BUBBLES, DROPS, AND PARTICLES

ROLAND CLIFT
Centre for Environmental Strategy
University of Surrey

JOHN GRACE
Professor and Canada Research Chair
Department of Chemical and Biological Engineering
University of British Columbia

MARTIN E. WEBER
Professor Emeritus
Department of Chemical Engineering
McGill University

DOVER PUBLICATIONS, INC.
Mineola, New York

Copyright

Bibliographical Note

This Dover edition, first published in 2005, is an unabridged republication of the work first published by Academic Press, Inc., New York, in 1978. A list of errata has been added to this edition.

International Standard Book Number: 0-486-44580-1

Manufactured in the United States of America
Dover Publications, Inc., 31 East 2nd Street, Mineola, N.Y. 11501

To the memory of

GERRY AND HEATHER RATCLIFF

Streams at some seasons
Wind their way through country lanes of beauty
And are dry.

Butterflies still hover
Down the rocky bed
And weeds grow strong and
Guard the pebbled way.

SARAH CHURCHILL

Contents

Chapter 5 Spheres at Higher Reynolds Numbers

Chapter 6 Nonspherical Rigid Particles at Higher Reynolds Numbers

Chapter 7 Ellipsoidal Fluid Particles

Chapter 8 Deformed Fluid Particles of Large Size

Chapter 9 Wall Effects

Chapter 10 Surface Effects, Field Gradients, and Other Influences

Chapter 11 Accelerated Motion without Volume Change

Chapter 12 Formation and Breakup of Fluid Particles

Preface

A vast body of literature dealing with bubbles, drops, and solid particles has grown up in engineering, physics, chemistry, geophysics, and applied mathematics. The principal objective of this book is to give a comprehensive critical review of this literature as it applies to the fluid dynamics, heat transfer, and mass transfer of single bubbles, drops, and particles. We have tried primarily to provide a reference text for research workers concerned with multiphase phenomena and a source of information, reference, and background material for engineers, students, and teachers who must deal with these phenomena in their work. In many senses, bubbles and drops are the chemical engineer's elementary particles. Inevitably the book has a bias toward the concerns of chemical engineers since each of the authors is a chemical engineer. However, we have attempted to keep our scope sufficiently broad to be of interest to readers from other disciplines. It became clear to us while preparing this book that workers in one area are commonly oblivious to advances in other fields. If this book does no more than bring literature from other fields to the attention of research workers, it will have accomplished part of our purpose.

A related objective of this book is to unify the treatment of solid particles, liquid drops, and gas bubbles. There are important similarities, as well as significant differences, that have often been overlooked among these three types of particle. Workers concerned with liquid drops, for example, sometimes fail to recognize the relevance of parallel work on bubbles or solid particles. Confusion has been created by differing—sometimes conflicting—nomenclature. To a large extent, we have written the book because we wished it had already existed.

An important limitation of this book is that we treat only phenomena in which particle–particle interactions are of negligible importance. Hence, direct application of the book is limited to single-particle systems or dilute suspensions.

Understanding the behavior of single particles is, however, a solid foundation upon which to build knowledge of multiple-particle systems. In addition, the literature on single particles is already so extensive that it warrants a book of its own.

Other limitations of our treatment should also be mentioned. Generally, we are concerned with bubbles, drops, and solid particles moving freely under the action of body forces (primarily gravity) in Newtonian fluids, but some work on stationary rigid bodies in flowing fluids is also applicable. We make little reference to direct applications or devices which use the phenomena under consideration, concentrating instead on the fundamentals of the phenomena. We make no mention of flexible and porous solid particles, and little mention of electrical and magnetic fields affecting particle motion and transfer processes. Coverage of static drops and bubbles, acoustical fields, phenomena involving change of phase, and noncontinuum effects is relatively scant, but reference is made in each of these cases to other works.

The fundamental principles and equations governing the behavior of bubbles, drops, and solid particles in Newtonian fluids are summarized in Chapter 1. Some readers may find the treatment too cursory here and in later chapters, so we provide extensive references where more detailed discussion may be found. Chapter 2 contains a summary of parameters used to characterize the shape of rigid particles, and of the factors which determine the shape of bubbles and drops. In Chapters 3–8 we treat the behavior of solid and fluid particles under steady incompressible flow in an external phase of very large extent. Since the sphere is of special importance in studies of bubbles, drops, and particles, two chapters are devoted to spherical particles. Bubbles and drops assume spherical shapes if either interfacial tension forces or continuous phase viscous forces are considerably larger than inertial effects. These conditions are obeyed in practice by small fluid particles. In addition, many solid particles may be approximated as spherical. An understanding of the behavior of spheres is therefore vital to the consideration of deformed fluid and solid particles. Chapter 3 deals with slow viscous flow past spherical particles, while Chapter 5 deals with flow at higher Reynolds numbers. Chapters 4 and 6 are devoted to nonspherical rigid particles at low and high Reynolds numbers, respectively. Nonspherical fluid particles are treated in Chapters 7 and 8.

The remaining chapters, 9–12, are devoted to effects which complicate the relatively simple case of a particle moving steadily through an unbounded fluid. Chapter 9 deals with the effects of rigid walls bounding the external fluid, with emphasis on ducts of circular cross section. Chapter 10 treats a series of factors which can influence motion or transfer rates, including turbulence, natural convection, surface roughness, and noncontinuum effects. However, effects of electrical charging at the interface are not included since they assume major significance only for particle–particle interactions. Chapters 11 and 12 relate to unsteady flows, the former to cases in which the particle volume is constant and the latter to processes whereby bubbles and drops grow, shrink, or divide.

Acknowledgments

We are indebted to many who have assisted us. We particularly acknowledge the help of T. H. Nguyen, L. Ng, P. Lefebvre, H. Bui, J. Brophy, D. Bhaga, T. Wairegi, K. V. Thambimuthu, Dr. L. Y.-L. Cheung, Dr. A. Mar, and Prof. J. M. Dealy for help in collecting and compiling original material, carrying out calculations, and performing tasks which were essential, though often tedious. Prof. N. Epstein, Prof. R. Guthrie, Dr. R. Nedderman, and Dr. M. Filla read some of the chapters and made useful comments. Pat Fong, Selma Abu-Merhy, and Helen Rousseau typed the manuscript. Students in two graduate courses at McGill University provided sounding boards for some of our ideas. The Graduate Faculty and the Department of Chemical Engineering of McGill University provided some financial assistance, for which we are grateful.

Each of us also owes personal thanks to members of our families, who cheerfully endured the trials of putting this volume together. Our thanks to David, Elizabeth, Gerry, Lawrence, Malcolm, Mark, Nora, Sherrill, and Vanessa.

BUBBLES, DROPS, AND PARTICLES

Chapter 1

Basic Principles

I. INTRODUCTION AND TERMINOLOGY

Bubbles, drops, and particles are ubiquitous. They are of fundamental importance in many natural physical processes and in a host of industrial and man-related activities. Rainfall, air pollution, boiling, flotation, fermentation, liquid–liquid extraction, and spray drying are only a few of the phenomena and operations in which particles play a primary role. Meteorologists and geophysicists study the behavior of raindrops and hailstones, and of solid particles transported by rivers. Applied mathematicians and applied physicists have long been concerned with fundamental aspects of fluid-particle interactions. Chemical and metallurgical engineers rely on bubbles and drops for such operations as distillation, absorption, flotation, and spray drying, while using solid particles as catalysts or chemical reactants. Mechanical engineers have studied droplet behavior in connection with combustion operations, and bubbles in electromachining and boiling. In all these phenomena and processes, there is relative motion between bubbles, drops, or particles on the one hand, and surrounding fluid on the other. In many cases, transfer of mass and/or heat is also of importance. Interactions between particles and fluids form the subject of this book.

Before turning to the principles involved, the reader should be aware of certain terminology which is basic to understanding the material presented in later chapters. Science is full of words which have very different connotations in the jargon of different disciplines. The present book is about particles and the term particle needs to be defined carefully within our context, to distinguish it from the way in which the nuclear physicist, for example, might use the word. For our purposes a "particle" is a self-contained body with maximum dimension between about 0.5 μm and 10 cm, separated from the surrounding medium by a recognizable interface. The material forming the particle will be termed the "dispersed phase." We refer to particles whose dispersed phases are composed of solid matter as "solid particles." If the dispersed phase is in the liquid state, the particle is called a "drop." The term "droplet" is often used to refer to small drops. The dispersed phase liquid is taken to be Newtonian. If the dispersed phase is a gas, the particle is referred to as a bubble. Together, drops and bubbles comprise "fluid particles." Following common usage, we use "continuous phase" to refer to the medium surrounding the particles. In this book we consider only cases in which the continuous phase is a Newtonian fluid (liquid or gas). In subsequent chapters we distinguish properties of the dispersed (or particle) phase by a subscript p from properties of the continuous phase which are unsubscripted. Occasionally the dispersed and continuous phases are referred to as the "inner" and "outer" phases, respectively.

Another distinction we use throughout the book is between rigid, noncirculating, and circulating particles. "Rigid particles," comprising most solid particles, can withstand large normal and shearing stresses without appreciable deformation or flow. "Noncirculating fluid particles" are those in which there is no internal motion relative to a coordinate system fixed to the particle. "Circulating particles" contain fluid which has motion of its own relative to any fixed coordinate system. We consider only cases in which the dispersed phase is continuous. Hence the scale of the particle must be large compared to the scale of molecular processes in the dispersed phase.

In this book we consider as particles only those bodies which are biologically inert and which are not self-propelling. To give some specific examples, raindrops, hailstones, river-borne gravel, and pockets of gas formed by cavitation or electrolysis are all considered to be particles. However, insects and microorganisms are excluded by their life, weather balloons and neutrons by their size, homogeneous vortices by the lack of a clearly defined interface, and rockets and airplanes by their self-propelling nature and size. Our attention is concentrated on particles which are free to move through the continuous phase under the action of some body force such as gravity. Thus heat exchanger tubes, for example, are not considered—not only because of their size but also because they are fixed in position. Some elements of our definitions are of necessity arbitrary. For example, a golf ball satisfies our definition of a particle while a football does not. In most cases, there is little ambiguity, however, so long as these general guidelines regarding terminology are borne in mind.

Other terms which can be defined quantitatively are introduced in the following sections. Some other terms, such as "turbulence," "viscosity," and "diffusivity" are used without definition. For a full explanation of these terms, we refer the reader to standard texts in fluid mechanics, heat transfer, and mass transfer.

II. THEORETICAL BASIS

The fundamental physical laws governing motion of and transfer to particles immersed in fluids are Newton's second law, the principle of conservation of mass, and the first law of thermodynamics. Application of these laws to an infinitesimal element of material or to an infinitesimal control volume leads to the Navier–Stokes, continuity, and energy equations. Exact analytical solutions to these equations have been derived only under restricted conditions. More usually, it is necessary to solve the equations numerically or to resort to approximate techniques where certain terms are omitted or modified in favor of those which are known to be more important. In other cases, the governing equations can do no more than suggest relevant dimensionless groups with which to correlate experimental data. Boundary conditions must also be specified carefully to solve the equations and these conditions are discussed below together with the equations themselves.

A. Fluid Mechanics

1. *The Navier–Stokes Equation*

Application of Newton's second law of motion to an infinitesimal element of an incompressible Newtonian fluid of density ρ and constant viscosity μ, acted upon by gravity as the only body force, leads to the Navier–Stokes equation of motion:

$$\rho\, D\mathbf{u}/Dt = \rho\mathbf{g} - \nabla p + \mu\nabla^2\mathbf{u}. \tag{1-1}$$

The term on the left-hand side, arising from the product of mass and acceleration, can be expanded using the expression for the substantial derivative operator

$$\frac{D}{Dt} = \frac{\partial}{\partial t} + \mathbf{u}\cdot\nabla, \tag{1-2}$$

where the first term, called the local derivative, represents changes at a fixed point in the fluid and the second term, the convective term, accounts for changes following the motion of the fluid. The $\rho\mathbf{g}$ term above is the gravity force acting on unit volume of the fluid. The final two terms in Eq. (1-1) represent the surface force on the element of fluid. If the fluid were compressible, additional terms would appear and the definition of p would require careful attention. For discussions of these matters, see Schlichting (S1), Bird *et al.* (B3), or standard texts on fluid dynamics. Equation (1-1) is written in scalar form in the most common coordinate systems in many texts [e.g. (B3)].

In the simplest incompressible flow problems under constant property conditions, the velocity and pressure fields ($\mathbf{u}$ and p) are the unknowns. In principle, Eq. (1-1) and the overall continuity equation, Eq. (1-9) below, are sufficient for solution of the problem with appropriate boundary conditions. In practice, solution is complicated by the nonlinearity of the Navier–Stokes equation, arising in the convective acceleration term $\mathbf{u} \cdot \nabla \mathbf{u}$.

In dimensionless form, Eq. (1-1) may be rewritten as

$$\frac{D\mathbf{u}'}{Dt'} = -\nabla' p_{\mathrm{m}}' + \frac{1}{\mathrm{Re}} (\nabla')^2 \mathbf{u}', \tag{1-3}$$

where the primes denote dimensionless quantities or operators formed using dimensionless variables. Reference quantities L, U_0 and p_0 are used together with the fluid properties to form the dimensionless quantities as follows:

$$\mathbf{u}' = \mathbf{u}/U_0 \tag{1-4}$$

$$x_1' = x_1/L; \; y_1' = y_1/L; \; z_1' = z_1/L; \; t' = tU_0/L \tag{1-5}$$

$$p_{\mathrm{m}}' = (p - p_0 - \rho g h_{\mathrm{v}})/\rho U_0^2 \tag{1-6}$$

$$\mathrm{Re} = \frac{\rho L U_0}{\mu} = \frac{L U_0}{\nu}, \tag{1-7}$$

where h_{v} is a coordinate directed vertically upwards. The Reynolds number, Re, is of enormous importance in fluid mechanics. From Eq. (1-3) it can be interpreted as an indication of the ratio of inertia to viscous forces. For convenience we have defined a dimensionless modified pressure, p_{m}', which gives the pressure field due to the flow (i.e., discounting hydrostatic pressure variations). Batchelor (B1) gives a good discussion of the modified pressure. It is useful in a wide range of problems where gravity effects can be isolated from the boundary conditions.

2. *Overall Continuity Equation*

Application of the principle of conservation of mass to a compressible fluid yields

$$(\partial\rho/\partial t) + \nabla \cdot \rho\mathbf{u} = 0, \tag{1-8}$$

which for an incompressible fluid reduces to

$$\nabla \cdot \mathbf{u} = 0. \tag{1-9}$$

In dimensionless form, Eq. (1-9) becomes simply

$$\nabla' \cdot \mathbf{u}' = 0. \tag{1-10}$$

3. *Velocity Boundary Conditions*

In order to solve the Navier–Stokes equations for the dispersed and continuous phases, relationships are required between the velocities on either side of an interface between the two phases. The existence of an interface assures that the normal velocity in each phase is equal at the interface, i.e.,

$$u_n = (u_n)_p \qquad \text{(everywhere on interface)}, \tag{1-11}$$

where the subscript n refers to motion normal to the interface. For a particle of constant shape and size the normal velocity is zero relative to axes fixed to the particle.

The condition on the tangential velocity at the interface is not as obvious as that on the normal velocity. There is now ample experimental evidence that the fluid velocity at the surface of a rigid or noncirculating particle is zero relative to the particle, provided that the fluid can be considered a continuum. This leads to the so-called "no-slip" condition, which for a fluid particle takes the form

$$u_t = (u_t)_p \qquad \text{(everywhere on interface)}, \tag{1-12}$$

where the subscript t refers to motion tangential to the surface.

Additional velocity boundary conditions are provided by the velocity field in the continuous phase remote from the particle and the existence of points, lines, and/or planes of symmetry. These conditions are set out in subsequent chapters for specific situations.

4. *Stress Boundary Conditions*

For solid particles a sufficient set of boundary conditions is provided by the no slip condition, the requirement of no flow across the particle surface, and the flow field remote from the particle. For fluid particles, additional boundary conditions are required since Eqs. (1-1) and (1-9) apply simultaneously to both phases. Two additional boundary conditions are provided by Newton's third law which requires that normal and shearing stresses be balanced at the interface separating the two fluids.

The interface between two fluids is in reality a thin layer, typically a few molecular dimensions thick. The thickness is not well defined since physical properties vary continuously from the values of one bulk phase to that of the other. In practice, however, the interface is generally treated as if it were infinitesimally thin, i.e., as if there were a sharp discontinuity between two bulk phases (L1). Of special importance is the surface or interfacial tension, σ, which is best viewed as the surface free energy per unit area at constant temperature. Many workers have used other properties, such as surface viscosity (see Chapter 3) to describe the interface.

A complete treatment of interfacial boundary conditions in tensor notation is given by Scriven (S2). If surface viscosities are ignored, the normal stress condition reduces to

$$p_p + (\tau_{nn})_p - p - \tau_{nn} = \sigma[(1/R_1) + (1/R_2)], \tag{1-13}$$

where R_1 and R_2 are the principal radii of curvature of the surface and the τ_{nn} are the deviatoric normal stresses (B1, S1). Under static conditions Eq. (1-13) reduces to the Laplace equation. The tangential stress condition corresponding to Eq. (1-13) is

$$\tau_{\mathrm{nt}} - (\tau_{\mathrm{nt}})_{\mathrm{p}} = \mathbf{\nabla}_{\mathrm{s}}\sigma, \tag{1-14}$$

where the τ_{nt} refer to the shearing stresses and $\mathbf{\nabla}_{\mathrm{s}}$ is the surface gradient (S2). For a spherical fluid particle with both bulk phases Newtonian and an incompressible axisymmetric flow field, Eqs. (1-13) and (1-14) become

$$p_{\mathrm{p}} - p - 2\mu_{\mathrm{p}}\left[\frac{\partial u_r}{\partial r}\right]_{\mathrm{p}} + 2\mu\frac{\partial u_r}{\partial r} = \frac{2\sigma}{a} \tag{1-15}$$

and

$$\mu_{\mathrm{p}}\left[r\frac{\partial}{\partial r}\left(\frac{u_\theta}{r}\right) + \frac{1}{r}\frac{\partial u_r}{\partial \theta}\right]_{\mathrm{p}} - \mu\left[r\frac{\partial}{\partial r}\left(\frac{u_\theta}{r}\right) + \frac{1}{r}\frac{\partial u_r}{\partial \theta}\right] = \frac{1}{a}\frac{d\sigma}{d\theta}. \tag{1-16}$$

The final term in Eq. (1-16) is especially important for cases in which σ varies around the surface of a fluid particle due to concentration or temperature gradients (see Chapters 3, 5, and 7).

5. *Stream Functions, Streamlines, and Vorticity*

From the definition of a particle used in this book, it follows that the motion of the surrounding continuous phase is inherently three-dimensional. An important class of particle flows possesses axial symmetry. For axisymmetric flows of incompressible fluids, we define a stream function, ψ, called Stokes's stream function. The value of $2\pi\psi$ at any point is the volumetric flow rate of fluid crossing any continuous surface whose outer boundary is a circle centered on the axis of symmetry and passing through the point in question. Clearly $\psi = 0$ on the axis of symmetry. Stream surfaces are surfaces of constant ψ and are parallel to the velocity vector, $\mathbf{u}$, at every point. The intersection of a stream surface with a plane containing the axis of symmetry may be referred to as a streamline. The velocity components, u_r and u_θ, are related to ψ in spherical-polar coordinates by

$$u_r = \frac{1}{r^2 \sin\theta}\frac{\partial\psi}{\partial\theta}; \qquad u_\theta = \frac{-1}{r\sin\theta}\frac{\partial\psi}{\partial r}. \tag{1-17}$$

The vorticity is defined as

$$\boldsymbol{\zeta} = \mathbf{\nabla} \times \mathbf{u}. \tag{1-18}$$

It can be shown that $\boldsymbol{\zeta}$ is twice the angular rotation of a fluid element. When $\boldsymbol{\zeta} = 0$ throughout a region of a fluid, the flow in that region is said to be irrotational. Flows which are initially irrotational remain irrotational if all the forces acting are conservative. Since gravity and pressure forces are conservative, vorticity generation in flow fields which are initially irrotational, such as around a particle accelerating in a stagnant fluid, arises from nonconservative viscous forces. For axisymmetric flows, vorticity can be treated as a scalar function. It is then often convenient to define surfaces of constant vorticity or lines of constant vorticity in a plane containing the axis of symmetry. Examples of streamlines and lines of constant vorticity are given in later chapters (for example, in Figs. 5.1 and 5.2).

It is often convenient to work in terms of a dimensionless stream function and vorticity defined, respectively, as

$$\Psi = \psi/U_0L^2 \tag{1-19}$$

and

$$Z = \zeta L/U_0. \tag{1-20}$$

6. *Inviscid Flow and Potential Flow Past a Sphere*

In practice all real fluids have nonzero viscosity so that the concept of an inviscid fluid is an idealization. However, the development of hydrodynamics proceeded for centuries neglecting the effects of viscosity. Moreover, many features (but by no means all) of certain high Reynolds number flows can be treated in a satisfactory manner ignoring viscous effects.

For $\mu = 0$ or $\mathrm{Re} \to \infty$, Eq. (1-1) may be rewritten

$$D\mathbf{u}/Dt = \mathbf{g} - \nabla p/\rho, \tag{1-21}$$

which is the well-known Euler equation. Integration of Eq. (1-21) along either a streamline or parallel to ζ for steady incompressible flows leads to Bernoulli's equation, i.e.,

$$(p/\rho g) + (|\mathbf{u}|^2/2g) + h_v = \text{constant}. \tag{1-22}$$

From Kelvin's theorem, inviscid motions in a gravity (conservative) field which are initially irrotational remain so. We may, therefore, write

$$\zeta = \nabla \times \mathbf{u} = 0. \tag{1-23}$$

Hence $\mathbf{u}$ may be written as the gradient of some scalar function, i.e.,

$$\mathbf{u} = \nabla\Phi, \tag{1-24}$$

where Φ is conventionally termed a "velocity potential." From this designation, irrotational motions derive the name "potential flow." For incompressible potential flows it can be shown that Bernoulli's equation, Eq. (1-22), applies throughout the flow field and that Φ satisfies Laplace's equation:

$$\nabla^2\Phi = 0 \tag{1-25}$$

If the flow is axisymmetric, ψ can be shown to obey the following equation in spherical polar coordinates (B1):

$$r^2(\partial^2\psi/\partial r^2) + \partial^2\psi/\partial\theta^2 - \cot\theta(\partial\psi/\partial\theta) = 0 \tag{1-26}$$

Since ψ by definition satisfies Eq. (1-9), potential flow solutions can be found by solving Eq. (1-26) for ψ subject to the required boundary conditions. The pressure field can then be found using Eq. (1-22).

Consider the case of a stationary sphere of radius a centered at the origin in a uniform stream of velocity $-U$. Equation (1-26) is second order and hence we require two boundary conditions. Remote from the sphere, the velocity must everywhere be $-U$, i.e.,

$$\psi = (-Ur^2/2)\sin^2\theta \qquad \text{as} \quad r \to \infty. \tag{1-27}$$

No fluid crosses the sphere boundary. Hence the surface is a stream surface and since this boundary also cuts the axis of symmetry

$$\psi = 0 \qquad \text{at} \quad r = a. \tag{1-28}$$

Equations (1-26) to (1-28) are satisfied by

$$\psi = -U\left(1 - \frac{a^3}{r^3}\right)\frac{r^2}{2}\sin^2\theta. \tag{1-29}$$

Application of Eq. (1-17) gives

$$u_r = -U\left(1 - \frac{a^3}{r^3}\right)\cos\theta; \qquad u_\theta = U\left(1 + \frac{a^3}{2r^3}\right)\sin\theta. \tag{1-30}$$

Since the pressure field depends only on the magnitude of the velocity (see Eq. (1-22)) and since the flow field has fore-and-aft symmetry, the modified pressure field forward from the equator of the sphere is the mirror image of that to the rear. This leads to d'Alembert's paradox: that the net force acting on the sphere is predicted to be zero. This paradox can only be resolved, and nonzero drag obtained, by accounting for the viscosity of the fluid. For inviscid flow, the surface velocity and pressure follow as

$$(u_\theta)_{r=a} = \tfrac{3}{2}U\sin\theta \tag{1-31}$$

$$(p)_{r=a} = p_0 + (\rho U^2/2)(1 - \tfrac{9}{4}\sin^2\theta). \tag{1-32}$$

These results are useful reference conditions for real flows past spherical particles. For example, comparisons are made in Chapter 5 between potential flow and results for flow past a sphere at finite Re. Other potential flow solutions exist for closed bodies, but none has the same importance as that outlined here for the motion of solid and fluid particles.

7. *Creeping Flow*

Whereas inviscid flow is a useful reference point for high Reynolds number flows, a different simplification known as the "creeping flow" approximation applies at very low Re. From Eq. (1-3), the terms on the right-hand side dominate as Re $\to 0$, so that the convective derivative may be neglected. In dimensional form, the resulting equation of motion is

$$\nabla p_{\mathrm{m}} = \mu \nabla^2 \mathbf{u}, \tag{1-33}$$

where p_{m} is the modified pressure introduced in dimensionless form in Eq. (1-6), i.e.,

$$p_{\mathrm{m}} = p - p_0 - \rho g h_{\mathrm{v}} \tag{1-34}$$

Comparison with the full Navier–Stokes equation, Eq. (1-1), shows that fluid inertia is completely neglected in Eq. (1-33). Problems arising from the nonlinearity of the convective acceleration term are thereby avoided. However, the order of the equation and hence the number of boundary conditions required are unchanged.

With this simplification, the equations governing incompressible fluid motion are Eq. (1-33) and the continuity equation, Eq. (1-9). Several important consequences follow from inspection of these equations. The fluid density does not appear in either equation. Both equations are "reversible" in the sense that they are still satisfied if $\mathbf{u}$ is replaced by $-\mathbf{u}$, whereas the nonlinearity of the Navier–Stokes equations prevents such "reversibility." If we take the divergence of Eq. (1-33) and apply Eq. (1-9), we obtain

$$\nabla^2 p_{\mathrm{m}} = 0, \tag{1-35}$$

so that the modified pressure is a harmonic function. For axisymmetric flows, we may write Eq. (1-33) in terms of the Stokes stream function as

$$E^4 \psi = 0, \tag{1-36}$$

where E^2 in spherical polar coordinates is

$$E^2 = \frac{\partial^2}{\partial r^2} + \frac{1}{r^2}\frac{\partial^2}{\partial \theta^2} - \frac{\cot\theta}{r^2}\frac{\partial}{\partial \theta}. \tag{1-37}$$

The creeping flow approximation has found wide application in problems such as lubrication, injection molding, and flow through porous media. Its application to rigid and fluid particles is discussed in Chapters 3 and 4. However, a fundamental difficulty, first recognized by Oseen, arises in applying Eq. (1-33) or (1-36) to particles in unbounded media. This difficulty, and Oseen's attempt to overcome it, are discussed in Chapter 3.

8. *Boundary Layer Theory*

As discussed above, no fluids are inviscid in practice. At high Reynolds number, viscous effects may be insignificant throughout large regions of the flow field and these regions may be treated as if the fluid were inviscid. However, the effect of viscosity must in general be taken into account in thin layers adjacent to boundaries in the flow. The essence of boundary layer theory in fluid mechanics, applicable only at high Re, is that viscous effects are considered to be restricted to thin layers called boundary layers and that certain simplifications can be made in the boundary layer because of its thinness. Usually derivatives with respect to the streamwise coordinate are neglected relative to those in the transverse direction. An analogous approach may be applied to heat and mass transfer at high Peclet numbers (see below) where we refer to temperature and concentration boundary layers. There are a number of excellent books on boundary layer theory [e.g. (S1)] to which the reader is referred.

B. Heat and Mass Transfer

1. *The Species Continuity Equation*

Application of the principle of conservation of mass to a binary system consisting of a non-reactive solute in dilute solution in an incompressible fluid yields

$$Dc/Dt = \mathscr{D}\nabla^2 c, \tag{1-38}$$

where $\mathscr{D}$, the diffusivity, is assumed constant. The driving force for diffusion is provided by molar concentration gradients. Hence Eq. (1-38) provides a good description of diffusion in most liquids, since the density is essentially constant, and in gases when the molecular weight of the solute is similar to that of the host gas. Alternate forms of the species continuity equation based on other driving forces are given by Bird *et al.* (B3) and Skelland (S4). Multicomponent diffusion is considered by Cussler (C1) and Bird *et al.* (B3).

In this book we limit our treatment to dilute solutions so that the diffusional mass flux is small. In this way the existence of diffusion does not appreciably alter the fluid motion, so that the velocity and stress boundary conditions can be considered to be unaltered. Treatments of diffusion with high mass fluxes appear elsewhere (B3, S3, S4).

Of the many possible boundary and initial conditions for Eq. (1-38), we consider in this book only uniform concentration at the particle surface, uniform concentration in the continuous phase far from the particle, and uniform initial concentrations in each phase. In addition, the interface is taken to be at an equilibrium described by a linear relationship between the concentrations in each phase:

$$c_p = Hc \qquad \text{(everywhere on interface).} \tag{1-39}$$

Equation (1-38) in dimensionless form becomes

$$\frac{D\phi}{Dt'} = \frac{1}{\mathrm{Pe}}(\nabla')^2\phi \tag{1-40}$$

where the dimensionless time and spatial coordinates are given by Eq. (1-5). A new dimensionless group, the Peclet number,

$$\mathrm{Pe} = LU_0/\mathscr{D} \tag{1-41}$$

appears in Eq. (1-40). Pe can also be written as the product of the Reynolds number, defined in Eq. (1-7), and the Schmidt number, $\mathrm{Sc} = \nu/\mathscr{D}$, i.e.,

$$\mathrm{Pe} = \mathrm{Re}\,\mathrm{Sc}. \tag{1-42}$$

The concentration is made dimensionless in one of several ways depending upon the situation considered. For example, for steady transfer to the continuous phase from a particle at constant concentration, the boundary conditions considered in this book are

$$\text{remote from particle:} \quad c = c_\infty, \tag{1-43}$$

$$\text{at particle surface:} \quad c = c_\mathrm{s}. \tag{1-44}$$

It is then convenient to define the dimensionless concentration as

$$\phi = (c - c_\infty)/(c_\mathrm{s} - c_\infty). \tag{1-45}$$

Other forms of ϕ appropriate to different physical situations are introduced in subsequent chapters.

2. *The Energy Equation*

Application of the first law of thermodynamics to an infinitesmal element of incompressible Newtonian fluid of uniform composition and constant properties yields

$$\rho C_\mathrm{t}(DT/Dt) = K_\mathrm{t}\nabla^2 T + \mu\Phi_\mathrm{v}, \tag{1-46}$$

where Φ_v, the dissipation function, represents the rate at which the tangential and deviatoric normal stresses do work on the element of fluid. Φ_v may also be viewed as the rate at which the internal energy of the fluid is increased due to viscous dissipation. Explicit forms for Φ_v are tabulated in standard texts [e.g. (B3, S1)]. The dissipation function becomes important in high-speed flows and in flows of fluids with extremely large viscosities (e.g., molten polymers). For almost all situations considered in this book, the simple form of the energy equation suffices with the dissipation term deleted, i.e.,

$$(DT/Dt) = (K_\mathrm{t}/\rho C_\mathrm{t})\nabla^2 T. \tag{1-47}$$

Equation (1-47) is identical in form to the species continuity equation, Eq. (1-38), and this leads to close analogies between heat and mass transfer as discussed in the next section.

Parallel to the boundary conditions discussed above for the species continuity equation, we consider in this book only uniform temperature on the surface of the particle, uniform temperature in the continuous phase remote from the particle and uniform initial temperatures in each phase. Hence

$$T = T_\mathrm{p} \qquad \text{(everywhere on interface)}. \tag{1-48}$$

Other types of boundary conditions are discussed in standard works on heat transfer [e.g. (E1, K1)].

Putting the energy equation into dimensionless form yields an equation identical to Eq. (1-40) with

$$\mathrm{Pe} = LU_0\rho C_t/K_t \tag{1-49}$$

or

$$\mathrm{Pe} = \mathrm{Re}\,\mathrm{Pr}. \tag{1-50}$$

Thus the Prandtl number, $\mathrm{Pr} = \mu C_t/K_t$, plays the same role in heat transfer as the Schmidt number, Sc, in mass transfer.

3. *Equivalence of Sherwood and Nusselt Numbers*

Since all properties have been assumed constant in Eqs. (1-1), (1-38), and (1-47), and the solute concentration has been assumed small, the Navier–Stokes equation may be solved independently of the species continuity and energy equations. We treat only one exception where the velocity field is considered to be affected by heat or mass transfer. This exception, natural convection, is covered in Chapter 10.

The formal analogy between heat and mass transfer under the conditions of no dissipation, low mass flux and constant properties can be completed as follows. Equations (1-38) and (1-39) and the boundary conditions considered in this book apply to heat transfer if one replaces c by $\rho C_t T$, c_p by $\rho_p C_{tp} T_p$, H by $\rho_p C_{tp}/\rho C_t$, $\mathscr{D}$ by the thermal diffusivity $\alpha = K_t/\rho C_t$, $\mathscr{D}_p$ by $\alpha_p = K_{tp}/\rho_p C_{tp}$ and the mass transfer coefficient k by $h/\rho C_t$.

Since the dimensionless equations and boundary conditions governing heat transfer and dilute-solution mass transfer are identical, the solutions to these equations in dimensionless form are also identical. Profiles of dimensionless concentration and temperature are therefore the same, while the dimensionless transfer rates, the Sherwood number ($\mathrm{Sh} = kL/\mathscr{D}$) for mass transfer, and the Nusselt number ($\mathrm{Nu} = hL/K_t$) for heat transfer, are identical functions of Re, Sc or Pr, and dimensionless time. Most results in this book are given in terms of Sh and Sc; the equivalent results for heat transfer may be found by simply replacing Sh by Nu and Sc by Pr.

4. *Thin Concentration Boundary Layer*

For transfer in the continuous phase, it is possible to simplify Eq. (1-38) when the continuous-phase Peclet number is large. For high Pe the concentration varies only in a thin layer adjacent to the particle surface. In this region the gradient of concentration normal to the surface is much larger than the gradient parallel to the surface. The thin concentration boundary layer approximation consists of neglecting diffusion parallel to the surface and retaining on the right-hand side of Eq. (1-38) only the term involving the derivative normal to the surface. Formally this requires $\mathrm{Pe} \to \infty$, which, for most practical situations, means $\mathrm{Sc} \to \infty$ for any finite Reynolds number. Surprisingly, this approximation is often reasonable down to Sc of order unity. Use of the thin concentration boundary layer approximation, sometimes called the asymptotic solution for $\mathrm{Sc} \to \infty$, does not require that Re be large or that the momentum boundary layer approximation (see above) be made.

Two particularly useful equations can be derived by applying the thin concentration boundary layer approximation to steady-state transfer from an axisymmetric particle (L2). The particle and the appropriate boundary layer coordinates are sketched in Fig. 1.1. The x coordinate is parallel to the surface ($x = 0$ at the front stagnation point), while the y coordinate is normal to the surface. The distance from the axis of symmetry to the surface is R. Equation (1-38), subject to the thin boundary layer approximation, then becomes

$$u_x(\partial c/\partial x) + u_y(\partial c/\partial y) = \mathscr{D}(\partial^2 c/\partial y^2) \tag{1-51}$$

with boundary conditions

$$\text{at} \quad y = 0 \qquad c = c_s, \tag{1-52}$$

$$\text{at} \quad y \to \infty \qquad c = c_\infty, \tag{1-53}$$

$$\text{at} \quad x = 0 \qquad c = c_\infty. \tag{1-54}$$

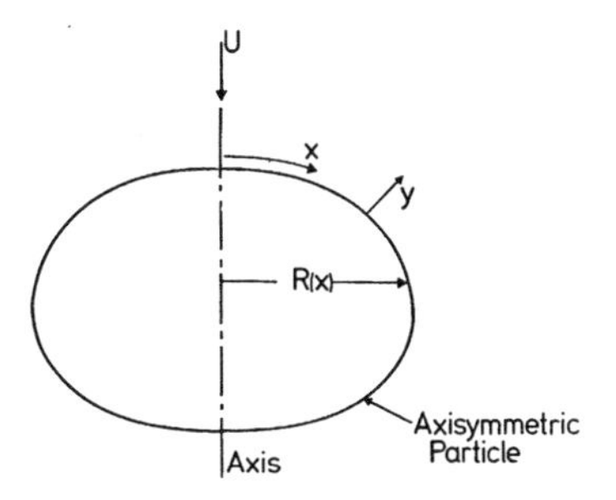

FIG. 1.1 Coordinates for the thin concentration boundary layer approximation.

In the thin layer adjacent to the particle surface the overall continuity equation may be written (S1)

$$\partial(u_x R)/\partial x + \partial(u_y R)/\partial y = 0. \tag{1-55}$$

Since we only require u_x and u_y near the surface, the following approximations may be used. For a solid particle we write

$$u_x = \zeta_s y, \tag{1-56}$$

where ζ_s, the surface vorticity, is given by

$$\zeta_s = (\partial u_x/\partial y)_{y=0}, \tag{1-57}$$

since $(\partial u_y/\partial x)_{y=0} = 0$ for all x from the normal velocity boundary condition, Eq. (1-11). For a fluid particle

$$u_x = u_t, \tag{1-58}$$

where u_t is the interfacial velocity discussed above. In general, ζ_s and u_t are functions of x. The normal velocity u_y is determined from u_x through the continuity equation, Eq. (1-55).

Combination of Eqs. (1-51) to (1-55) with either Eq. (1-56) or (1-58) yields an equation which may be solved to give concentration profiles from which mass transfer rates may be found. For a solid particle the average Sherwood number is

$$\mathrm{Sh_e} = \frac{(kA/A_e)d_e}{\mathscr{D}} = 0.641\left\{\int_0^{X_M}\left[\frac{\zeta_s d_e}{2U}\mathscr{R}^3\right]^{1/2} dX\right\}^{2/3}\mathrm{Pe}^{1/3}, \tag{1-59}$$

where

$$X = 2x/d_e \tag{1-60}$$

$$\mathscr{R} = 2R/d_e \tag{1-61}$$

$$\mathrm{Pe} = d_e U/\mathscr{D}. \tag{1-62}$$

Here X_M is the maximum value of X and A_e is the surface area of the volume equivalent sphere. For a fluid particle the average Sherwood number is

$$\mathrm{Sh_e} = \frac{(kA/A_e)d_e}{\mathscr{D}} = 0.798\left[\int_0^{X_M}\left(\frac{u_t}{U}\right)\mathscr{R}^2\, dX\right]^{1/2}\mathrm{Pe}^{1/2}. \tag{1-63}$$

Equation (1-63) is valid as long as the x direction velocity is essentially equal to the tangential velocity throughout the concentration boundary layer. This requires (L2) that

$$\mathrm{Sh_e} \gg \frac{d_e}{2u_t}\left(\frac{\partial u_x}{\partial y}\right)_{y=0}. \tag{1-64}$$

As we shall see in Chapter 3, this places severe restrictions on the range of $\kappa = \mu_p/\mu$ for which Eq. (1-63) can be applied. Equations equivalent to Eq. (1-63) have been derived for fluid particles from another point of view (B2, S5).

REFERENCES

B1. Batchelor, G. K., "An Introduction to Fluid Dynamics." Cambridge Univ. Press, London and New York, 1967.

B2. Beek, W. J., and Kramers, H., *Chem. Eng. Sci.* **17**, 909–921 (1962).

B3. Bird, R. B., Stewart, W. E., and Lightfoot, E. N., "Transport Phenomena." Wiley, New York, 1960.

C1. Cussler, E. L., "Multicomponent Diffusion." Elsevier, Amsterdam, 1976.

E1. Eckert, E. R. G., and Drake, R. M., "Analysis of Heat and Mass Transfer." McGraw-Hill, New York, 1972.
K1. Kays, W. M., "Convective Heat and Mass Transfer." McGraw-Hill, New York, 1966.
L1. Levich, V. G., "Physicochemical Hydrodynamics." Prentice-Hall, New York, 1962.
L2. Lochiel, A. C., and Calderbank, P. H., *Chem. Eng. Sci.* **19**, 471–484 (1964).
S1. Schlichting, H., "Boundary Layer Theory," 6th Ed. McGraw-Hill, New York, 1968.
S2. Scriven, L. E., *Chem. Eng. Sci.* **12**, 98–108 (1960).
S3. Sherwood, T. K., Pigford, R. L., and Wilke, C. R., "Mass Transfer." McGraw-Hill, New York, 1975.
S4. Skelland, A. H. P., "Diffusional Mass Transfer." Wiley, New York, 1974.
S5. Stewart, W. E., Angelo, J. B., and Lightfoot, E. N., *AIChE J.* **16**, 771–786 (1970).

Chapter 2

Shapes of Rigid and Fluid Particles

I. INTRODUCTION

Natural and man-made solid particles occur in almost any imaginable shape from roughly spherical pollen and fly ash through cylindrical asbestos fibers to irregular mineral particles. Bubbles and drops, on the other hand, adopt a smaller range of shapes, and although they are often axisymmetric, they are spherical only under special circumstances. A nonspherical particle may have planes and axes of symmetry, but it cannot possess the unique point-symmetry of the sphere. Thus a nonspherical particle presents problems which are more complex than those arising for the sphere. This chapter presents a summary of methods of classifying and quantifying the shapes of particles. A method is also presented for distinguishing which of three overall shape regimes a fluid particle adopts in unhindered motion under the influence of gravity.

II. CLASSIFICATION OF PARTICLE SHAPES

A. Symmetry

It is convenient to classify symmetric particles into several general groups. A shape may belong to more than one group, and this overlap generally makes it easier to predict flow properties, motion in free fall or rise, etc. The most useful divisions are as follows.

1. *Axisymmetric Particles*

This group comprises bodies generated by rotating a closed curve around an axis. Spheroidal particles (also called ellipsoids of revolution) are of particular interest, since they correspond closely to the shapes adopted by many drops and bubbles and to the shapes of some solid particles. A spheroid is an ellipsoid of revolution, generated by rotating an ellipse about one of its principal axes. If this is the minor axis, the body is said to be oblate; otherwise the spheroid is prolate.

Axisymmetric shapes are conveniently described by the "aspect ratio" E, defined as the ratio of the length projected on the axis of symmetry to the maximum diameter normal to the axis. Thus, E is the ratio of semiaxes for a spheroid, with $E < 1$ for an oblate spheroid and $E > 1$ for a prolate spheroid.

2. *Orthotropic Particles*

A body has a plane of symmetry if the shape is unchanged by reflection in the plane. Orthotropic particles have three mutually perpendicular planes of symmetry. An axisymmetric particle is symmetric with respect to all planes containing its axis, so that it is orthotropic if it has a plane of symmetry normal to the axis, i.e., if it has fore-and-aft symmetry.

3. *Spherically Isotropic Particles*

This group comprises regular polyhedra and all shapes obtained by symmetrically smoothing or cutting pieces from these bodies. It includes isometric orthotropic particles, i.e., shapes for which the half-body obtained by cutting the particle along a plane of symmetry is the same whichever plane is chosen for the cut. Particles obtained by symmetrical deformation of a regular tetrahedron are "spherically isotropic" (see Chapter 4), even though they are not orthotropic.

Some simple examples may help to clarify these classes of symmetry. Circular cylinders, disks, and spheroids are axisymmetric and orthotropic; cones are axisymmetric but not orthotropic; none of these are strictly spherically isotropic. Parallelepipeds are orthotropic, but the cube is the only spherically isotropic parallelepiped. Regular octahedra and tetrahedra are spherically isotropic; octahedra are orthotropic whereas tetrahedra are not.

B. Shape Factors

Most particles of practical interest are irregular in shape, and so do not fall into the above categories. A variety of empirical factors have been proposed to describe nonspherical particles and correlate their flow behavior. Empirical description of particle shape is provided by identifying two characteristic parameters from the following (B3):

(i) volume, V
(ii) surface area, A
(iii) projected area, A_p
(iv) projected perimeter, P_p

The projected area and perimeter must be determined normal to some specified axis. For axisymmetric bodies, the reference direction is taken parallel or normal to the axis of symmetry (see Chapter 4). Many naturally occurring particles have an oblate or lenticular form. In this case, the reference direction is usually taken parallel to the particle thickness t, the minimum distance between two parallel planes tangential to opposite surfaces. This choice has some immediate practical advantages. If the particle is observed or photographed at rest on a flat horizontal surface (such as a microscope slide), then the outline usually defines A_p and P_p. In the intermediate Reynolds number range (see Chapter 6), an oblate particle tends to fall with its greatest area horizontal, so that correlations based on these A_p and P_p values are useful. They are less reliable in creeping flow or at high Re where other orientations may be adopted.

An "equivalent sphere" is defined as the sphere with the same value of one of the above measures. The commonest referent is the volume-equivalent sphere, which many authors describe as the equivalent sphere without further definition. The "particle shape factor" is defined as the ratio of another measure from the above list to the corresponding value for the equivalent sphere. Of the many possible shape factors, those which have proved most useful are described below. All shape factors are open to the criticism that a range of bodies with different forms may have the same shape factor, but this is inevitable if regular or complex shapes are to be described by a single parameter.

1. *Volumetric Shape Factor*

Heywood (H5) proposed a widely used empirical parameter based on the projected profile of a particle. The "volumetric shape factor" is defined as

$$k = V/d_A{}^3, \tag{2-1}$$

where $d_A = \sqrt{4A_p/\pi}$ is the "projected area diameter," the diameter of the sphere with the same projected area as the particle.

A number of methods have been suggested for obtaining an estimate for d_A without determining A_p:

(i) The diameter may be estimated by comparison with a graticule superimposed on the image of the particle. This method has the disadvantage that it is open to subjective operator error. Generally it leads to overestimation of d_A, especially for elongated particles (H6).

(ii) Two images of the particle are displaced until they just touch, as shown in Fig. 2.1a. The displacement gives the "image-shearing diameter." This method greatly reduces operator error. Moreover, a number of values can be obtained for a given particle, corresponding to different orientations of the image relative to the direction of displacement. The mean of these values gives a good estimate for d_A (H6).

(iii) A line with random orientation is drawn to bisect the projected area. The intercept of the outline on this line, shown in Fig. 2.1b, gives the value of the "statistical intercept diameter" proposed by Martin *et al.* (M1). This method is

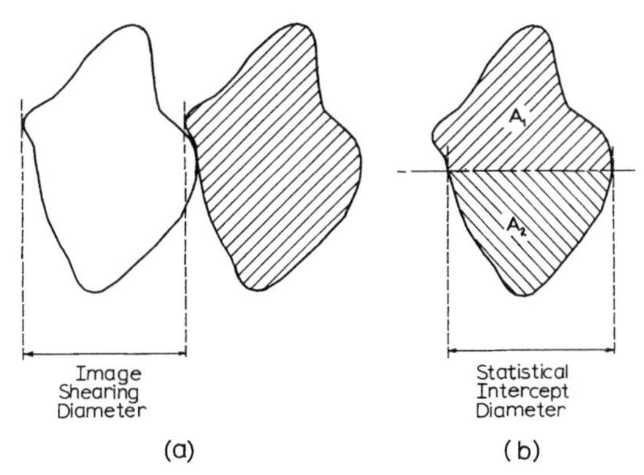

FIG. 2.1 Methods of estimating d_A, the projected area diameter: (a) image shearing method [after Heywood (H6)]; (b) statistical intercept method [after Martin *et al.* (M1)].

subject to operator error since the position of the line bisecting the area is judged subjectively. A number of such measurements can, however, be made with different orientations; the geometric mean gives d_A (H6).

Automatic techniques for characterizing particle shape without operator error are also under development, based primarily on fiber optics with automatic signal processing. Kaye (K1) has given a useful review of recent developments.

Even if an estimate for d_A is available, the volumetric shape factor can only be evaluated if the particle volume is known, and this may not be readily available for naturally occurring particles, or if a distribution of particle sizes or shapes is present. Heywood (H4) suggested that k may be estimated from the corresponding value, k_e, of an isometric particle of similar form by the relationship

$$k = k_e/(e_1\sqrt{e_2}). \tag{2-2}$$

The parameters e_1 and e_2 are obtained from:

(i) the thickness, t, defined above.
(ii) the breadth, b, defined as the minimum distance between two parallel planes which are perpendicular to the planes defining the thickness and tangential to opposite surfaces.
(iii) the length, l, projected on a plane normal to the planes defining t and b.
The "flatness ratio" is then

$$e_1 = b/t \tag{2-3}$$

and the "elongation ratio" is

$$e_2 = l/b. \tag{2-4}$$

Values of k_e for some regular shapes and approximate values for irregular shapes are given in Table 6.3. Equation (2-2) is exact for regular shapes such as spheroids and cylinders, and the group $(e_1\sqrt{e_2})^{-1}$ has itself been proposed as a simple shape factor (C1, M2).

2. *Sphericity*

Wadell (W1) proposed that the "degree of true sphericity" be defined as

$$\psi = \frac{A_e}{A} = \frac{\text{surface area of volume-equivalent sphere}}{\text{surface area of particle}}, \tag{2-5}$$

so that $\psi = 1$ for a true sphere. Although the sphericity was first introduced simply as a measure of particle shape, it was subsequently claimed to be useful for correlating drag coefficients (W3). There is some theoretical justification for the use of ψ as a correlating parameter for creeping flow past bodies whose geometric proportions resemble a sphere, but for other circumstances its use is purely empirical. The more the aspect ratio departs from unity, the lower the sphericity. For irregular particles, it is difficult to determine ψ directly.

3. *Circularity*

Wadell (W1) also introduced the "degree of circularity":

$$\not{c} = \frac{P_A}{P_p} = \frac{\text{perimeter of projected-area-equivalent sphere}}{\text{projected perimeter of particle}} = \frac{\pi d_A}{P_p} \tag{2-6}$$

Unlike the sphericity, $\not{c}$ can be determined from microscopic or photographic observation. Use of $\not{c}$ is only justified on empirical grounds, but it has the potential advantage of allowing correlation of the dependence of flow behavior on particle orientation. For an axisymmetric particle projected parallel to its axis, $\not{c}$ is unity.

Determination of the perimeter may be avoided if Feret's "statistical projected

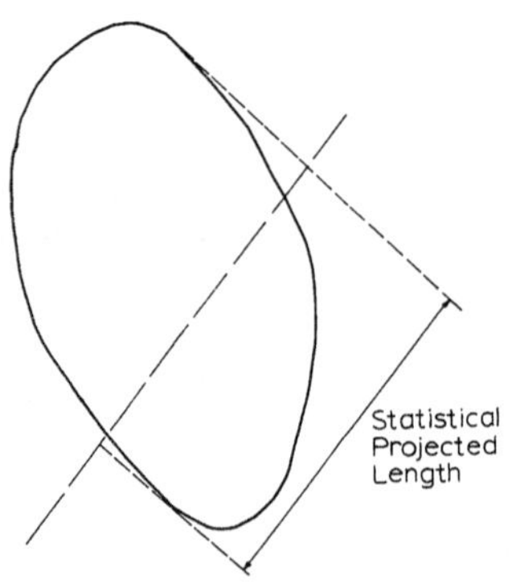

FIG. 2.2 Method of estimating d_p, the projected perimeter diameter [after Feret (F1)].

length" (F1) is employed. As for Martin's intercept diameter, a line with random orientation is drawn across the projected area. The length of the particle projected onto this line as shown in Fig. 2.2 gives the statistical projected length. Provided that the profile is not re-entrant, the mean of a number of such determinations gives the "projected perimeter diameter," d_p, the diameter of a sphere with the same projected perimeter as the particle (W6). The circularity is then given by

$$¢ = d_A/d_p. \tag{2-7}$$

4. *Operational Sphericity and Circularity*

Since the sphericity and circularity are so difficult to determine for irregular particles, Wadell (W1) proposed that ψ and $¢$ be approximated by "operational shape factors":

$$\psi_{op} = \left[\frac{\text{volume of particle}}{\text{volume of smallest circumscribing sphere}}\right]^{1/3}, \tag{2-8}$$

$$¢_{op} = \left[\frac{\text{projected area of particle}}{\text{area of smallest circumscribing circle}}\right]^{1/2}. \tag{2-9}$$

For rounded particles the operational sphericity, ψ_{op}, is well approximated (K2, P3) by $(e_2 e_1)^{-1/3}$, which is exact for ellipsoids. However, ψ_{op} is not generally a good approximation to ψ. Aschenbrenner (A2) showed that a better approximation to ψ is given by a "working sphericity" obtained from the flatness and elongation ratios by a result derived for a specific truncated polygonal form:

$$\psi_w = \frac{12.8(e_1 e_2^2)^{1/3}}{1 + e_2(1 + e_1) + 6\sqrt{1 + e_2^2(1 + e_1^2)}}. \tag{2-10}$$

This parameter has been found to correlate well with the settling behavior of naturally occurring mineral particles (B4).

The operational circularity, $¢_{op}$, is sometimes called the "projection sphericity," since Wadell (W2) suggested that $¢_{op}$ provides an estimate of ψ based on a two-dimensional projection. However, $¢_{op}$ does not approximate ψ for regular bodies, and is virtually uncorrelated with settling behavior for natural irregular particles (B4). Quick methods for evaluating $¢_{op}$ are available. It may be determined as

$$¢_{op} = d_A/(\text{diameter of circumscribing circle}), \tag{2-11}$$

where d_A is the projected-area diameter determined, for example, by image-shearing or from the Martin intercept diameter previously described. Rittenhouse (R1) has given a series of calibrated outlines for irregular particles covering the range $0.45 \le ¢_{op} \le 0.97$. Pye and Pye (P3) showed that for rounded particles $¢_{op}$ is approximated closely by $e_2^{-1/2}$, an exact relation for ellipsoids.

5. *Perimeter-Equivalent Factor*

For axisymmetric bodies with creeping flow parallel to the axis of symmetry, Bowen and Masliyah (B3) found that the most useful shape parameter was based on the sphere with the same perimeter, P', projected normal to the axis. Their shape factor is given by

$$\Sigma = \frac{\text{surface area of particle}}{\text{surface area of perimeter-equivalent sphere}} = \frac{A}{A_{\text{p}'}}. \tag{2-12}$$

It is shown in Chapter 4 that Σ is also a valuable correlating parameter for motion normal to the axis, and for diffusional mass and heat transfer.

III. SHAPE REGIMES FOR FLUID PARTICLES

Bubbles and drops tend to deform when subject to external fluid fields until normal and shear stresses balance at the fluid–fluid interface. When compared with the infinite number of shapes possible for solid particles, fluid particles at steady state are severely limited in the number of possibilities since such features as sharp corners or protuberances are precluded by the interfacial force balance.

A. Static Bubbles and Drops

Bubbles or drops which are prevented from moving under the influence of gravity by a flat plate are termed "sessile" (see Fig. 2.3a and 2.3b). When bubbles or drops remain attached to a surface with gravity acting to pull them away, they are called "pendant" (see Fig. 2.3c and 2.3d). Floating bubbles or drops, shown in Fig. 2.3e, are those which sit at the interface between two fluids.

The profiles of pendant and sessile bubbles and drops are commonly used in determinations of surface and interfacial tensions and of contact angles. Such methods are possible because the interfaces of static fluid particles must be at equilibrium with respect to hydrostatic pressure gradients and increments in normal stress due to surface tension at a curved interface (see Chapter 1). It is simple to show that at any point on the surface

$$(\rho_{\text{p}} - \rho)gy = \sigma\left(\frac{2}{R_0} - \frac{1}{R_1} - \frac{1}{R_2}\right), \tag{2-13}$$

where y is measured vertically upwards from point O on the axis of symmetry where the radius of curvature of the surface is R_0, while R_1 and R_2 are the principal radii of curvature at the point of interest. The above equation shows why the radius of curvature must increase on proceeding away from O for pendant drops or bubbles while decreasing for sessile bubbles or drops. When substitutions are made for R_1 and R_2, a second-order ordinary differential equation results which must be solved numerically (B1). The recent book by Hartland and Hartley (H3) provides complete and accurate tabulations as well

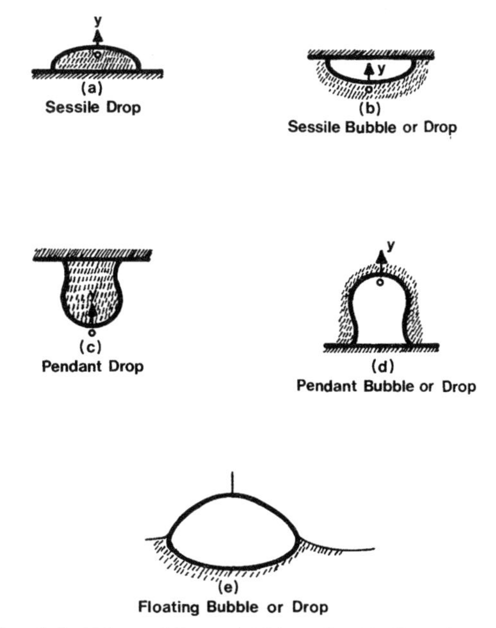

FIG. 2.3 Shapes of static bubbles and drops: (a),(b) sessile; (c),(d) pendant; (e) floating. (Shading denotes more dense fluid in each case.)

as a general review, approximate solutions, numerical methods, and treatment of menisci and stationary particles subject to applied forces. Another useful review of this subject, including also the case of fluid particles at equilibrium in centrifugal fields, has been prepared by Princen (P1). Many standard texts on surface chemistry [e.g. (A1)] also contain discussions of the use of pendant and sessile drops in determining interfacial tensions.

B. BUBBLES AND DROPS IN FREE MOTION

Bubbles and drops in free rise or fall in infinite media under the influence of gravity are generally grouped under the following three categories:

(a) "Spherical": Generally speaking, bubbles and drops are closely approximated by spheres if interfacial tension and/or viscous forces are much more important than inertia forces. For our purposes, fluid particles will be termed "spherical" if the minor to major axis ratio lies within 10% of unity. Spherical fluid particles in free rise or fall are discussed in Chapters 3 and 5.

(b) "Ellipsoidal": The term "ellipsoidal" is generally used to refer to bubbles and drops which are oblate with a convex interface (viewed from inside) around the entire surface. Photographs of bubbles and drops in this regime are given in Fig. 2.4a, b, and d. It must be noted that actual shapes may differ considerably from true ellipsoids and that fore-and-aft symmetry must not be assumed.

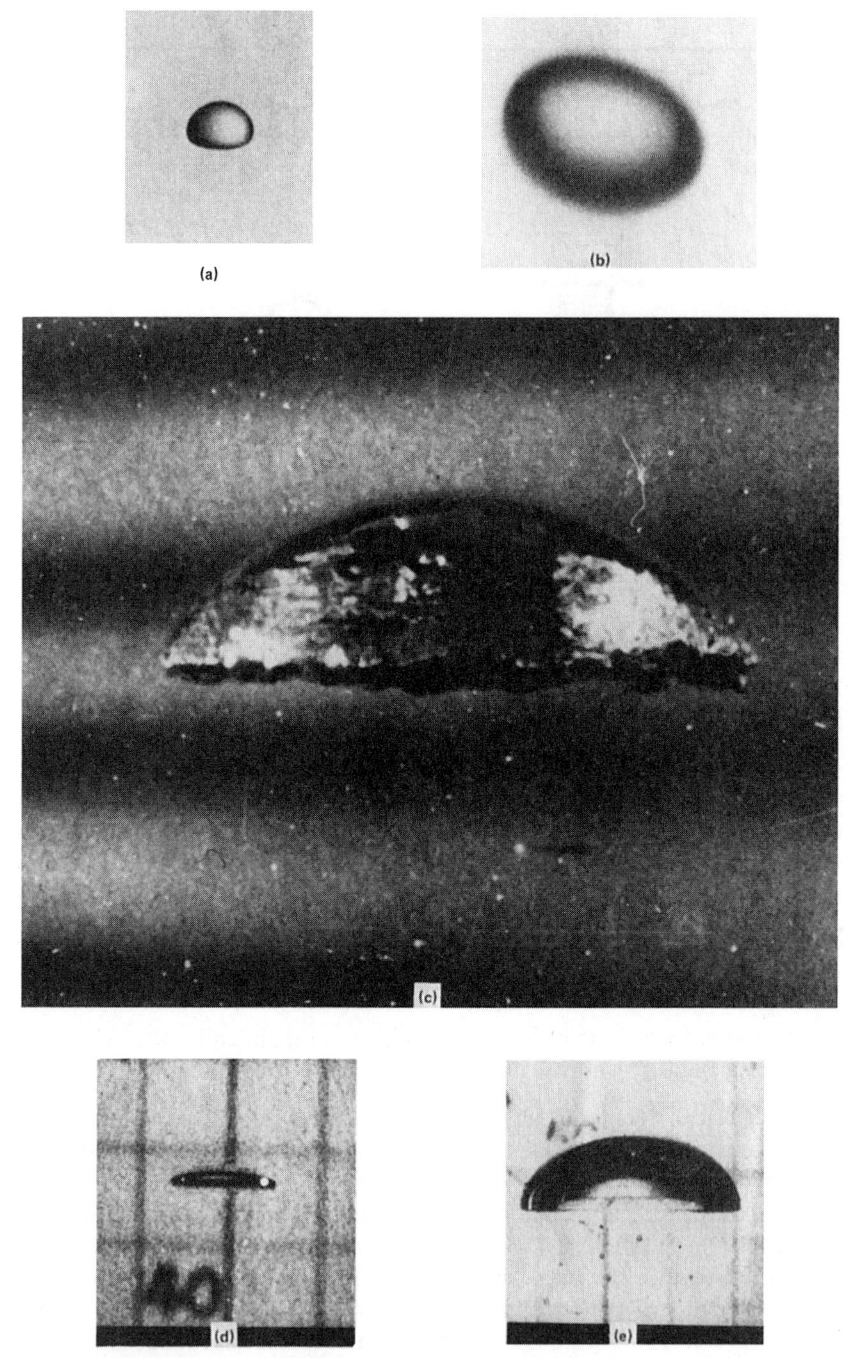

FIG. 2.4 Photographs of bubbles and drops in different shape regimes. All photographs reproduced with the permission of the publisher and author, if previously published.

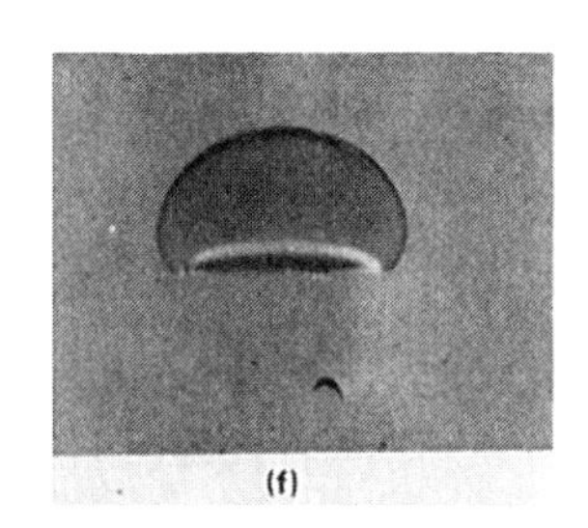
(f)

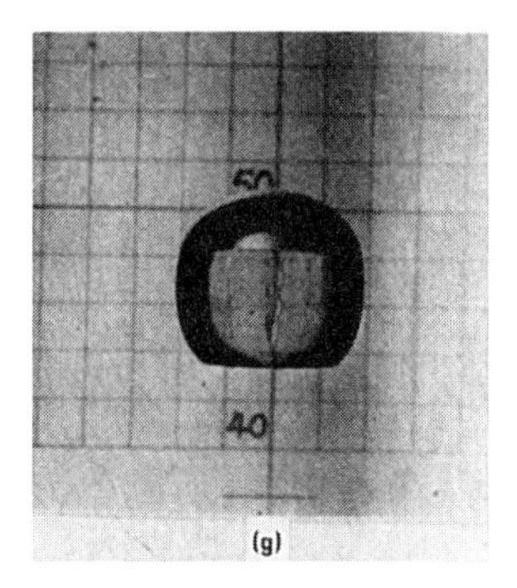

(g)

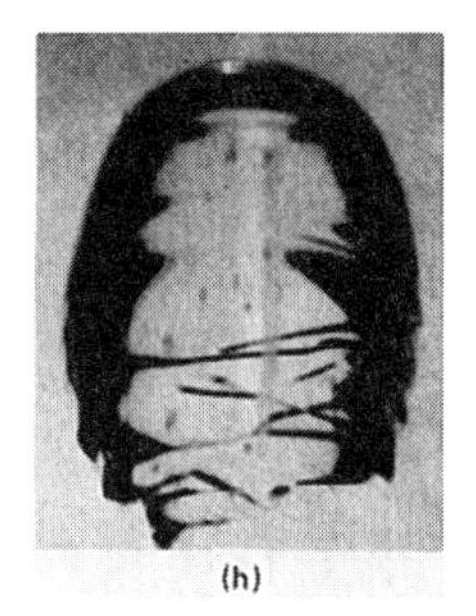
(h)

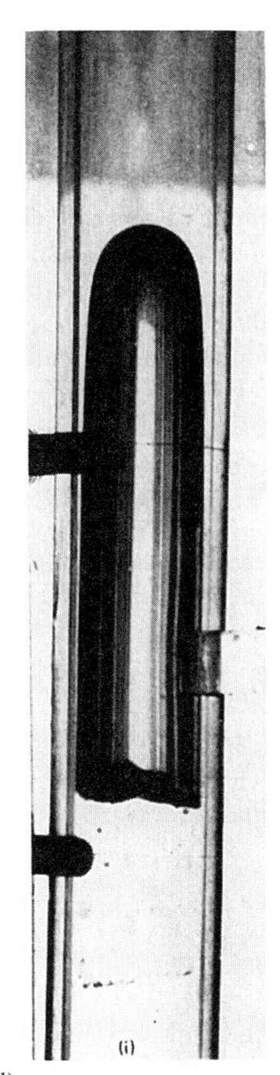
(i)

FIG. 2.4 (continued)

Fig.	Ref.	Dispersed fluid	Continuous fluid	M	d_e (cm)	Eo	Comments
a	(P2)	water	air	1.7×10^{-12}	0.58	4.5	see Chapter 7
b		air	water	3.1×10^{-11}	0.6	5	see Chapter 7; note wobble
c	(D1)	air	water	3.1×10^{-11}	4.2	240	see Chapter 8
d	(B2)	air	aqueous sugar solution	8.2×10^{-4}	1.4	32	very flat ellipsoid
e	(B2)	air	aqueous sugar solution	5.5	4.1	290	see Chapter 8
f	(W5)	chloroform	aqueous sugar solution	1.2×10^{3}	5.1	70	see Chapter 8
g	(B2)	air	aqueous sugar solution	45	4.3	320	smooth skirt; see Chapter 8
h	(W5)	air	paraffin oil	0.34	5.8	780	wavy skirt; see Chapter 8
i	(B5)	air	mineral oil	2.6×10^{-5}	$\gg D$	N.A.	$Fr_D = 0.33$, $D = 2.6$ cm; see Chapter 9

Moreover, ellipsoidal bubbles and drops commonly undergo periodic dilations or random wobbling motions which make characterization of shape particularly difficult. Chapter 7 is devoted to this regime.

(c) "Spherical-cap" or "ellipsoidal-cap": Large bubbles and drops tend to adopt flat or indented bases and to lack any semblance of fore-and-aft symmetry. Such fluid particles may look very similar to segments cut from spheres or from oblate spheroids of low eccentricity; in these cases the terms "spherical-cap" and "ellipsoidal-cap" are used. If the fluid particle has an indentation at the rear, it is often said to be "dimpled." Large spherical- or ellipsoidal-caps may also trail thin envelopes of dispersed fluid referred to as "skirts." Photographs of freely rising fluid particles in this regime are shown in Fig. 2.4c, e, f, g and h. Spherical- and ellipsoidal-caps with and without skirts are treated in Chapter 8.

When bubbles and drops rise or fall in bounded media their shape is affected by the walls of the container. If the bubble or drop is sufficiently large, it fills most of the container cross section and the "slug flow" regime results. A photograph of a slug flow bubble is shown in Fig. 2.4i. The effect of bounding walls is treated in Chapter 9.

For bubbles and drops rising or falling freely in infinite media it is possible to prepare a generalized graphical correlation in terms of the Eötvös number,† Eo; Morton number,‡ M; and Reynolds number, Re (G1, G2):

$$\mathrm{Eo} = g\,\Delta\rho\, d_e^{\,2}/\sigma, \tag{2-14}$$

$$\mathrm{M} = g\mu^4\,\Delta\rho/\rho^2\sigma^3, \tag{2-15}$$

$$\mathrm{Re} = \rho d_e U/\mu. \tag{2-16}$$

The resulting plot is shown in Fig. 2.5. Figure 2.5 does not apply to the extreme values of density ratio, $\gamma = \rho_p/\rho$, or viscosity ratio, $\kappa = \mu_p/\mu$, found for liquid drops falling through gases. Drops in gases are considered explicitly in Chapter 7. Aside from this exclusion, the range of fluid properties and particle volumes covered by Fig. 2.5 is very broad indeed. Since Re is the only one of the three groups to contain the terminal velocity, Fig. 2.5 may be used to estimate terminal velocities as well as the shape regime, although more accurate predictive correlations are usually available. It is notable that μ_p does not play an important role in determining terminal velocities and shape regimes since it does not appear

† When the Hungarian alphabet was reformed in the 1920's, Eötvös (pronounced Ertversh) was given special dispensation to keep the archaic spelling. For convenience, we drop the umlauts from now on. In the present context, the name appears to have originated with Harmathy (H2). This group is sometimes referred to as the Bond number.

‡ We have called this group the Morton number throughout this book, although it was used prior to Haberman and Morton (H1) by Rosenberg (R2) who refers to an even earlier user. In the literature, the group is often simply referred to as the M-group or property group, and its inverse as the P-group.

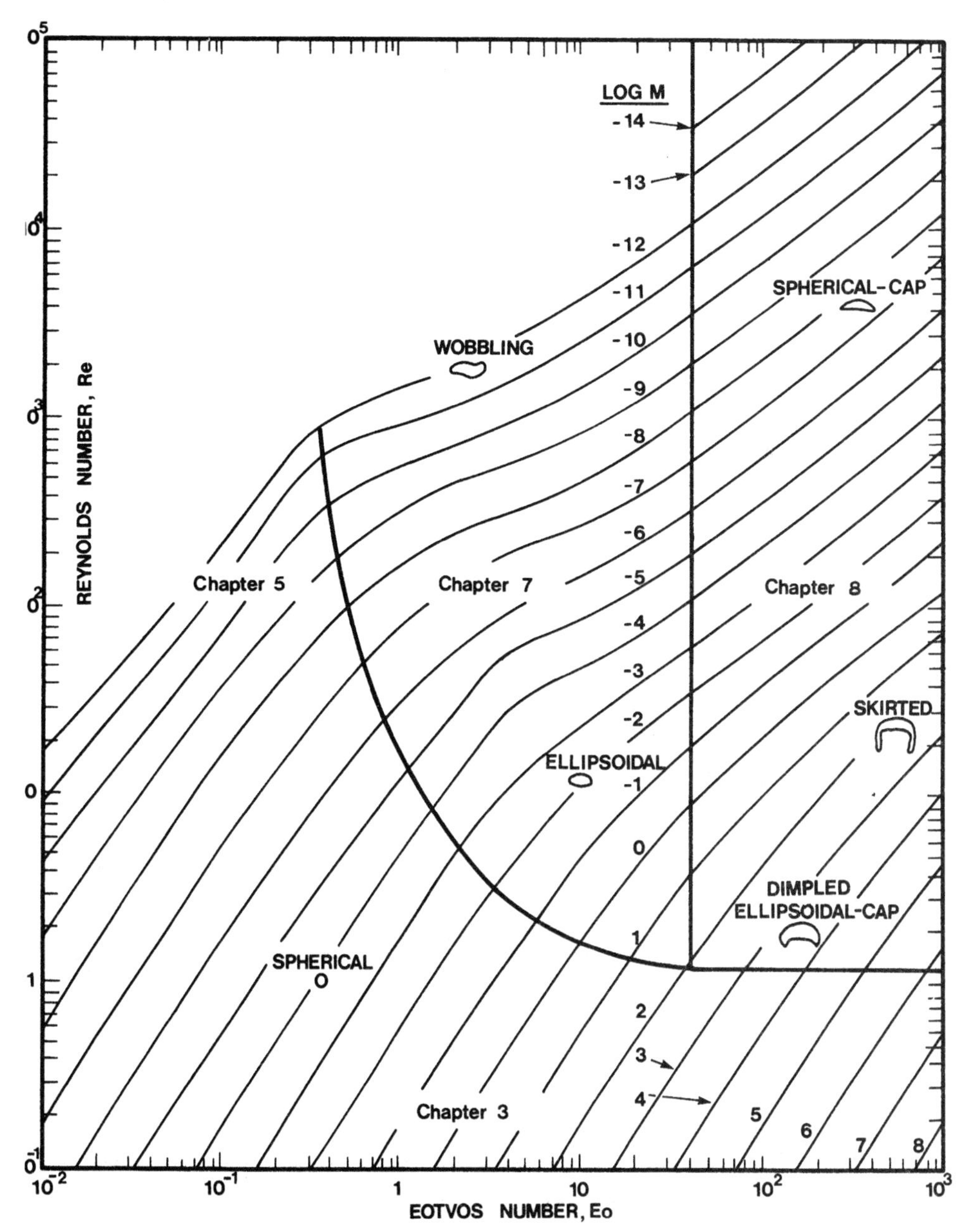

FIG. 2.5 Shape regimes for bubbles and drops in unhindered gravitational motion through liquids.

in any of the three groups used to construct Fig. 2.5. The role of μ_p may be significant, however, for very pure (surfactant-free) systems or for large fluid particles in high M liquids. These cases are considered in Chapters 3 and 8.

Figure 2.5 shows boundaries between the three principal shape regimes described above, as given by Grace *et al.* (G2). While the boundaries between the principal shape regimes are somewhat arbitrary, it is clear that bubbles and drops are ellipsoidal at relatively high Re and intermediate Eo while the spherical- or ellipsoidal-cap regime requires that both Eo and Re be large. Various subregimes may also be mapped (B2, W4), and some of these are included in Fig. 2.5. Again the boundaries are somewhat arbitrary. Nevertheless, Fig. 2.5 is a useful tool for demonstrating the wide range of bubble and drop behavior considered in the following chapters.

REFERENCES

A1. Adamson, A. W., "Physical Chemistry of Surfaces," 2nd Ed., Wiley (Interscience), New York, 1967.

A2. Aschenbrenner, B., *J. Sediment. Petrol.* **26**, 15–31 (1956).

B1. Bashforth, F., and Adams, J. C., "An Attempt to Test the Theories of Capillary Action." Cambridge Univ. Press, London, 1883.

B2. Bhaga, D., Ph.D. Thesis, McGill Univ., Montreal, 1976.

B3. Bowen, B. D., and Masliyah, J. H., *Can. J. Chem. Eng.* **51**, 8–15 (1973).

B4. Briggs, L. I., McCulloch, D. S., and Moser, F., *J. Sediment. Petrol.* **32**, 645–656 (1962).

B5. Brown, R. A. S., *Can. J. Chem. Eng.* **43**, 217–223 (1965).

C1. Corey, A. T., M. S. Thesis, Colorado A&M Coll., Fort Collins, 1949.

D1. Davenport, W. G., Bradshaw, A. V., and Richardson, F. D., *J. Iron Steel Inst.* **205**, 1034–1042 (1967).

F1. Feret, L. R., *Assoc. Int. Essai Mater., Zurich* **2**, Group D, 428–436 (1931).

G1. Grace, J. R., *Trans. Inst. Chem. Eng.* **51**, 116–120 (1973).

G2. Grace, J. R., Wairegi, T., and Nguyen, T. H., *Trans. Inst. Chem. Eng.* **54**, 167–173 (1976).

H1. Haberman, W. L., and Morton, R. K., *David Taylor Model Basin Rep.* No. 802 (1953).

H2. Harmathy, T. Z., *AIChE J.* **6**, 281–288 (1960).

H3. Hartland, S., and Hartley, R. W., "Axisymmetric Fluid–Liquid Interfaces." Elsevier, Amsterdam, 1976.

H4. Heywood, H., *Powder Metall.* **7**, 1–28 (1961).

H5. Heywood, H., *Symp. Interaction Fluids and Parts., Inst. Chem. Eng., London* pp. 1–8 (1962).

H6. Heywood, H., Cambridge, L. A., Corbett, J. E., Brown, M., and Gratton, A., *Loughborough Univ. Chem. Eng. J.* **6**, 22–26 (1971).

K1. Kaye, B. H., *Powder Technol.* **8**, 293–306 (1973).

K2. Krumbein, W. C., *J. Sediment. Petrol.* **11**, 64–72 (1941).

M1. Martin, G., Blyth, C. E., and Tongue, H., *Trans. Br. Ceram. Soc.* **23**, 61–118 (1923–4).

M2. McNown, J. S., and Malaika, J., *Trans. Am. Geophys. Union* **31**, 74–82 (1950).

P1. Princen, H. M., *in* "Surface and Colloid Science" (E. Matijevic, ed.), Vol. 2, pp. 1–84, Wiley, New York, 1969.

P2. Pruppacher, H. R., and Beard, K. V., *Q. J. R. Meteorol. Soc.* **96**, 247–256 (1970).

P3. Pye, W. D., and Pye, M. H., *J. Sediment. Petrol.* **13**, 28–34 (1943).

R1. Rittenhouse, G., *J. Sediment. Petrol.* **13**, 179–181 (1943).

R2. Rosenburg, B., *David Taylor Model Basin Rep.* No. 727 (1950).

W1. Wadell, H., *J. Geol.* **41**, 310–331 (1933).
W2. Wadell, H., *J. Geol.* **43**, 250–280 (1935).
W3. Wadell, H., *J. Franklin Inst.* **217**, 459–490 (1934).
W4. Wairegi, T., Ph.D. Thesis, McGill Univ., Montreal, 1974.
W5. Wairegi, T., and Grace, J. R., *Int. J. Multiphase Flow* **3**, 67–77 (1976).
W6. Walton, W. H., *Nature (London)* **162**, 329–330 (1948).

Chapter 3

Slow Viscous Flow Past Spheres

I. INTRODUCTION

The system considered in this chapter is a rigid or fluid spherical particle of radius a moving relative to a fluid of infinite extent with a steady velocity U. The Reynolds number is sufficiently low that there is no wake at the rear of the particle. Since the flow is axisymmetric, it is convenient to work in terms of the Stokes stream function ψ (see Chapter 1). The starting point for the discussion is the "creeping flow approximation," which leads to Eq. (1-36). It was noted in Chapter 1 that Eq. (1-36) implies that the flow field is "reversible," so that the flow field around a particle with fore-and-aft symmetry is also symmetric. Extensions to the creeping flow solutions which lack fore-and-aft symmetry are considered in Sections II, E and F.

II. FLUID MECHANICS

A. Fluid Spheres: Hadamard–Rybczynski Solution

One of the most important analytic solutions in the study of bubbles, drops, and particles was derived independently by Hadamard (H1) and Rybczynski (R5). A fluid sphere is considered, with its interface assumed to be completely free from surface-active contaminants, so that the interfacial tension is constant. It is assumed that both Re and Re_p are small so that Eq. (1-36) can be applied to both fluids, i.e.,

$$E^4\psi = E^4\psi_p = 0. \tag{3-1}$$

The boundary conditions require special attention. Taking a reference frame fixed to the particle with origin at its center, they are

(a) Uniform stream flow in the $-z$ direction at large distances from the sphere:

$$\psi/r^2 \to -(U/2)\sin^2\theta \qquad \text{as} \quad r \to \infty; \tag{3-2}$$

(b) No flow across the interface:

$$\psi = \psi_p = 0 \qquad \text{at} \quad r = a; \tag{3-3}$$

(c) Continuity of tangential velocity across the interface:

$$\partial\psi/\partial r = \partial\psi_p/\partial r \qquad \text{at} \quad r = a; \tag{3-4}$$

(d) Continuity of tangential stress across the interface:

$$\frac{\partial}{\partial r}\left(\frac{1}{r^2}\frac{\partial\psi}{\partial r}\right) = \kappa\,\frac{\partial}{\partial r}\left(\frac{1}{r^2}\frac{\partial\psi_p}{\partial r}\right) \qquad \text{at} \quad r = a, \tag{3-5}$$

where $\kappa = \mu_p/\mu$ is the viscosity ratio.

(e) Continuity of normal stress across the interface:

$$p - 2\mu\frac{\partial}{\partial r}\left(\frac{1}{r^2\sin\theta}\frac{\partial\psi}{\partial\theta}\right) + \frac{2\sigma}{a} = p_p - 2\mu_p\frac{\partial}{\partial r}\left(\frac{1}{r^2\sin\theta}\frac{\partial\psi_p}{\partial\theta}\right) \qquad \text{at} \quad r = a, \tag{3-6}$$

where p and p_p are the modified pressures in each phase (see Chapter 1) and the term $2\sigma/a$ results from the pressure increment associated with interfacial tension.

The solution of Eq. (3-1) with boundary conditions (3-2) to (3-5) may be found in a number of standard texts [e.g. (B2, L3)], and is

$$\psi = -\frac{Ur^2\sin^2\theta}{2}\left[1 - \frac{a(2 + 3\kappa)}{2r(1 + \kappa)} + \frac{\kappa a^3}{2r^3(1 + \kappa)}\right], \tag{3-7}$$

$$\psi_p = \frac{Ur^2\sin^2\theta}{4(1 + \kappa)}\left[1 - \frac{r^2}{a^2}\right]. \tag{3-8}$$

The internal motion given by Eq. (3-8) is that of Hill's spherical vortex (H6). Streamlines are plotted in Figs. 3.1 and 3.2 for $\kappa = 0$ and $\kappa = 2$, and show the fore-and-aft symmetry required by the creeping flow equation. It may also be noted in Fig. 3.2 that the streamlines are not closed; for any value of κ, the solution predicts that outer fluid is entrained along with the moving sphere. This entrainment, sometimes known as "drift," is infinite in creeping flow. This problem is discussed further in Chapter 4.

The solution given by Eqs. (3-7) and (3-8) is derived using only the first four boundary conditions (L3); i.e. without considering the normal stress condition, Eq. (3-6). The modified pressures can be obtained from Eq. (1-33) and are given by

$$p = p_0 + [\mu Ua\cos\theta(2 + 3\kappa)/2r^2(1 + \kappa)], \tag{3-9}$$

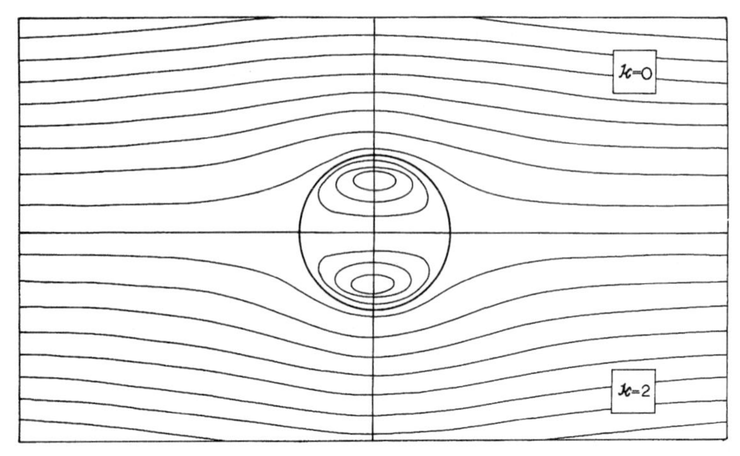

FIG. 3.1 Streamlines relative to spherical fluid particle at low Re: Hadamard–Rybczynski solution.

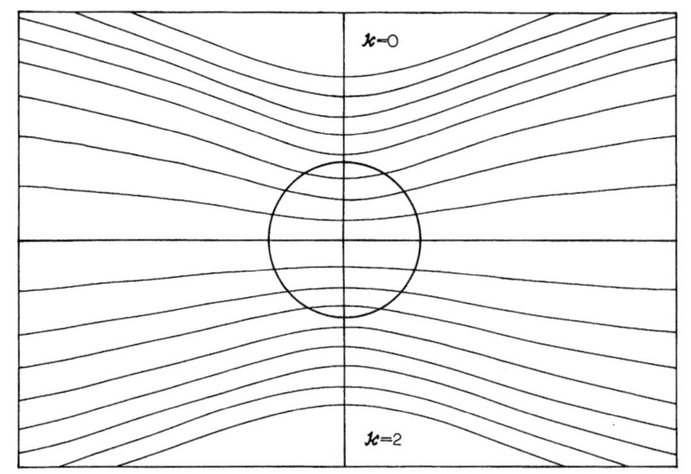

FIG. 3.2 Streamlines for motion of fluid sphere through stagnant fluid at low Re caused by translation of spherical fluid particle: Hadamard–Rybczynski solution.

$$p_p = p_{0p} - [5\mu_p U r \cos\theta / a^2(1 + \kappa)], \tag{3-10}$$

where p_0 and p_{0p} are constants. Even though Eqs. (3-9) and (3-10) are derived without considering Eq. (3-6), the latter is also satisfied if $p_{0p} - p_0 = 2\sigma/a$. Thus the problem is not overspecified, and the assumed spherical shape is consistent with the other assumptions in the derivation. This leads to the important conclusion that bubbles and drops are spherical when the creeping flow approximation is valid, and only deform from a spherical shape when inertial terms become significant. A further corollary is that it is not necessary for surface tension forces to be predominant for a bubble or drop to be spherical. Moreover, as we shall see below, surface-active contaminants may cause marked changes in internal circulation and drag for a bubble or drop, but the effect on shape is negligible at low Reynolds numbers. Thus if Reynolds numbers are very low, bubbles and drops remain spherical no matter how small the surface tension forces. The onset of deformation of fluid particles is discussed in Chapter 7.

The pressure distribution given by Eq. (3-9) is an odd function of θ, so that the particle experiences a net pressure force or "form drag." Integration of the pressure over the surface of the particle leads to a drag component given by

$$C_{D1} = \frac{8}{3\mathrm{Re}}\left(\frac{2 + 3\kappa}{1 + \kappa}\right). \tag{3-11}$$

This result may be contrasted with potential flow past a sphere, where the streamlines again have fore-and-aft symmetry but p is an even function of θ so that there is no net pressure force (see Chapter 1). Additional drag components arise from the deviatoric normal stress:

$$C_{D2} = 32/[3\mathrm{Re}(1 + \kappa)] \tag{3-12}$$

and from the shear stress:

$$C_{D3} = \frac{16}{\mathrm{Re}}\frac{\kappa}{(1 + \kappa)}. \tag{3-13}$$

The overall drag coefficient is the sum of these three contributions:

$$C_D = \frac{2F_D}{\pi\rho U^2 a^2} = \frac{8}{\mathrm{Re}}\left(\frac{2 + 3\kappa}{1 + \kappa}\right), \tag{3-14}$$

so that $C_{D1} = C_D/3$ for all κ. For a gas bubble ($\kappa = 0$) the shear component, C_{D3}, is zero and the deviatoric normal stress plays a very important role in determining the overall drag.†

The terminal velocity of a fluid particle in creeping flow is obtained by equating the total drag to the net gravity force, $4\pi a^3 \Delta\rho\, g/3$, giving:

$$U_T = \frac{2}{3}\frac{ga^2\,\Delta\rho}{\mu}\left(\frac{1 + \kappa}{2 + 3\kappa}\right). \tag{3-15}$$

Finally, the vorticity at the interface is

$$\zeta_s = \frac{U\sin\theta}{2a}\left(\frac{2 + 3\kappa}{1 + \kappa}\right). \tag{3-16}$$

† It is an interesting semantic question whether C_{D2} should be regarded as a component of form drag or of skin friction.

B. Rigid Spheres: Stokes's Solution

Stokes's solution (S9) for steady creeping flow past a rigid sphere may be obtained directly from the results of the previous section with $\kappa \to \infty$. The same results are obtained by solving Eq. (3-1) with Eqs. (3-4) to (3-6) replaced by the single condition that $u_\theta = 0$ at $r = a$. The corresponding streamlines are shown in Figs. 3.3a and 3.4a. As for fluid spheres, the particle causes significant

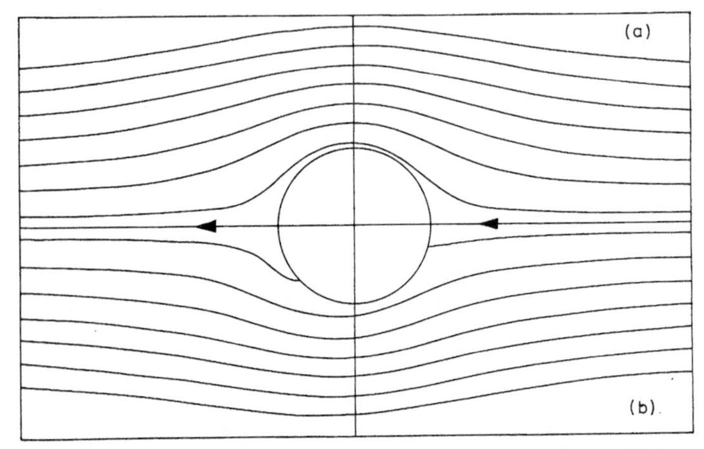

FIG. 3.3 Streamlines relative to rigid sphere at low Re: (a) Stokes's solution; (b) Oseen approximation.

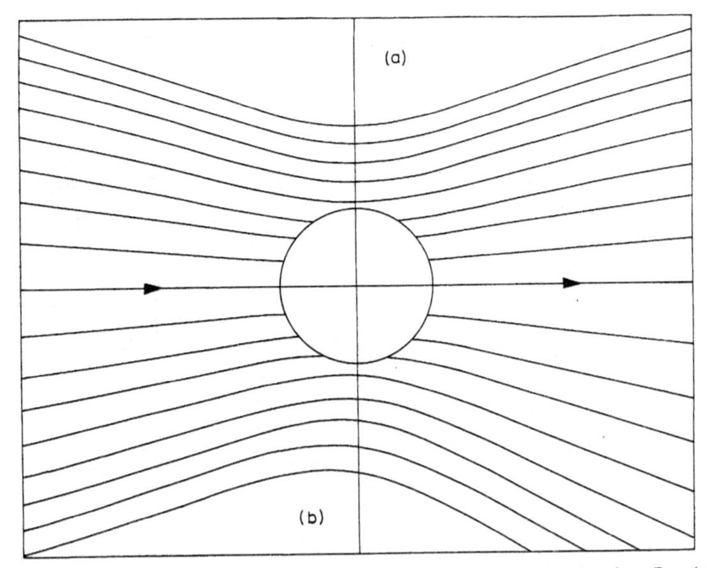

FIG. 3.4 Streamlines for motion of rigid sphere through stagnant fluid at low Re: (a) Stokes's solution; (b) Oseen approximation.

streamline curvature over a very extensive region, and there is infinite drift. On the axis of symmetry, the fluid velocity falls to half the sphere velocity almost two radii from the surface. The corresponding distance for potential flow is 0.7 radii.

From Eqs. (3-11) to (3-14), the total drag coefficient is given by "Stokes's law":

$$C_{\mathrm{DSt}} = 24/\mathrm{Re}. \tag{3-17}$$

Two thirds of this drag arises from skin friction, one third from form drag, and the component due to deviatoric normal stress is zero. The corresponding terminal velocity follows from Eq. (3-15) as:

$$U_{\mathrm{TS}} = 2ga^2\,\Delta\rho/9\mu = gd^2\,\Delta\rho/18\mu. \tag{3-18}$$

The surface vorticity obtained from Eq. (3-16) is

$$\zeta_{\mathrm{s}} = 3U \sin\theta/2a. \tag{3-19}$$

C. Experimental Results: Fluid Spheres

The Hadamard–Rybczynski theory predicts that the terminal velocity of a fluid sphere should be up to 50% higher than that of a rigid sphere of the same size and density. However, it is commonly observed that small bubbles and drops tend to obey Stokes's law, Eq. (3-18), rather than the corresponding Hadamard–Rybczynski result, Eq. (3-15). Moreover, internal circulation is essentially absent. Three different mechanisms have been proposed for this phenomenon, all implying that Eq. (3-5) is incomplete.

Bond and Newton (B3) found that small bubbles and drops followed Stokes's law while, with increasing diameter, there was a rather sharp increase in velocity toward the Hadamard–Rybczynski value. They suggested that a circulating particle requires energy locally to stretch interfacial area elements over the leading hemisphere, while these shrink over the rear surface. It was hypothesized that this process caused additional tangential stresses to retard the particle and that surface tension played the dominant role in determining whether U_{T} followed Eq. (3-15) or (3-18). They proposed that internal circulation could only occur for $\mathrm{Eo} > 4$. This has come to be known as the "Bond criterion." That it gave fair agreement with observed bubble or drop sizes at which the terminal velocity was midway between the Stokes and Hadamard–Rybczynski values probably reflects the fact that the degree of contamination by surface active substances is often roughly proportional to the surfactant-free interfacial tension, σ_{SF} (D3, G7). However, subsequent experimental work [e.g. (G1, G2, G3, L5)] has shown that the Bond criterion is not always applicable. Harper *et al.* (H5) and Kenning (K1), on the other hand, have shown that the surface energy argument is valid if tangential gradients of temperature and hence surface tension are considered. However, the effect is much too small to account for the immobile interfaces of small bubbles and drops.

Boussinesq (B4) proposed that the lack of internal circulation in bubbles and drops is due to an interfacial monolayer which acts as a viscous membrane. A constitutive equation involving two parameters, surface shear viscosity and surface dilational viscosity, in addition to surface tension, was proposed for the interface. This model, commonly called the "Newtonian surface fluid model" (W2), has been extended by Scriven (S3). Boussinesq obtained an exact solution to the creeping flow equations, analogous to the Hadamard–Rybczinski result but with surface viscosity included. The resulting terminal velocity is

$$U_{\mathrm{T}} = \frac{2}{3}\frac{ga^2\,\Delta\rho}{\mu}\left[\frac{1 + \kappa + C/\mu}{2 + 3\kappa + 3C/\mu}\right], \tag{3-20}$$

where C is equal to the surface dilational viscosity divided by 1.5 times the radius. Although Eq. (3-20) reduces to Eqs. (3-15) and (3-18) for $C = 0$ and $C = \infty$, respectively, the transition between these results with decreasing radius is in practice much sharper than predicted [e.g. (B3)]. A further difficulty with surface viscosity is that it is very difficult to obtain reliable measurements (O1, W2).

The most reasonable explanation for the absence of internal circulation for small bubbles and drops was provided by Frumkin and Levich (F1, L3). Surface-active substances tend to accumulate at the interface between two fluids, thereby reducing the surface tension. When a drop or bubble moves through a continuous medium, adsorbed surface-active materials are swept to the rear, leaving the frontal region relatively uncontaminated. The concentration gradient results in a tangential gradient of surface tension which in turn causes a tangential stress (see Eq. (1-14)) tending to retard surface motion. These gradients are most pronounced for small bubbles and drops, in agreement with the tendency for small fluid particles to be particularly subject to retardation. Models relating to surface contamination are discussed in the next section.

The surface contamination theory implies that all bubbles and drops, no matter how small, will show internal circulation if the system is sufficiently free of surface-active contaminants. Experimental evidence tends to support this view. For example, Redfield and Houghton (R2) took considerable pains to purify systems in which air bubbles rose in aqueous dextrose solutions, and reported excellent agreement with the Hadamard–Rybczynski drag relationship. Similarly, Levich (L3) reports that mercury drops falling through pure glycerine have velocities which are 50% greater than the Stokes value. Observations at higher Reynolds numbers also confirm this theory qualitatively (B1, E1, E2, L5).

Internal circulation patterns have been observed experimentally for drops by observing striae caused by the shearing of viscous solutions (S7) or by photographing non-surface-active aluminum particles or dyes dispersed in the drop fluid [e.g. (G2, G3, J2, L5, M1, S1)]. A photograph of a fully circulating falling drop is shown in Fig. 3.5a. Since the internal flow pattern for the Hadamard–Rybczynski analysis satisfies the complete Navier–Stokes equation

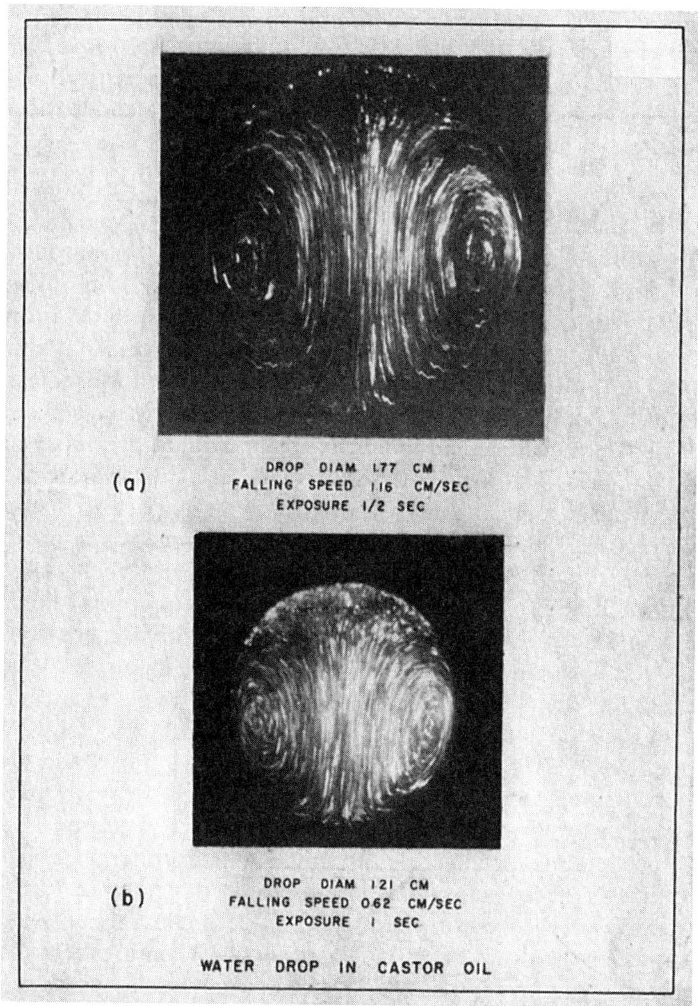

FIG. 3.5 Internal circulation in a water drop falling through castor oil [from Savic (S1), reproduced by permission of the National Research Council of Canada]: (a) $d = 1.77$ cm, $U_T = 1.16$ cm/s, exposure 1/2 s, fully circulating; (b) $d = 1.21$ cm, $U_T = 0.62$ cm/s, exposure 1 s, stagnant cap at top of drop.

(H3, T1), it is unimportant that the Reynolds number of the internal motion was rather large for many flow visualization studies which set out to verify the Hadamard–Rybczynski predictions, so long as the Reynolds number based on the continuous fluid properties was small and the fluid particle spherical. The observed streamlines show excellent qualitative agreement with theory, although quantitative comparison is difficult in view of refractive mdex differences and the possibility of surface contamination. When a trace of surface-active contaminant is present, the motion tends to be damped out first at the rear of the bubble or drop (G3, H7, M1, S1). A photograph reproduced in Fig. 3.5b demonstrates that the internal vortex for a falling drop is pushed forward, leaving a stagnant region at the rear where the contaminant tends to accumulate. Similar asymmetry has been noted by others (G9, L5).

D. Effect of Surface Contaminants

Traces of surface-active contaminants may have a profound effect on the behavior of drops and bubbles. Even though the amount of impurity may be so small that there is no measurable change in the bulk fluid properties, a contaminant can eliminate internal circulation, thereby significantly increasing the drag and drastically reducing overall mass- and heat-transfer rates. Systems which exhibit high interfacial tensions, including common systems like air/water, liquid metals/air, and aqueous liquids/nonpolar liquids, are most subject to this effect (D2, L5). The measures required to purify such systems and the precautions needed to ensure no further contamination are so stringent that one must accept the presence of surface-active contaminants in most systems of practical importance. For this reason, the Hadamard–Rybczynski theory is not often obeyed in practice, although it serves as an important limiting case.

Accounting for the influence of surface-active contaminants is complicated by the fact that both the amount and the nature of the impurity are important in determining its effect (G7, L5, R1). Contaminants with the greatest retarding effect are those which are insoluble in either phase (L5) and those with high surface pressures (G7). A further complication is that bubbles and drops may be relatively free of surface-active contaminants when they are first injected into a system, but internal circulation and the velocity of rise or fall decrease with time as contaminant molecules accumulate at the interface (G3, L5, R3). Further effects of surface impurities are discussed in Chapters 7 and 10. For a useful synopsis of theoretical work on the effect of contaminants on bubbles and drops, see the critical review by Harper (H3). Attention here is confined to the practically important case of a surface-active material which is insoluble in the dispersed phase. The effects of ions in solution or in double layers adjacent to the interface are not considered.

The first attempt to account for surface contamination in creeping flow of bubbles and drops was made by Frumkin and Levich (F1, L3) who assumed that the contaminant was soluble in the continuous phase and distributed over the interface. The form of the concentration distribution was controlled by one of three rate limiting steps: (a) adsorption-desorption kinetics, (b) diffusion in the continuous phase, (c) surface diffusion in the interface. In all cases the terminal velocity was given by an equation identical to Eq. (3-20) where C, now called the "retardation coefficient", is different for the three cases. The analysis has been extended by others (D6, D7, N2).

Since the Frumkin–Levich approach predicts symmetrical internal circulation, various workers have tried to account for the asymmetry clearly shown in photographs such as Fig. 3.5. Savic was the first to attempt an analysis by assuming that the contaminant was strongly surface active and insoluble in both phases. The equations solved and the boundary conditions imposed were Eqs. (3-1) to (3-4) with the tangential stress condition replaced by:

$$\text{for } r = a; \qquad 0 < \theta < \theta_0,\ \tau_{r\theta} = 0, \tag{3-21}$$

$$\text{for } r = a; \qquad \theta_0 < \theta < \pi,\ u_\theta = 0, \tag{3-22}$$

where θ_0 is the "stagnant cap angle," measured from the nose of the bubble or drop. Note that Eq. (3-21) restricts the direct applicability of Savic's analytic results to cases in which $\kappa \to 0$. The terminal velocity is

$$U_T = \frac{2}{9}\frac{ga^2\,\Delta\rho}{\mu}\,Y(\theta_0) = U_{TS}Y(\theta_0). \tag{3-23}$$

Savic's calculated values of Y, along with values obtained subsequently (D5, H8), are plotted in Fig. 3.6. Also shown is an asymptotic solution (H4) for a small stagnant cap ($3\pi/4 < \theta < \pi$):

$$Y = 1.5/(1 + 2(\pi - \theta_0)^3/3\pi). \tag{3-24}$$

Savic estimated cap angles from his photographs and the resulting predictions using Fig. 3.6 showed good agreement with experimental terminal velocities.

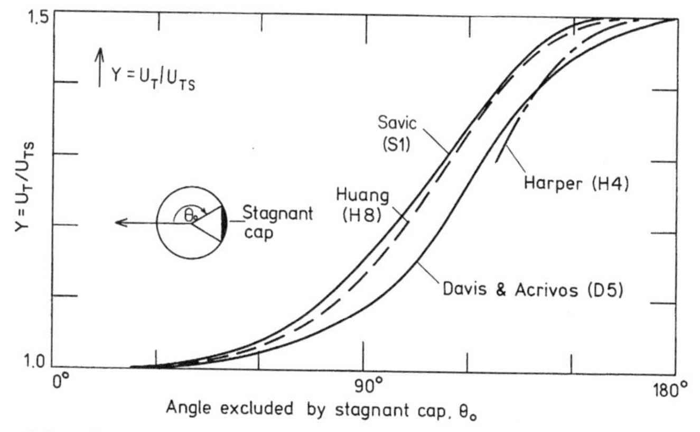

FIG. 3.6 Effect of stagnant cap on terminal velocity of a bubble or inviscid drop.

By assuming that the surface tension on the surface of a fluid sphere varied from the surfactant-free value, σ_{sf}, at the nose to zero at the rear, Savic also deduced a relationship between velocity and Eotvos number, shown in Fig. 3.7, which agrees qualitatively with the experimental results of Bond and Newton. Modifications of this approach for cases where the maximum change in local interfacial tension is less than σ_{sf} have been devised for bubbles (D5, G7) and for drops (U1). Griffith (G7) treated a surface-active monolayer distributed under the influence of the shear at the interface. The difference between the interfacial tension between the pure fluids and the equilibrium value with the surfactant present may then be denoted by $\Delta\sigma'$ and the corresponding modified Eotvos number by Eo′. Davis and Acrivos (D5) assumed that the supply of surfactant is unlimited, so that the minimum surface tension corresponds to the condition at which the surface film collapses. The difference between this value and the surfactant-free value, assumed to prevail at the nose, may be denoted by $\Delta\sigma^*$, and the corresponding Eotvos number by Eo*. The resulting curves are shown in Fig. 3.7.

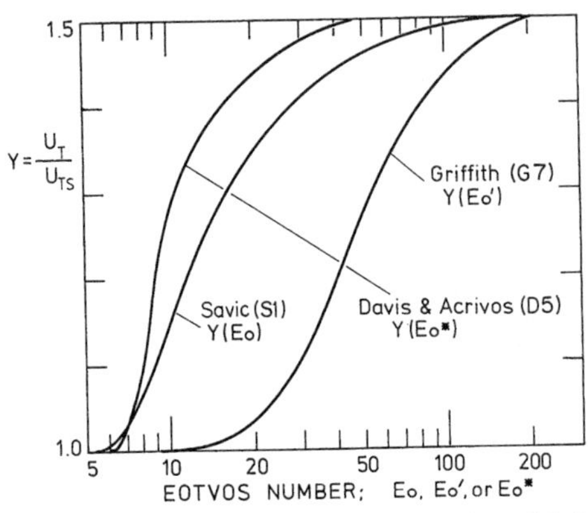

FIG. 3.7 Effect of surfactant on the terminal velocity of small bubbles and drops.

Subsequent theoretical work has allowed the contaminant to distribute itself over the interface under the influence of fluid shear as expressed through Eq. (1-16). Schecter and Farley (S2) showed that a drop or bubble would remain spherical and Eq. (1-15) could be satisfied only for a particular variation of interfacial tension around the periphery. Using this variation led to symmetrical circulation. Wasserman and Slattery (W2) assumed that the surface contaminant diffused to the particle from the continuous phase and was convected along the interface. A perturbation solution was obtained for an air bubble of 0.0022 cm diameter in water containing a trace of isoamyl alcohol. The contaminant had little effect on bubble shape, while it drastically reduced the terminal velocity. Noting these results Levan and Newman (L2) derived stream functions for creeping flow of a spherical fluid particle with an arbitrary variation of interfacial tension. These stream functions can be used for any mechanism of contaminant transport to and along the interface. Applying their stream functions to Wasserman and Slattery's example showed that the interfacial contaminant concentration was highest and the interfacial velocity lowest near the rear stagnation point.

Following a suggestion made by Davies (D2, D4), we define a "degree of circulation" Z such that the terminal velocity of a spherical bubble or drop in slow viscous flow is given by

$$U_T = U_{TS}(1 + Z/2), \tag{3-25}$$

where U_{TS} is the Stokes terminal velocity defined by Eq. (3-18). Griffith assumed that the effects of viscosity and surface contamination could, to a first approximation, be separated, so that

$$Z = [2/(2 + 3\kappa)]\{2(Y - 1)\}, \tag{3-26}$$

where the first bracketed term accounts for the influence of the viscosity of the dispersed phase and follows directly from the Hadamard–Rybczynski analysis, and the term in Y accounts for surface effects. Griffith's values for Y, plotted in Fig. 3.7, give good agreement with experimental results except for one or two anomalous cases.

Figure 3.7 together with Eqs. (3-25) and (3-26) provide an approximate but rational means of estimating the effect of surface-active impurities for bubbles and drops at low Reynolds numbers. If the contaminant is strong (i.e., for a large surface excess of surfactant) and its type and amount can be characterized in terms of $\Delta\sigma'$, the difference in interfacial tension between the pure and equilibrated phases, Griffith's curve can be used to estimate Y. If the amount of surfactant is relatively large and the value of σ at which the surface film collapses is known, the Davis and Acrivos curve should be used. When the amount and types of contaminant are unknown, the Savic curve in Fig. 3.7 describes the limiting case where the surface is so fully contaminated that the surface tension varies from its value for a pure system at the front to zero at the rear. For the other limit of a very pure system, Y should be taken as 1.5. For cases of intermediate but unknown purity, transition from rigid to circulating behavior occurs for Eo lower than the Savic values, and there is presently no alternative to using the Bond criterion, which corresponds to a maximum reduction in interfacial tension of 10% to 45% (D5).

E. Oseen's Approximation

There is a fundamental difficulty, first noted by Oseen (O2), in applying the creeping flow equations to particles in unbounded media. In the creeping flow solution given by Eqs. (3-7) and (3-8), the ratio of neglected inertia terms to retained viscous terms is $O[\mathrm{Re}(r/a)]$. For any finite Re, the neglected terms dominate at large distances from the sphere, and the creeping flow approximation is only valid for distances less than order a/Re. To remove this inconsistency, Oseen suggested that the Navier–Stokes equation should be linearized by simplifying, rather than neglecting, the inertia term. The continuous phase velocity $\mathbf{U}$ is written as $(\mathbf{v} - U\mathbf{i})$ so that $\mathbf{v}$ represents the deviation from the uniform stream, $-U\mathbf{i}$. For steady flow

$$\frac{D\mathbf{U}}{Dt} \equiv \mathbf{v} \cdot \nabla\mathbf{v} - U\mathbf{i} \cdot \nabla\mathbf{v}. \tag{3-27}$$

The final term in Eq. (3-27) dominates at large distances from the body. Since this is the region in which inertial effects are significant, Oseen suggested that the nonlinear term $\mathbf{v} \cdot \nabla\mathbf{v}$ be neglected. Equation (1-33) then becomes

$$\nabla p = \mu \nabla^2 \mathbf{v} + \rho U\mathbf{i} \cdot \nabla\mathbf{v}, \tag{3-28}$$

which is generally called Oseen's equation. The additional term, $\rho U\mathbf{i} \cdot \nabla\mathbf{v}$, removes the property of "reversibility," so that solutions no longer possess fore-and-aft symmetry.

For a rigid sphere, the boundary conditions are

$$\mathbf{v} \to 0 \qquad \text{as} \qquad r \to \infty, \tag{3-29}$$

$$\mathbf{v} = U\mathbf{i} \qquad \text{at} \qquad r = a. \tag{3-30}$$

Lamb (L1) has given an approximate solution:

$$\psi = -\frac{Ur^2 \sin^2 \theta}{2}\left[1 + \frac{a^3}{2r^3}\right] + \frac{3Ua^2}{\text{Re}}(1 - \cos\theta)\left[1 - \exp\left\{-\frac{r\,\text{Re}}{4a}(1 + \cos\theta)\right\}\right]. \tag{3-31}$$

Corresponding streamlines are shown in Figs. 3.3b and 3.4b. Like the creeping flow result, the Oseen solution predicts infinite "drift." For large r the velocity is unbounded, but the divergent terms are $O[\text{Re}^2]$ and formally beyond the range of the Oseen approximation. For $r \ll a/\text{Re}$, the stream function may be approximated as

$$\psi = -\frac{Ur^2 \sin^2 \theta}{2}\left[1 - \frac{3a}{2r} + \frac{a^3}{2r^3} + O\left(\frac{r\,\text{Re}}{a}\right)\right]. \tag{3-32}$$

The modified pressure at the surface is:

$$p = \frac{\mu U}{2a}\left[3\cos\theta + \frac{\text{Re}}{4}(3\cos^2\theta - 1)\right]. \tag{3-33}$$

Equations (3-32) and (3-33) differ from the Stokes solution only in the Re terms. The contribution to p is symmetrical about the equator, so that the form drag is the same for the two solutions.

The vorticity at the surface is given by

$$\zeta_s = \frac{3U \sin\theta}{2a}\left[1 + \frac{\text{Re}}{4}\right]\exp\left\{-\frac{\text{Re}}{4}(1 - \cos\theta)\right\}, \tag{3-34}$$

and is shown in Fig. 3.8. By comparison with Stokes's solution, vorticity is increased over the leading hemisphere and reduced over the rear. In the outer

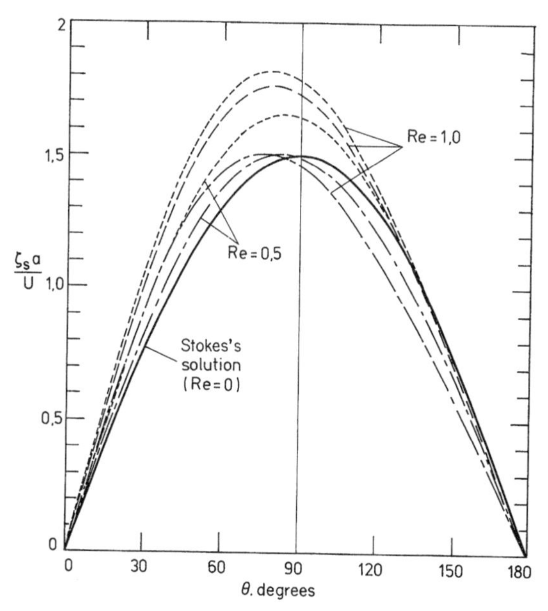

FIG. 3.8 Dimensionless vorticity, $\zeta_5 a/U$, at surface of rigid sphere: --- Oseen, Eq. (3-34);--- Proudman and Pearson (P3), Eq. (3-38); — Woo (W5) (numerical).

part of the flow, vorticity is small except for a wake-like region behind the particle. The vorticity distribution leads to a drag coefficient greater† than the Stokes law value (C_{DSt}):

$$C_D = \frac{24}{Re}\left[1 + \frac{3}{16}\,Re\right] \tag{3-35}$$

or

$$\frac{C_D}{C_{DSt}} - 1 = \frac{3}{16}\,Re. \tag{3-36}$$

A plot of $(C_D/C_{DSt} - 1)$ against Re gives a particularly sensitive indication of departures from Stokes law (M3). Figure 3.9 compares the recommended correlation of reliable drag data with Oseen's solution and other approximations. As Re → 0, the drag approaches zero via the Oseen drag, Eq. (3-35), rather than via the Stokes' drag, emphasizing that the Stokes solution is strictly invalid for Re ≠ 0. This implies that there is never true fore-and-aft symmetry in the flow field, and is particularly important for the motion of interacting particles (S8).

† The difference is of smaller order than the error in either solution and Eq. (3-35) is exact to $O(Re)$ (P3). In fact, the Re term in $(C_D/C_{DSt} - 1)$ can be deduced from the Stokes drag alone for any three-dimensional body symmetrical about a plane normal to the direction of motion (C6).

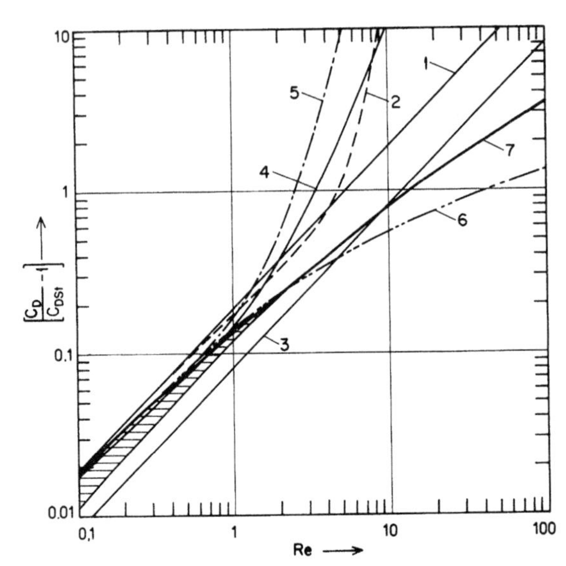

FIG. 3.9 Fractional departure from Stokes's law. Numbers 1-6 correspond to expressions in Table 3.1. Curve numbered 7 is a correlation of available data (see Chapter 5). Shaded region represents range of experimental scatter.

Equation (3-33) shows how the inertia term changes the pressure distribution at the surface of a rigid particle. The same general conclusion applies for fluid spheres, so that the normal stress boundary condition, Eq. (3-6), is no longer satisfied. As a result, increasing Re causes a fluid particle to distort towards an oblate ellipsoidal shape (T1). The onset of deformation of fluid particles is discussed in Chapter 7.

F. Further Extensions to the Creeping Flow Solutions

Figure 3.9 shows that Eq. (3-36) is applicable only for Re ≤ 0.1. Several more complete series solutions to Eq. (3-28) have been obtained (G5, S5) including one to 24 terms (V2). The expression of Goldstein involving terms to Re^5 is shown in Table 3.1. Figure 3.9 shows that this series diverges rapidly from the experimental correlation for Re > 4. Although series solutions are more accurate representatives of the Oseen drag, the Oseen drag itself is only an approximation to the true drag. To improve the approximation Lewis and Carrier (L4) proposed a semiempirical modification of the Oseen equation in which the final term in Eq. (3-28) is multiplied by a parameter c which may be a function of Re. The drag result is given in Table 3.1 and plotted in Fig. 3.9 with their suggested value of $c = 0.43$. Although the approximation is better, the form of the dependence on Re is not improved.

Rather than obtaining accurate solutions to Oseen's approximate equation, Proudman and Pearson (P3) suggested a technique to obtain successive approxi-

TABLE 3.1

Drag Coefficient Expressions at Low Reynolds Number

No.	Approximation from	$(C_D/C_{DST}) - 1$	Reference
1	Oseen equation, Eq. (3-28)	$\frac{3}{16}\mathrm{Re}$	Oseen (O2)
2	Oseen equation, Eq. (3-28)	$\frac{3}{16}\mathrm{Re} - \frac{19}{1280}\mathrm{Re}^2 + \frac{71}{20408}\mathrm{Re}^3 - \frac{30179}{34406400}\mathrm{Re}^4 + \frac{122519}{550502400}\mathrm{Re}^5$	Goldstein (G5)
3	Modified Eq. (3-28) (see text)	$\frac{3c}{16}\mathrm{Re}$ $c = 0.43$ (C3)	Lewis and Carrier (L4)
4	Navier–Stokes equation, Eq. (1-1)	$\frac{3}{16}\mathrm{Re} + \frac{9}{160}\mathrm{Re}^2 \ln\frac{\mathrm{Re}}{2} + O[\mathrm{Re}^2]$	Proudman and Pearson (P3)
5	Navier–Stokes equation, Eq. (1-1)	$\frac{3}{16}\mathrm{Re} + \frac{9}{160}\mathrm{Re}^2\left[\ln \mathrm{Re} + \gamma + \frac{2}{3}\ln 2 - \frac{323}{360}\right] + \frac{27}{640}\mathrm{Re}^3 \ln\left(\frac{\mathrm{Re}}{2}\right) + O[\mathrm{Re}^3]$ where γ = Euler's constant = 0.5772157...	Chester and Breach (C7)
6	Navier–Stokes equation, Eq. (1-1)	$\frac{3}{16}\varepsilon + \frac{9}{160}\varepsilon^2\left[\ln \varepsilon + \gamma + \frac{2}{3}\ln 2 - \frac{548}{360} + \frac{5m}{8}\right] + \frac{27}{640}\varepsilon^3 \ln\left(\frac{\varepsilon}{2}\right) + O[\varepsilon^3]$ where $\varepsilon = \mathrm{Re}(C_D/C_{DSt})^{-m}$, $m = 5$ (P4)	Proudman (P2)

mations to the Navier–Stokes equation. Since different forms for the stream function are appropriate in the region near the sphere and in the outer part of the flow field, two separate expansions were used. The "inner" or "Stokes" series was chosen to satisfy the boundary condition at $r = a$, while the "outer" or "Oseen" series satisfied the condition for $r \to \infty$. Alternate terms in each expansion were generated by "matching" the series in a region of supposed common validity. For a rigid sphere, the two-term inner series gives (V1):

$$\psi = -\frac{U}{4}(r-a)^2 \sin^2\theta\left[2 + \frac{a}{r} + \frac{3\mathrm{Re}}{16}\left\{\left(2 + \frac{a}{r}\right)(1 + \cos\theta) + \frac{a^2}{r^2}\cos\theta\right\}\right]. \tag{3-37}$$

Figure 3.8 shows the corresponding vorticity at the surface of the sphere:

$$\zeta_s = \frac{3U \sin\theta}{2a}\left[1 + \frac{3}{16}\mathrm{Re}\left\{1 + \frac{4}{3}\cos\theta\right\}\right]. \tag{3-38}$$

For Re > 16, the vorticity from Eq. (3-38) becomes negative over part of the rear hemisphere, indicating a recirculatory wake. In practice, recirculation starts at Re = 20 (see Chapter 5). Moreover, the predicted length of the wake at higher Re agrees well with experiment, although the width is less well predicted. Unfortunately, these predictions are fortuitous. Wake formation occurs at Re beyond the range where perturbations to Stokes's solution are valid, and inclusion of higher terms in the inner series eliminates the recirculatory wake (V1). Comparison with the numerical solution of Woo (W5) shows that Eq. (3-38) gives a close representation of the surface vorticity for Re < 0.5. Even at Re = 1.0, the error is less than 4%.

The drag coefficient corresponding to the two-term approximation is the Oseen value, Eq. (3-35). The addition of a further term yields expression 4 in Table 3.1. Figure 3.9 shows that this expression fits the data within about 1.5% for Re < 0.7, but divergence is rapid at higher Re. The series was extended to terms of order Re^3 (C7); see Table 3.1. Figure 3.9 shows that the additional terms make the fit worse. Similar conclusions apply for fluid spheres (A3, G6). Proudman suggested that the divergence might result from the unsuitability of Re as the expansion parameter. He proposed instead expansion in terms of a semiempirical parameter ε; see Table 3.1. His result, with the value $m = 5$ suggested by Pruppacher *et al.* (P4), is plotted in Fig. 3.9. Agreement with the data is better than for any of the other analytic results, but deviation is still marked for Re > 3.

Thus, analytic solutions for flow around a spherical particle have little value for Re > 1. For Re somewhat greater than unity, the most accurate representation of the flow field is given by numerical solution of the full Navier–Stokes equation, while empirical forms should be used for C_D. These results are discussed in Chapter 5.

III. HEAT AND MASS TRANSFER

There are no solutions for transfer with the generality of the Hadamard–Rybczynski solution for fluid motion. If resistance within the particle is important, solute accumulation makes mass transfer† a transient process. Only approximate solutions are available for this situation with internal and external mass transfer resistances included. The following sections consider the resistance in each phase separately, beginning with steady-state transfer in the continuous phase. Section B contains a brief discussion of unsteady mass transfer in the continuous phase under conditions of steady fluid motion. The resistance within the particle is then considered and methods for approximating the overall resistance are presented. Finally, the effect of surface-active agents on external and internal resistance is discussed.

† See Chapter 1 for discussion of the equivalence of heat and mass transfer.

A. External Resistance—Steady State

For axisymmetric flow with constant properties, the diffusion equation may be written (see Eq. (1-38)) as:

$$u_r \frac{\partial c}{\partial r} + \frac{u_\theta}{r} \frac{\partial c}{\partial \theta} = \frac{\mathscr{D}}{r^2}\left[\frac{\partial}{\partial r}\left(r^2 \frac{\partial c}{\partial r}\right) + \frac{1}{\sin\theta} \frac{\partial}{\partial \theta}\left(\sin\theta \frac{\partial c}{\partial \theta}\right)\right]. \tag{3-39}$$

For a rigid sphere or a fluid particle with negligible internal resistance and constant concentration, the boundary conditions are:

$$c = c_s \qquad \text{at} \quad r = a, \tag{3-40}$$

$$c \to c_\infty \qquad \text{as} \quad r \to \infty, \tag{3-41}$$

$$\partial c/\partial\theta = 0 \qquad \text{at} \quad \theta = 0 \quad \text{and} \quad \theta = \pi. \tag{3-42}$$

Section II shows that the dimensionless external velocity field (u_r/U, u_θ/U) is a function of dimensionless position (r/a, θ) and κ for creeping flow. The dimensionless concentration defined in Eq. (1-45) is a function of these quantities and of the Peclet number, $\text{Pe} = 2aU/\mathscr{D}$. Hence the Sherwood number, $\text{Sh} = 2ka/\mathscr{D}$, is a function of κ and Pe (with additional dependence on Re unless the creeping flow approximation is valid). The exact solution of Eqs. (3-39) to (3-42) with the Hadamard–Rybczynski velocity field is not available for all values of Pe and κ, but several special cases have been treated.

1. *Stagnant Continuous Phase*

When the velocity is everywhere zero, diffusion is in the radial direction only. Equation (3-39) reduces to

$$\frac{\partial}{\partial r}\left(r^2 \frac{\partial c}{\partial r}\right) = 0 \tag{3-43}$$

with boundary conditions given by Eqs. (3-40) and (3-41). Since there is no dependence on θ, local and average values of Sh are equal and

$$\text{Sh}_0 = 2, \tag{3-44}$$

where the subscript denotes stagnant fluid. Equation (3-43) may also be regarded as the limiting form of Eq. (3-39) for $\mathscr{D} \to \infty$ ($\text{Pe} \to 0$); i.e., convective terms in the diffusion equation are neglected, just as inertia terms in the Navier–Stokes equation are neglected in the creeping flow approximation. Thus, Sh_0 may be considered analogous to the drag coefficient in creeping flow.

2. *Rigid Sphere in Creeping Flow*

Equation (3-39) has been solved for steady Stokes flow past a rigid sphere (B6, M2). The resulting values of Sh, obtained numerically for a wide range of Pe, are shown as the $\kappa = \infty$ curve in Fig. 3.10. For small Pe, Sh approaches Sh_0, while for large Pe, Sh becomes proportional to $\text{Pe}^{1/3}$. The numerical solution

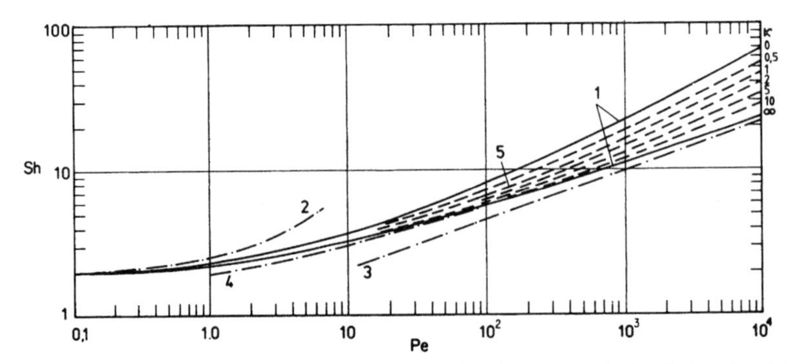

FIG. 3.10 External Sh for spheres in Stokes flow: (1) Exact numerical solution for rigid and circulating spheres; (2) Brenner (B6): rigid sphere, Pe → 0, Eq. (3-45); (3) Levich (L3): rigid sphere, Pe → ∞, Eq. (3-47): (4) Acrivos and Goddard (A1): rigid sphere, Pe → ∞, Eq. (3-48): (5) Approximate values: fluid spheres.

provides a standard for assessing the validity of asymptotic solutions for Pe → 0 and Pe → ∞.

Brenner (B6) pointed out that similar problems arise in obtaining Eq. (3-44) as in the low Re approximation for fluid flow. The neglected convection terms dominate far from the particle, since the ratio of convective to diffusive terms is $O[\mathrm{Pe}(r/a)]$. An asymptotic solution to Eq. (3-39) with Pe → 0 was therefore obtained by the matching procedure of Proudman and Pearson discussed above. Brenner's result for the first term in a series expansion for Sh may be written:

$$(\mathrm{Sh}/\mathrm{Sh}_0) - 1 = \tfrac{1}{4}\mathrm{Pe}. \tag{3-45}$$

Equation (3-45) is analogous to the Oseen correction to the Stokes drag, and is accurate to $O[\mathrm{Pe}]$.† It applies for any rigid or fluid sphere at any Re, provided that Pe → 0 and the velocity remote from the particle is uniform. Figure 3.10 shows that Eq. (3-45) is accurate for Pe $\dot{<}$ 0.5. Acrivos and Taylor (A2) extended the solution to higher terms, but, as for drag, the additional terms only yield slight improvement at Pe < 1.

Levich (L3) obtained an asymptotic solution to Eq. (3-39) for Pe → ∞, using the thin concentration boundary layer assumption discussed in Chapter 1. Curvature of the boundary layer and angular diffusion are neglected (i.e., the last term in Eq. (3-39) is deleted), so that the solution does not hold at the rear of the sphere where the boundary layer thickens and angular diffusion is significant. The asymptotic boundary layer formula, Eq. (1-59), reduces for a sphere to:

$$\mathrm{Sh} = 0.641\ \mathrm{Pe}^{1/3}\left[\int_0^{\pi}\left(\frac{\zeta_s a}{U}\sin^3\theta\right)^{1/2} d\theta\right]^{2/3}. \tag{3-46}$$

† Furthermore, just as for drag, the Pe term in $(\mathrm{Sh}/\mathrm{Sh}_0 - 1)$ can be deduced from Sh_0 alone for any particle symmetric about a plane normal to the direction of motion (B6).

Substitution of the Stokes surface vorticity, ζ_s, from Eq. (3-19) yields

$$\mathrm{Sh} = 0.991\ \mathrm{Pe}^{1/3}. \tag{3-47}$$

Figure 3.10 shows that Eq. (3-47) gives Sh approximately 10% too low for $\mathrm{Pe} = 10^3$, while the deviation becomes worse at lower Re. Acrivos and Goddard (A1) used a perturbation method to obtain the first-order correction to Eq. (3-47):

$$\mathrm{Sh} = 0.991\ \mathrm{Pe}^{1/3} + 0.92. \tag{3-48}$$

Figure 3.10 shows that Eq. (3-48) lies within 3% of the numerical solution for $\mathrm{Pe} > 30$.

It is convenient to have a relationship for Sh valid for all Pe in creeping flow. The following equation agrees with the numerical solution within 2%:

$$\mathrm{Sh} = 1 + (1 + \mathrm{Pe})^{1/3}. \tag{3-49}$$

Equation (3-49) can be used for $\mathrm{Re} \leq 1$ even though the Stokes surface vorticity is not accurate for $\mathrm{Re} > 0.1$. This fortuitous result follows because mass transfer is much less sensitive than drag to errors in ζ_s.

Figure 3.11 shows the local Sherwood number, $\mathrm{Sh}_{\mathrm{loc}}$, for the limits of high and low Pe. Values for $\mathrm{Pe} = 0.1$ are not symmetrical about the equator, and show the greatest transfer rates over the leading surface indicating that the

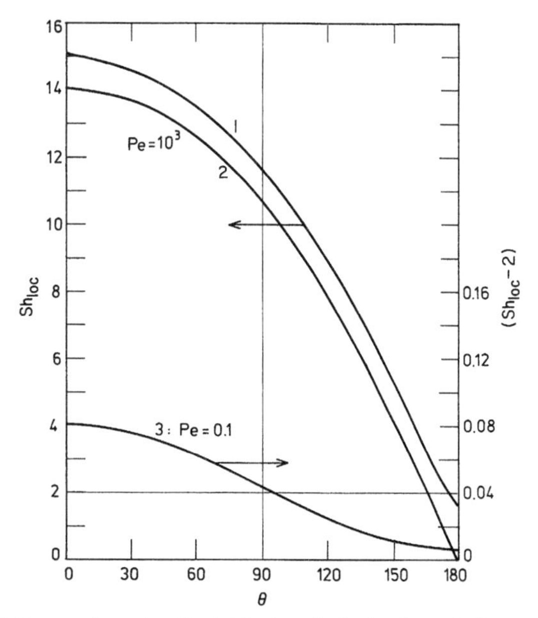

FIG. 3.11 Local Sherwood number for rigid sphere in Stokes flow: (1) Exact numerical solution: $\mathrm{Pe} = 10^3$; (2) High Pe asymptotic solution (L3): $\mathrm{Pe} = 10^3$; (3) Low Pe asymptotic solution (A2): $\mathrm{Pe} = 0.1$.

concentration field lacks fore-and-aft symmetry. The local Sherwood number always exceeds $Sh_0 = 2$, indicating that convection increases the transfer rate at all points on the surface. For high Pe, on the other hand, the solute is swept to the rear so that the local concentration gradient is reduced and $Sh_{loc} < Sh_0$ (H2, S6), as shown by the curves for $Pe = 10^3$. The boundary layer solution unrealistically predicts that Sh_{loc} falls to zero at the rear stagnation point due to neglect of angular diffusion.

3. *Fluid Sphere in Creeping Flow*

For fluid spheres with $\kappa = 0$, Eq. (3-39) has been solved numerically with the Hadamard–Rybczynski velocity field (O1), and the resulting variation of Sh with Pe is shown in Fig. 3.10. The values are approximated within 6% for all Pe by

$$Sh = 1 + (1 + 0.564\, Pe^{2/3})^{3/4}. \tag{3-50}$$

For $Pe \to 0$ an asymptotic solution through the matching procedure has been obtained for all κ (B6). As for solid spheres its range of applicability is limited to $Pe \lessdot 1$.

For a fluid sphere with $Pe \to \infty$ the thin concentration boundary layer approximation, Eq. (1-63), becomes

$$Sh = 0.798\, Pe^{1/2}\left[\int_0^{\pi} \frac{(u_\theta)_{r=a}}{U} \sin^2\theta\, d\theta\right]^{1/2}. \tag{3-51}$$

Inserting the Hadamard–Rybczynski form for u_θ yields†

$$Sh = 0.651\sqrt{Pe/(1 + \kappa)} \tag{3-52}$$

which agrees with the numerical solution for $\kappa = 0$ within 10% for $Pe \gtrdot 100$. From Eqs. (1-64) and (3-52), this approximation applies if

$$Pe \gg 2.4(3\kappa + 1)^2(1 + \kappa). \tag{3-53}$$

The Hadamard–Rybczynski results are applicable if $Re \lessdot 1$, so that

$$Sc \gg 2.4(3\kappa + 1)^2(1 + \kappa). \tag{3-54}$$

For liquids of low viscosity, Sc is of order 10^3, so that Eq. (3-54) is satisfied for $\kappa \lessdot 2$; thus Eq. (3-52) is valid for bubbles or drops of low viscosity. Experimental data on dissolution of small low-viscosity drops (W1) and bubbles (C1) with

† The dependence of Sh on $Pe/(1 + \kappa)$ at high Pe results because the Hadamard–Rybczynski analysis gives dimensionless velocities (u_r/U, u_θ/U) proportional to $(1 + \kappa)^{-1}$ within and close to the particle (Eqs. (3-7) and (3-8)). Similar dependence is encountered for unsteady external transfer (Section B.2), and for internal transfer at all Pe (Section C.4). These results do not give the rigid sphere values as $\kappa \to \infty$, because of fundamental differences between the boundary layer approximations for the two cases (see Chapter 1), and are only valid for $\kappa \lessdot 2$.

$\kappa \dot{<} 2$ agree with Eq. (3-52) if the particle is spherical and there is negligible interfacial contamination.

Although no exact solution valid for all κ and Pe is available, an approximate solution using the integral boundary layer approach has been given (B5). The curves for intermediate κ in Fig. 3.10 were prepared by locating them between the exact solutions for $\kappa = 0$ and $\kappa = \infty$, using the relative spacings from this approximate solution. As κ increases, the variation of Sh at high Pe changes from $\mathrm{Pe}^{1/2}$ to $\mathrm{Pe}^{1/3}$. Although this procedure is not exact, the curves in Fig. 3.10 are recommended for predicting Sh for any κ with $\mathrm{Re} \dot{<} 1$.

4. *Extension to Larger* Re

The preceding results can be extended beyond the creeping flow regime by using any of the flow fields discussed in Sections II.E and II.F. For moderate to high Sc, say $\mathrm{Sc} > 1$, the layer of variable concentration lies near the sphere, and only stream functions accurate in this region give improved results. The Oseen stream function only differs significantly from the Stokes stream function in the outer field, and is not useful for extending the theoretical prediction of Sh to finite Re. However, the Proudman and Pearson solution can be used to extend the range of the solutions for rigid spheres. Since the inner stream function contains Re, the value of Sh is a function of a set of two dimensionless groups from among Re, Sc, and Pe.

Gupalo and Ryazantsev (G10) followed the analysis of Acrivos and Taylor (A2) with the Proudman–Pearson stream function rather than Stokes flow. For $\mathrm{Sc} > 10$, the two predictions for Sh agree within 1%, while for $\mathrm{Sc} = 1$ they differ by at most 8% for $\mathrm{Pe} < 1$. The results of Gupalo and Ryazantsev, although valid to higher Re, are still restricted to $\mathrm{Pe} \to 0$, so that this extension is of little practical value.

The asymptotic solution for $\mathrm{Pe} \to \infty$ embodied in Eq. (3-46) can be extended to finite Re in a similar way. The Oseen value for surface vorticity, ζ_s, predicts little effect of Re. However, the Proudman and Pearson expression for ζ_s, Eq. (3-38), yields:

$$\mathrm{Sh} = 0.991\ \mathrm{Pe}^{1/3}[1 + (\mathrm{Re}/4)]^{0.27} \tag{3-55}$$

where the integral has been approximated by a simpler form which agrees within 2%. Equation (3-55) extends the range of the boundary layer solution up to $\mathrm{Re} \doteq 1$ but, as with drag predictions, the Proudman and Pearson approach has little value at higher Re.

B. Unsteady External Resistance

If a particle is suddenly exposed to a step change in the composition of the continuous phase, or if the surface composition undergoes a step change to a new constant value, the rate of mass transfer becomes a function of time even if the fluid motion is steady and the fluid properties are constant. During unsteady transfer, the concentration field is governed by:

$$\frac{\partial c}{\partial t} + u_r \frac{\partial c}{\partial r} + \frac{u_\theta}{r} \frac{\partial c}{\partial \theta} = \frac{\mathscr{D}}{r^2}\left[\frac{\partial}{\partial r}\left(r^2 \frac{\partial c}{\partial r}\right) + \frac{1}{\sin\theta}\frac{\partial}{\partial \theta}\left(\sin\theta \frac{\partial c}{\partial \theta}\right)\right]. \tag{3-56}$$

The boundary conditions of Eqs. (3-40) to (3-42) apply for $t > 0$, with the additional condition

$$c = c_\infty \quad \text{at} \quad t = 0, r \geq a. \tag{3-57}$$

Considering the order of magnitude of terms in Eq. (3-56), we see that for any finite $\mathscr{D}$ there will be times short enough that the terms in u_r, u_θ, and $(\partial/\partial\theta)$ are small relative to the others. Thus, at very short times, unsteady transfer is not affected by convection, and the time variation of Sh is identical to that in a stagnant medium. It is convenient to express the results in terms of the dimensionless time $\tau = \mathscr{D}t/a^2$, sometimes called the Fourier number,† which may be regarded as the ratio of real time to the time for diffusion to become established. For long times, Sh approaches the steady values in Fig. 3.10.

1. *Stagnant Continuous Phase*

For a stagnant medium or with $\mathscr{D} \to 0$, Pe $\to 0$ and Eq. (3-56) reduces to

$$\frac{\partial c}{\partial t} = \frac{\mathscr{D}}{r^2}\frac{\partial}{\partial r}\left(r^2 \frac{\partial c}{\partial r}\right). \tag{3-58}$$

The instantaneous Sherwood number follows as

$$\text{Sh} = 2(1 + 1/\sqrt{\pi\tau}). \tag{3-59}$$

The dimensionless time, τ_x, for Sh to come within $100x\%$ of the steady value indicates the duration of the unsteady state; for Pe $= 0$, $\tau_{0.1} = 31.8$, and $\tau_{1/e} = 2.35$. Diffusivities in gases are of order 10^4 times diffusivities in liquids; hence, for particles with equal size and equal exposure, transient effects in a stagnant medium are much more significant in liquids.

2. *Solutions for Larger* Pe

For a rigid or circulating sphere in creeping flow, Sh may be written as a series expansion in Pe and τ, valid for small Pe and τ (C8, K2, K5). The first term in Pe is $O[\text{Pe}^2]$. Hence, Eq. (3-59) for small τ and Eq. (3-45) for long times are both valid to $O[\text{Pe}]$. The results of Konopliv and Sparrow (K2) and Choudhury and Drake (C8) for rigid or circulating spheres with Pe < 0.5 are approximated within 5% for all times by:

$$\text{Sh} = 2 + \left[\frac{\text{Pe}^2}{4} + \frac{4}{\pi\tau}\right]^{1/2}. \tag{3-60}$$

† For heat transfer the Fourier number is $\alpha t/a^2$. The heat transfer analogs of the mass transfer dimensionless groups can be found by making the substitutions described in Chapter 1.

Thus the effect of fluid motion is to reduce the unsteady period; e.g., $\tau_{0.1}$ is reduced by a factor of 3 on raising Pe from zero to 0.5.

Unsteady transfer with $\mathrm{Pe} \to \infty$ has been treated using the thin concentration boundary layer assumptions. With this approximation, the last term in Eq. (3-56) is deleted. Hence, for small τ where the convection term is negligible, the transfer rate for rigid or circulating spheres is identical to that for diffusion from a plane into a semi-infinite region:

$$\mathrm{Sh}_{\tau\to 0} = 2/\sqrt{\pi\tau}. \tag{3-61}$$

The range over which Eq. (3-61) provides a useful approximation may be evaluated by comparison with more detailed solutions.

Complete solutions are available for $\mathrm{Pe} \to \infty$ for rigid spheres (K4) and for fluid spheres (C5, R4) subject to the limitation of Eq. (3-54). Approximations good within 3% are, for rigid spheres:

$$\mathrm{Sh} = \mathrm{Pe}^{1/3}\left[0.956 + \left(\frac{2}{\sqrt{\pi\tau}\mathrm{Pe}^{1/3}}\right)^5\right]^{1/5} \tag{3-62}$$

and for fluid spheres:

$$\mathrm{Sh} = \sqrt{\frac{\mathrm{Pe}}{1+\kappa}}\left[0.117 + \left(2\sqrt{\frac{1+\kappa}{\pi\tau\,\mathrm{Pe}}}\right)^5\right]^{1/5} \tag{3-63}$$

The duration of the unsteady period, denoted by t_x, the time required for Sh to come within $100x\%$ of the steady value, is different for rigid and fluid spheres. For a rigid sphere at high Pe, $\tau_x \propto \mathrm{Pe}^{-2/3}$. From Stokes's law, Eq. (3-18), $U_\mathrm{T} \propto a^2$; hence t_x is independent of particle size for a given fluid. However, for a fluid sphere, $\tau_x \propto (1+\kappa)/\mathrm{Pe}$; thus $U_\mathrm{T} t_x/a$ is a constant, and a given fractional approach to steady state is achieved when the particle has moved a fixed number of radii, e.g.,

$$U_\mathrm{T} t_{0.1}/a = 1.8(1+\kappa). \tag{3-64}$$

Approximate values of Sh for intermediate Pe may be obtained by using Eq. (3-61) until Sh equals the steady-state value of Fig. 3.10. For larger τ this steady-state value is used. Although this approximation underestimates the duration of the unsteady period, the error in Sh is not large. In terms of the time-averaged Sh or the total mass transferred, the error is less than 15% for all times.

C. Transfer with Variable Particle Concentration

When mass diffuses into or out of a fluid particle, the concentration within the particle changes with time. Therefore the time derivatives must be retained in the diffusion equations for both internal and external fluids. The internal and external concentration fields are related at the interface. If there is no chemical reaction at the interface, the species mass fluxes on each side are equal. A second condition on interfacial concentrations is given by Eq. (1-39). If we assume constant properties in each phase and axial symmetry, the concentration fields are described by Eq. (3-56) with $\mathscr{D}$ replaced by $\mathscr{D}_p$ within the particle. Assuming that each phase is initially at uniform concentration, the boundary conditions are:

$$c_p = c_{p0} \text{ and } c = c_\infty \quad \text{at} \quad t = 0, \tag{3-65}$$

$$\partial c_p/\partial r = 0 \quad \text{at} \quad r = 0, \tag{3-66}$$

$$c \to c_\infty \quad \text{as} \quad r \to \infty, \tag{3-67}$$

$$c_p = Hc \quad \text{and} \quad \mathscr{D}_p \frac{\partial c_p}{\partial r} = \mathscr{D} \frac{\partial c}{\partial r} \quad \text{at} \quad r = a \quad \text{for} \quad t > 0, \tag{3-68}$$

$$\partial c_p/\partial \theta = \partial c/\partial \theta = 0 \quad \text{at} \quad \theta = 0 \quad \text{and} \quad \theta = \pi. \tag{3-69}$$

The profiles of the dimensionless concentrations:

$$\Phi_p = (c_p - Hc_\infty)/(c_{p0} - Hc_\infty), \qquad \Phi = (c - c_\infty)/(c_{p0}/H - c_\infty) \tag{3-70}$$

are then governed by Pe, H, $\mathscr{D}_p/\mathscr{D}$, and τ, as well as by Re and κ which determine the dimensionless velocity fields.

Since the concentration within the particle varies with time, instantaneous mass transfer rates are difficult to measure. Experimental data are frequently presented in terms of the fractional approach to equilibrium:

$$F = 1 - \bar{\Phi}_p = (c_{p0} - \bar{c}_p)/(c_{p0} - Hc_\infty), \tag{3-71}$$

where $\bar{c}_p$ is the average concentration within the particle at time t, i.e., the concentration obtained by mixing the dispersed fluid. As t increases, F increases from zero to unity. This group is sometimes termed the "extraction efficiency," but the definition of Eq. (3-71) applies for transfer both into and out of the particle.

1. *Approximation for Short Times*

It was noted in Section B that, for finite Pe and short times, Eq. (3-56) is dominated by the first term on each side. Mass transfer is then determined by unsteady diffusion, and fluid motion has no effect on F. Only the region near the interface is affected, and diffusion occurs as if it were between two semi-infinite media, giving† (C4):

$$F = 6\sqrt{\frac{\tau_p}{\pi}}\left[\frac{1}{1 + H\sqrt{\mathscr{D}_p/\mathscr{D}}}\right], \tag{3-72}$$

† For heat transfer the group $H\sqrt{\mathscr{D}_p/\mathscr{D}}$ becomes $\sqrt{\rho_p C_{tp} K_{tp}/\rho C_t K_t}$.

where τ_p is $\mathscr{D}_p t/a^2$. Immediately after $\tau = \tau_p = 0$, the concentration outside the interface changes to:

$$c = (c_\infty + c_{p0}\sqrt{\mathscr{D}_p/\mathscr{D}})/(1 + H\sqrt{\mathscr{D}_p/\mathscr{D}}). \tag{3-73}$$

For semi-infinite media, this interface concentration remains constant, but for a particle it changes with time towards c_∞. Equations (3-72) and (3-73) are compared with more complete solutions below.

2. *Stagnant Phases*

If the fluids are stagnant (i.e., $\mathrm{Pe} = \mathrm{Pe_p} = 0$), the concentration profiles display angular symmetry and the fractional approach to equilibrium is a function only of H, $\mathscr{D}_p/\mathscr{D}$, and τ or τ_p. The corresponding solution for F (K3, P1) is shown in Figs. 3.12–3.14, for a wide range of values of these parameters. Fluid motion always increases F for given τ_p, so that these solutions give a lower limit for the fractional approach to equilibrium.

Figure 3.12 shows the variation of F with τ_p when $H = 1$. As $\mathscr{D}_p/\mathscr{D}$ decreases, the curves approach a limiting case solved much earlier, (N1), often called the Newman solution. This corresponds to negligible external resistance, and Eqs.

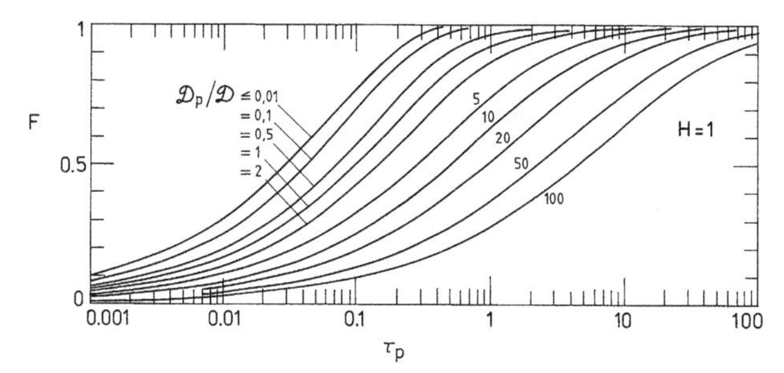

FIG. 3.12 Variation of fractional approach to equilibrium F with dimensionless time, $\tau_p = \mathscr{D}_p t/a^2$, for a sphere in stagnant surroundings with $H = 1$.

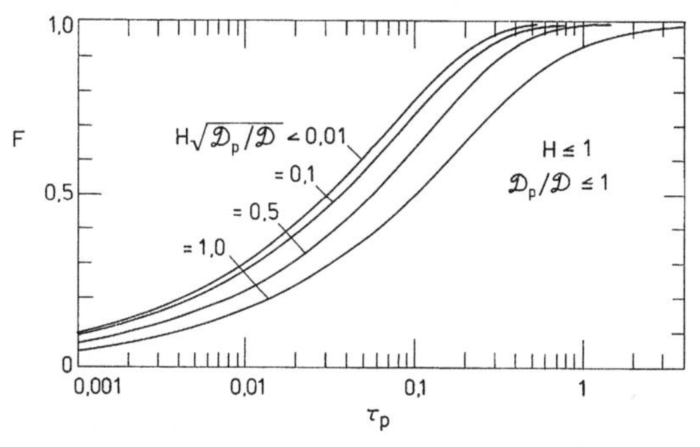

FIG. 3.13 Variation of fractional approach to equilibrium F with dimensionless time, $\tau_p = \mathscr{D}_p t/a^2$, for a sphere in stagnant surroundings with $H \leq 1$ and $\mathscr{D}_p/\mathscr{D} \leq 1$.

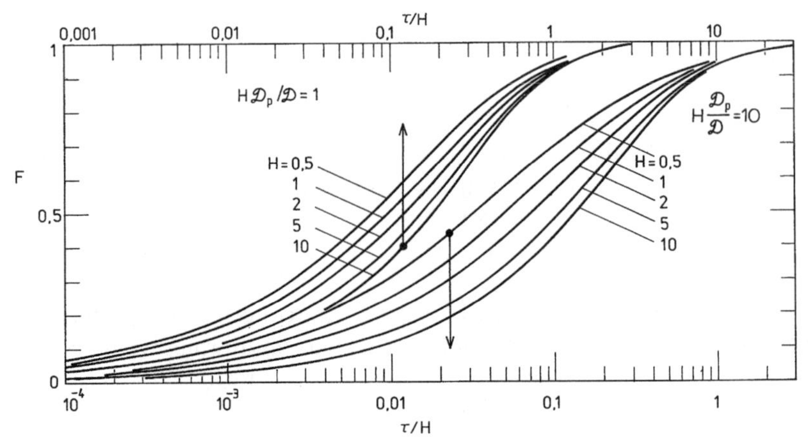

FIG. 3.14 Variation of fractional approach to equilibrium F with dimensionless time, $\tau = \mathscr{D}t/a^2$, for a sphere in stagnant surroundings with $H\mathscr{D}_p/\mathscr{D} \geq 1$.

(1-39) and (3-73) show that, as $H\sqrt{\mathscr{D}_p/\mathscr{D}} \to 0$, the concentration inside the interface approaches Hc_∞. At the other extreme, $H\sqrt{\mathscr{D}_p/\mathscr{D}} \gg 1$, the external resistance controls and the concentration within the particle is nearly uniform. The short-time solution, Eq. (3-72), gives a good approximation for $F < 0.2$. Figure 3.13 shows $F(\tau_p)$ for $H \leq 1$ and $\mathscr{D}_p/\mathscr{D} < 1$. In this case the results are brought closer together by using $H\sqrt{\mathscr{D}_p/\mathscr{D}}$ as parameter, as suggested by Eq. (3-72), and calculations for different combinations of H and $\mathscr{D}_p/\mathscr{D}$ lie within 3% of the curves shown. Typical results for larger H and $\mathscr{D}_p/\mathscr{D}$ are shown in Fig. 3.14.

3. *Limiting Cases*

It was noted above that the external resistance may sometimes be neglected relative to the internal resistance. Criteria for the importance of the external resistance can be developed from Eq. (3-72) for short times, and for long times from the steady-state external resistance taking the internal resistance as roughly $a/\mathscr{D}_p$. The external resistance is negligible for short times when

$$H\sqrt{\mathscr{D}_p/\mathscr{D}} \ll 1 \tag{3-74}$$

and for long times when

$$H\mathscr{D}_p/\mathscr{D} \ll \mathrm{Sh}. \tag{3-75}$$

The short-time criterion is the more stringent except when $H \gg 1$ and $\mathscr{D}_p/\mathscr{D} \ll 1$. External resistance controls and the concentration within the particle is uniform when the inequalities in Eqs. (3-74) and (3-75) are reversed. Even if the external resistance is not negligible relative to the internal resistance, it may be possible to assume constant external resistance, i.e., quasi-steady behavior. Comparison between the external transient time from Section B and the time constant for a particle with uniform concentration shows that the quasi-steady assumption is justified if

$$H(\mathrm{Sh}) \gg 1. \tag{3-76}$$

Table 3.2 summarizes these criteria, and indicates the section in this chapter where each limiting case is discussed.

TABLE 3.2

Transient Transfer to Spheres: Criteria for Limiting Cases

	Negligible external resistance	Significant external resistance: Transient	Significant external resistance: Quasi-steady (H Sh $\gg 1$)
Negligible internal resistance (particle concentration uniform)		Section C.5.a Short times: $H\sqrt{\mathscr{D}_p/\mathscr{D}} \gg 1$ Long times: $H\mathscr{D}_p/\mathscr{D} \gg \mathrm{Sh}$	Section C.5.b
Significant internal resistance	Section C.4 Short times: $H\sqrt{\mathscr{D}_p/\mathscr{D}} \ll 1$ Long times: $H\mathscr{D}_p/\mathscr{D} \ll \mathrm{Sh}$	General case (Section C.2 for Pe $\to 0$)	Section C.6

4. *Negligible External Resistance*

If the external resistance is negligible, it is only necessary to solve Eq. (3-56) for the dispersed fluid with boundary conditions given by Eqs. (3-65), (3-66), (3-69) and

$$c_p = Hc_\infty \quad \text{at} \quad r = a \quad \text{for} \quad t > 0. \tag{3-77}$$

These equations have been solved for rigid (N1) and circulating spheres (J1, K6, W3, W4) in creeping flow. Since the dimensionless velocities within the particle are proportional to $(1 + \kappa)^{-1}$ (see Eq. (3-8)), F is a function only of τ_p and $\mathrm{Pe}_p/(1 + \kappa)$. In presenting the results, it is instructive to consider the instantaneous overall Sherwood number, Sh_p, as well as F. The driving force is taken as the difference between the concentration inside the interface, Hc_∞, and the mixed mean particle concentration, $\bar{c}_p$, giving

$$\mathrm{Sh}_p = \frac{a}{\bar{c}_p - Hc_\infty} \int_0^\pi \left(\frac{\partial c_p}{\partial r}\right)_{r=a} \sin\theta \, d\theta = \frac{-2}{3(\bar{c}_p - Hc_\infty)} \frac{d\bar{c}_p}{d\tau_p} = \frac{2}{3(1 - F)} \frac{dF}{d\tau_p}. \tag{3-78}$$

Hence the time-averaged Sherwood number is

$$\overline{\mathrm{Sh}}_p = -2 \ln(1 - F)/3\tau_p. \tag{3-79}$$

Figure 3.15 shows the variation of F with τ_p for several values of $\mathrm{Pe}_p/(1+\kappa)$. For a stagnant sphere, $\mathrm{Pe}_p/(1+\kappa) = 0$ and Newman (N1) obtained

$$F = 1 - \frac{6}{\pi^2} \sum_{n=1}^{\infty} \frac{1}{n^2} \exp(-n^2\pi^2\tau_p), \tag{3-80}$$

$$\mathrm{Sh}_p = \frac{2\pi^2}{3} \sum_{n=1}^{\infty} \exp(-n^2\pi^2\tau_p) \Bigg/ \sum_{n=1}^{\infty} \frac{1}{n^2} \exp(-n^2\pi^2\tau_p). \tag{3-81}$$

For a circulating sphere with $\mathrm{Pe}_p/(1+\kappa) \to \infty$, the time required for diffusion is much greater than that for fluid circulation, so that surfaces of uniform concentration coincide with the Hadamard–Rybczynski streamlines. Kronig and Brink (K6) showed that the solution is then

$$F = 1 - \frac{3}{8} \sum_{n=1}^{\infty} A_n^2 \exp(-16\lambda_n\tau_p) \tag{3-82}$$

$$\mathrm{Sh}_p = \frac{32}{3} \sum_{n=1}^{\infty} A_n^2\lambda_n^2 \exp(-16\lambda_n\tau_p) \Bigg/ \sum_{n=1}^{\infty} A_n^2 \exp(-16\lambda_n\tau_p). \tag{3-83}$$

Corresponding values for F, evaluated by finite differences from the governing equations, are shown in Fig. 3.15. As Pe_p increases, circulation causes F to rise more rapidly, but the effect is not large: τ_p for a given F decreases by at most a factor of three as $\mathrm{Pe}_p/(1+\kappa)$ increases from zero to infinity. In fact, the Kronig–Brink curve in Fig. 3.17 is closely approximated by Eq. (3-80) with $\mathscr{D}_p$ replaced by $2.5\mathscr{D}_p$. Thus circulation causes an effective diffusivity at most 2.5 times the molecular value. For negligible external resistance, the short-time approximation given by Eq. (3-72) becomes

$$F = 6\sqrt{\tau_p/\pi}. \tag{3-84}$$

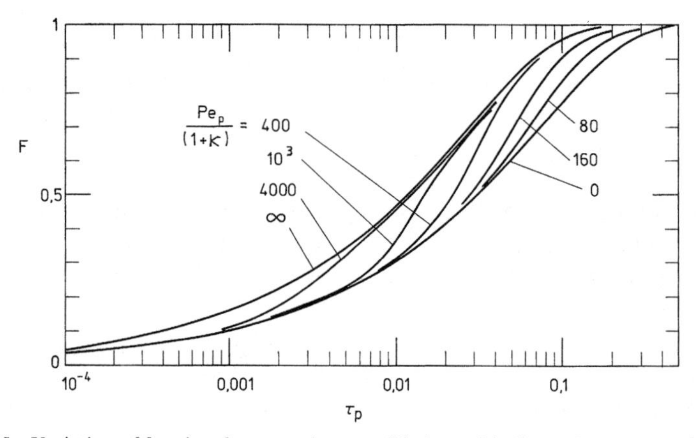

Fig. 3.15 Variation of fractional approach to equilibrium with dimensionless time for spheres in creeping flow with negligible external resistance.

Equation (3-84) lies within 10% of the Newman solution for $\tau_p < 10^{-2}$, and within 10% of the Kronig–Brink curve for $\tau_p < 10^{-4}$.

Figure 3.16 shows the time variation of Sh_p. Although Sh_p cannot easily be measured, it is useful for displaying the interaction of diffusion and circulation. The period of the local maxima and minima shown in Fig. 3.16 is inversely proportional to $Pe_p/(1 + \kappa)$. As a fluid element circulates along the surface of the particle and up through its center, solute diffuses to it from the region of the stagnation ring. A fluid element originally near the surface of the drop and depleted in solute is enriched in solute before it returns to the neighborhood of the surface. Thus the flux remains higher than it would have been if there were no diffusion from the stagnant regions of the drop. This is reflected by an increase in $dF/d\tau_p$ (Fig. 3.15) and a maximum in Sh_p. For long times, Sh_p approaches an asymptotic value, shown in Fig. 3.17. For the Newman solution the steady asymptotic value is

$$(Sh_p)_{\tau_p \to \infty} = 2\pi^2/3 = 6.58, \tag{3-85}$$

while for the Kronig–Brink solution

$$(Sh_p)_{\tau_p \to \infty} = \tfrac{32}{3}\lambda_1 = 17.66. \tag{3-86}$$

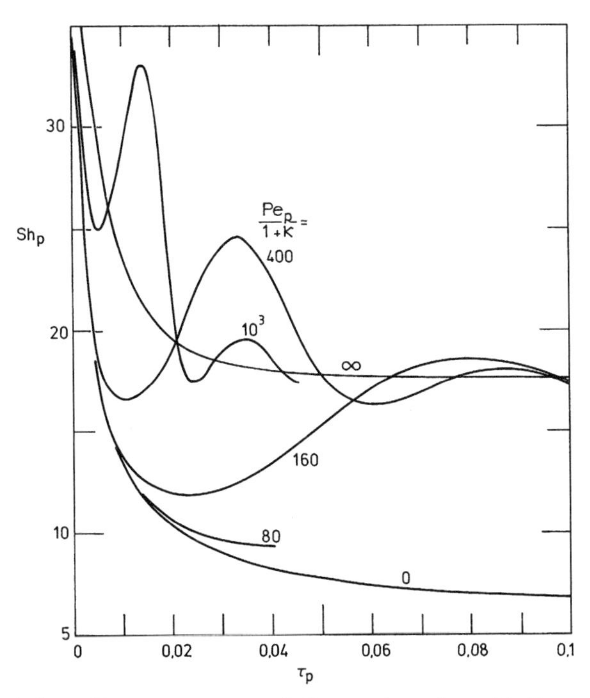

FIG. 3.16 Variation of instantaneous overall Sherwood number with dimensionless time for spheres in creeping flow with negligible external resistance.

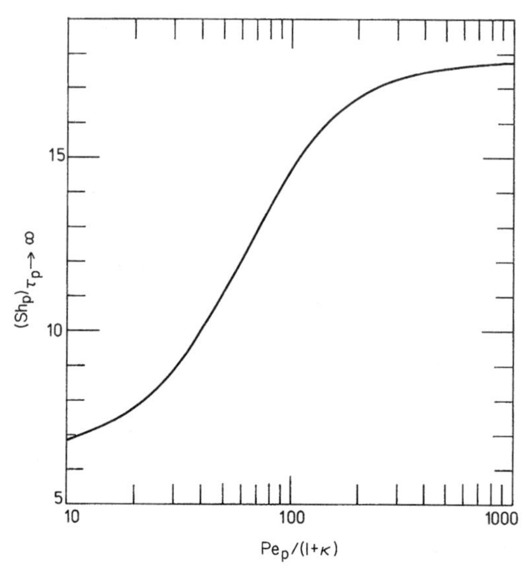

FIG. 3.17 Asymptotic value of overall Sherwood number at long times for spheres in creeping flow with negligible external resistance.

The asymptote is within 5% of the Newman value for $\mathrm{Pe_p}/(1+\kappa) < 10$ and within 5% of the Kronig–Brink value for $\mathrm{Pe_p}/(1+\kappa) > 250$. Figure 3.16 shows that the steady state is reached sooner for higher $\mathrm{Pe_p}/(1+\kappa)$. However, for $\tau_p > 0.15$, $\mathrm{Sh_p}$ is close to its steady asymptotic value for all $\mathrm{Pe_p}/(1+\kappa)$. For $\mathrm{Pe_p}/(1+\kappa) > 1000$ little error is incurred by using the Kronig–Brink result, since $\mathrm{Sh_p}$ is within 15% of the Kronig–Brink value even with oscillation. Experimental data at low Re for heat transfer (C2) and extraction from single drops (B7, G4, J3) agree with the Kronig–Brink analysis if care is taken to eliminate the external resistance, to exclude surfactants, and to correct for end effects.

5. *Negligible Internal Resistance*

When the internal resistance is negligible, the particle concentration is uniform and its time variation can be related to the external concentration gradient by a mass balance on the diffusing species:

$$dF/d\tau = (3\mathrm{Sh}/2H)(1 - F) \tag{3-87}$$

with the initial condition

$$F = 0 \quad \text{at} \quad \tau = 0. \tag{3-88}$$

a. *Transient External Resistance* With the time variation of the external resistance unspecified, the problem posed by Eq. (3-87) reduces to diffusion and

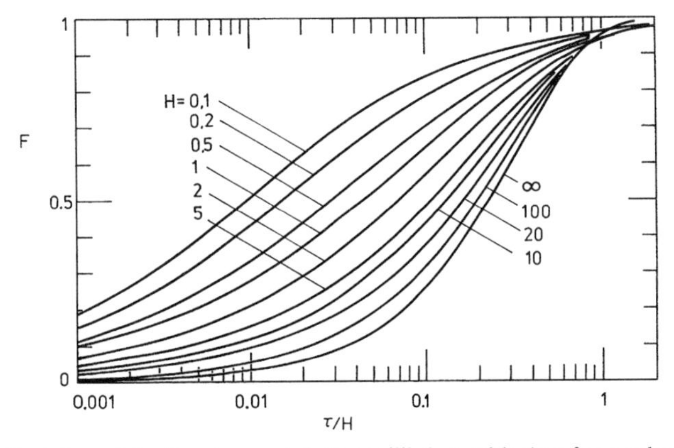

FIG. 3.18 Variation of fractional approach to equilibrium with time for a sphere in stagnant surroundings (Pe = 0) with negligible internal resistance ($H\mathscr{D}_p/\mathscr{D} > 25$).

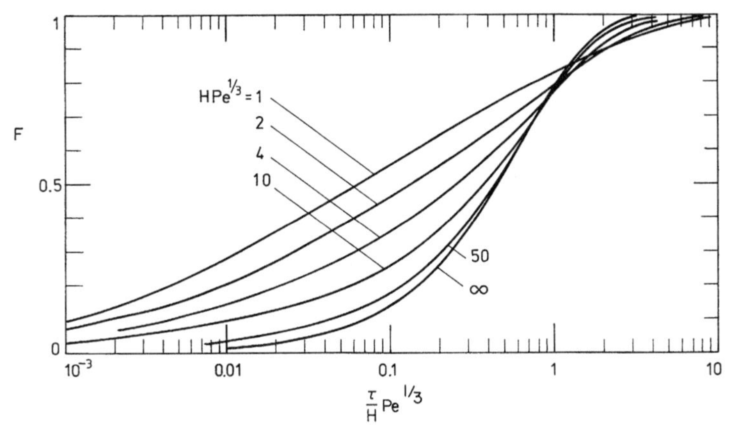

FIG. 3.19 Variation of fractional approach to equilibrium with time for rigid spheres with negligible internal resistance in creeping flow at high Pe.

convection in the external phase subject to a time-varying boundary condition. It can be solved with any of the step function solutions in Section B using Duhamel's theorem, to give the variation of F with τ (A4, C4, K4, K5). The solution for a sphere in stagnant surroundings, Eq. (3-59), yields the results in Fig. 3.18, valid for $H\mathscr{D}_p/\mathscr{D} > 25$† (K3). Figures 3.19 and 3.20 show corresponding results for rigid (K4) and circulating (D1) spheres in creeping flow at high Pe, obtained from the step function solutions. A solution is also available (K5) for rigid spheres with Pe < 1.

† For heat transfer the group $H\mathscr{D}_p/\mathscr{D}$ becomes K_{tp}/K_t.

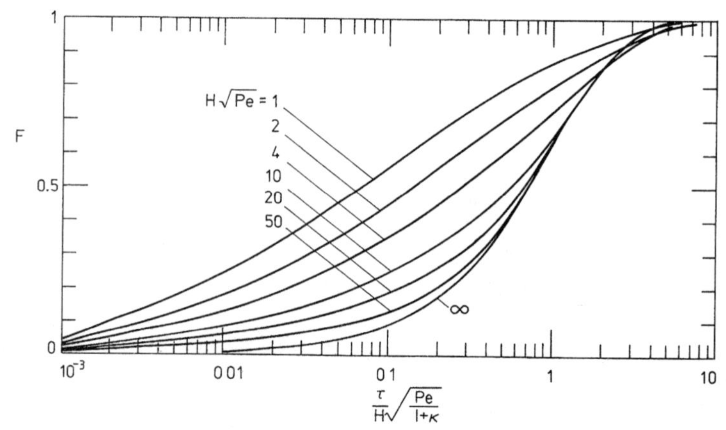

FIG. 3.20 Variation of fractional approach to equilibrium with time for circulating spheres with negligible internal resistance in creeping flow at high Pe.

b. *Quasi-Steady External Resistance* In the cases shown in Figs. 3.18 to 3.20, the curves approach a limit corresponding to the quasisteady case as H, $H\mathrm{Pe}^{1/3}$, or $H\mathrm{Pe}^{1/2}$ becomes very large. If the external resistance is assumed constant at its steady value, the solution to Eqs. (3-87) and (3-88) is

$$F = 1 - \exp(-3\tau\ \mathrm{Sh}/2H) = 1 - \exp(-3\tau_p\ \mathrm{Bi}), \tag{3-89}$$

where Bi, the Biot number, is $ka/H\mathscr{D}_p$.† The appropriate value of Sh is given by Eq. (3-44) for stagnant fluids (Fig. 3.18), by Eq. (3-47) for rigid particles (Fig. 3.19), and by Eq. (3-52) for circulating spheres (Fig. 3.20).

6. *Comparable Resistance in Each Phase*

Except for stagnant fluids, discussed in Section C.2, there are no general solutions for the case where the transient resistances in both phases are significant. If the external resistance is assumed constant, Eq. (3-56) must be solved with boundary conditions given by Eqs. (3-65), (3-66), (3-69), and

$$\partial c_p/\partial r = \mathrm{Bi}(Hc_\infty - c_p)/a \tag{3-90}$$

Solutions have been obtained for a rigid sphere with $\mathrm{Pe}_p = 0$ (G8), and the results are shown in Fig. 3.21. We have complemented these with solutions for a sphere circulating with the Hadamard–Rybczynski velocities at $\mathrm{Pe}_p/(1+\kappa) \rightarrow \infty$, assuming Sh_{loc} proportional to $\sin\theta$ and with the overall mean Sh used to define Bi. These results are shown in Fig. 3.22. For $\mathrm{Bi} \rightarrow \infty$ (i.e., negligible external resistance), the limiting curves are the Newman solution in Fig. 3.21 and the Kronig–Brink solution in Fig. 3.22. For $\mathrm{Bi} < 0.2$ the internal resistance

† For heat transfer, $\mathrm{Bi} = ha/K_{tp}$.

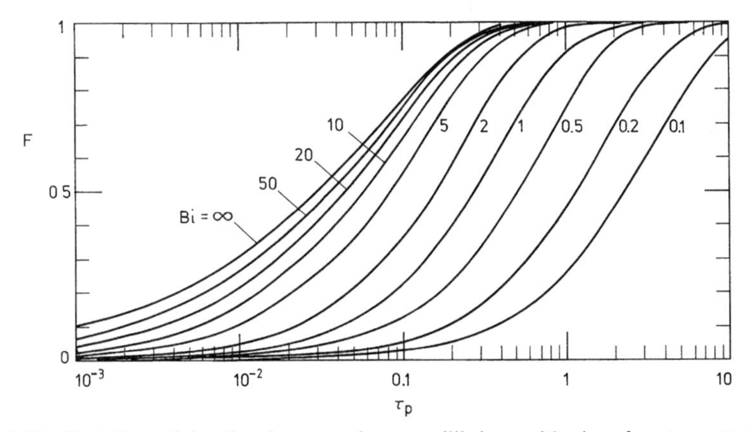

FIG. 3.21 Variation of fractional approach to equilibrium with time for stagnant spheres ($Pe_p = 0$) with constant external resistance.

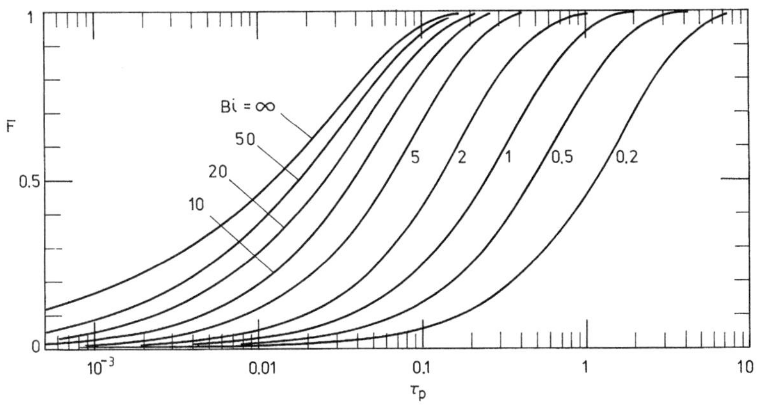

FIG. 3.22 Variation of fractional approach to equilibrium with time for circulating spheres with $Pe_p/(1 + \kappa) \rightarrow \infty$ and constant external resistance.

is negligible, so that the particle concentration is uniform and Eq. (3-89) can be used to predict F.

D. Effect of Surface Contaminants

Surface contaminants affect mass transfer via hydrodynamic and molecular effects, and it is convenient to consider these separately. Hydrodynamic effects include two phenomena which act in opposition. In the absence of mass transfer, contaminants decrease the mobility of the interface as discussed in Section II.D. In the presence of mass transfer, however, motion at the interface may be enhanced through the action of local surface tension gradients caused by small differences in concentration along the interface. This enhancement of surface motion, often called the Marangoni effect (S4), is considered in more detail in Chapter 10. The molecular effects are interfacial resistances to mass transfer which may arise from the interaction of surface contaminants with the species being transferred. The magnitude of the interfacial resistance depends upon the nature of the transferring substances and the contaminants.† Here we assume that the contaminants cause no additional resistance to transfer. Finite interfacial resistances are considered briefly in Chapter 10.

1. *External Resistance—Steady State*

The effect of surface contaminants on mass transfer has been calculated using the models of Frumkin and Levich (F1, L6), Schecter and Farley (S2), and Savic (G7, S1) (see Section II.D). The following crude but systematic method of estimating the effect of surface contamination on mass transfer is based on Savic's stagnant-cap approach, analogous to the treatment of terminal velocity. The values of the velocity ratio, Y, and angle excluded by the stagnant cap, θ_0, are estimated from Fig. 3.7 and Huang's curve in Fig. 3.6. The Sherwood number for the mobile interface, $\mathrm{Sh_M}$, is obtained by treating it as part of a fully circulating sphere, and is therefore taken from Fig. 3.10 at the appropriate κ and Pe. An approximate upper limit for transfer through the stagnant cap is obtained by treating it as a portion of a rigid sphere at the same Pe, so that the appropriate Sherwood number, $\mathrm{Sh_S}$, is obtained from Fig. 3.10 or Eq. (3-49). The overall Sherwood number is then estimated as

$$\mathrm{Sh} = Y_\mathrm{M}\,\mathrm{Sh_M} + Y_\mathrm{S}\,\mathrm{Sh_S}. \tag{3-91}$$

Use of Savic's surface velocities in Eq. (3-51) yields for high Pe:

$$\mathrm{Sh} = 0.651\,\mathrm{Pe}^{1/2}\sqrt{3(Y-1)/Y}, \tag{3-92}$$

which suggests that the weighting factor for mass transfer through the mobile interface can be approximated by

$$Y_\mathrm{M} = \sqrt{3(Y-1)/Y}. \tag{3-93}$$

Similarly, Y_S follows from the Levich solution for a rigid sphere in creeping flow at high Pe as

$$Y_\mathrm{S} = 1 - \sqrt{(\theta_0 - \sin\theta_0\cos\theta_0)/\pi}. \tag{3-94}$$

2. *Transfer with Variable Particle Concentration*

Dispersed phase resistances are increased when surface contaminants reduce interfacial mobility. Huang and Kintner (H9) used Savic's stagnant-cap theory in a semiempirical model for this resistance. A simpler quasi-steady model is proposed here, analogous to that for continuous phase resistance. The Sherwood

† It is unlikely that appreciable molecular resistance to heat transfer across fluid–solid or fluid–fluid interfaces can be caused by surface contamination.

numbers for the cap and the mobile portion of the sphere are obtained as functions of τ_p from the Newman and Kronig–Brink solutions described in Section C.4. It is assumed that the cap angle is constant with time and that the spherical segment bounded by the cap is stagnant and occupies a fraction f_{VS} of the sphere volume:

$$f_{VS} = (1 + \cos\theta_0)[2 + \cos\theta_0(1 - \cos\theta_0)]/4. \tag{3-95}$$

Mass balances for the mobile and stagnant portions of the particle then give

$$(1 - f_{VS})(dF_M/d\tau_p) = \tfrac{3}{2}\text{Sh}_M(1 - f_{AS})(1 - F_M) - \tfrac{3}{2}\text{Sh}_I f_{AI}(F_M - F_S), \tag{3-96}$$

$$f_{VS}(dF_S/d\tau_p) = \tfrac{3}{2}\text{Sh}_S f_{AS}(1 - F_S) + \tfrac{3}{2}\text{Sh}_I f_{AI}(F_M - F_S), \tag{3-97}$$

where the subscripts M and S indicate the mobile and stagnant portions of the particle, I indicates the plane separating these portions, f_{AS} is the fraction of the particle surface occupied by the stagnant cap, and f_{AI} is the area of contact between the two portions of the particle expressed as a fraction of the surface area of the sphere. The mean approach to equilibrium is

$$F = (1 - f_{VS})F_M + f_{VS}F_S. \tag{3-98}$$

It is further assumed that the resistances between the two portions of the sphere are additive:

$$\text{Sh}_I^{-1} = \text{Sh}_M^{-1} + \text{Sh}_S^{-1}, \tag{3-99}$$

with Sh_M and Sh_S given by Eqs. (3-81) and (3-83). For uniform initial composition, the initial condition is

$$F_M = F_S = 0 \quad \text{at} \quad \tau_p = 0. \tag{3-100}$$

These equations have been solved numerically to give the variation of F with τ_p shown in Fig. 3.23 for several values of the angle excluded by the

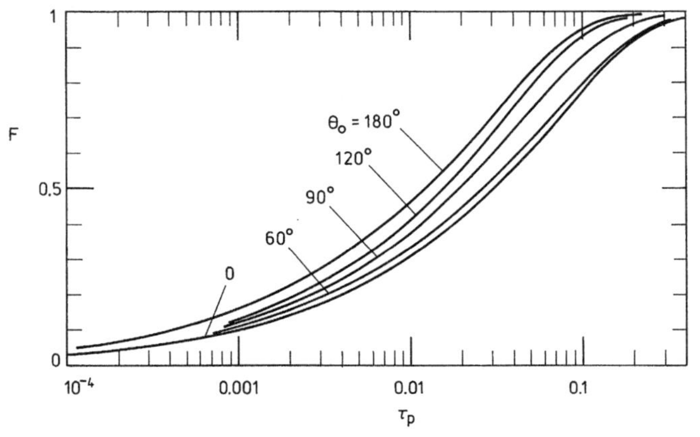

FIG. 3.23 Variation of fractional approach to equilibrium with time for fluid particles with contaminated interface and $\text{Pe}_p/(1 + \kappa) \to \infty$; θ_0 = angle excluded by stagnant cap.

stagnant cap, θ_0. Results lie between the solutions of Newman and Kronig and Brink. This model can easily be extended to include changes of cap angle with time.

REFERENCES

A1. Acrivos, A., and Goddard, J. D., *J. Fluid Mech.* **23**, 273–291 (1965).
A2. Acrivos, A., and Taylor, T. D., *Phys. Fluids* **5**, 387–394 (1962).
A3. Akiyama, T., and Yamaguchi, K., *Can. J. Chem. Eng.* **53**, 695–698 (1975).
A4. Arpaci, V. S., "Conduction Heat Transfer." Addison-Wesley, Reading, Massachusetts, 1966.
B1. Bachhuber, C., and Sanford, C., *J. Appl. Phys.* **45**, 2567–2569 (1974).
B2. Batchelor, G. K., "An Introduction to Fluid Dynamics." Cambridge Univ. Press, London and New York, 1967.
B3. Bond, W. N., and Newton, D. A., *Philos. Mag.* **5**, 794–800 (1928).
B4. Boussinesq, J., *C. R. Acad. Sci.* **156**, 1124–1130 (1913).
B5. Bowman, C. W., Ward, D. M., Johnson, A. I., and Trass, O., *Can. J. Chem. Eng.* **39**, 9–13 (1961).
B6. Brenner, H., *Chem. Eng. Sci.* **18**, 109–122 (1963).
B7. Brounshtein, B. I., Zheleznyak, A. S., and Fishbein, G. A., *Int. J. Heat Mass Transfer* **13**, 963–973 (1970).
C1. Calderbank, P. H., Johnson, D. S. L., and Loudon, J., *Chem. Eng. Sci.* **25**, 235–256 (1970).
C2. Calderbank, P. H., and Korchinski, I. J. O., *Chem. Eng. Sci.* **6**, 65–78 (1956).
C3. Carrier, G. F., Techn. Rep., Math. Sci. Div., Off. Nav. Res. nonR-653-00/1. Brown Univ., Providence, Rhode Island (1953).
C4. Carslaw, H. S., and Jaeger, J. C., "Conduction of Heat in Solids," 2nd Ed. Oxford Univ. Press, London and New York, 1959.
C5. Chao, B. T., *J. Heat Transfer* **91**, 273–280 (1969).
C6. Chester, W., *J. Fluid Mech.* **13**, 557–569 (1962).
C7. Chester, W., and Breach, D. R., *J. Fluid Mech.* **37**, 751–760 (1969).
C8. Choudhury, P. N., and Drake, D. G., *Q. J. Mech. Appl. Math.* **24**, 23–36 (1971).
D1. Dang, V. D., Ruckenstein, E., and Gill, W. N., *Chem. Eng. (London)* **241**, 248–251 (1970).
D2. Davies, J. T., *Adv. Chem. Eng.* **4**, 1–50 (1963).
D3. Davies, J. T., and Mayers, G. R. A., *Chem. Eng. Sci.* **16**, 55–66 (1961).
D4. Davies, J. T., and Rideal, E. K., "Interfacial Phenomena." Academic Press, New York, 1963.
D5. Davis, R. E., and Acrivos, A., *Chem. Eng. Sci* **21**, 681–685 (1966).
D6. Dukhin, S. S., and Buikov, M. V., *Russ. J. Phys. Chem.* **39**, 482–485 (1965).
D7. Dukhin, S. S., and Deryagin, B. V., *Russ. J. Phys. Chem.* **35**, 715–717 (1961).
E1. Edge. R. M., and Grant, C. D., *Chem. Eng. Sci.* **26**, 1001–1012 (1971).
E2. Edge, R. M., and Grant, C. D., *Chem. Eng. Sci.* **27**, 1709–1721 (1972).
F1. Frumkin, A., and Levich, V. G., *Zh. Fiz. Khim.* **21**, 1183–1204 (1947).
G1. Garner, F. H., and Hammerton, D., *Chem. Eng. Sci.* **3**, 1–11 (1954).
G2. Garner, F. H., and Haycock, P. J., *Proc. R. Soc., Ser. A* **252**, 457–475 (1959).
G3. Garner, F. H., and Skelland, A. H. P., *Chem. Eng. Sci.* **4**, 149–158 (1955).
G4. Garner, F. H., and Tayeban, M., *An. R. Soc. Esp. Fis. Quim., Ser. B* **56**, 479–498 (1960).
G5. Goldstein, S., *Proc. R. Soc., Ser. A* **123**, 216–235 (1929).
G6. Golovin, A. M., and Ivanov, M. F., *J. Appl. Mech. Tech. Phys. (USSR)* **12**, 91–94 (1973).
G7. Griffith, R. M., *Chem. Eng. Sci.* **17**, 1057–1070 (1962).
G8. Gröber, H., *VDIZ.* **69**, 705–712 (1925).
G9. Groothuis, H., and Kramers, H., *Chem. Eng. Sci.* **4**, 17–25 (1955).

G10. Gupalo, Y. P., and Ryazantsev, Y. S., *Chem. Eng. Sci.* **27**, 61–68 (1972).
H1. Hadamard, J. S., *C. R. Acad. Sci.* **152**, 1735–1738 (1911).
H2. Hales, H. B., Sc.D. Thesis, Mass. Inst. Technol., Cambridge, 1967.
H3. Harper, J. F., *Adv. Appl. Mech.* **12**, 59–129 (1972).
H4. Harper, J. F., *J. Fluid Mech.* **58**, 539–545 (1973).
H5. Harper, J. F., Moore, D. W., and Pearson, J. R. A., *J. Fluid Mech.* **27**, 361–366 (1967).
H6. Hill, M. J. M., *Philos. Trans. R. Soc., London, Ser. A* **185**, 213–245 (1894).
H7. Horton, T. J., Fritsch, T. R., and Kintner, R. C., *Can. J. Chem. Eng.* **43**, 143–146 (1965).
H8. Huang, W. S., Ph.D. Thesis, Illinois Inst. of Technol., Chicago, 1968.
H9. Huang, W. S., and Kintner, R. C., *AIChE J.* **15**, 735–744 (1969).
J1. Johns, L. E., and Beckmann, R. B., *AIChE J.* **12**, 10–16 (1966).
J2. Johnson, A. I., and Braida, L., *Can. J. Chem. Eng.* **35**, 165–172 (1957).
J3. Johnson, A. I., and Hamielec, A. E., *AIChE J.* **6**, 145–149 (1960).
K1. Kenning, D. B. R., *Chem. Eng. Sci.* **24**, 1385–1386 (1969).
K2. Konopliv, N., and Sparrow, E. M., *Heat Transfer 1970*, **3**, FC7.4 (1970).
K3. Konopliv, N., and Sparrow, E. M., *Wärme-Stoffübertrag.* **3**, 197–210 (1970).
K4. Konopliv, N., and Sparrow, E. M., *J. Heat Transfer* **94**, 266–272 (1972).
K5. Konopliv, N., and Sparrow, E. M., *Q. Appl. Math.* **29**, 225–235 (1972).
K6. Kronig, R., and Brink, J. C., *Appl. Sci. Res., Sect. A* **2**, 142–154 (1950).
L1. Lamb, H., *Philos. Mag.* **21**, 112–121 (1911).
L2. Levan, M. D., and Newman, J., *AIChE J.* **22**, 695–701 (1976).
L3. Levich, V. G., "Physicochemical Hydrodynamics." Prentice-Hall, New York, 1962.
L4. Lewis, J. A., and Carrier, F. G., *Q. Appl. Math.* **7**, 228–234 (1949).
L5. Linton, M., and Sutherland, K. L., *Proc. Int. Congr. Surf. Act., 2nd, London* **1**, 494–502 (1957).
L6. Lochiel, A. C., *Can. J. Chem. Eng.* **43**, 40–44 (1965).
M1. Magarvey, R. H., and Kalejs, J., *Nature (London)* **198**, 377–378 (1963).
M2. Masliyah, J. H., and Epstein, N., *Prog. Heat Mass Transfer* **6**, 613–632 (1972).
M3. Maxworthy, T., *J. Fluid Mech.* **23**, 369–372 (1965).
N1. Newman, A. B., *Trans. Am. Inst. Chem. Eng.* **27**, 203–220 (1931).
N2. Newman, J., *Chem. Eng. Sci.* **22**, 83–85 (1967).
O1. Oellrich, L., Schmidt-Traub, H., and Brauer, H., *Chem. Eng. Sci.* **28**, 711–721 (1973).
O2. Oseen, C. W., *Ark. Mat., Astron. Fys.* **6**, No. 29 (1910).
P1. Plöcher, U. J., and Schmidt-Traub, H., *Chem.-Ing.-Tech.* **44**, 313–319 (1972).
P2. Proudman, I., *J. Fluid Mech.* **37**, 759–760 (1969).
P3. Proudman, I., and Pearson, J. R. A., *J. Fluid Mech.* **2**, 237–262 (1957).
P4. Pruppacher, H. R., LeClair, B. P., and Hamielec, A. E., *J. Fluid Mech.* **44**, 781–790 (1970).
Q1. Quinn, J. A., and Scriven, L. E., *Interfacial Phenomena, Adv. Semin. AIChE, 14th, 1970.*
R1. Raymond, D. R., and Zieminski, S. A., *AIChE J.* **17**, 57–65 (1971).
R2. Redfield, J. A., and Houghton, G., *Chem. Eng. Sci.* **20**, 131–139 (1965).
R3. Robinson, J. V., *J. Phys. Colloid Chem.* **51**, 431–437 (1947).
R4. Ruckenstein, E., *Int. J. Heat Mass Transfer* **10**, 1785–1792 (1967).
R5. Rybczynski, W., *Bull. Int. Acad. Pol. Sci. Lett., Cl. Sci. Math. Nat., Ser. A* pp. 40–46 (1911).
S1. Savic, P., *Natl. Res. Counc. Can., Rep.* No. MT-22 (1953).
S2. Schecter, R. S., and Farley, R. W., *Can. J. Chem. Eng.* **41**, 103–107 (1963).
S3. Scriven, L. E., *Chem. Eng. Sci.* **12**, 98–108 (1960).
S4. Scriven, L. E., and Sternling, C. V., *Nature (London)* **187**, 186–188 (1960).
S5. Shakespear, G. A., *Philos. Mag.* **28**, 728–734 (1914).
S6. Sih, P. S., and Newman, J., *Int. J. Heat Mass Transfer* **10**, 1749–1756 (1967).
S7. Spells, K. E., *Proc. Phys. Soc., London, Sect. B* **65**, 541–546 (1952).
S8. Steinberger, E. H., Pruppacher, H. R., and Neiburger, M., *J. Fluid Mech.* **34**, 809–819 (1968).

S9. Stokes, G. G., *Trans. Cambridge Philos. Soc*, **9**, 8–27 (1851). (*Math. Phys. Pap.* **3**, 1.)
T1. Taylor, J. D., and Acrivos, A. J., *J. Fluid Mech.* **18**, 466–476 (1964).
U1. Ueyama, K., and Hatanaka, J. I., *J. Chem. Eng. Jpn.* **9**, 17–22 (1976).
V1. Van Dyke, M. D., "Perturbation Methods in Fluid Mechanics." Academic Press, New York, 1964.
V2. Van Dyke, M. D., *J. Fluid Mech.* **44**, 365–372 (1970).
W1. Ward, D. M., Trass, O., and Johnson, A. I., *Can. J. Chem. Eng.* **40**, 164–168 (1962).
W2. Wasserman, M. L., and Slattery, J. C., *AIChE J.* **15**, 533–547 (1969).
W3. Watada, H., Hamielec, A. E., and Johnson, A. I., *Can. J. Chem. Eng.* **48**, 255–260 (1970).
W4. Wellek, R. M., Andoe, W. V., and Brunson, R. J., *Can. J. Chem. Eng.* **48**, 645–655 (1970).
W5. Woo, S.-W., Ph.D. Thesis, McMaster Univ., Hamilton, Ontario, 1971.

Chapter 4

Slow Viscous Flow Past Nonspherical Rigid Particles

I. INTRODUCTION

In this chapter, we extend the discussion of the previous chapter to non-spherical shapes. Only solid particles are considered and the discussion is limited to low Reynolds number flows. The flow pattern and heat and mass transfer for a nonspherical particle depend on its orientation. This introduces complications not present for spherical particles. For example, the net drag force is parallel to the direction of motion only if the particle has special shape properties or is aligned in specific orientations.

II. FLUID MECHANICS

A. General Considerations

It is convenient to define an "equivalent sphere" as in Chapter 2. Drag is then related to that on the equivalent sphere either by a "drag ratio" defined as

$$\Delta = \frac{\text{drag on particle}}{\text{drag on equivalent sphere at same velocity}} \tag{4-1}$$

or by a "settling factor,"

$$S = \frac{\text{terminal velocity of particle}}{\text{terminal velocity of equivalent sphere with same density}}. \tag{4-2}$$

When these factors are based on the volume-equivalent sphere,

$$S_e = \Delta_e^{-1} \tag{4-3}$$

in creeping flow because of the linear dependence of drag on relative velocity.

Since the net drag on an arbitrary particle is generally not parallel to the direction of motion, a particle falls† vertically without rotation only if it possesses a certain symmetry or a specific orientation. The following guidelines for solid particles with uniform density are derived from general results for creeping flow (H3):

(i) Orthotropic particles (see Chapter 2) have no preferred orientation and always fall without rotation. Motion is vertical only if a plane of symmetry is horizontal.

(ii) Axisymmetric particles fall vertically if the axis is vertical. If the particle has fore-and-aft symmetry, it is orthotropic. It therefore falls vertically also if the axis of symmetry is horizontal, and always moves without rotation. Otherwise, it falls without rotating only when its axis is vertical; it is only stable, however, in one of the two directions.

(iii) Spherically isotropic particles always fall vertically without rotation, and the settling velocity is independent of orientation. This is the origin of the name for this class of shapes.

Particles subject to Brownian motion tend to adopt random orientations, and hence do not follow these rules. A particle without these symmetry properties may follow a spiral trajectory, and may also rotate or wobble. In general, the drag and torque on an arbitrary particle translating and rotating in an unbounded quiescent fluid are determined by three second-order tensors which depend on the shape of the body:

(i) A symmetric translation tensor which describes the resistance to translational motion.

(ii) A symmetric rotation tensor giving the torques resulting from rotation.

(iii) An asymmetric coupling tensor which defines torques resulting from translation and drag forces resulting from rotation.

The use of these resistance tensors is developed in detail by Happel and Brenner (H3). While enabling compact formulation of fundamental problems, these tensors have limited application since their components are rarely available even for simple shapes. Here we discuss specific classes of particle shape without recourse to tensor notation, but some conclusions from the general treatment are of interest. Because the translation tensor is symmetric, it follows that every particle possesses at least three mutually perpendicular axes such that, if the particle is translating without rotation parallel to one of these axes, the total

† The same guidelines apply to rising particles with density less than that of the surrounding fluid.

drag force is also parallel to the axis (H3). These axes are usually called "principal axes of translation." If the particle is translating with velocity **U** parallel to principal axis *i*, then the drag is given by

$$\mathbf{F_D} = -\mu c_i \mathbf{U}, \tag{4-4}$$

where the three values of c_i are termed the "principal translational resistances." For an orthotropic particle, the principal axes are normal to the planes of symmetry. For an axisymmetric particle, the axis of symmetry is one of the principal axes.

Particles which are orthotropic, axisymmetric, or spherically isotropic possess a point about which the coupling tensor is zero. In this case, pure translation in creeping flow never causes a torque component of drag. The resistance to any translation can then be estimated by a simple procedure described by Dahneke (D1), relying on the linearity of the governing equations. The total drag is obtained by adding the drag components due to the components of velocity parallel to each of the principal axes of translation. Thus, if the principal axes are defined by the three orthogonal unit vectors **i**, **j**, **k**, the total drag resulting from translation at velocity **U** is given by

$$\mathbf{F_D} = -\mu[\mathbf{i}c_1 U_1 + \mathbf{j}c_2 U_2 + \mathbf{k}c_3 U_3]. \tag{4-5}$$

For a large number of identical particles with random orientations, the mean resistance is obtained by integrating Eq. (4-5) over the range of orientations (H3). The mean resistance follows as

$$\bar{c} = 3/(c_1{}^{-1} + c_2{}^{-1} + c_3{}^{-1}) \tag{4-6}$$

while the mean direction of settling is parallel to the gravity field. The resistance given by Eq. (4-6) is used to describe the translational motion of dilute suspensions of small particles of arbitrary shape and random orientation (e.g. as a result of Brownian motion).

B. Axisymmetric Particles

1. *General Considerations*

a. *Resistance to Translation* Consider an axisymmetric particle translating with steady velocity **U** through a stationary unbounded viscous fluid. The orientation of the particle is defined as shown in Fig. 4.1 by the angle θ between its axis of symmetry and the direction of motion. In the plane of Fig. 4.1, the principal axes of translation are parallel and perpendicular to the axis of symmetry. Therefore the components of drag along the principal axes follow from Eq. (4-5):

parallel to the axis of symmetry:

$$F_{D1} = -\mu c_1 U \cos\theta; \tag{4-7}$$

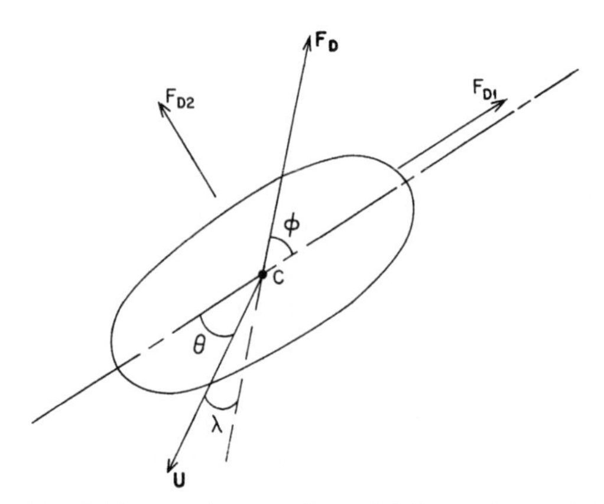

FIG. 4.1 Arbitrary axisymmetric particle in steady translation.

perpendicular to the axis of symmetry:

$$F_{D2} = -\mu c_2 U \sin\theta, \tag{4-8}$$

where c_1 and c_2 are the principal resistances for translation parallel and normal to the axis of symmetry. The net drag is then

$$F_D = -\mu U[c_1{}^2 \cos^2\theta + c_2{}^2 \sin^2\theta]^{1/2} \tag{4-9}$$

at an angle ϕ to the axis of symmetry such that

$$\tan\phi = (c_2/c_1)\tan\theta. \tag{4-10}$$

Thus the drag resulting from any translation can be determined if the two principal resistances are known. The principal resistances of common axisymmetric shapes are given in subsequent sections.

b. *Motion in Free Fall or Rise* For the particle to move steadily in free fall or rise, two conditions must be met:

(i) The total drag $\mathbf{F_D}$ must be directed vertically to counterbalance the net gravity force acting on the particle.
(ii) The point on the axis of symmetry through which $\mathbf{F_D}$ acts, C in Fig. 4.1, must coincide with the center of mass (assuming the particle has uniform density).

For a particle without fore-and-aft symmetry, condition (ii) is generally met only when the axis is vertical: hence such particles fall with a "tumbling" motion. However, if the particle has fore-and-aft symmetry of shape and density, both $\mathbf{F_D}$ and immersed weight must act through the point where the plane of symmetry cuts the axis; condition (ii) is automatically satisfied, and the particle falls without rotation. Condition (i) then determines the direction of motion. The angle ϕ becomes the inclination of the axis from the vertical, so that the particle falls at an angle to the vertical given by

$$\lambda = \phi - \theta = \tan^{-1}[(c_2 - c_1)\tan\phi/(c_2 + c_1\tan^2\phi)]. \tag{4-11}$$

As a general guide, c_1 is usually less than c_2 for a prolate particle, so that $\theta < \phi$ and the direction of motion is between the axis and the vertical. On the other hand, an oblate body usually has $c_1 > c_2$ so that the direction of fall is between the vertical and the equator. The settling velocity follows from Eq. (4-9):

$$U = \frac{g\,\Delta\rho\,V}{\mu c_1}\left[\frac{1 + (c_1\tan\phi/c_2)^2}{1 + \tan^2\phi}\right]^{1/2}, \tag{4-12}$$

where V is the particle volume and $\Delta\rho$ the density difference between the particle and the fluid. The component parallel to the axis of the particle is

$$U\cos\theta = \frac{g\Delta\rho\,V}{\mu c_1\sqrt{1 + \tan^2\phi}}, \tag{4-13}$$

while the vertical component is

$$U\cos\lambda = \frac{g\,\Delta\rho\,V}{\mu c_1 c_2}\left[\frac{c_1 + c_2\tan^2\phi}{1 + \tan^2\phi}\right]. \tag{4-14}$$

For a dilute suspension of identical particles oriented randomly, the mean resistance follows from Eq. (4-6):

$$\bar{c} = 3/(c_1{}^{-1} + 2c_2{}^{-1}), \tag{4-15}$$

so that the mean settling velocity is

$$\bar{U} = \frac{g\,\Delta\rho\,V}{3\mu}\left[\frac{1}{c_1} + \frac{2}{c_2}\right]. \tag{4-16}$$

c. *Translation Parallel to the Axis of Symmetry* Many more results have been reported in the literature for the axial resistance c_1 than for the normal resistance c_2 of axisymmetric particles, since axisymmetric flows are more tractable than three-dimensional flows. The equation of motion for creeping flow parallel to the axis of symmetry, Eq. (1-36), may be expressed in various orthogonal curvilinear coordinate systems (H3). For a frame of reference fixed to the particle with origin on the axis of symmetry, the boundary conditions are Eq. (1-27) and

$$\psi = 0 \qquad \text{on the surface of the particle,} \tag{4-17}$$

$$\partial\psi/\partial n = 0 \qquad \text{on the surface of the particle,} \tag{4-18}$$

where n is a coordinate normal to the surface.

A useful theorem due to Payne and Pell (P3) enables the drag on an axisymmetric body to be calculated directly from the stream function ψ' for steady motion of the body through stagnant fluid[†]:

$$F_D = 8\pi\mu \lim_{r\to\infty} \{\psi'/r \sin^2\theta\}. \tag{4-19}$$

This theorem could have been used to obtain the drag for fluid and solid spheres in Chapter 3. Explicit analytic solutions are available for bodies whose boundaries are easily described in relatively simple coordinate systems. Results for spheroids and disks (O1, P3, S1) are discussed below. Solutions are also available for lenses and hemispheres (P3), hollow spherical caps (D3, C3, P3), toroids (P4), long spindles or needles (P5), and pairs of identical spheres (S7).

Techniques have also been developed for obtaining approximate solutions in axisymmetric creeping flow. The general approach is to use a series expansion for the drag or stream function, truncate the series after a number of terms, and use the boundary conditions to evaluate the terms retained. The "point-force approximation technique" developed by Burgers (B10) is applicable to particles (e.g., needles or fibres) which have a large aspect ratio. The total drag is approximated by a distributed line force along the axis of symmetry. The force is represented by a polynomial approximation in which the constants are determined by satisfying Eq. (4-18) as closely as possible. O'Brien (O2) expanded the stream function as an infinite series in general spherical coordinates and truncated after a finite number of terms. The remaining coefficients were obtained by satisfying Eq. (4-18) at the same number of points on the surface of the particle. Bowen and Masliyah (B4) improved this approach by fitting the solution to the boundary condition in the least-squares sense over the entire surface. Gluckman *et al.* (G2, G3) developed a "multipole representation technique," by which any convex axisymmetric body is represented as an array of oblate spheroids. Again, individual terms in the resulting series for ψ are determined by satisfying Eq. (4-17) and Eq. (4-18) at a finite number of points on the surface. These approximate techniques allow reliable results to be obtained for bodies as deformed as cylinders and cones. However, care is required in handling plane surfaces normal to flow, e.g., the ends of a cylinder. Results obtained by these techniques are discussed below.

2. *Spheroids*

Spheroidal particles can be treated analytically, and allow study of shapes ranging from slightly deformed spheres to disks and needles. Moreover, a spheroid often provides a useful approximation for the drag on a less regular

[†] This theorem leads to an interesting result concerning "drift." For an axisymmetric body, the drift volume (D2), the volume enclosed between the initial and final position of a horizontal layer of tracer fluid, is given by $V_d = \lim_{r\to\infty}(\psi')$. If the body is to have finite drag, then $\lim_{r\to\infty}(\psi'/r)$ $(0 < \theta < \pi)$ must be finite from Eq. (4-19). Hence V_d must be unbounded. Therefore any axisymmetric body with finite drag in creeping flow through an infinite medium must cause infinite drift. It seems likely that this result should apply to bodies without axisymmetry, but no proof of this appears to have been given.

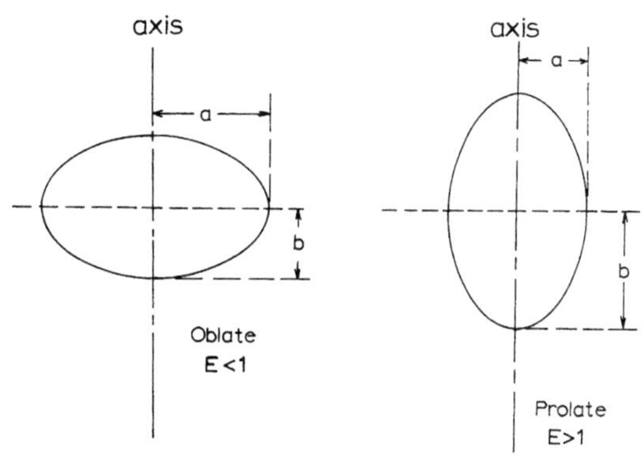

FIG. 4.2 Dimensions of spheroids.

particle (see Section E below). Figure 4.2 shows the notation used here: the axial dimension of the particle is $2b$ while the maximum dimension normal to the axis of symmetry is $2a$. The aspect ratio, b/a, of the particle is denoted by E.

a. *Creeping Flow* Table 4.1 gives expressions for the principal resistances of spheroids, first obtained (G1) from Oberbeck's general results for ellipsoids

TABLE 4.1

Resistances of Spheroids in Creeping Flow[a]

	Oblate ($E < 1$)	Prolate ($E > 1$)
Principal Resistances		
1. Axial, c_1		
Exact:	$\dfrac{8\pi a(1 - E^2)}{[(1 - 2E^2)\cos^{-1} E/\sqrt{1 - E^2}] + E}$	$\dfrac{8\pi a(E^2 - 1)}{[(2E^2 - 1)\ln(E + \sqrt{E^2 - 1})/\sqrt{E^2 - 1}] - E}$
Approx:	$1.2\pi a(4 + E)$	$1.2\pi a(4 + E)$
2. Normal, c_2		
Exact:	$\dfrac{16\pi a(1 - E^2)}{[(3 - 2E^2)\cos^{-1} E/\sqrt{1 - E^2}] - E}$	$\dfrac{16\pi a(E^2 - 1)}{[(2E^2 - 3)\ln(E + \sqrt{E^2 - 1})/\sqrt{E^2 - 1}] + E}$
Approx.:	$1.2\pi a(3 + 2E)$	$1.2\pi a(3 + 2E)$
Mean Resistance (random orientation), $\bar{c}$:		
Exact:	$6\pi a\sqrt{1 - E^2}/\cos^{-1} E$	$6\pi a\sqrt{E^2 - 1}/\ln(E + \sqrt{E^2 - 1})$
Ratio of form drag to skin friction (axisymmetric flow), R:		
Exact:	$\dfrac{E\cos^{-1} E - \sqrt{1 - E^2}}{E^2\sqrt{1 - E^2} - E\cos^{-1} E}$	$\dfrac{E\ln(E + \sqrt{E^2 - 1}) - \sqrt{E^2 - 1}}{E^2\sqrt{E^2 - 1} - E\ln(E + \sqrt{E^2 - 1})}$

[a] After Aoi (A1), Gans (G1), Happel and Brenner (H3), Oberbeck (O1), and Payne and Pell (P3).

(O1). Results for thin disks are obtained in the limit as $E \to 0$. Approximate relationships, obtained by treating the spheroid as a slightly deformed sphere (H3, S1), are also given. The drag ratio may conveniently be expressed as the ratio of the resistance of the spheroid to that of the sphere with the same equatorial radius a:

$$\Delta_{a1} = c_1/6\pi a, \tag{4-20}$$

$$\Delta_{a2} = c_2/6\pi a, \tag{4-21}$$

$$\bar{\Delta}_a = \bar{c}/6\pi a, \tag{4-22}$$

where $\bar{c}$ is the mean resistance for a large number of spheroids with random orientation, obtained from Eq. (4-15). The drag ratio is thus equal to the radius of the sphere with the same resistance in creeping flow, expressed as a multiple of a.

Figure 4.3 shows exact and approximate values for Δ_{a1}, the axial drag ratio. Corresponding curves for Δ_{a2} and $\bar{\Delta}_a$ appear in Fig. 4.4 with the exact values of Δ_{a1} for comparison. Due to the dependence of surface area on the axial dimension, $2b$, drag increases with E. For flow parallel to the axis, the polar regions contribute least to the total drag, so that Δ_{a1} depends less strongly than Δ_{a2} on E. The approximate results give good estimates for the resistances. The maximum deviation for oblate particles, approximately 6% for both Δ_{a1} and Δ_{a2}, occurs for disks. For prolate particles the deviation increases with aspect ratio; for $E = 5$ the error is almost 10% in Δ_{a2} but less than 1% in Δ_{a1}.

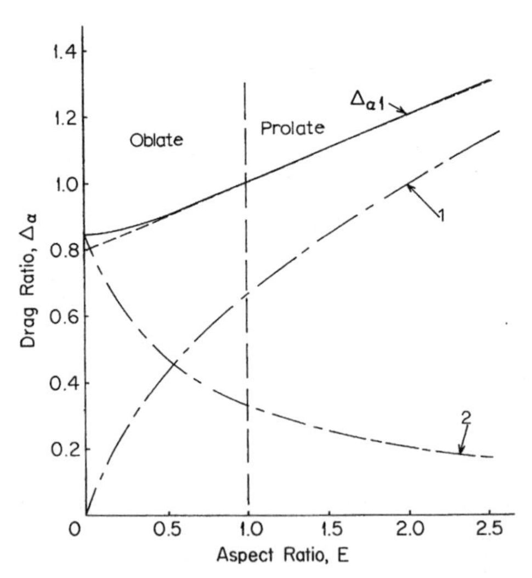

FIG. 4.3 Drag ratios for spheroids in axisymmetric flow. Drag ratio Δ_{a1}: — Exact; ---- Approximate; —·—· Drag Component; (1) Friction; (2) Form.

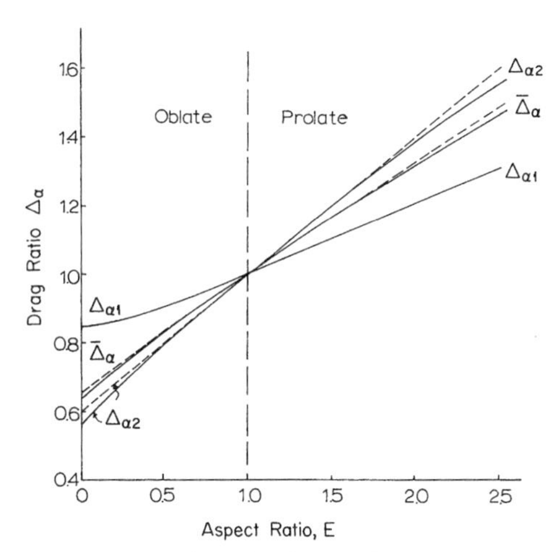

FIG. 4.4 Drag ratios for spheroids: — Exact; --- Approximate.

It is common practice to define a "hydraulic equivalent" sphere as the sphere with the same density and terminal settling velocity as the particle in question. For a spheroid in creeping flow, the hydraulic equivalent sphere diameter is $2a\sqrt{E/\Delta_a}$ and thus depends on orientation.

It was noted in Chapter 3 that the ratio R of form drag to skin friction for a rigid sphere in Stokes flow is 1:2. Table 4.1 gives expressions for R due to Aoi (A1) for flow parallel to the axis of spheroids. The ratios of form drag and skin friction to the total drag on a sphere of radius a are $R\Delta_{a1}/(1 + R)$ and $\Delta_{a1}/(1 + R)$, respectively. These two terms are plotted in Fig. 4.3. The two components of drag depend strongly on aspect ratio for oblate spheroids. However, the changes are largely compensating so that the dependence of total drag on E is weak.

The drag ratio based on the sphere with equal volume is

$$\Delta_e = \Delta_a E^{-1/3}. \tag{4-23}$$

Figure 4.5 shows the variation of Δ_e with E for flow parallel and normal to the axis, and averaged over random orientations. Except for disk-like particles, the dependence of Δ_e on aspect ratio is rather weak. In axial motion, a somewhat prolate spheroid experiences less drag than the volume-equivalent sphere: Δ_{e1} passes through a minimum of 0.9555 for $E = 1.955$. For motion normal to the axis of symmetry, Δ_{e2} takes a minimum of 0.9883 at $E = 0.702$. However, the average resistance $\bar{\Delta}_e$ is a minimum for a sphere.

b. *Axisymmetric Motion at Somewhat Higher Reynolds Numbers* The inconsistency noted by Oseen (see Chapter 3) is also present in creeping flow solutions for nonspherical bodies. Extensions to the Stokes solution similar to those for a sphere in Chapter 3 have been investigated for flow parallel to the axis of

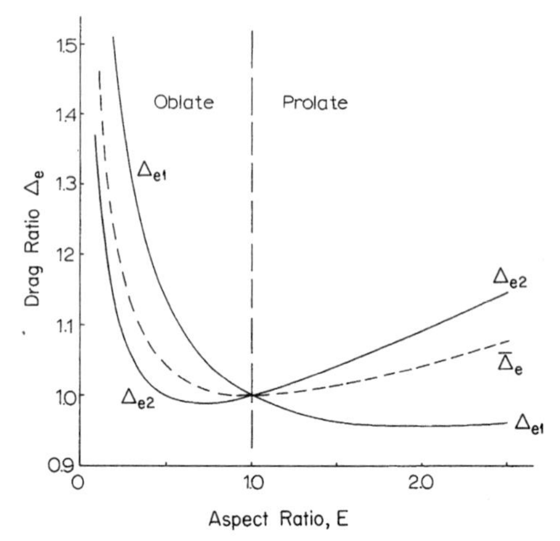

FIG. 4.5 Drag ratios for spheroids compared to volume-equivalent spheres.

a spheroid. Motion with any other orientation is significantly harder to analyze due to the need for three spatial coordinates. Breach (B5) applied Proudman and Pearson's method of inner and outer expansions to obtain:

(i) Oblate:

$$\frac{C_D}{C_{DSt}} - 1 = \frac{3\Delta_{a1}\,\mathrm{Re}}{16} + \frac{9\Delta_{a1}{}^2\,\mathrm{Re}^2 \ln(\mathrm{Re}/2)}{160} + O(\mathrm{Re}^2), \tag{4-24}$$

(ii) Prolate:

$$\frac{C_D}{C_{DSt}} - 1 = \frac{3\Delta_{a1}\,\mathrm{Re}}{16} + \frac{9\Delta_{a1}{}^2\,\mathrm{Re}^2 \ln(E\,\mathrm{Re}/2)}{160} + O(\mathrm{Re}^2), \tag{4-25}$$

where Δ_{a1} is the drag ratio defined by Eq. (4-20) with c_1 from Table 4.1 and Re is based on the equatorial diameter, $2a$. The first term on the right of each equation gives the Oseen drag (Al, O3). The term Δ_{a1} Re is the Reynolds number for the sphere with the same Stokes resistance as the spheroid. Within the Oseen range, the ratio of form to friction drag is independent of Δ_{a1} Re (Al).

Figure 4.6 compares Eqs. (4-24) and (4-25) with selected experimental and numerical results for spheroids. When plotted in this form, $(C_D/C_{DSt} - 1)$ is only weakly dependent on E for Δ_{a1} Re less than about unity. The drag is then very close to the Oseen value, and Eqs. (4-24) and (4-25) are accurate. Above this range, the equations predict that the drag should exceed the Oseen value, whereas the reverse occurs in practice. Thus, as for spheres in Chapter 3, analytic results have little value for $\Delta_{a1}\,\mathrm{Re} \gtrsim 1$.

For higher Re, departure from the Oseen drag increases with increasing aspect ratio. It is common practice to determine a hydraulic equivalent diameter

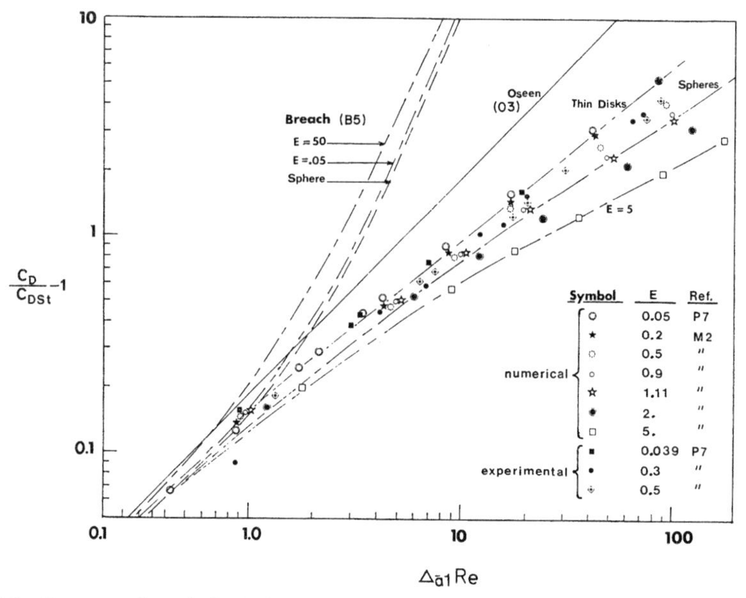

FIG. 4.6 Departure from Stokes's drag for spheroids. Breach curves from Eqs. (4-24) and (4-25).

for an irregular particle by measuring its terminal settling velocity under a specific set of conditions, usually in creeping flow. Figure 4.6 shows that, even for simple shapes, the dependence of drag on particle shape prevents the hydraulic equivalent diameter determined at one Re from being used to predict the settling velocity reliably at another Re. This problem is aggravated by particle orientation effects, discussed in Chapter 6.

3. *Cylinders*

Analytic results for cylinders comparable to those discussed for spheroids are not available. However, Heiss and Coull (H4) reported accurate experimental determinations for cylinders, spheroids, and rectangular parallelepipeds, and developed a general correlation for settling factors. In terms of the volume drag ratio, Δ_e, their results may be written:

(i) for motion parallel to the axis:

$$\Delta_{e1} = \frac{1}{\chi_1\sqrt{\psi}} \exp\left[\frac{0.622(\chi_1 - 1)}{\sqrt{\psi}\chi_1^{0.345}}\right], \tag{4-26}$$

(ii) for motion normal to the axis:

$$\Delta_{e2} = \frac{1}{\chi_2\sqrt{\psi}} \exp[0.576\sqrt{\psi\chi_2}(\chi_2 - 1)], \tag{4-27}$$

where ψ is the sphericity defined in Chapter 2, and χ is a shape factor similar to the circularity, defined as:

$$\chi = \frac{d_e}{d_A} = \frac{\text{diameter of volume-equivalent sphere}}{\text{diameter of projected-area-equivalent sphere}}. \tag{4-28}$$

The area defining d_A is projected parallel to the direction of motion. The modified circularity χ is related to Heywood's shape factor (see Chapter 2) by

$$k = \pi\chi^3/6, \tag{4-29}$$

provided that k is evaluated for the same projected area.†

For spheroids with aspect ratios between 0.1 and 10, Eqs. (4-26) and (4-27) agree closely with the analytic results in Table 4.1 (H4). For cylinders, these results may be written explicitly in terms of the aspect ratio, $E = L/d$, using

$$\psi = (18E^2)^{1/3}/(2E + 1), \tag{4-30}$$

$$\chi_1 = (3E/2)^{1/3}, \tag{4-31}$$

$$\chi_2 = (3/16)^{1/3}\sqrt{\pi}E^{-1/6}. \tag{4-32}$$

The principal resistances may be obtained from the drag ratios as

$$c = 3\pi d(3E/2)^{1/3}\,\Delta_e. \tag{4-33}$$

Figures 4.7 and 4.8 show experimental and numerical results for the resistance of cylinders. The drag values predicted by the multipole representation technique of Gluckman *et al.* (G3) lie closer to the experimental values (B2, H4) than do the series truncation approximations of Bowen and Masliyah (B4). Equation (4-26) gives a reasonable approximation for $0.1 < E < 10$, but is unreliable outside this range. If the drag on a cylinder is approximated by that on a spheroid of the same aspect ratio, the value of Δ_e for the spheroid must be multiplied by $1.5^{-1/3} = 0.874$, since the volume of the cylinder is 1.5 times that of the spheroid. Figure 4.7 shows the resulting curve, obtained from the exact results for spheroids in Table 4.1. The drag on a cylinder approaches that on the spheroid as $E \to 0$ or $E \to \infty$. For $E < 0.1$, the result for a spheroid gives a close estimate for the drag on a cylindrical disk. For $E > 9$, the closest approximation is given by Cox's result from slender-body theory, Eq. (4-36) below.

Figure 4.8 shows that Eq. (4-27) gives a good approximation for the drag on a cylinder with motion normal to the axis for the range in which experimental results are available. The curve obtained from the exact results for spheroids can be used to estimate Δ_e for very small or large E. The slender-body result, Eq. (4-37), appears to be applicable for $E > 3$.

† Singh and Chowdhury (S4) proposed alternative correlations in terms of ψ and χ for cylinders and square bars. Their equations are simpler in form than Eqs. (4-26) and (4-27). However, the fit to available data is no better for $0.5 < E < 5.0$, and the trend is wrong outside this range.

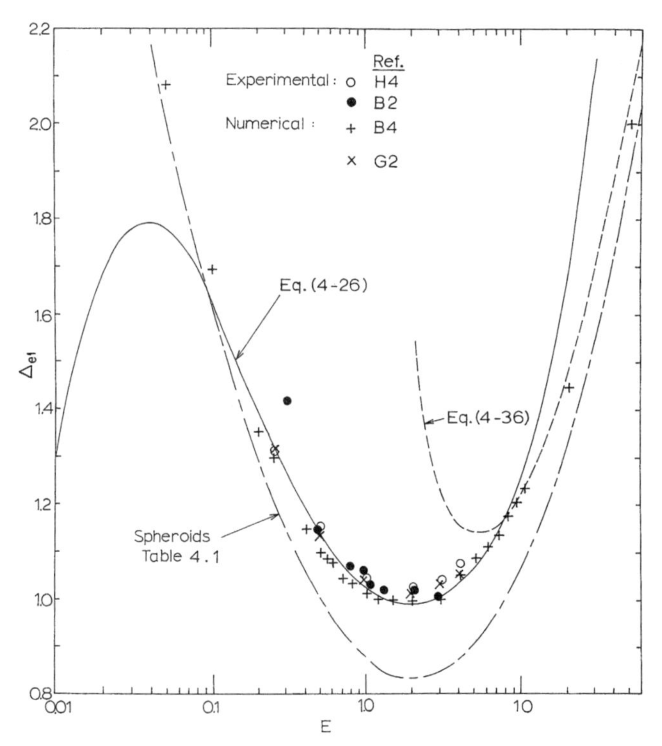

FIG. 4.7 Drag ratio for cylinders in axial motion.

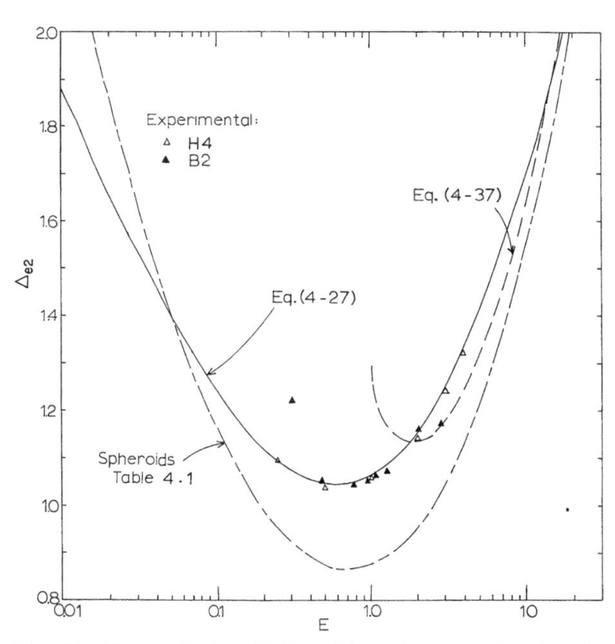

FIG. 4.8 Drag ratio for cylinders with motion normal to the axis.

4. *Slender Bodies*

In the "point force approximation" technique (see Section 1c), Burgers (B10) suggested a polynomial approximation for the distributed line force along the axis of a body of large aspect ratio:

$$f(x) = -8\pi\mu U[B_1 + B_2(x/L)^2 + B_3(x/L)^4] \qquad (-L/2 \le x \le L/2), \quad (4\text{-}34)$$

where the x axis is the axis of symmetry, the body has length L, and B_1, B_2, and B_3 are constants determined by requiring the fluid velocity induced by the drag to counterbalance the free stream velocity as closely as possible on the surface of the particle. The total drag is then

$$F_D = \int_{-L/2}^{L/2} f(x)\,dx = -8\pi\mu UL(B_1 + B_2/12 + B_3/80). \qquad (4\text{-}35)$$

This approach leads to expressions for the resistance to axial motion of the form

$$c_1 = 2\pi L/[\ln(2E) - \kappa]. \qquad (4\text{-}36)$$

For prolate spheroids, κ is predicted to be 0.50, in agreement with the result in Table 4.1 when $E \to \infty$. For cylinders, Burgers obtained $\kappa = 0.72$. Broersma (B9) improved Burgers' estimate for cylinders by taking further terms and solving numerically. The value obtained, $\kappa = 0.80685$, was subsequently confirmed by Cox (C5) using an asymptotic expansion. Cox's treatment has the advantages of leading to an estimate of the error in the approximation and of enabling results to be obtained for curved slender bodies and for cases in which the axis of the body is not parallel to the direction of motion. Cox showed that the principal resistance for motion normal to the axis of symmetry is

$$c_2 = 4\pi L/[\ln(2E) + 1 - \kappa]. \qquad (4\text{-}37)$$

For prolate spheroids, Eq. (4-37) with $\kappa = 0.5$ again agrees with the limiting exact result for $E \to \infty$. The validity of these equations for cylinders is demonstrated in Figs. 4.7 and 4.8. Comparison of Eqs. (4-36) and (4-37) shows that the ratio of c_2 to c_1 tends to 2 as $E \to \infty$. This result holds for any axisymmetric particle, while $c_2 < 2c_1$ for finite aspect ratios (W2). Consequently a needle-like particle falls twice as fast when oriented vertically at low Re than when its axis is horizontal.

Cox (C5) and Tchen (T1) also obtained expressions for the drag on slender cylinders and ellipsoids which are curved to form rings or half circles. The advantages of prolate spheroidal coordinates in dealing with slender bodies have been demonstrated by Tuck (T2). Batchelor (B1) has generalized the slender body approach to particles which are not axisymmetric and Clarke (C2) has applied it to twisted particles by considering a surface distribution rather than a line distribution.

5. *Arbitrary Axisymmetric Particles*

Bowen and Masliyah (B4) give a useful discussion of the axial resistance of various axisymmetric bodies. For particles which may be regarded as spheres with axisymmetric deformations, simple estimates for the resistance are available. Suppose that a particle of volume V with principal resistance c_1 is obtained by deforming a sphere of volume V_s and resistance c_{1s}. It is convenient to use two factors introduced by O'Brien (O2):

$$\Delta c = (c_1 - c_{1s})/c_{1s}, \tag{4-38}$$

$$\Delta V = (V - V_s)/V_s. \tag{4-39}$$

For deformations with fore-and-aft symmetry,

$$\Delta c = 0.2\,\Delta V + O[(\Delta V)^2] \tag{4-40}$$

while "μ-deformations" (lacking fore-and-aft symmetry) give (B3)

$$\Delta c = 0.25\,\Delta V + O[(\Delta V)^2] \tag{4-41}$$

Equation (4-40) is equivalent to the approximate results for spheroids given in Table 4.1; Figs. 4.3 and 4.4 demonstrate that the approximation is useful even for grossly deformed spheres.

Bowen and Masliyah examined the axial resistance of cylinders with flat, hemispherical and conical ends, and of double-headed cones and cones with hemispherical caps, together with the established results for spheroids. Widely used shape factors (including sphericity) did not give good correlations, while Eqs. (4-26) and (4-27) were found to be inapplicable to particles other than cylinders and spheroids. The best correlation was provided by the perimeter-equivalent factor $\sum$ defined in Chapter 2. With this parameter, the equivalent sphere has the same perimeter as the particle viewed normal to the axis. Based on their numerical results, Bowen and Masliyah obtained the correlation

$$\Delta_{P'} = 0.244 + 1.035\textstyle\sum - 0.712\sum^2 + 0.441\sum^3. \tag{4-42}$$

Figure 4.9 shows that Eq. (4-42) also gives a good correlation of results of other workers. For shapes where experimental data are lacking, Eq. (4-42) is likely to give the best estimate for resistance to axial translation. Figure 4.9 also shows experimental results for rectangular parallelepipeds, which may be regarded as analogous to axisymmetric particles (see Section C below). The shape factor and drag ratio are evaluated from the arithmetic mean of the maximum and minimum perimeters, viewed normal to the corresponding axes. Equation (4-42) also correlates these results within about 6%, suggesting that this form of correlation may prove to be useful for nonaxisymmetric particles.

Results for translation normal to the axis are more limited and all experimental. Figure 4.10 shows available data plotted employing Bowen and Masliyah's

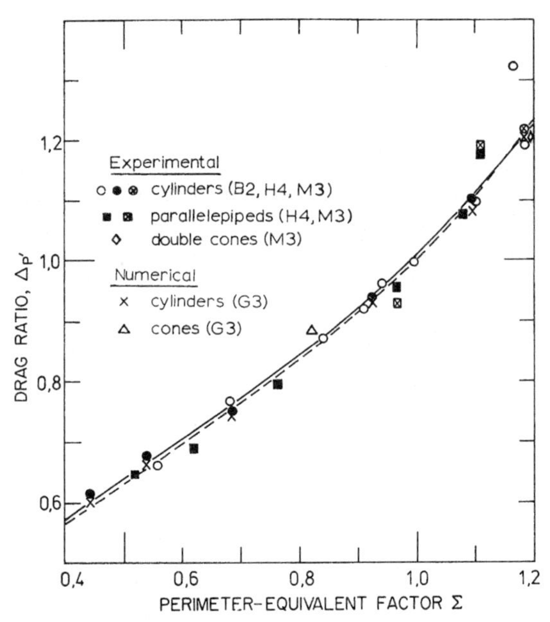

FIG. 4.9 Bowen and Masliyah correlation for axial resistance of axisymmetric particles: — Eq. (4-42); --- Spheroids (exact).

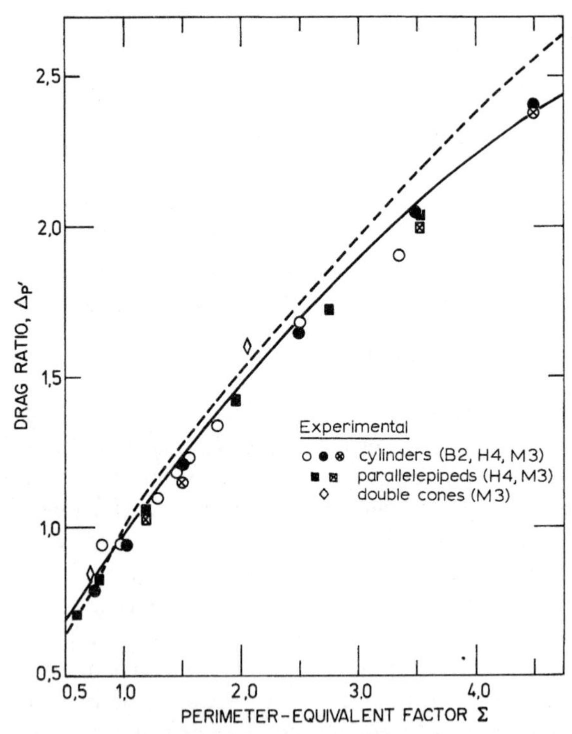

FIG. 4.10 Resistance of axisymmetric particles to translation normal to the axis: — Eq. (4-43); --- Spheroids (exact).

groups, where the perimeter viewed along the axis (or "equivalent axis" discussed below for rectangular parallelepipeds) has been used. The parameters $\sum$ and $\Delta_{\mathrm{P}'}$ also appear to give a useful correlation for the second principal resistance, since the data fall generally below but within 10% of the exact curve for spheroids. The following correlation is obtained from the experimental data together with an equal number of points from the exact curve for spheroids:

$$\Delta_{\mathrm{P}'} = 0.392 + 0.621\sum - 0.040\sum^2. \tag{4-43}$$

Equation (4-43) may be used to estimate the normal resistance of particles for which no experimental results are available.

C. Orthotropic Particles

For an orthotropic particle in steady translation through an unbounded viscous fluid, the total drag is given by Eq. (4-5). In principle, it is possible to follow a development similar to that given in Section II.B.1 for axisymmetric particles, to deduce the general behavior of orthotropic bodies in free fall. This is of limited interest, since no analytic results are available for the principal resistances of orthotropic particles which are not bodies of revolution. General conclusions from the analysis were given in II.A.

The only orthotropic particles for which comprehensive experimental results are available are "square bars," rectangular parallelepipeds with one pair of square faces. Symmetry then shows that the two principal resistances corresponding to translation with square faces parallel to the direction of motion are equal. These resistances will be denoted by c_2, while the resistance for translation normal to the square faces will be called c_1. Consider such a particle in arbitrary translation at velocity $\mathbf{U}$. Figure 4.11 shows a section of the particle parallel to the square faces; U_2 is the component of $\mathbf{U}$ in this plane, and the angle between U_2 and principal axis 2 is θ. From Eq. (4-5), the drag components are as shown in Fig. 4.11. Hence the drag component parallel to U_2 is

$$F_{\mathrm{D}||} = \mu U_2(c_2 \cos^2\theta + c_3 \sin^2\theta), \tag{4-44}$$

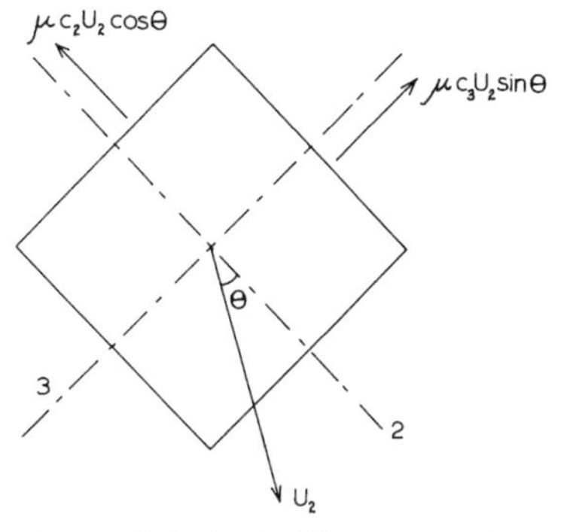

Fig. 4.11 Rectangular parallelepiped with square section in steady translation.

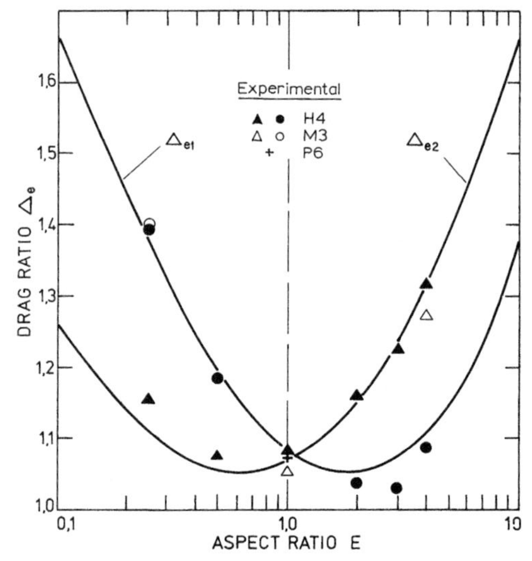

FIG. 4.12 Drag ratios for rectangular parallelepipeds with square section.

while that perpendicular to U_2 is

$$F_{D\perp} = \mu U_2(c_2 \cos\theta \sin\theta - c_3 \sin\theta \cos\theta). \tag{4-45}$$

Since $c_3 = c_2$ for this class of shapes, $F_{D\perp}$ is zero while $F_{D\|}$ is $\mu U_2 c_2$. Thus the drag component in the plane of Fig. 4.11 is always in the $-U_2$ direction with magnitude independent of the direction of U_2. Thus, for drag in steady translation or motion in free fall, these particles may be treated like axisymmetric particles.† The axis through the centres of the square faces is like an axis of symmetry and may be termed the "equivalent axis."

Heiss and Coull (H4) measured the drag on rectangular parallelepipeds. Results were correlated by Eqs. (4-26) and (4-27). For a particle with dimensions $l \times l \times El$

$$\psi = (9\pi E^2/2)^{1/3}/(1 + 2E). \tag{4-46}$$

For motion parallel to the equivalent axis

$$\chi_1 = (0.75E\sqrt{\pi})^{1/3}, \tag{4-47}$$

while for motion normal to the equivalent axis

$$\chi_2 = (0.75\sqrt{\pi/E})^{1/3}. \tag{4-48}$$

† Symmetry arguments show that the same conclusion applies to bodies, such as bars with regular polygonal cross sections, whose shapes are unchanged on rotation through an angle of $2\pi/n$ ($n > 2$) about the equivalent axis. Unfortunately, this simplification does not apply to rotational motion, or in general to heat or mass transfer.

The principal resistances may be obtained from the drag ratios as

$$c = 3l(6E\pi^2)^{1/3}\,\Delta_e \tag{4-49}$$

Corresponding values for Δ_e are shown in Fig. 4.12. Agreement with available experimental data is reasonable, but no better than for the more general correlations, Eqs. (4-42) and (4-43).

D. Spherically Isotropic Particles

For a particle which is spherically isotropic (see Chapter 2), the three principal resistances to translation are all equal. It may then be shown (H3) that the net drag is $-\mu c\mathbf{U}$ regardless of orientation. Hence a spherically isotropic particle settling through a fluid in creeping flow falls vertically with its velocity independent of orientation.

Settling velocities of such particles have been measured by Pettyjohn and Christiansen (P6) and Chowdhury and Fritz (Cl). Correlation of both sets of data with the results of Heiss and Coull for cubes gives:

$$\Delta_e = 1/(1 + 0.367 \ln \psi). \tag{4-50}$$

For a cube of side l, Eq. (4-50) gives the resistance as $12.70l$, compared with experimental values of $12.58l$ (P6), $12.63l$ (H4), and $12.71l$ (Cl). To the accuracy of the determinations, the resistance can be taken as $4\pi l$ (D1). It is noteworthy that Eqs. (4-26) and (4-27) predict that a spherically isotropic cylinder with aspect ratio 0.812 should have a drag ratio of 1.050, while Eq. (4-50) gives $\Delta_e = 1.054$. Agreement is so favorable that Eq. (4-50) may be useful for spherically isotropic particles other than the simple shapes for which it was developed.

E. Particles of Arbitrary Shape

No fully satisfactory method is available for correlating the drag on irregular particles. Settling behavior has been correlated with most of the more widely used shape factors. Settling velocity may be entirely uncorrelated with the "visual sphericity" obtained from the particle outline alone (B8). General correlations for nonspherical particles are discussed in Chapter 6.

For creeping flow, a few simple general results often lead to useful estimates for the resistance or settling velocity of arbitrary particles. Sharp edges have little effect on drag, the most significant features being areas where the tangential stress is parallel to the direction of motion. For a sphere which has been slightly deformed, the average resistance $\bar{c}$ is equal to that of the sphere with the same volume (H3). However, the average resistance should be used with care, since even slight asymmetry causes a particle to adopt a preferred orientation (M3). Hill and Power (H6) showed that the Stokes drag on an arbitrary particle is less than or equal to that on a body which encloses it and greater than or equal to that on a body contained within it. Judicious choice of circumscribed and inscribed bodies can give close bounds on the resistance or settling velocity.

Weinberger (W2) showed that the sphere has the largest average Stokes settling velocity of all bodies of a given volume. Keller *et al.* (K1) showed that creeping flow solutions always underestimate drag at nonzero Re. The results for spheroids discussed above illustrate these general principles.

III. HEAT AND MASS TRANSFER

Very few solutions have been obtained for heat or mass transfer to nonspherical solid particles in low Reynolds number flow. For Re = 0 the species continuity equation has been solved for a number of axisymmetric shapes, while for creeping flow only spheroids have been studied.

A. External Resistance—Steady State

For constant-property steady flow the species continuity equation, Eq. (1-38), becomes

$$\frac{\mathrm{Pe}}{2}\frac{\mathbf{u}}{U}\cdot\nabla' c = (\nabla')^2 c, \tag{4-51}$$

where Pe is based on some characteristic length and c is the species concentration. The boundary conditions are

$$c = c_s \qquad \text{on surface of particle,} \tag{4-52}$$

$$c = c_\infty \qquad \text{far from particle.} \tag{4-53}$$

Brenner (B7) has shown that, whatever the particle shape, the total mass or heat transferred from a particle in creeping flow is the same if the flow infinitely far from the particle is reversed. Although the variation of the rate of transfer over the surface of the particle may differ under forward and reverse flow, the total rate of transfer is the same.

1. *Stagnant External Phase*

For Pe = 0, Eq. (4-51) reduces to Laplace's equation

$$\nabla^2 c = 0, \tag{4-54}$$

with the boundary conditions given by Eqs. (4-52) and (4-53). Diffusion from a finite particle into a stagnant external medium is analogous to the electrostatic problem of a charged conductor located in a charge-free homogeneous dielectric medium. This problem has been treated thoroughly (S5, V1, W1) and the capacitance, the ratio of charge to potential, has been obtained for a number of shapes. These results can be utilized directly in the present application by

TABLE 4.2

External Conductances of Particles in a Stagnant Medium[a]

Particle shape	Conductance, $k_0 A/\mathscr{D}$
Sphere (radius $= a$)	$4\pi a$
Spheroid	
oblate ($E < 1$)	$\dfrac{4\pi a\sqrt{1-E^2}}{\cos^{-1}(E)}$
prolate ($E > 1$)	$\dfrac{4\pi a\sqrt{E^2-1}}{\ln(E+\sqrt{E^2-1})}$
Circular Cylinder ($0 \leq L/d \leq 8$, radius $= a$)	$\left[8 + 6.95\left(\dfrac{L}{d}\right)^{0.76}\right]a$
Thin rectangular plate (side $L_1 \geq$ side L_2)	$2\pi L_1/\ln(4L_1/L_2)$
Cube (edge $= a$)	$0.656(4\pi a)$
Touching spheres (equal size, radius $= a$)	$(2\ln 2)(4\pi a)$
Intersecting spheres, radii a_1 and a_2, with orthogonal intersection	$4\pi\left[a_1 + a_2 - \dfrac{a_1 a_2}{\sqrt{a_1{}^2 + a_2{}^2}}\right]$

[a] After Smythe (S5, S6), Weber (W1), Schneider (S2), Reitan and Higgins (R1), and Hahne and Grigul (H1).

noting the equivalence

$$k_0 A/\mathscr{D} \equiv C_e/\varepsilon, \tag{4-55}$$

where C_e is the capacitance of the conducting particle and ε is the permittivity of the medium. The quantity $k_0 A/\mathscr{D}$ may be called the "conductance" and has dimensions of length.[†] The subscript zero denotes the absence of external flow. The distribution of conductance over the surface of a particle is identical to the distribution of surface charge in the geometrically similar electrostatic problem. With the exception of spheres, the local conductance is not uniform over the surface because edges and corners, where the curvature is high, have higher conductances. There is no solution to Eq. (4-54) for an infinite cylinder of any cross section. Instead, the steady-state rate of transfer to an infinite body is zero. Conductances obtained from solutions to Eq. (4-54) for finite bodies of various shapes are given in Table 4.2. Values of Sh_0 can be obtained from the tabulated conductances by dividing by the surface area of the particle and multiplying by a characteristic length.

[†] In the heat transfer literature the corresponding quantity, $h_0 A/K_t$, is sometimes called the conduction shape factor.

The conductance of arbitrary axisymmetric particles may be approximated using the correlation given in Fig. 4.13. By analogy with the drag ratio, a "conductance factor" is defined as

$$\Delta' = \frac{\text{conductance of particle}}{\text{conductance of equivalent sphere}}. \tag{4-56}$$

The graphical correlation is presented in terms of the perimeter equivalent factor $\sum$ used in Figs. 4.9 and 4.10. The points have been calculated for axisymmetric shapes.

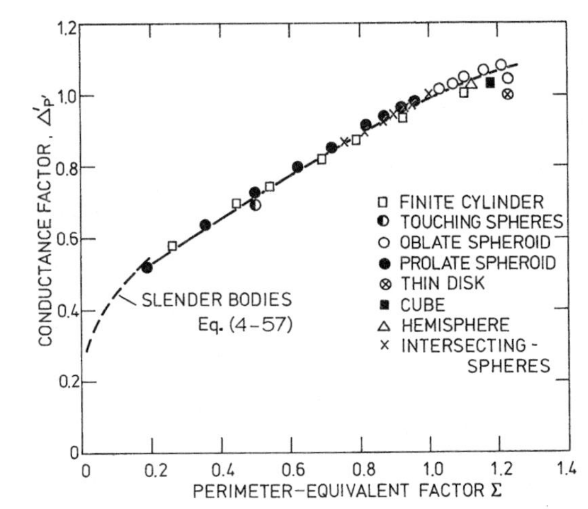

FIG. 4.13 Correlation for conductance factor of axisymmetric particles in stagnant media (based on perimeter-equivalent sphere).

For needle-like bodies an electrostatic slender body theory is available (M4) which yields

$$k_0 A/\mathscr{D} = 2\pi L/[\ln(4E) - 1]. \tag{4-57}$$

Comparison with the conductance for cylinders indicates that Eq. (4-57) is accurate within 5% for $E \geq 10$.

For shapes whose boundaries are not simply described in a single coordinate system, numerical solution of Eq. (4-54) is required. However, it is possible to provide upper and lower bounds for the conductance (P8) in much the same way as for the drag. A lower bound for an arbitrary particle is the conductance of the sphere of the same volume, i.e.,

$$k_0 A/\mathscr{D} \geq 2\pi d_e. \tag{4-58}$$

Another lower bound is given by

$$k_0 A/\mathscr{D} \geq 8\sqrt{A_p/\pi}, \tag{4-59}$$

where A_p is the area of the maximum orthogonal projection of the body onto a plane. The equality is achieved for a disk. An upper bound is given by the conductance of a shape circumscribing the particle, spheres and spheroids being frequently used. More precise bounds can be obtained with extra effort (P1, P8).

2. *Creeping Flow*

a. *Particles of Arbitrary Shape* For Pe → 0, an asymptotic solution for a particle of arbitrary shape has been obtained using matched asymptotic expansions (B6). To first order in velocity, the solution is

$$\frac{k}{k_0} = 1 + \frac{1}{8\pi}\left(\frac{k_0 A}{\mathscr{D}}\right)\frac{U}{\mathscr{D}}. \tag{4-60}$$

The effect of flow depends solely on a Peclet number formed using the conductance in a stagnant medium, $k_0A/\mathscr{D}$, as the characteristic length. Equation (4-60) has wider generality; it is valid for a fluid or solid particle of any shape at any Re so long as Pe → 0 and the stream far from the particle is uniform. This expression gives a good prediction of the conductance ratio for $k/k_0 < 1.2$. Equation (3-45) is the special case of Eq. (4-60) for spheres. The next term in the series expansion depends explicitly upon the shape and the orientation of the particle.

b. *Spheroids* For creeping flow at finite Pe, Eq. (4-51) has been solved numerically for oblate and prolate spheroids of axis ratio 0.2 (M1). Solutions were obtained up to Pe = 70 with equatorial diameter as characteristic length in both Pe and Sh. An asymptotic solution for Pe → ∞ has also been obtained (S3) for spheroids of any aspect ratio using the thin concentration boundary layer approach (see Chapter 1). With the equatorial diameter as characteristic length, this solution is

$$\mathrm{Sh} = 0.991K(\mathrm{Pe})^{1/3}, \tag{4-61}$$

where K is plotted in Fig. 4.14. For oblate spheroids the following formula holds asymptotically:

$$K = 4(E/3\pi)^{1/3} \qquad (E \to 0). \tag{4-62}$$

The corresponding asymptotic formula for prolate spheroids is

$$K = \frac{4}{\pi}\left(\frac{4}{3}\right)^{1/3}[E\ln 2E]^{-1/3} \qquad (E \to \infty). \tag{4-63}$$

Equations (4-62) and (4-63) yield good predictions for $E < 0.1$ and $E > 10$, respectively.

The asymptotic formulae, Eqs. (4-62) and (4-63), predict that $K \to 0$ at the respective limits. However, Sh does not go to zero because the assumption of a thin concentration boundary layer breaks down for extreme values of E.

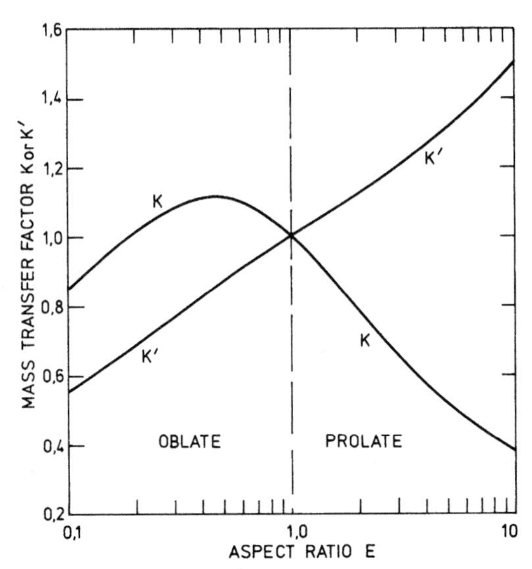

FIG. 4.14 Factors to be used with Eqs. (4-61), (4-68), (4-69) and (4-70) for predicting heat and mass transfer to spheroids in creeping flow.

It was assumed that the concentration boundary layer was thin relative to the shorter axis of the spheroid. The order of magnitude of the boundary layer thickness can be approximated by the thickness δ of a fictitious film

$$\delta = \mathscr{D}/k. \tag{4-64}$$

Combining Eqs. (4-64) and (4-61) yields

$$\delta/a \propto K^{-1}\,\mathrm{Pe}^{-1/3}. \tag{4-65}$$

For the oblate spheroid

$$\delta/a \propto (E\,\mathrm{Pe})^{-1/3}, \tag{4-66}$$

and thus the analysis leading to Eq. (4-62) applies only if $E\,\mathrm{Pe} \to \infty$. Similarly for the prolate spheroid the analysis leading to Eq. (4-63) applies only if $\mathrm{Pe}/E \to \infty$.

The asymptotic solution can be recast in a variety of forms using different characteristic lengths. Pasternak and Gauvin (P2) proposed a length which is useful at higher Re (see Chapter 6):

$$L' = \frac{\text{surface area of particle}}{\text{maximum perimeter projected on a plane normal to the flow}} \tag{4-67}$$

If we denote the Sherwood and Peclet numbers based on this length by primes, Eq. (4-61) becomes

$$\mathrm{Sh}' = 0.991K'(\mathrm{Pe}')^{1/3}, \tag{4-68}$$

where K' is plotted in Fig. 4.14.

The correlation presented earlier for spheres, Eq. (3-60), suggests the following form for spheroids

$$\mathrm{Sh} = (\mathrm{Sh}_0/2) + [(\mathrm{Sh}_0/2)^3 + K^3\,\mathrm{Pe}]^{1/3}. \tag{4-69}$$

With K-values from Fig. 4.14 and Sh_0 derived from Table 4.2, Eq. (4-69) predicts Sh within 10% of the numerical values of Masliyah and Epstein (M1) for $\mathrm{Pe} < 70$ and $E = 0.2$ for oblate spheroids and $E = 5$ for prolate spheroids. The analogous correlation with L' as the characteristic length is

$$\mathrm{Sh}' = (\mathrm{Sh}_0'/2) + [(\mathrm{Sh}_0'/2)^3 + (K')^3\,\mathrm{Pe}']^{1/3}. \tag{4-70}$$

c. *Other Shapes* Flow normal to an infinite cylinder at low Re and Pe has been treated by the method of matched asymptotic expansions (H5). The first two terms in the expansion are

$$\mathrm{Sh} = \frac{2}{\ln\dfrac{8}{\mathrm{Pe}} - \gamma}\left[1 - \frac{a_3}{\left(\ln\dfrac{8}{\mathrm{Pe}} - \gamma\right)^2}\right], \tag{4-71}$$

where γ is Euler's constant. The first term represents transfer from a line source into a uniform stream. The coefficient a_3, a function of Sc, must be evaluated numerically. It increases with increasing Sc from zero at $\mathrm{Sc} = 0$ to 1.38 at $\mathrm{Sc} = 0.72$. Experimental data for heat transfer from fine wires to air (C4) agree well with Eq. (4-71) for $\mathrm{Re} \lessdot 0.4$.

For flow parallel to a cylinder the rate of mass or heat transfer decreases with axial distance. Far from the leading end, the transfer at low Pe may be considered as transfer from a line source into a uniform stream and the local Sherwood number becomes

$$\mathrm{Sh} = \frac{4}{\ln\left(\dfrac{16x}{d\,\mathrm{Pe}}\right) - \gamma} \tag{4-72}$$

Near the leading end a more complex analysis is necessary (B3).

It would seem that no theoretical calculations have been made for shapes other than spheroids. In addition, no experimental measurements have been reported for shapes other than spheres or circular cylinders in creeping flow. Equation (4-60) is useful for cases in which Pe is small.

B. Transfer with Variable Particle Concentration

The variation of particle concentration with time has been determined only for quasi-steady external resistance in two cases.

1. *Negligible Internal Resistance*

For a particle of arbitrary shape a mass balance yields

$$V\,dF/dt = (kA/H)(1 - F), \tag{4-73}$$

subject to the initial condition

$$F = 0 \qquad \text{at} \quad t = 0. \tag{4-74}$$

Solution yields the fractional approach to equilibrium as

$$F = 1 - \exp[-(kA/HV)t]. \tag{4-75}$$

If the characteristic length is taken as

$$d_e/2 = 3V/A, \tag{4-76}$$

Eq. (4-75) can be rewritten as

$$F = 1 - \exp(-3\,\mathrm{Bi}\,\tau_p), \tag{4-77}$$

where $\mathrm{Bi} = kd_e/2\mathscr{D}_pH$ and $\tau_p = 4t\mathscr{D}_p/d_e^2$ are the Biot and Fourier numbers. Equation (4-75) is expected to apply when $\mathrm{Bi} \lessdot 0.1$.

2. *Comparable Resistance in Each Phase*

Diffusion within a solid particle with convection at the boundary is described by

$$\partial c_p/\partial\tau_p = (\nabla')^2 c_p. \tag{4-78}$$

At the surface

$$\partial c_p/\partial n = -\mathrm{Bi}\, c_p \tag{4-79}$$

where n is a coordinate normal to the particle surface. These equations have been solved to yield c_p as a function of position and time for simple geometries: spheres (see Chapter 3), semi-infinite slabs, and infinite cylinders which can be described using a single coordinate (L1, S2). Values of c_p as a function of position and time for certain two- and three-dimensional shapes can be constructed from these simple cases. The basic requirement is that the boundaries must be described by constant values of the coordinate parameters used in the one-dimensional solutions. For example, the concentration history of rectangular parallelepipeds and finite cylinders can be determined in this way. Luikov (L1) outlines the method and gives equations for F derived from volume integration of c_p.

The time variation of concentration at the center and at the foci of prolate spheroids has been calculated for negligible external resistance, $\mathrm{Bi} \to \infty$ (H2). These appear to be the only calculations for shapes other than those mentioned above.

REFERENCES

A1. Aoi, T., *J. Phys. Soc. Jpn.* **10**, 119–129 (1955).
B1. Batchelor, G. K., *J. Fluid Mech.* **44**, 419–440 (1970).
B2. Blumberg, P. N., and Mohr, C. M., *AIChE J.* **14**, 331–334 (1968).

B3. Bourne, D. E., and Davies, D. R., *Q. J. Mech. Appl. Math.* **11**, 52–66 (1958).
B4. Bowen, B. D., and Masliyah, J. H., *Can. J. Chem. Eng.* **51**, 8–15 (1973).
B5. Breach, D. R., *J. Fluid Mech.* **10**, 300–314 (1961).
B6. Brenner, H., *Chem. Eng. Sci.* **18**, 109–122 (1963).
B7. Brenner, H., *J. Math. Phys. Sci.* **1**, 173–179 (1967).
B8. Briggs, L. I., McCulloch, D. S., and Moser, F., *J. Sediment. Petrol.* **32**, 645–656 (1962).
B9. Broersma, S., *J. Chem. Phys.* **32**, 1632–1635 (1960).
B10. Burgers, J. M., *in* "Second Report on Viscosity and Plasticity," pp. 113–184. North-Holland Publ., Amsterdam, 1938.
C1. Chowdhury, K. C. R., and Fritz, W., *Chem. Eng. Sci.* **11**, 92–98 (1959).
C2. Clarke, N. S., *J. Fluid Mech.* **52**, 781–793 (1972).
C3. Collins, W. D., *Mathematika* **10**, 72–78 (1963).
C4. Collis, D. C., and Williams, M. J., *J. Fluid Mech.* **6**, 357–384 (1959).
C5. Cox, R. G., *J. Fluid Mech.* **44**, 791–810 (1970).
D1. Dahneke, B. E., *J. Aerosol Sci.* **4**, 139–145 (1973).
D2. Darwin, Sir C., *Proc. Cambridge Philos. Soc.* **49**, 342–354 (1953).
D3. Dorrepaal, J. M., O'Neill, M. E., and Ranger, K. B., *J. Fluid Mech.* **75**, 273–286 (1976).
G1. Gans, R., *Sitzungsber. Math-Phys. Kl. Akad. Wiss. München* **41**, 191–203 (1911).
G2. Gluckman, M. J., Pfeffer, R., and Weinbaum, S., *J. Fluid Mech.* **50**, 705–740 (1971).
G3. Gluckman, M. J., Weinbaum, S., and Pfeffer, R., *J. Fluid Mech.* **55**, 677–709 (1972).
H1. Hahne, E., and Grigul, U., *Int. J. Heat Mass Transfer* **18**, 751–767 (1975).
H2. Haji-Sheikh, A., and Sparrow, E. M., *J. Heat Transfer* **88**, 331–333 (1966).
H3. Happel, J., and Brenner, H., "Low Reynolds Number Hydrodynamics," 2nd ed. Noordhoff, Leyden, Netherlands, 1973.
H4. Heiss, J. F., and Coull, J., *Chem. Eng. Prog.* **48**, 133–140 (1952).
H5. Hieber, C. A., and Gebhart, B., *J. Fluid Mech.* **32**, 21–28 (1968).
H6. Hill, R., and Power, G., *Q. J. Mech. Appl. Math.* **9**, 313–319 (1956).
K1. Keller, J. B., Rubenfeld, L. A., and Molyneux, J. E., *J. Fluid Mech.* **30**, 97–125 (1967).
L1. Luikov, A. V., "Analytical Heat Diffusion Theory." Academic Press, New York, 1968.
M1. Masliyah, J. H., and Epstein, N., *Prog. Heat Mass Transfer* **6**, 613–632 (1972).
M2. Masliyah, J. H.,and Epstein, N., *J. Fluid Mech.* **44**, 493–512 (1970).
M3. McNown, J. S., and Malaika, J., *Trans. Am. Geophys. Union* **31**, 74–82 (1950).
M4. Miles, J. W., *J. Appl. Phys.* **38**, 192–196 (1967).
O1. Oberbeck, H. A., *Crelles J.* **81**, 62–87 (1876).
O2. O'Brien, V., *AIChE J.* **14**, 870–875 (1968).
O3. Oseen, C. W., *Ark. Mat., Astron. Fys.* **6**, No. 29 (1910).
P1. Parr, W. E., *J. Soc. Ind. Appl. Math.* **9**, 334–386 (1961).
P2. Pasternak, I. S., and Gauvin, W. H., *Can. J. Chem. Eng.* **38**, 35–42 (1960).
P3. Payne, L. E., and Pell, W. H., *J. Fluid Mech.* **7**, 529–549 (1960).
P4. Pell, W. H., and Payne, L. E., *Mathematika* **7**, 78–92 (1960).
P5. Pell, W. H., and Payne, L. E., *Q. Appl. Math.* **18**, 257–262 (1960–1961).
P6. Pettyjohn, E. A., and Christiansen, E. B., *Chem. Eng. Prog.* **44**, 157–172 (1948).
P7. Pitter, R. L., Pruppacher, H. R., and Hamielec, A. E., *J. Atmos. Sci.* **30**, 125–134 (1973).
P8. Polya, G., and Szego, G., "Isoperimetric Inequalities in Mathematical Physics," Ann. Math. Stud., No. 27. Princeton Univ. Press, Princeton, New Jersey, 1951.
R1. Reitan, D. K., and Higgins, T. J., *J. Appl. Phys.* **22**, 223–226 (1951).
S1. Sampson, R. A., *Phil. Trans. R. Soc. London, Ser. A* **182**, 449–518 (1891).
S2. Schneider, P. J., *in* "Handbook of Heat Transfer" (W. H. Rohsenhow and J. P. Hartnett, eds.), Section 3. McGraw-Hill, New York, 1973.
S3. Sehlin, R. C., M. S. Thesis, Carnegie-Mellon Univ., Pittsburgh, Pennsylvania, 1969.
S4. Singh, A. H., and Chowdhury, K. C. R., *Chem. Eng. Sci.* **24**, 1185–1186 (1969).
S5. Smythe, W. R., "Static and Dynamic Electricity," 3rd Ed., McGraw-Hill, New York, 1968.

S6. Smythe, W. R., *J. Appl. Phys.* **33**, 2966–2967 (1962).
S7. Stimson, M., and Jeffrey, G. B., *Proc. R. Soc., Ser. A* **111**, 110–116 (1926).
T1. Tchen, C., *J. Appl. Phys.* **25**, 463–473 (1954).
T2. Tuck, E. O., *J. Fluid Mech.* **18**, 619–635 (1964).
V1. Van Bladel, J., "Electromagnetic Fields." McGraw-Hill, New York, 1964.
W1. Weber, E., "Electromagnetic Fields, Vol. I, Mapping of Fields." Wiley, New York, 1950.
W2. Weinberger, H. F., *J. Fluid Mech.* **52**, 321–344 (1972).

Chapter 5

Spheres at Higher Reynolds Numbers

I. INTRODUCTION

Analytic solutions for flow around and transfer from rigid and fluid spheres are effectively limited to Re < 1 as discussed in Chapter 3. Phenomena occurring at Reynolds numbers beyond this range are discussed in the present chapter. In the absence of analytic results, sources of information include experimental observations, numerical solutions, and boundary-layer approximations. At intermediate Reynolds numbers when flow is steady and axisymmetric, numerical solutions give more information than can be obtained experimentally. Once flow becomes unsteady, complete calculation of the flow field and of the resistance to heat and mass transfer is no longer feasible. Description is then based primarily on experimental results, with additional information from boundary layer theory.

II. RIGID SPHERES

A. Fluid Mechanics

1. *Theoretical Approaches*

a. *Numerical Solution of Governing Equations* For numerical solution of the Navier–Stokes and continuity equations in axisymmetric flow, it is useful to introduce the dimensionless stream function, Ψ, and vorticity, $Z = \zeta a/U$ (see Chapter 1). The Navier–Stokes equation for steady flow becomes

$$E^2\Psi = ZR\sin\theta \qquad (5\text{-}1)$$

and

$$\frac{\partial \Psi}{\partial R}\frac{\partial}{\partial \theta}\left(\frac{Z}{R\sin\theta}\right) - \frac{\partial \Psi}{\partial \theta}\frac{\partial}{\partial R}\left(\frac{Z}{R\sin\theta}\right) = \frac{2E^2}{\mathrm{Re}}(ZR\sin\theta), \tag{5-2}$$

where

$$E^2 = \frac{\partial}{\partial R^2} + \frac{\sin\theta}{R^2}\frac{\partial}{\partial\theta}\left(\frac{1}{\sin\theta}\frac{\partial}{\partial\theta}\right) \quad \text{and} \quad R = r/a. \tag{5-3}$$

The boundary conditions for a sphere of radius a are:

(a) at $\theta = 0$ and π: $\Psi = Z = 0,$ (5-4)

(b) at $R = 1$: $\Psi = 0;\ \partial\Psi/\partial R = 0;\ Z = E^2\Psi/\sin\theta,$ (5-5)

(c) at $R \to \infty$: $\Psi/R^2 \to \frac{1}{2}\sin^2\theta;\ Z \to 0.$ (5-6)

Useful results have been obtained by solving finite-difference equations obtained from Eqs. (5-1) and (5-2) by Taylor-series expansion. These algebraic equations are solved by iteration to give Ψ and Z at a number of discrete points forming the nodes of a grid. Because Ψ and Z vary most rapidly near the particle surface, the intervals in the grid are commonly taken to increase exponentially with R. The outer boundary condition, Eq. (5-6), is satisfied on some outer envelope; care is required to ensure that this is sufficiently remote from the particle (L8). The basic finite-difference scheme was developed by Jenson (J1), but the grid used was too coarse to give accurate results. Subsequent studies (H1, L5, L8, I1, M2, W9) have used the same technique with digital computers to give accurate results. The results of the various workers generally agree closely.

An alternative approach is to solve the time-dependent problem in which the development of the flow is calculated from some arbitrary initial state. Eq. (5-1) is unchanged, but $\partial(ZR\sin\theta)/\partial t^*$ must be added to the left side of Eq. (5-2), where

$$t^* = tU/a \tag{5-7}$$

is a dimensionless time. The usual initial condition is an "impulsive start," in which there is no relative motion for $t^* < 0$, while the relative velocity between particle and fluid is constant for $t^* \geq 0$. Solution is continued until the flow is effectively steady. This method sometimes requires less computation than the iterative approach. Rimon and Cheng (R8) and Rafique (R1) have obtained results by this method. However, the former may be unreliable (M4, R1) due to use of a different condition at $R \to \infty$ which makes the solution sensitive to the outer boundary of numerical calculation.

Dennis and Walker (D3) expanded Ψ and Z as a series of Legendre functions in the position coordinates. Equations (5-1) and (5-2) were reduced to a set of ordinary differential equations, solved numerically. This approach is inconvenient for high Re since the number of terms which must be included becomes prohibitive. Solutions to the steady equations were obtained for Re < 40 (D3) and for impulsively started motion for Re < 100 (D4).

Form drag and skin friction drag coefficients are obtained from the numerical results by integrating the distributions of surface pressure and vorticity:

$$C_{DP} = \int_0^\pi \frac{p_s}{\rho U^2/2} \sin 2\theta \, d\theta \tag{5-8}$$

and

$$C_{DF} = \frac{4}{\mathrm{Re}} \int_0^\pi \frac{a}{U}\left(\frac{\partial}{\partial\theta} + \cot\theta\right)\zeta_s \sin 2\theta \, d\theta. \tag{5-9}$$

The total drag coefficient, C_D, is the sum of C_{DP} and C_{DF}.

The validity of the numerical solutions has often been justified by comparison with experimental values for drag and wake dimensions (see below). However, these are not very sensitive to detailed changes in the flow field. Seeley *et al.* (S7) have given a more precise comparison based on measurements of stream function and surface vorticity at Re $\doteq 300$. Near the front stagnation point significant discrepancies were found, attributable to the relative sparseness of grid points in this region in all numerical solutions. Agreement was much better in the vicinity of separation.

b. *Boundary Layer Theory* Boundary layer theory has been applied to predict fluid velocities with some success for Re > 3000, but with less success at lower Re. The main difficulties are that the pressure distribution only follows potential flow up to about 30° from the front stagnation point, that the boundary layer thickness is only small relative to the sphere radius at very high Re, and that the tangential velocity in the boundary layer shows a maximum which is greater than the free stream value. Although an exact solution is available (F4, S1a) using the potential flow solution as the outer boundary condition, it gives velocity and vorticity distributions which are only realistic within 20° of the front stagnation point. Separation is predicted at 109.6° which corresponds to observed separation at Re $\doteq 400$, whereas at very high Re where boundary layer theory should be more reliable, separation occurs at about 81°. A somewhat more reliable treatment was given by Tomotika (T3) using Pohlhausen's method [see (S1a)] and an experimental pressure distribution (F1). This approach predicts separation correctly at 81°, but the predicted velocity distribution is again only accurate over the leading part of the sphere (S7). A full evaluation of boundary layer solutions in the range $300 < \text{Re} < 3000$ is given by Seeley *et al.* (S7).

For $\text{Re} < 100$ the pressure distribution departs from the ideal distribution even over the leading surface. This is because the boundary layer thickness is too large for conventional boundary layer theory to be applicable. Gluckman *et al.* (G8) attempted to overcome this limitation by a modified boundary layer theory accounting for the effect of the displacement thickness on the outer potential flow and by allowing for a pressure gradient across the boundary layer. Their predictions for the separation point are shown in Fig. 5.6 below. Generally the wake size is overestimated. However, the dependence of θ_s on Re shows roughly the right form and the approach warrants further development.

2. *Development of Flow Field with Reynolds Number*

a. *Unseparated Flow* ($1 < \text{Re} < 20$) As shown in Chapter 3, steady flow past spheres has fore-and-aft symmetry only in the limit of zero Reynolds number. Asymmetry becomes progressively more marked as Re increases. Figures 5.1 and 5.2 show streamlines and vorticity contours calculated numerically (M2). For $\text{Re} = 1$, asymmetry is most apparent in the vorticity distribution. The surface vorticity has a maximum forward of the equator (Fig. 5.3), while

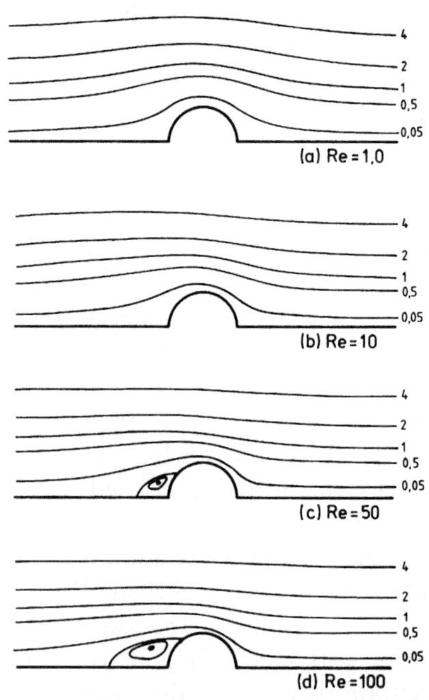

FIG. 5.1 Streamlines for flow past a sphere. Numerical results of Masliyah (M2). Flow from right to left. Values of Ψ indicated. (a) $\text{Re} = 1.0$; (b) $\text{Re} = 10$; (c) $\text{Re} = 50$; (d) $\text{Re} = 100$.

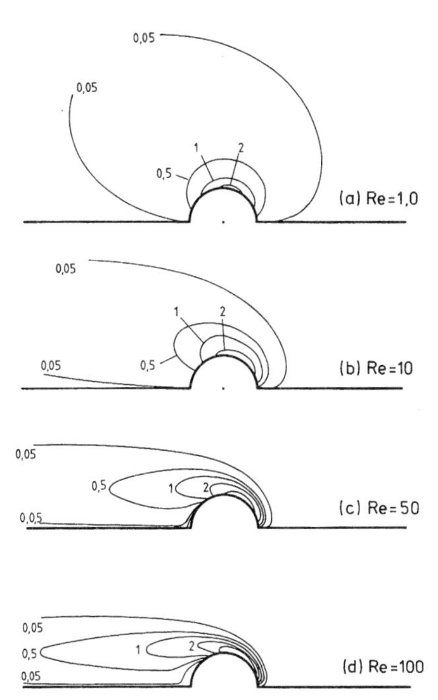

FIG. 5.2 Vorticity contours for flow past a sphere. Numerical results of Masliyah (M2). Flow from right to left. Values of Z indicated. (a) Re = 1.0; (b) Re = 10; (c) Re = 50; (d) Re = 100.

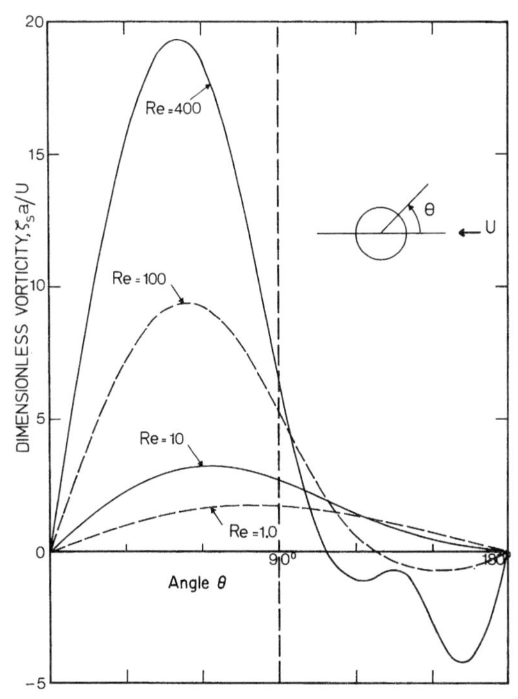

FIG. 5.3 Dimensionless vorticity at surface of sphere. Numerical results of Woo (W9).

contours remote from the body (Fig. 5.2) show convection of vorticity downstream. By Re = 10, asymmetry is also apparent in the streamlines (Fig. 5.1) while the position of the maximum surface vorticity has moved further forward (Fig. 5.3). The excess modified surface pressure shows some recovery at the rear (Fig. 5.4).

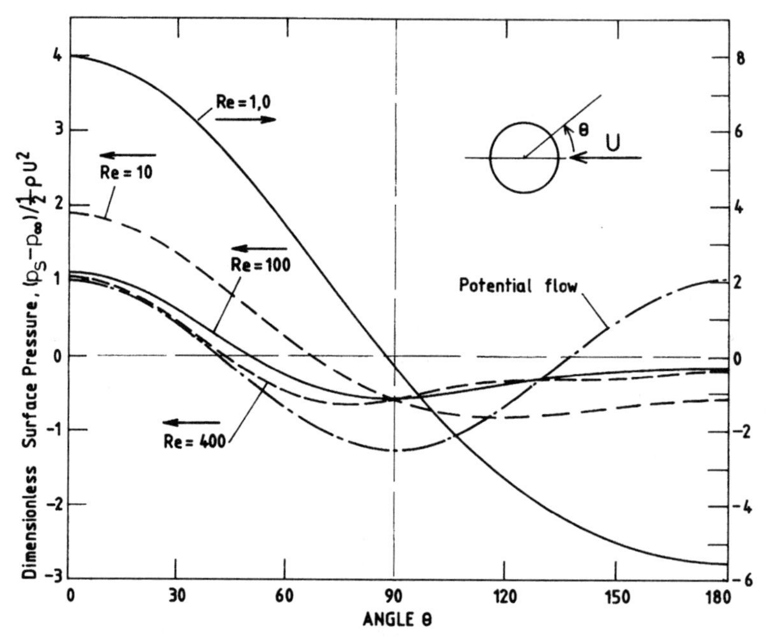

FIG. 5.4 Dimensionless excess modified pressure at surface of sphere. Numerical results of Woo (W9). (Note different scale for Re = 1.0 curve.)

b. *Onset of Separation* (Re ≐ 20) Flow separation is indicated by a change in the sign of the vorticity and first occurs at the rear stagnation point. The precise Re at which recirculation begins has been the subject of debate [e.g., see (G3)]. Some experimental and numerical results (N1, N7, R8) suggest separation at Re = 10, but this evidence is questionable (L8, M5). Taneda (T2) gave the onset of wake formation as Re = 24, but the difficulty of observing a very small eddy probably makes this figure slightly high. By extrapolating observed wake lengths to zero, Kalra and Uhlherr (K2) concluded that separation first occurs at Re = 20, in close agreement with the most reliable numerical solutions (D3, L5, L8, M2, W9). Drag determinations also indicate a change in flow regime at Re ≐ 20 (P7). The best estimate for the onset of recirculation is therefore Re = 20.

c. *Steady Wake Region* (20 < Re ⋖ 130) As Re increases beyond 20, the separation ring moves forward so that the attached recirculating wake widens and lengthens. The outer streamlines also curve less and vorticity is convected further downstream. Development of the wake is evident in photographs of flow past a rigid sphere (T2) reproduced in Fig. 5.5. The wake changes from a convex to a concave shape at Re $\doteq$ 35 (N1).

The dimensions of the attached wake are shown in Figs. 5.6, 5.7, and 5.8. The various numerical solutions agree closely with flow visualization results of Taneda (T2), although other workers (K2) report separation slightly closer to the rear. The separation angle, measured in degrees from the front stagnation point, is well approximated by

$$\theta_s = 180 - 42.5[\ln(\mathrm{Re}/20)]^{0.483} \qquad (20 < \mathrm{Re} \lessdot 400). \qquad (5\text{-}10)$$

Predicted and observed wake lengths and wake volumes agree closely for Re = 100 (Figs. 5.7 and 5.8). For Re $\gtrdot$ 100, the excess pressure over the leading surface of the sphere approaches that for an ideal fluid, but there is little recovery in the wake. As Re increases, the importance of skin friction decreases relative to form drag.

d. *Onset of Wake Instability* (130 $\lessdot$ Re $\lessdot$ 400) As Re is increased beyond about 130, diffusion and convection of vorticity no longer keep pace with vorticity generation. Instead, discrete pockets of vorticity begin to be shed from the wake. The Re at which vortex shedding begins is often called the "lower critical Reynolds number," although the transition is much more gradual than this label would imply.

At Re $\doteq$ 130, a weak long-period oscillation appears in the tip of the wake (T2). Its amplitude increases with Re, but the flow behind the attached wake remains laminar to Re above 200. The amplitude of oscillation at the tip reaches 10% of the sphere diameter at Re = 270 (G10). At about this Re, large vortices, associated with pulsations of the fluid circulating in the wake, periodically form and move downstream (S6). Vortex shedding appears to result from flow instability, originating in the free surface layer and moving downstream to affect the position of the wake tip (R11, R12, S6).

The relative importance of form drag continues to increase in this region with skin friction becoming inferior once Re > 150 (M2), and C_D begins to level out. The separation angle is still given by Eq. (5-10).

e. *High Subcritical Reynolds Number Range* (400 < Re < 3.5 × 10^5) Unsteadiness and asymmetry, originating in wake instability and shedding, limit the range of applicability of numerical results, based as they are on axisymmetric and often steady flow equations (see above). Predictions of the separation angle (Fig. 5.6) appear to be reliable to higher Re than predictions of wake length (Fig. 5.7) or wake volume (Fig. 5.8). This suggests that unsteadiness downstream has little effect upstream near the particle surface, at least for rigidly supported or heavy particles, and this has been confirmed by flow visualization (A4, S6). The surface vorticity distribution in the wake (see curve

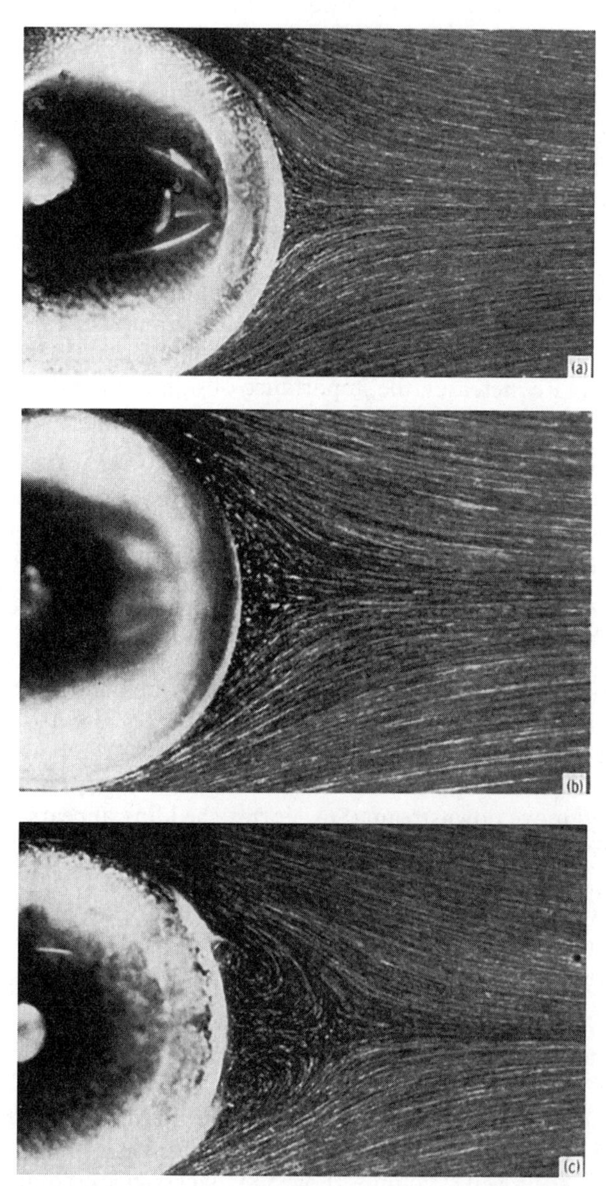

FIG. 5.5 Photographs of Taneda (T2) showing the development of the attached wake behind rigid spheres. (a) Re = 17.9; (b) Re = 26.8; (c) Re = 37.7; (d) Re = 73.6; (e) Re = 118; (f) Re = 133. (Reproduced with permission.)

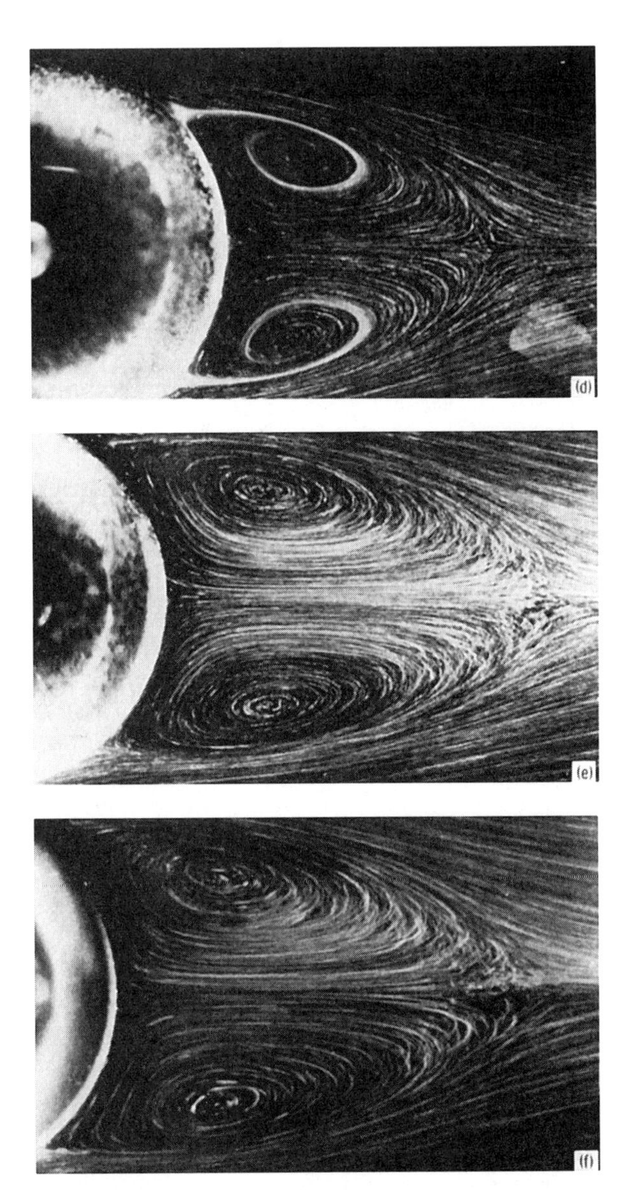

FIG. 5.5 (*Cont.*)

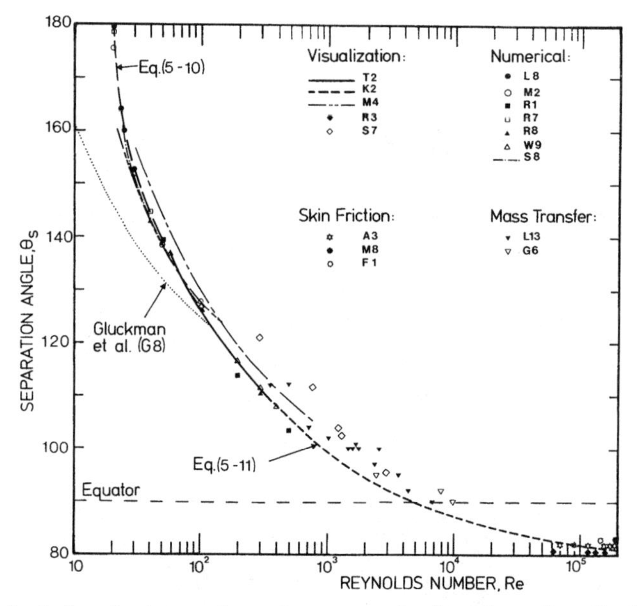

FIG. 5.6 Angle from front stagnation point to separation from the surface of a rigid sphere.

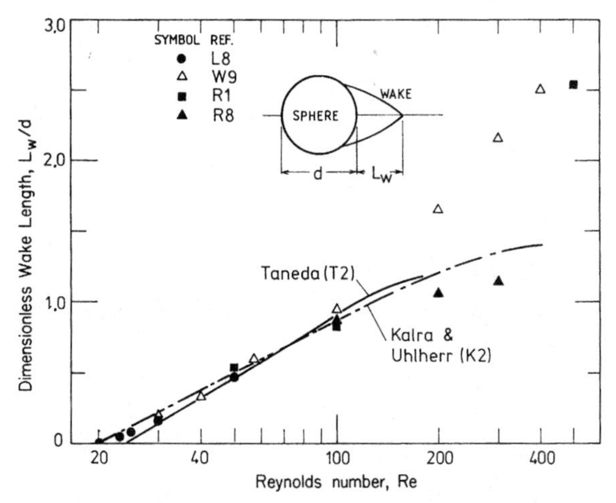

FIG. 5.7 Length of sphere wake.

for Re = 400 in Fig. 5.3) has been interpreted as indicating secondary eddies, but these do not appear to have been observed experimentally.

As Re increases beyond about 400, vortices are shed as a regular succession of loops from alternate sides of a plane which precesses slowly about the axis (A4, K6, M11). As shown in Fig. 5.9, the Strouhal number Sr for vortex shedding increases. At the same time, the point at which the detached shear layer

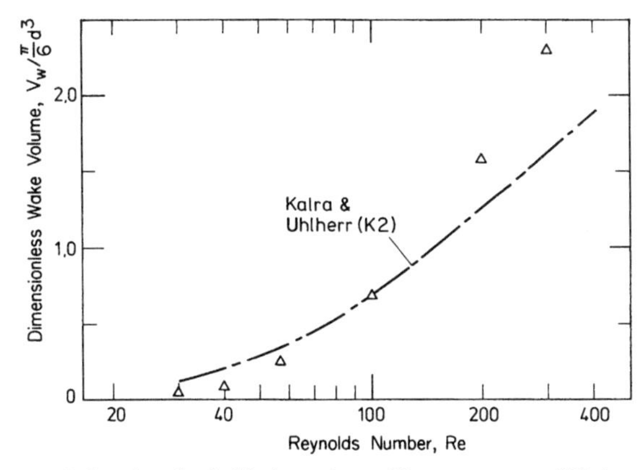

FIG. 5.8 Volume of closed wake behind a sphere. Measurements of Kalra and Uhlherr (K2) and numerical predictions of Woo (W9).

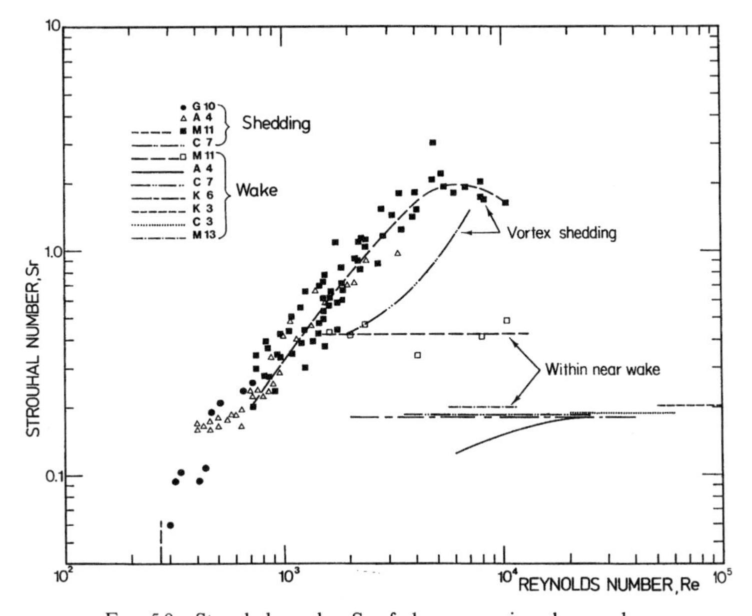

FIG. 5.9 Strouhal number Sr of phenomena in sphere wake.

rolls up to form shed vortices moves closer to the sphere. Shed loops progressively lose their character (A4) and may combine to form "vortex balls" (M11). By Re = 1300, the wake shows three-dimensional rotation, while velocities near the rear surface of the sphere fluctuate in direction and magnitude due to vortex shedding (S6). At Re $\doteq$ 6000, Sr reaches a maximum, and the point at which the shear layer rolls up approaches the sphere surface (A4). From here until the critical transition (Re $\doteq 3 \times 10^5$, see below), separation occurs at a

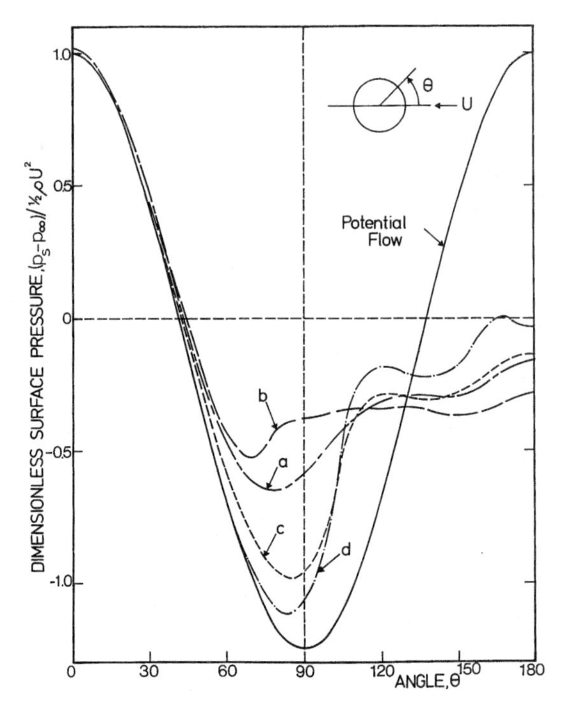

FIG. 5.10 Dimensionless pressure at surface of sphere: (a) Numerical results of Woo (W9): Re = 400; (b)–(d) Measurements of Achenbach (A3): (b) Re = 1.62×10^5; (c) Re = 3.18×10^5; (d) Re = 1.14×10^6.

point which rotates around the sphere at the shedding frequency (A4). The wake may appear like a pair of helical vortex filaments (F2, K6), although the structure cannot be so regular in detail (A4). Hot-wire measurements in the near-wake show strong periodicity right up to the critical transition, with Sr, ranging between 0.18 and 0.2, virtually independent of Re (A4, C3, C7, K3, K6, M13). Möller (M11) reported Sr = 0.42 for "vortex balls," but this is inconsistent with subsequent measurements. Because of the periodicity, the wake should not be considered turbulent. As discussed later, wake shedding can cause appreciable fluctuations in the motion of freely falling particles, thereby affecting mean drag.

Figure 5.10 shows the surface pressure distribution at different values of Re. The distribution changes remarkably little between Re = 400 and 1.6×10^5. Since form drag now predominates as noted above, C_D is also insensitive to Re. For $750 \leq \text{Re} \leq 3.5 \times 10^5$, the "Newton's law" range,† C_D varies by only $\pm 13\%$

† Newton proposed a law equivalent to $C_D = 0.5$ (N4), and confirmed this experimentally by timing the fall of spheres from the dome of St. Paul's Cathedral (N5). However, his explanation was based on ideas which bear little resemblance to current concepts of fluid mechanics.

about a value of 0.445. An alternative label for this range, the "turbulent flow" range, is inaccurate and misleading. The drag force in this and other ranges of Re is treated in sections 3 and 4.

Throughout the Newton's law range, the separation ring continues to move forward as Re increases. At Re $\doteq$ 5000, separation moves in front of the equator towards a limit of 81–83° (A3, F1, M8, R4). Direct observations of the separation ring are scant for $800 < \text{Re} < 6 \times 10^4$. Several workers [e.g., B14, L10, L13, N3, W1) have determined the point of minimum heat or mass transfer in this range, but, as discussed below, this occurs aft of separation. Seeley *et al.* (S7) report some flow visualization results, but they found separation closer to the rear than observed by other workers, perhaps due to wall effects. As shown in Fig. 5.6, a realistic interpolation is provided by

$$\theta_s = 78 + 275\,\text{Re}^{-0.37} \qquad (400 < \text{Re} < 3 \times 10^5). \tag{5-11}$$

f. *Critical Transition and Supercritical Flow* ($\text{Re} > 3.5 \times 10^5$) As Re increases beyond 2×10^5, changes in the flow pattern occur which are so marked that they are termed "critical transition." Figure 5.11 shows the separation point in this range determined from direct visualization (R3) and inferred from pressure and skin friction measurements (A3). On increasing Re above 2×10^5, separation begins to move aft, while fluctuations in the position of the separation point and in pressure and skin friction become more marked. The detached free shear layer becomes turbulent soon after separation and, for $\text{Re} \gtrsim 2.8 \times 10^5$, reattaches to the surface (A3). As a result of enhanced momentum transfer, the turbulent boundary layer is able to withstand the adverse pressure gradient longer without separation. Final separation therefore shifts abruptly downstream. In the same range the surface pressure minimum decreases towards the

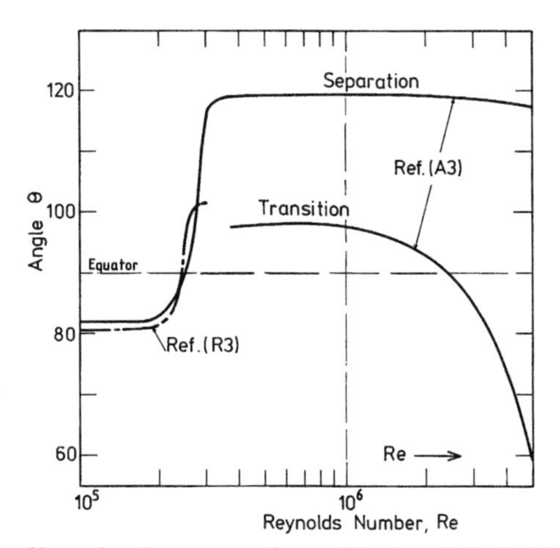

FIG. 5.11 Position of boundary layer separation and laminar/turbulent transition in the critical region and beyond. Experimental results of Achenbach (A3) and Raithby and Eckert (R3).

potential flow value, and more pressure is recovered in the wake (see Fig. 5.10). Similar changes can be induced at lower Re by "tripping" the boundary layer with an irregularity such as a fine wire attached to the sphere surface [e.g., see (M8)].

As a result of the changes in pressure distribution, form drag drops sharply in the critical range. The drag coefficient C_D falls from 0.5 at $\mathrm{Re} = 2 \times 10^5$ to 0.07 at $\mathrm{Re} = 4 \times 10^5$ (see Fig. 5.12), while the proportion of the total drag resulting from skin friction rises from 1.3 to 12.5% (A3). Critical transition is sensitive to free stream turbulence as discussed in Chapter 10. Thus drag measurements in this range show considerable scatter (A3, M8). The results least affected by turbulence (A3) appear to be those of Millikan and Klein (M10) who determined the drag on a sphere towed by an aircraft. Definition of a "critical Reynolds number" is arbitrary; for convenience, it is taken as the Re at which C_D reaches 0.3 (C6, D7), $\mathrm{Re}_c = 3.65 \times 10^5$ for turbulence-free flow (M10).

Above the critical range, further increases in Re cause the "separation bubble" between laminar separation and turbulent reattachment to shrink, although the positions of laminar/turbulent transition and final separation remain essentially fixed (see Fig. 5.11). For $\mathrm{Re} \gtrsim 10^6$, transition from laminar to turbulent flow occurs without a separation bubble (A3). At still higher Re, both transition and separation move forward on the sphere. As the pressure recovery in the wake declines, C_D increases slightly and tends towards a constant value of approximately 0.19 at very high Re (A3). Appreciable fluctuating lift forces occur in the supercritical range, with an r.m.s. lift coefficient of approximately 0.06, accompanied by fluctuating moments (W6). The fluctuations appear to be due to shedding of large turbulent eddies, with corresponding random changes in wake configuration.

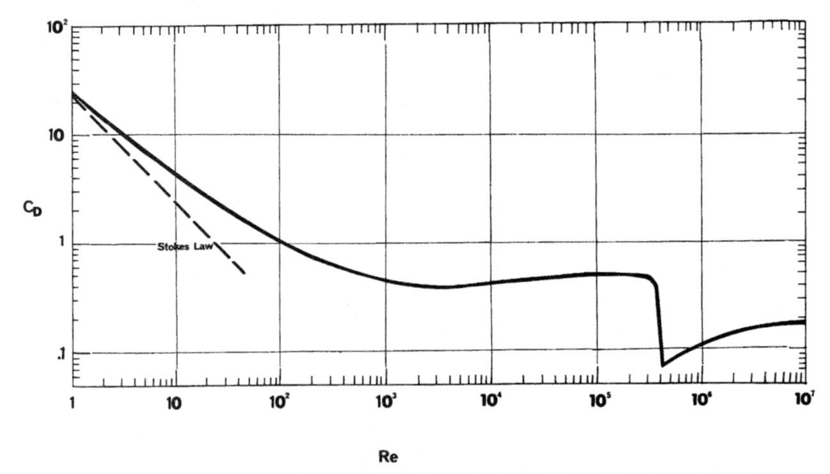

FIG. 5.12 Drag coefficient of a sphere as a function of Reynolds number (standard drag curve).

3. *Drag: Standard Drag Curve*

The conventional correlation for the drag on a sphere in steady motion is presented as a graph, see Fig. 5.12, called the "standard drag curve", where C_D is plotted as a function of Re. Many empirical or semiempirical equations have been proposed to approximate this curve. Some of the more popular are listed in Table 5.1. None of these correlations appears to consider all available data.

TABLE 5.1

Relationships for Sphere Drag

Author(s)	Range	Relationship for C_D	Range of deviation in C_D (%)
1. Schiller and Nauman (S1)	$\mathrm{Re} < 800$	$\frac{24}{\mathrm{Re}}(1 + 0.15\,\mathrm{Re}^{0.687})$	$+5$ to -4
2. Lapple (L3)	$\mathrm{Re} < 1000$	$\frac{24}{\mathrm{Re}}(1 + 0.125\,\mathrm{Re}^{0.72})$	$+5$ to -8
3. Langmuir and Blodgett (L2)	$1 < \mathrm{Re} < 100$	$\frac{24}{\mathrm{Re}}(1 + 0.197\,\mathrm{Re}^{0.63} + 2.6 \times 10^{-4}\,\mathrm{Re}^{1.38})$	$+6$ to $+1$
4. Allen (A5)	(a) $2 < \mathrm{Re} < 500$ (b) $1 < \mathrm{Re} < 1000$	$10\,\mathrm{Re}^{-1/2}$ $30\,\mathrm{Re}^{-0.625}$	-8 to -52 $+70$ to -15
5. Gilbert *et al.* (G7)	$0.2 < \mathrm{Re} < 2000$	$0.48 + 28\,\mathrm{Re}^{-0.85}$	$+24$ to -11
6. Kurten *et al.* (K8)	$0.1 < \mathrm{Re} < 4000$	$0.28 + \frac{6}{\mathrm{Re}^{1/2}} + \frac{21}{\mathrm{Re}}$	$+7$ to -6
7. Abraham (A2)	$\mathrm{Re} < 6000$	$0.2924(1 + 9.06\,\mathrm{Re}^{-1/2})^2$	$+9$ to -6
8. Ihme *et al.* (I1)	$\mathrm{Re} < 10^4$	$0.36 + \frac{5.48}{\mathrm{Re}^{0.573}} + \frac{24}{\mathrm{Re}}$	$+10$ to -10
9. Rumpf [see (K8)]	(a) $\mathrm{Re} < 10$ (b) $\mathrm{Re} < 100$ (c) $\mathrm{Re} < 10^5$	$2 + 24/\mathrm{Re}$ $1 + 24/\mathrm{Re}$ $0.5 + 24/\mathrm{Re}$	-3 to -5 $+14$ to -20 $+30$ to -39
10. Clift and Gauvin (C6)	$\mathrm{Re} < 3 \times 10^5$	$\frac{24}{\mathrm{Re}}(1 + 0.15\,\mathrm{Re}^{0.687}) + 0.42/(1 + 4.25 \times 10^4\,\mathrm{Re}^{-1.16})$	$+6$ to -4
11. Brauer (Bl1)	$\mathrm{Re} < 3 \times 10^5$	$0.40 + \frac{4}{\mathrm{Re}^{1/2}} + \frac{24}{\mathrm{Re}}$	$+20$ to -18
12. Tanaka and Iinoya (T1)	$\mathrm{Re} < 7 \times 10^4$	$\log_{10} C_D = a_1 w^2 + a_2 w + a_3$ where $w = \log_{10}\mathrm{Re}$ and a_1, a_2, and a_3 are given for 7 intervals of Re	$+6$ to -9

Table 5.2 gives a new correlation, based on a critical examination of available data for spheres (N6). Results in which wall effects, compressibility effects, noncontinuum effects, support interference, etc. are significant have been excluded. The whole range of Re has been divided into 10 subintervals, with a distinct correlation for each interval. Adjacent equations for C_D match within 1% at the boundaries between subintervals, but the piecewise fit shows slight gradient discontinuities there. The Re = 20 boundary corresponds to onset of wake formation as discussed above, the remaining boundaries being chosen for convenience.

For Re < 0.01, the Oseen result is reliable (see Chapter 3). Equation B was originally proposed by Beard (B7) as a fit to two specific sets of data (B5, P8)

TABLE 5.2[a,b]

Recommended Drag Correlations: Standard Drag Curve, $w = \log_{10} \mathrm{Re}$

	Range	Correlation
(A)	$\mathrm{Re} < 0.01$	$C_D = 3/16 + 24/\mathrm{Re}$
(B)	$0.01 < \mathrm{Re} \le 20$	$\log_{10}\left[\frac{C_D \mathrm{Re}}{24} - 1\right] = -0.881 + 0.82w - 0.05w^2$
		i.e., $C_D = \frac{24}{\mathrm{Re}}\left[1 + 0.1315\,\mathrm{Re}^{(0.82 - 0.05w)}\right]$
(C)	$20 \le \mathrm{Re} \le 260$	$\log_{10}\left[\frac{C_D \mathrm{Re}}{24} - 1\right] = -0.7133 + 0.6305w$
		i.e., $C_D = \frac{24}{\mathrm{Re}}\left[1 + 0.1935\,\mathrm{Re}^{0.6305}\right]$
(D)	$260 \le \mathrm{Re} \le 1500$	$\log_{10} C_D = 1.6435 - 1.1242w + 0.1558w^2$
(E)	$1.5 \times 10^3 \le \mathrm{Re} \le 1.2 \times 10^4$	$\log_{10} C_D = -2.4571 + 2.5558w - 0.9295w^2 + 0.1049w^3$
(F)	$1.2 \times 10^4 < \mathrm{Re} < 4.4 \times 10^4$	$\log_{10} C_D = -1.9181 + 0.6370w - 0.0636w^2$
(G)	$4.4 \times 10^4 < \mathrm{Re} \le 3.38 \times 10^5$	$\log_{10} C_D = -4.3390 + 1.5809w - 0.1546w^2$
(H)	$3.38 \times 10^5 < \mathrm{Re} \le 4 \times 10^5$	$C_D = 29.78 - 5.3w$
(I)	$4 \times 10^5 < \mathrm{Re} \le 10^6$	$C_D = 0.1w - 0.49$
(J)	$10^6 < \mathrm{Re}$	$C_D = 0.19 - 8 \times 10^4/\mathrm{Re}$

[a] Sources of data: Achenbach (A3); Arnold (A7); Bailey and Hiatt (B1); Beard and Pruppacher (B5); Davies (D2); Dennis and Walker (D3); Goin and Lawrence (G9); Goldburg and Florsheim (G10); Gunn and Kinzer (G14); Hoerner (H14); Ihme *et al.* (I1); LeClair (L5); Liebster (L12); Masliyah (M2); Maxworthy (M7, M8); Millikan and Klein (M10); Möller (M11); Pettyjohn and Christiansen (P4); Pruppacher and Steinberger (P8); Rafique (R1); Rimon and Cheng (R8); Roos and Willmarth (R10); Schmiedel (S2); Shakespear (S9); Vlajinac and Covert (V3); Wieselsberger (W4); Woo (W9).

[b] Number of data points: C—149; D—74; E—61; F—52; G—142.

and agrees closely with all reliable experimental and numerical data in its range. Correlations C to G were obtained by least-squares regression. Correlation H fits the data of Millikan and Klein (M10) while I and J correspond to Achenbach's results (A3). The correlations in Table 5.2 may be regarded as the best available approximation to the standard drag curve. The standard curve, calculated from these equations, is shown in Fig. 5.12.

The recommended standard drag curve of Fig. 5.12 differs from the curve originally given by Lapple and Shepherd (L4) and widely reproduced [e.g., (P3)]. They underestimate C_D by up to 5% for Re < 100 and also place the critical Re too low. The revised curve of Bailey (B2) is in close agreement with the one recommended here except near the critical transition where there is considerable spread in the measurements and he used only a single set of free-flight data. Deviations of other empirical relations from the recommended ones are listed in Table 5.1. The high errors for Allen's equations are noteworthy in view of their common use [e.g., (G11)] for calculating terminal settling velocities.

4. *Terminal Velocity in Free Fall or Rise*

For a particle moving with steady terminal velocity U_T in a gravitational field, the drag force balances the difference between the weight and buoyancy:

$$F_D = g\,\Delta\rho(\pi/6)d^3, \tag{5-12}$$

so that the drag coefficient becomes

$$C_D = 4\Delta\rho\, gd/3\rho\, U_T^2 = 4\rho\,\Delta\rho\, gd^3/3\mu^2 \mathrm{Re}_T^2, \tag{5-13}$$

where Re_T is the Reynolds number at the terminal velocity. As noted above, $C_D \doteq 0.445$ for $750 < \mathrm{Re} < 3.5 \times 10^5$, so that for this range

$$U_T \doteq 1.73(gd\,\Delta\rho/\rho)^{1/2} \qquad \text{or} \qquad \mathrm{Re}_T \doteq 1.73 N_D^{1/2} \qquad (750 < \mathrm{Re} < 3.5 \times 10^5) \tag{5-14}$$

where

$$N_D = C_D \mathrm{Re}_T^2 = 4\rho\,\Delta\rho\, gd^3/3\mu^2 \tag{5-15}$$

The term N_D is sometimes called the "Best number."† An analytic expression for the terminal velocity corresponding to Stokes' law is also available at low Re [Eq. (3-18)]. Outside these ranges of Re, or when more accurate predictions are required, C_D vs. Re relationships are inconvenient for determining terminal velocities since both groups involve U_T. Hence an iterative procedure is needed. It is more convenient to express Re as a function of N_D, the latter being independent of U_T. Empirical correlations of this form, based on the same data

† The group $\pi N_D/8$ is often termed the "Archimedes number," while $3N_D/4$ is sometimes called the Galileo number or Archimedes number.

as in Table 5.2, are presented in Table 5.3. Adjacent correlations agree within 1% at the arbitrary boundaries of the ranges. Values of Re calculated from the correlations in Tables 5.2 and 5.3 agree within 4%. Davies (D2) gave similar correlations for $N_D < 4.5 \times 10^7$, $\mathrm{Re} < 10^4$. Although his expressions are based on pre-1945 data, they differ by at most 5% from the results in Table 5.3. Re_T is tabulated as a function of $N_D^{1/3}$ in Appendix A.

TABLE 5.3

Correlations for Re as a Function of N_D, $W = \log_{10} N_D$

Range	Correlation
(A) $N_D \leq 73$; $\mathrm{Re} \leq 2.37$	$\mathrm{Re} = N_D/24 - 1.7569 \times 10^{-4} N_D^2 + 6.9252 \times 10^{-7} N_D^3 - 2.3027 \times 10^{-10} N_D^4$
(B) $73 < N_D \leq 580$; $2.37 < \mathrm{Re} \leq 12.2$	$\log_{10} \mathrm{Re} = -1.7095 + 1.33438W - 0.11591W^2$
(C) $580 < N_D \leq 1.55 \times 10^7$; $12.2 < \mathrm{Re} \leq 6.35 \times 10^3$	$\log_{10} \mathrm{Re} = -1.81391 + 1.34671W - 0.12427W^2 + 0.006344W^3$
(D) $1.55 \times 10^7 < N_D \leq 5 \times 10^{10}$; $6.35 \times 10^3 < \mathrm{Re} \leq 3 \times 10^5$	$\log_{10} \mathrm{Re} = 5.33283 - 1.21728W + 0.19007W^2 - 0.007005W^3$

It is also useful to define a dimensionless terminal velocity:

$$N_U = \mathrm{Re}_T/C_D = 3\rho^2 U_T^3/4\Delta\rho g\mu. \tag{5-16}$$

Here $N_U^{1/3}$ is plotted versus $N_D^{1/3}$ in Fig. 5.13 and tabulated in Appendix B. This tabulation is particularly convenient for estimation of terminal velocities or diameters since $N_D^{1/3}$ is independent of U_T and proportional to d, while $N_U^{1/3}$ is proportional to U_T and independent of d.

Figure 5.13 shows that there is a range, $2.3 \times 10^3 \lessdot N_D^{1/3} \lessdot 3.8 \times 10^3$, for which three terminal velocities are possible. This range is of practical interest for meteorological balloons (S4) and large hailstones (B8, W5). The intermediate value, portion AB of the curve, corresponds to the critical range. The terminal velocity corresponding to this part of the curve is unstable in the sense that, if U_T increases, the drag decreases. Thus terminal velocities in the critical range are not observed experimentally unless there is significant free stream turbulence (see Chapter 10). Instead, a sphere can show two stable terminal velocities, and may even alternate between them giving a mean velocity close to the unstable value (M1). The curve beyond B represents supercritical motion.

Fluctuations in speed and direction also occur in the subcritical range (down to $\mathrm{Re} \doteq 270$) (G10). A sphere shows a rocking motion and follows a zigzag or spiral trajectory† in this range (C5, M1, P5) with wavelength about $12d$ and

† Newton encountered this problem in experiments to determine the drag on spheres falling through liquids (N4).

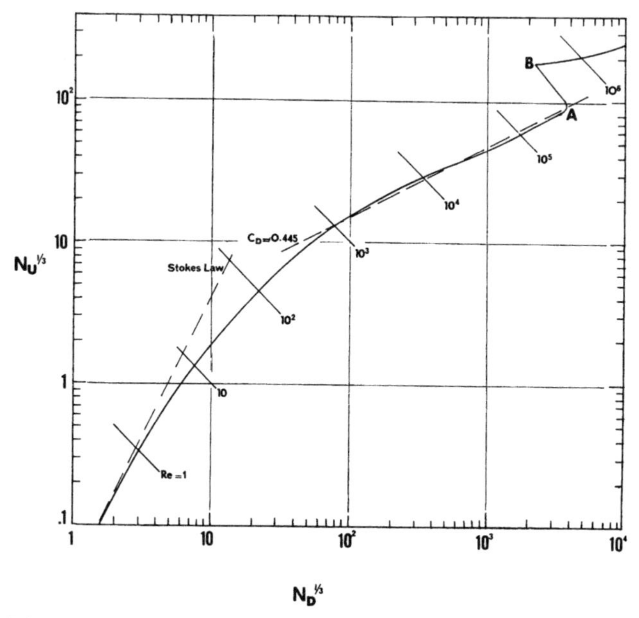

FIG. 5.13 Dimensionless terminal velocity ($N_U^{1/3}$) for sphere as a function of dimensionless diameter ($N_D^{1/3}$).

lateral amplitude approximately $0.37d/(1 + 2\gamma)$ (M1). At least for the lower Reynolds numbers, this phenomenon is associated with wake-shedding, which induces secondary motion of the particle at the same frequency.

Variations in vertical velocity are typically 5% of the mean (S10); horizontal velocities are of the same order and decrease as the density ratio, $\gamma = \rho_p/\rho$, increases (P5, S10). Wandering is enhanced if the centre of mass is displaced from the geometric centre of the particle (V2). Secondary motion increases the mean drag, i.e., a particle undergoing secondary motion tends to have a vertical terminal velocity less than that calculated from the drag on a fixed sphere (G10, P5, S10). This retardation appears to become more significant as γ is reduced (P5, V2), and to be negligible for $\mathrm{Re} < 10^3$ (M1). The following correlations are proposed, based on data reported by Sheth† (S10) for $10^3 < \mathrm{Re} < 2 \times 10^5$ and $\gamma \geq 1$:

$$C_D' = C_D[1 + 0.13/(2.8\gamma - 1)], \tag{5-17}$$

$$N_U' = N_U[1 - 0.2/(2.8\gamma - 1)], \tag{5-18}$$

where the prime denotes the value appropriate to a sphere with density ratio γ in free motion, and C_D and N_U may be calculated from the standard correlations above. From Eq. (5-18), the terminal velocity is reduced by 3.5% on reducing γ

† Sheth proposed a different correlation which shows anomalous behaviour for large γ.

from a very large value to nearly unity. In supercritical flow, horizontal motion is more marked, with erratic changes of speed and direction rather than periodic motion (M1, M14, S4, S5, W6), resulting from the fluctuating lift noted in section 2. Secondary motion is more important for nonspherical and fluid particles, and is discussed further in Chapters 6 and 7.

Figure 5.14 shows terminal velocities of spheres of various densities in air and water at 20°C calculated from the correlations in Tables 5.2 and 5.3, incorporating corrections for secondary motion, Eq. (5-18), and slip (see Chapter 10).

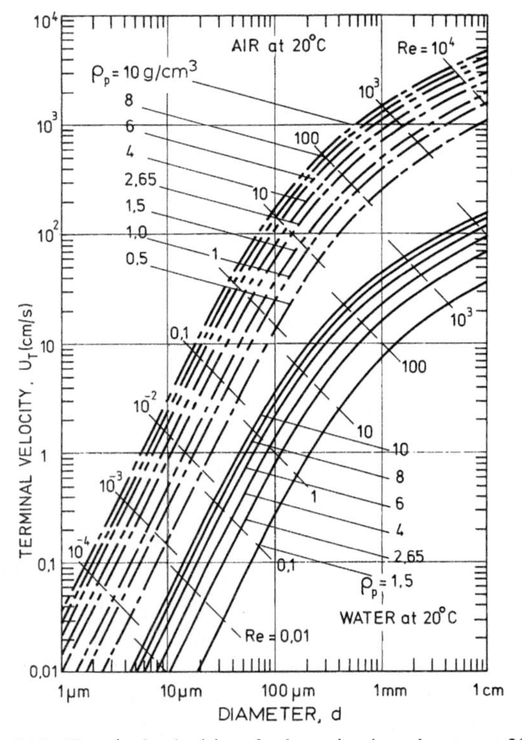

FIG. 5.14 Terminal velocities of spheres in air and water at 20°C.

B. Heat and Mass Transfer

1. *Numerical Solutions*

For axisymmetric flow the species continuity equation, Eq. (1-38), written in terms of the dimensionless concentration ϕ and stream function Ψ (see Chapter 1) is

$$\frac{\partial \Psi}{\partial R}\frac{\partial \phi}{\partial \theta} - \frac{\partial \Psi}{\partial \theta}\frac{\partial \phi}{\partial R} = \frac{2R^2 \sin\theta}{\text{Pe}}\left[\frac{1}{R^2}\frac{\partial}{\partial R}\left(R^2 \frac{\partial \phi}{\partial R}\right) + \frac{1}{R^2 \sin\theta}\frac{\partial}{\partial \theta}\left(\sin\theta \frac{\partial \phi}{\partial \theta}\right)\right]. \quad (5\text{-}19)$$

The boundary conditions are

$$\text{(a) at} \quad \theta = 0 \text{ and } \pi: \quad \frac{\partial \phi}{\partial \theta} = 0, \tag{5-20}$$

$$\text{(b) at} \quad R = 1: \quad \phi = 1, \tag{5-21}$$

$$\text{(c) at} \quad R \to \infty: \quad \phi \to 0. \tag{5-22}$$

Since the stream function depends upon Reynolds number, the rate of transfer will depend upon both Re and Sc except in the limit Re $\to$ 0 treated in Chapter 3. Solutions to Eq. (5-19) have been obtained using the techniques discussed earlier, i.e., finite-difference schemes (A6, D5, I1, M6, W9), solution to the time-dependent problem (H11), and series expansions (D5).

The local and mean Sherwood numbers are obtained from the numerical results using the equations

$$\text{Sh}_{\text{loc}} = -2(\partial \phi / \partial R)_{R=1} \tag{5-23}$$

and

$$\text{Sh} = \frac{1}{2} \int_0^{\pi} \text{Sh}_{\text{loc}} \sin \theta \, d\theta. \tag{5-24}$$

2. *Mechanism of Transfer*

Figure 5.15 shows streamlines and concentration contours calculated by Masliyah and Epstein (M6). Even in creeping flow, Fig. 5.15a, the concentration contours are not symmetrical. The concentration gradient at the surface, and thus Sh_{loc}, is largest at the front stagnation point and decreases with polar angle; see also Fig. 3.11. The diffusing species is convected downstream forming a region of high concentration at the rear (often referred to as a "concentration wake") which becomes narrower at higher Peclet number.

We consider the changes which occur at increasing Reynolds number and at a constant Schmidt number of 0.7, typical of evaporation of liquids into air or of heat transfer to air (Pr = 0.7). Figure 5.15b shows streamlines and concentration contours at Re = 20 where a steady wake first appears. Although there is no flow separation, a concentration wake is evident downstream from the sphere. At Re = 100, where separation occurs at $\theta = 126°$ and a large recirculating wake exists, the downstream concentration wake has narrowed and the concentration contours are distorted by the recirculatory flow in the wake; see Fig. 5.15c. The variation of Sh_{loc} with polar angle θ for the same Sc and various Re is shown in Fig. 5.16 (W9). For Re < 20, Sh_{loc} decreases monotonically from front to rear, but between Re = 30 and 57 a minimum first appears even though separation occurs for Re > 20. This minimum moves forward with increasing Re; however, as shown in Fig. 5.16, it occurs aft of the separation point due to the presence of angular diffusion. The increased Sh_{loc}

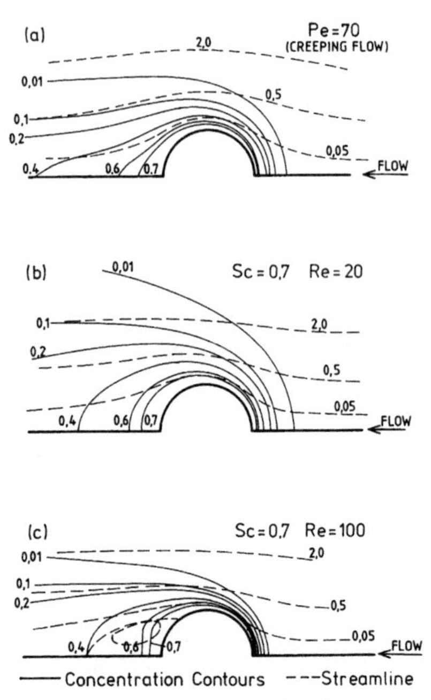

FIG. 5.15 Streamlines and concentration contours for flow past a sphere. Numerical results of Masliyah and Epstein (M6). Flow from right to left. Values of Ψ and ϕ indicated. (a) Creeping flow, Pe = 70; (b) Re = 20, Sc = 0.7; (c) Re = 100, Sc = 0.7.

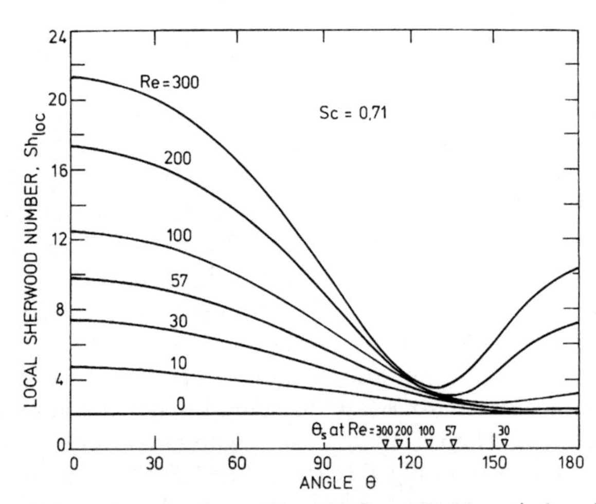

FIG. 5.16 Local Sherwood number for a sphere with Sc = 0.71. Numerical results of Woo (W9).

at the rear of the sphere at high Re is caused by the action of the recirculating wake (M3). Mass transferred from that portion of the sphere covered by the recirculating wake is ultimately transferred to the external flow by diffusion across the separating streamline, $\Psi = 0$. Elements of fluid in the wake near the separating streamline move away from the sphere losing mass to the external fluid. On their return toward and over the rear surface of the sphere the concentration increases. Thus these wake elements of fluid "carry" mass (or heat) from the rear of the sphere to the external stream which then carries it away. Due to the recirculatory motion in the wake, fluid approaching the rear stagnation point does not have zero concentration and the approach velocity is less than the free stream velocity. Therefore, Sh_{loc} is lower at the rear than at the front stagnation point, at least until vortices are shed.

As Re increases further and vortices are shed, the local rate of mass transfer aft of separation should oscillate. Although no measurements have been made for spheres, mass transfer oscillations at the shedding frequency have been observed for cylinders (B9, D6, S12). At higher Re the forward portion of the sphere approaches boundary layer flow while aft of separation the flow is complex as discussed above. Figure 5.17 shows experimental values of the local Nusselt number Nu_{loc} for heat transfer to air at high Re. The vertical lines on each curve indicate the values of the separation angle. It is clear that the transfer rate at the rear of the sphere increases more rapidly than that at the front and that even at very high Re the minimum Nu_{loc} occurs aft of separation. Also shown in Fig. 5.17 is the thin concentration boundary layer

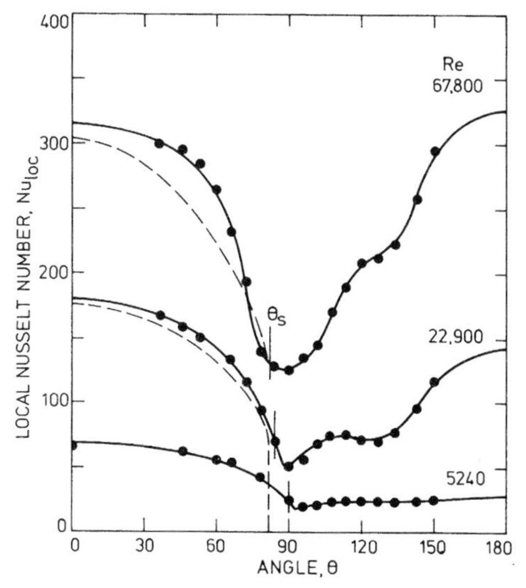

FIG. 5.17 Local Nusselt number for heat transfer from a sphere to air ($Pr = 0.71$). Experimental results of Galloway and Sage (G1). Dashed lines are predictions of boundary layer theory by Lee and Barrow (L10).

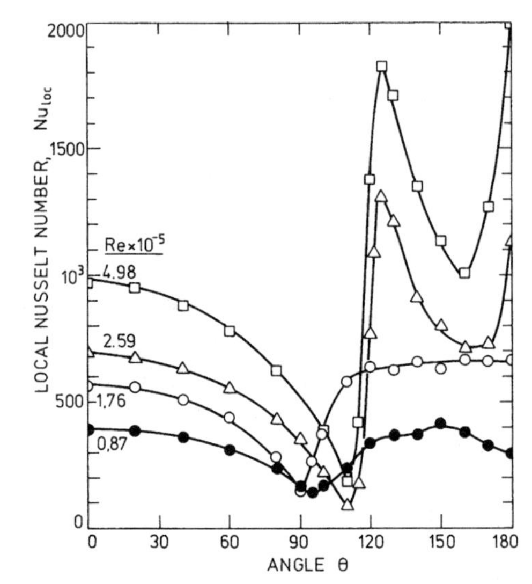

FIG. 5.18 Local Nusselt number for heat transfer from a sphere to air ($\mathrm{Pr} = 0.71$). Experimental results of Xenakis *et al.* (X1).

theory prediction for the forward portion of the sphere (L10).† The theory, which predicts that Nu_{loc} is zero at the separation point, fails due to the neglect of angular diffusion which becomes increasingly important as the separation ring is approached. Theoretical predictions lie beneath the data for two reasons: first, the velocity profile outside the boundary layer differs from that assumed (S7); second, in an experiment the approaching stream is usually turbulent (for example, the intensity of turbulence was 1.3% for the data in Fig. 5.17)—see Chapter 10.

Figure 5.18 shows the only reliable Nu_{loc} data available near the critical Reynolds number (X1). Since the data were taken with a side support, there is some effect on the separation and transition angles. Thus the values of Nu_{loc} are probably subject to error (R2, R3) although the trend with Re should be correct. At $\mathrm{Re} = 0.87 \times 10^5$ the Sh_{loc} variation is similar to that shown at lower Re in Fig. 5.17. At $\mathrm{Re} = 1.76 \times 10^5$ the critical transition has already occurred, with the separation bubble accounting for the minimum in Nu_{loc} at $\theta = 110°$. The maximum in Nu_{loc} at $\theta = 125°$ reflects the increased transfer rate in the attached turbulent boundary layer. The local minimum at $\theta = 160°$ is due to final separation. These angles do not agree exactly with those in Fig. 5.11 because of the crossflow support and the fact that angular diffusion shifts the

† Several results may be derived from the use of boundary layer theory depending upon the velocity profile assumed to exist outside the boundary layer. Lee and Barrow (L10) used the velocity profile of Tomotika (T3) which was, in turn, fitted to the surface pressure data of Fage (F1) at $\mathrm{Re} = 157{,}000$. This profile predicts separation at $\theta = 81°$ as noted in Section A.1.

minima rearward. As Re increases, the rate of transfer over the rear hemisphere increases more rapidly with Re than the rate over the forward hemisphere because $\mathrm{Nu}_{\mathrm{loc}} \propto \mathrm{Re}^{0.8}$ for a turbulent boundary layer while $\mathrm{Nu}_{\mathrm{loc}} \propto \mathrm{Re}^{0.5}$ for a laminar boundary layer.

We now consider the effect of Schmidt number. At constant Reynolds number, increasing Sc narrows the concentration wake. Figure 5.19 shows the results of numerical solutions (H11, W9) for $\mathrm{Sh}_{\mathrm{loc}}$ at several Re and Sc. As Sc increases from zero at Re = 100, the local Sherwood number increases, its minimum value shifting forward toward the separation point. In the limit as $\mathrm{Sc} \to \infty$, angular diffusion is negligible and the minimum occurs at the separation point. Thus determinations of the separation angle from the minimum value of $\mathrm{Sh}_{\mathrm{loc}}$ are reliable only for experiments at large Sc. Also shown in Fig. 5.19 are the data of Frössling (F3) for sublimation of naphthalene spheres in air. Although the values of Re and Sc do not match exactly, the data and the numerical solutions agree well.

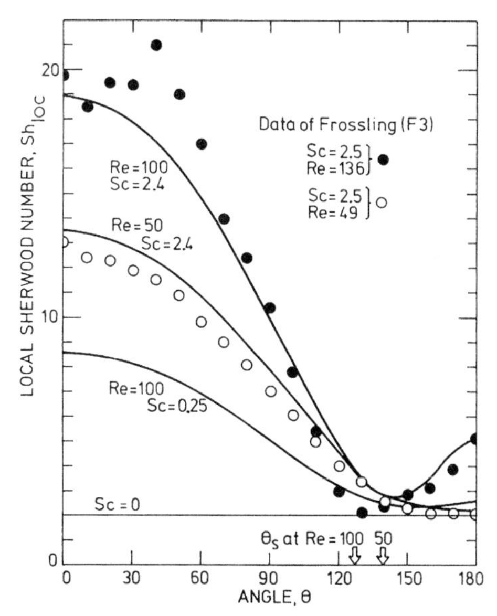

FIG. 5.19 Local Sherwood number for a sphere: Solid lines are the numerical results of Woo (W9) and Hatim (H11) at the values of Re and Sc indicated. Points are the data of Frössling (F3) for sublimation of naphthalene into air.

3. *Correlation of Average Sherwood Number*

Available numerical solutions for $1 \leq \mathrm{Re} \leq 400$ and $0.25 \leq \mathrm{Sc} \leq 100$ (A6, D5, H11, I1, M6, W9) can be correlated within 3% by the expression

$$(\mathrm{Sh} - 1)/\mathrm{Sc}^{1/3} = [1 + (1/\mathrm{ReSc})]^{1/3}\,\mathrm{Re}^{0.41} \tag{5-25}$$

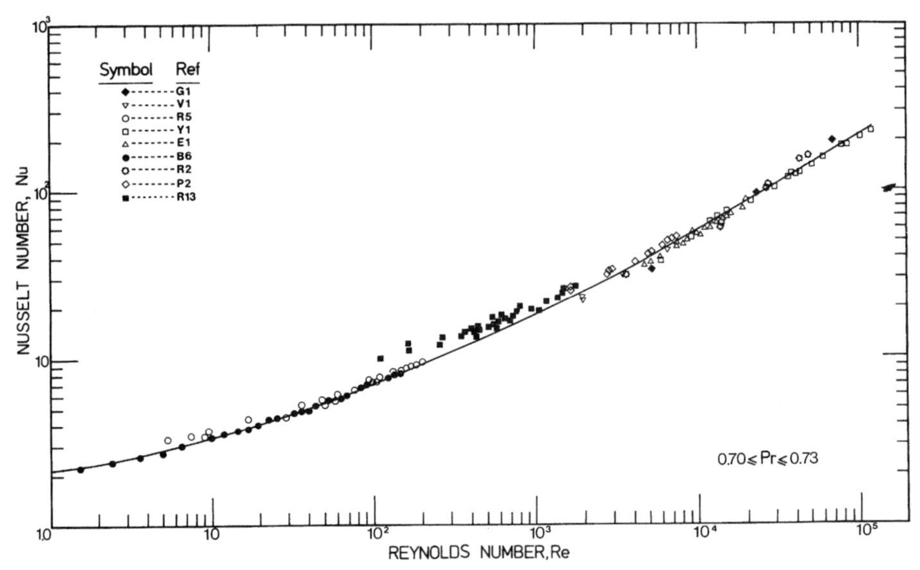

FIG. 5.20 Nusselt number for heat transfer from a sphere to air (0.70 < Pr < 0.73). Lines calculated from Eqs. A and B of Table 5.4 and Eq. (5-25).

The form of Eq. (5-25) was suggested by noting that the first order curvature corrections to Eqs. (3-47) and (5-35) are near unity and by matching the expression to the creeping flow result, Eq. (3-49), at $\text{Re} = 1$. Equation (5-25) also represents the results of the application of the thin concentration boundary layer approach ($\text{Sc} \to \infty$) through Eq. (3-46), using numerically calculated surface vorticities.† Thus the Schmidt number dependence is reliable for any $\text{Sc} > 0.25$.

Experimental data on heat transfer from spheres to an air stream are shown in Fig. 5.20. Despite the large number of studies over the years, the amount of reliable data is limited. The data plotted correspond to a turbulence intensity less than 3%, negligible effect of natural convection (i.e., $\text{Gr}/\text{Re}^2 < 0.1$; see Chapter 10), rear support or freefloating, wind tunnel area blockage less than 10%, and either a guard heater on the support or a correction for conduction down the support. Only recently has the effect of support position and guard heating been appreciated: a side support causes about a 10% increase in Nu

† Up to $\text{Re} = 20$ there is no difficulty in using the thin concentration boundary layer method with the calculated surface vorticities. For larger Re two methods of calculating transfer to the wake were pursued: first, neglect transfer aft of separation; second, consider transfer aft of separation as if it were a forward stagnation point, i.e., apply the theory starting from $\theta = 180°$ and work forward to separation. The true value of Sh should lie between these two limits and probably closer to the first as the discussion in Section B.2 suggests. The limits on Sh for the entire sphere were within 3% at $\text{Re} = 100$ and 10% at $\text{Re} = 400$. Equation (5-25) is within 3% of the values calculated neglecting transfer aft of separation.

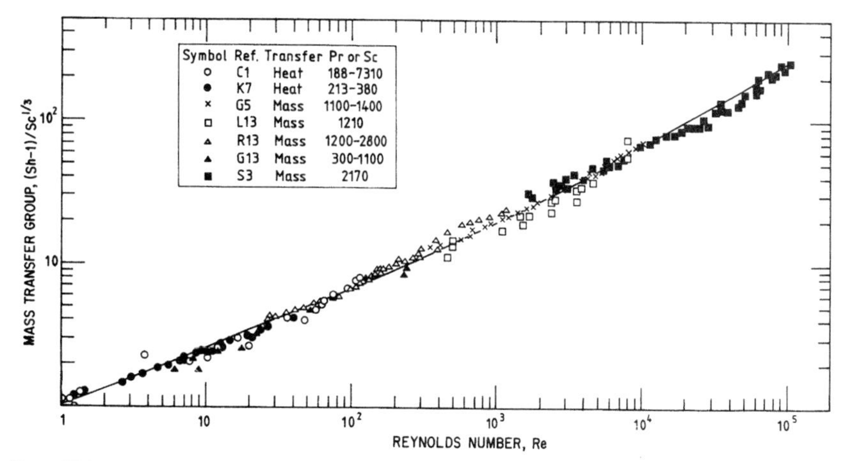

FIG. 5.21 Heat or mass transfer group for a sphere at high Pr or Sc. Lines calculated from Eqs. C and D of Table 5.4 and Eq. (5-25).

as does the lack of a guard heater (P2, R2). More data are needed, especially for $200 \lessdot \mathrm{Re} \lessdot 2000$ where the available data match poorly with the remaining results. Comparable experimental data for heat and mass transfer at high Pr and Sc are shown in Fig. 5.21 with the mass transfer group used in Eq. (5-25) as the ordinate. The least reliable data here are those for $\mathrm{Re} > 10^4$. Equations correlating the air and high Sc data are given in Table 5.4. All data for $1 < \mathrm{Re} < 100$ are well correlated by Eq. (5-25). Separate equations are given for the data in Figs. 5.20 and 5.21. All the data are also correlated by Eqs. (E) and (F) of

TABLE 5.4

Correlations for Transfer from Stationary Spheres

Heat transfer to air (Pr = 0.7)—Fig. 5.20		
(A)	$100 < \mathrm{Re} \leq 4000$	$\mathrm{Nu} = 1 + 0.677\,\mathrm{Re}^{0.47}$
(B)	$4 \times 10^3 < \mathrm{Re} \leq 1 \times 10^5$	$\mathrm{Nu} = 1 + 0.272\,\mathrm{Re}^{0.58}$
Mass transfer at high Sc (Sc > 200)—Fig. 5.21		
(C)	$100 < \mathrm{Re} \leq 2000$	$\mathrm{Sh} = 1 + 0.724\,\mathrm{Re}^{0.48}\,\mathrm{Sc}^{1/3}$
(D)	$2 \times 10^3 < \mathrm{Re} \leq 1 \times 10^5$	$\mathrm{Sh} = 1 + 0.425\,\mathrm{Re}^{0.55}\,\mathrm{Sc}^{1/3}$
All data		
(E)	$100 < \mathrm{Re} \leq 2000$	$(\mathrm{Sh} - 1)/Q\,\mathrm{Sc}^{1/3} = 0.752\,\mathrm{Re}^{0.472}$
(F)	$2 \times 10^3 < \mathrm{Re} \leq 1 \times 10^5$	$(\mathrm{Sh} - 1)/Q\,\mathrm{Sc}^{1/3} = 0.44\,\mathrm{Re}^{1/2} + 0.034\,\mathrm{Re}^{0.71}$

where $Q = \left(1 + \dfrac{1}{\mathrm{Re\,Sc}}\right)^{1/3}$

Table 5.4, so that these equations are recommended for general use. The $\frac{1}{2}$-power term in Eq. (F) can be viewed as the contribution from the portion of the sphere with a laminar boundary layer forward of separation, while the 0.71-power term corresponds to the section aft of separation. Justification for the latter power is found from local Sh values as discussed in the next chapter.

4. *Spheres in Free Fall or Free Rise*

Figures 5.22 and 5.23 present the result of combining the equations in Table 5.4 with the correlations of Table 5.3 to predict heat transfer for spheres falling in air at 20°C and mass transfer for spheres in water at 20°C with $Sc = 10^3$. The decrease in terminal velocity due to secondary motion has not been taken into account because the transfer rate depends on the overall relative velocity between the sphere and the fluid, not the vertical velocity component alone.

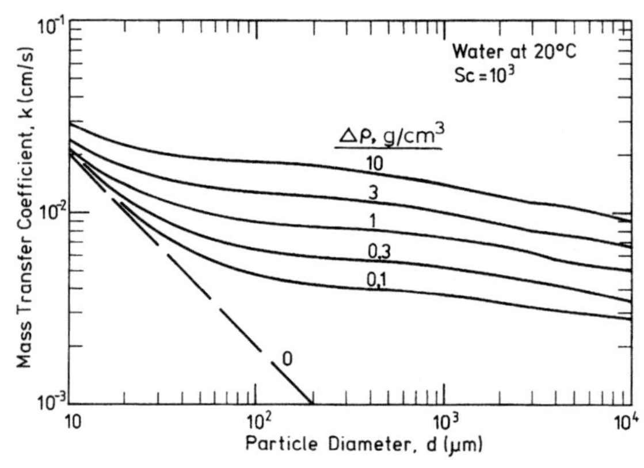

FIG. 5.22 Mass transfer coefficients for a sphere in free rise ($\rho_p < \rho$) or free fall ($\rho_p > \rho$) in water at 20°C with $Sc = 10^3$.

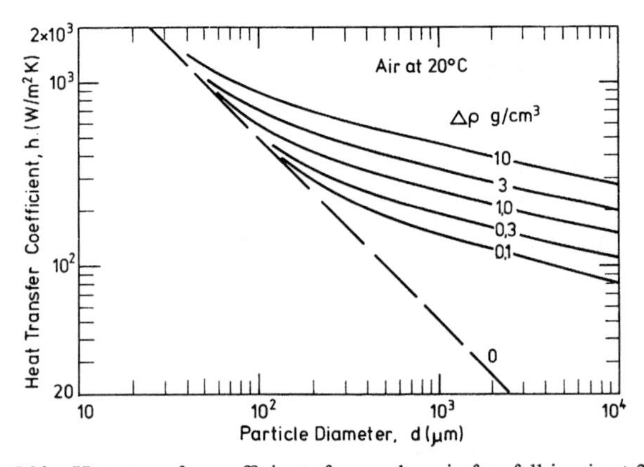

FIG. 5.23 Heat transfer coefficients for a sphere in free fall in air at 20°C.

The straight line for $\Delta\rho = 0$ represents diffusion in a stagnant medium [Eq. (3-44)]. In air spheres with diameters less than about 30 μm have transfer rates essentially equal to those in a stagnant medium, while in water the diameter for this to occur must be less than about 3 μm. In water the mass transfer coefficient is only weakly dependent on diameter, a prediction which has been verified experimentally (C2). For free fall in air, the transfer coefficient exhibits a larger decrease with diameter. The following expressions fit the predictions of Figs. 5.22 and 5.23 over the ranges indicated:

for free fall or rise in water with $d > 100\ \mu$m:

$$k(\mathrm{Sc})^{2/3} = 0.45(\Delta\rho/\rho)^{0.3} g^{0.3} \nu^{0.4} d^{-0.1}, \tag{5-26}^\dagger$$

for free fall in air with $d > 300\ \mu$m:

$$k(\mathrm{Sc})^{2/3} = 0.83(\Delta\rho/\rho)^{1/4} g^{1/4} \nu^{1/2} d^{-1/4}. \tag{5-27}^\dagger$$

III. FLUID SPHERES

A. Introduction and General Considerations

As noted in Chapter 2, bubbles and drops remain nearly spherical at moderate Reynolds numbers (e.g., at Re = 500) if surface tension forces are sufficiently strong. For drops and bubbles rising or falling freely in systems of practical importance, significant deformations from the spherical occur for all Re $\gtrsim$ 600 (see Fig. 2.5). Hence the range of Re covered in this section, roughly 1 < Re < 600, is more restricted than that considered in Section II for solid spheres. Steady motion of deformed drops and bubbles at all Re is treated in Chapters 7 and 8.

When a fluid sphere exhibits little internal circulation, either because of high $\kappa = \mu_p/\mu$ or because of surface contaminants, the external flow is indistinguishable from that around a solid sphere at the same Re. For example, for water drops in air, a plot of C_D versus Re follows closely the curve for rigid spheres up to a Reynolds number of 200, corresponding to a particle diameter of approximately 0.85 mm (B5). In fact, many of the experimental points used in Section II to determine the "standard drag curve" refer to spherical drops in gas streams, where high values of κ ensure negligible internal circulation.

Here we consider three theoretical approaches. As for rigid spheres, numerical solutions of the complete Navier–Stokes and transfer equations provide useful quantitative and qualitative information at intermediate Reynolds numbers (typically Re $\lesssim$ 300). More limited success has been achieved with approximate techniques based on Galerkin's method. Boundary layer solutions have also been devised for Re $\gtrsim$ 50. Numerical solutions give the most complete and

† These equations can be used for heat transfer by replacing Sc by Pr and k by $h/\rho C_\mathrm{t}$.

probably the most reliable results, but Galerkin's method has the advantage of giving analytic expressions. The boundary layer theories also lead to analytic forms for the drag coefficient and Sherwood (or Nusselt) number.

B. Fluid Mechanics

1. *Numerical Solutions*

Numerical solutions of the flow around and inside fluid spheres are again based on the finite difference forms of Eqs. (5-1) and (5-2) (B10, H6, L5, L9). The necessity of solving for both internal and external flows introduces complications not present for rigid spheres. The boundary conditions are those described in Chapter 3 for the Hadamard–Rybczynski solution; i.e., the internal and external tangential fluid velocities and shear stresses are matched at $R = 1$ ($r = a$), while Eq. (5-6) applies as $R \to \infty$. Most reported results refer to the limits in which κ is either very small (B10, H5, H7, L7) or large (L9). For intermediate κ, solution is more difficult because of the coupling between internal and external flows required by the surface boundary conditions, and only limited results have been published (A1, R7). Details of the numerical techniques themselves are available (L5, R7).

The major qualitative results of the numerical work are as follows:

a. *Wake Formation* Internal circulation delays the onset of flow separation and wake formation in the external fluid. This is not surprising, since a well-known (if rarely used) method of delaying boundary layer separation on solid bodies is to cause the surface to move in the same direction as the passing fluid (S1a). Table 5.5 shows the increase in wake angle, measured from the front stagnation point, by comparison with rigid spheres for the special case of spherical raindrops in air ($\kappa = 55$, $\gamma = 790$). A curious feature of such wakes is that the recirculating eddy may be completely detached from the sphere surface (L5); for example, this condition occurs for water drops in air in the

TABLE 5.5

Wake Characteristics of a Spherical Raindrop Compared with a Rigid Sphere[a]

	Raindrop			Rigid sphere	
Re	θ_{s1}	θ_{s2}	L_w/d	θ_s	L_w/d
30	180	164	0.15	153	0.15
57	180	147	0.53	138	0.53
100	170	136	0.85	127	0.94
300	157	124	1.90	111	2.17

[a] From LeClair *et al.* (L9).

range $20 \lessdot \mathrm{Re} \lessdot 100$. Thus separation of flow outside the sphere does not necessarily imply formation of a secondary internal vortex (S13). Two different angles are required to characterize the wake and both appear in Table 5.5. The position on the surface at which separation occurs is indicated as θ_{s1}, whereas θ_{s2} is measured to the furthest upstream extension of the recirculating eddy. The wake length, measured from the rear of the sphere, is slightly less for the water drop than for a corresponding rigid sphere. For a gas bubble in a liquid, and for a droplet ($\kappa = 1$, $\gamma = 0.5$) uncontaminated with surfactants, no separation is predicted even for Reynolds numbers as high as 200 (H1, H5, R9).

Figure 5.24 shows predicted surface vorticity distributions at $\mathrm{Re} = 100$ and for $\kappa = 0$ (gas bubble), $\kappa = 1$ (liquid drop in liquid of equal viscosity), and $\kappa = 55$ (water drop in air), and for a rigid sphere. The results for the raindrop are very close to those for a rigid sphere. The bubble shows much lower surface vorticity due to higher velocity at the interface, while the $\kappa = 1$ drop is intermediate. The absence of separation for the bubble and $\kappa = 1$ drop is indicated by the fact that vorticity does not change sign.

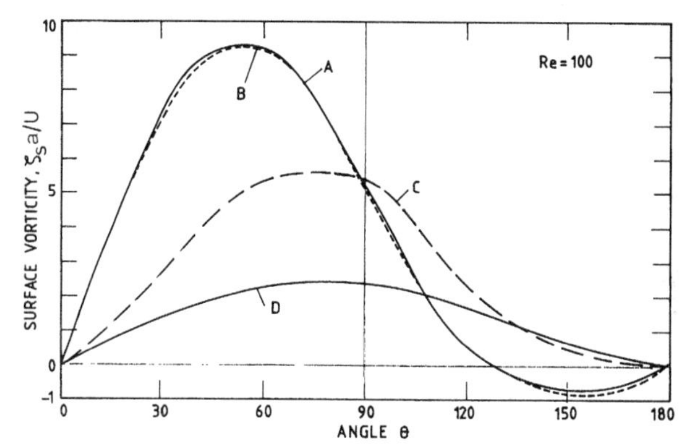

FIG. 5.24 Vorticity distribution at surface of sphere for $\mathrm{Re} = 100$ (numerical results): (A) Rigid sphere (L5); (B) Water drop in air; $\kappa = 55$, $\gamma = 790$ (L5, L9); (C) Liquid drop; $\kappa = 1$, $\gamma = 0.5$ (R9); (D) Gas bubble; $\kappa = \gamma = 0$ (H6).

b. *Internal Circulation* As discussed in Chapter 3, creeping flow around a fluid sphere is symmetrical about the equatorial plane. At higher Re, the stagnation ring in the internal fluid shifts forward of the equator.† Under some circumstances, e.g., $\mathrm{Re} > 300$ for water drops in air (L9), a small secondary internal vortex of opposite sense may occur inside the fluid sphere near the rear stagnation point. Experimental evidence for this secondary vortex is scant, but positive (P6).

† Experimenters who have observed asymmetry of internal circulation patterns have generally attributed this to accumulation of surface-active materials at the rear, causing a stagnant cap (see Chapter 3). It seems likely that at least part of the asymmetry results from the forward shift of the internal vortex at nonzero Re, as predicted numerically.

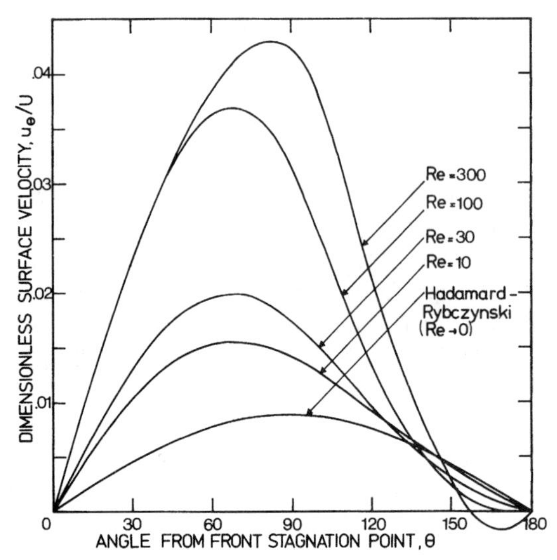

FIG. 5.25 Dimensionless fluid velocities for water drops in air ($\kappa = 55$, $\gamma = 790$). Numerical predictions of LeClair *et al.* (L9).

Figure 5.25 shows surface velocities for water drops in air with Re in the range 10 to 300, together with the Hadamard–Rybczinski solution for the same κ. Increasing asymmetry and a progressive increase in surface velocity with Re are evident. Experimental measurements (G4, H15, P6) generally give significantly lower velocities, presumably due to surface contamination. Internal and external streamlines and vorticity contours are shown in Figs. 5.26 and 5.27 for Re = 100 and $\kappa = 55$ (corresponding to a 0.6 mm diameter raindrop

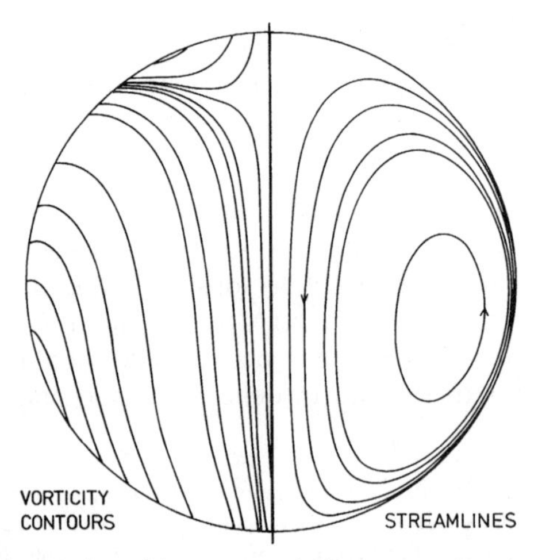

FIG. 5.26 Streamlines and vorticity contours inside a water drop in air at Re = 100 ($\kappa = 55$, $\gamma = 790$). Numerical predictions of LeClair (L5).

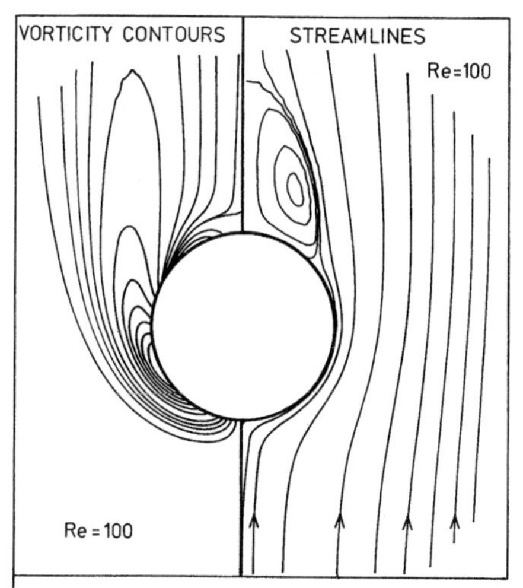

FIG. 5.27 Streamlines and vorticity contours outside a water drop in air at Re = 100 ($\kappa = 55$, $\gamma = 790$). Numerical predictions of LeClair (L5).

at its terminal velocity in air). Note that both internal and external flows show asymmetry and regions of negative vorticity near the rear of the sphere.

c. *Surface Pressure and Drag* Figure 5.28 shows numerical results for surface pressure distributions at Re = 100, together with those for the reference cases of potential flow and of a rigid sphere at the same Re. The curve for the

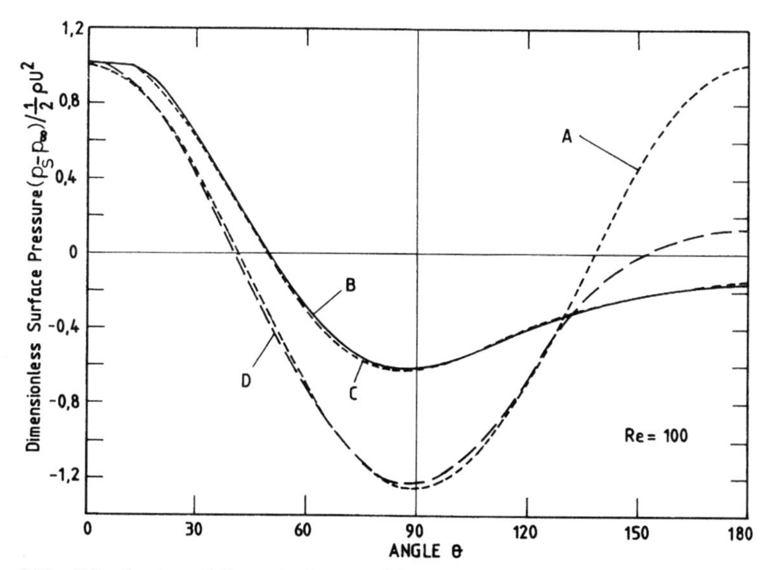

FIG. 5.28 Distribution of dimensionless modified pressure at surface of spheres at Re = 100, compared with potential flow distribution. (A) Potential flow: $(p_s - p_\infty)/\frac{1}{2}\rho U^2 = 1 - 2.25 \sin^2 \theta$ (B) Rigid sphere (L5); (C) Water drop in air; $\kappa = 55$, $\gamma = 790$ (L9); (D) Gas bubble; $\kappa = \gamma = 0$ (H6).

water drop ($\kappa = 55$) lies close to that for a rigid sphere. The pressure distribution for a bubble ($\kappa = 0$) follows the potential flow distribution very closely up to about 130° from the front stagnation point, much further than for a rigid sphere.

Values of C_{DP}, C_{DF}, and C_D are presented in Table 5.6 for bubbles in liquids (B10, H6, L7) and for water drops in air (L5, L9), with corresponding results for rigid spheres (L9). The viscous sphere ($\kappa = 55$) has essentially the same drag as a rigid sphere. The bubble ($\kappa = 0$) has much lower values of both form drag and skin friction. However, the ratio of form drag to skin friction is insensitive to κ. An equation which gives a good fit to numerical predictions of drag on spherical bubbles (H1) is:

$$C_D = 14.9\,\mathrm{Re}^{-0.78} \qquad (\kappa \to 0, \mathrm{Re} \gtrsim 2) \tag{5-28}$$

2. *Error Distribution Solutions*

Error distribution (or Galerkin) methods are based on choosing a polynomial for the stream function which is made to satisfy all the boundary conditions together with an integral form of the Navier–Stokes equation. Snyder *et al.* (S11) surveyed the application of this technique in fluid mechanics. Its success depends strongly on the form of polynomial chosen (H4). Kawaguti (K4, K5) applied this technique to flow around a rigid sphere, but the results are of limited interest since even the total drag predictions are inaccurate. Hamielec *et al.* (H3, H5, H7) applied Galerkin's method to fluid spheres up to Re = 500. Since inertia terms for the internal fluid were neglected, their solutions are restricted to small $\mathrm{Re_p}$. For $4 < \mathrm{Re} < 100$, the following correlation was suggested for the total drag:

$$C_D = \frac{3.05(783\kappa^2 + 2142\kappa + 1080)}{(60 + 29\kappa)(4 + 3\kappa)} \mathrm{Re}^{-0.74}. \tag{5-29}$$

Nakano and Tien (N2) investigated the effect of increasing $\mathrm{Re_p}$ by including inertia terms for both phases. Changes in $\mathrm{Re_p}$ had little effect on the external streamlines or on overall drag. On the other hand, internal circulation velocities increased significantly as $\mathrm{Re_p}$ increased, and the internal vortex was displaced forward. These results are in qualitative agreement with the numerical treatments and with experimental observations. However, there are substantial quantitative discrepancies, especially in the wake region and in local values of surface pressure (H4).

3. *Boundary Layer Theories*

Consider a circulating spherical bubble ($\kappa \ll 1$, $\gamma \ll 1$) for which $\mathrm{Re} \gg 1$, and compare this with a rigid sphere at $\mathrm{Re} \gg 1$. For the latter case, the boundary layer is perceived as a thin layer at the particle surface where viscous forces

TABLE 5.6

Drag Coefficients for Rigid and Fluid Spheres[a]

Re	C_{DP}				C_{DF}				C_D			
	$\kappa = \infty$	$\kappa = 55$	$\kappa = 0.3$	$\kappa = 0$	$\kappa = \infty$	$\kappa = 55$	$\kappa = 0.3$	$\kappa = 0$	$\kappa = \infty$	$\kappa = 55$	$\kappa = 0.3$	$\kappa = 0$
0.1	80.913			63.3	163.16			128.5	244.07			191.8
1.0	9.066		6.33	6.14	18.25		12.87	12.23	27.315		19.20	18.3
5.0	2.412	2.4[b]	1.63	1.69[b]	4.617	4.60[b]	3.18	3.0[b]	7.029	7.0	4.81	4.69
10.0	1.52	1.51	0.99	0.98	2.77	2.71	1.90	1.67	4.29	4.23	2.89	2.64
20	1.008	1.0[b]		0.54[b]	1.703	1.69[b]		0.86[b]	2.711	2.69		1.40[b]
30	0.81	0.81	0.41[b]	0.42[b]	1.30	1.29	0.79[b]	0.65[b]	2.11	2.10	1.20	1.07[b]
40	0.72[b]	0.71[b]		0.33[b]	1.08[b]	1.07[b]		0.50[b]	1.80[b]	1.78[b]		0.83[b]
50	0.65[b]	0.64[b]	0.28[b]	0.288	0.92[b]	0.92[b]	0.58[b]	0.435	1.57[c]	1.56	0.86[b]	0.723
57	0.63[b]	0.63		0.27[b]	0.88[b]	0.88		0.37[b]	1.51[b]	1.51		0.64[b]
100	0.51	0.49		0.181	0.59	0.59		0.224	1.096	1.08		0.405
200	0.40	0.39[b]		0.134	0.372	0.37[b]		0.132	0.772	0.76[b]		0.266
300	0.35	0.34		0.11[b]	0.28	0.29		0.094[b]	0.632	0.63		0.204[b]
400	0.320	0.31[b]		0.09[b]	0.233	0.23[b]		0.075[b]	0.552	0.54[b]		0.165[b]
500				0.068				0.057	0.555[c]			0.125
1000				0.062				0.031	0.471[c]			0.093

[a] From Abdel-Alim and Hamielec (A1), Brabston and Keller (B10), Hamielec *et al.* (H6), LeClair (L5), LeClair and Hamielec (L7), and LeClair *et al.* (L9).
[b] Interpolated or extrapolated.
[c] From standard drag relationships, Table 5.2.

play a dominant role and across which the velocity variation is of order U; outside this layer, the flow departs little from the irrotational pattern. For the bubble on the other hand, it is not necessary for the outer fluid to come to rest at the sphere surface. Flow deviates significantly less from irrotational motion. At first sight it might appear that potential flow could be a valid solution for the entire external flow field about a circulating bubble. However, the velocity derivatives in that case would not satisfy the tangential stress boundary condition. Thus a boundary layer must still exist on the surface, but it is of a rather different kind from that on a rigid body. In particular, the velocity variation across the boundary layer is only of order $U/\mathrm{Re}^{1/2}$. Moreover, the boundary layer is much thinner, and remains attached to the surface longer than on a comparable rigid body. These features are discussed at length by Levich (L11), Batchelor (B4), and Harper (H8). Harper has given a particularly thorough review of boundary layer solutions for circulating particles, and has pointed out a number of errors and misconceptions in the literature.

Since the flow is only slightly perturbed from irrotational, a first approximation for the drag on a spherical bubble may be obtained by calculating the viscous energy dissipation for potential flow past a sphere. This gives (L11):

$$C_D = 48/\mathrm{Re}. \tag{5-30}$$

Moore (M12) extended Eq. (5-30) by solving the boundary layer equations analytically, except in the vicinity of the rear of the bubble where the velocity and pressure fields were found to have singularities. The drag on the bubble was calculated using a momentum argument (L1) and by extending the energy dissipation calculation to include the contribution from the boundary layer and wake. Moore's improved drag estimate is:

$$C_D = \frac{48}{\mathrm{Re}}\left[1 - \frac{2.21}{\mathrm{Re}^{1/2}} + O(\mathrm{Re}^{-5/6})\right]. \tag{5-31}$$

Equations (5-30) and (5-31) are plotted in Fig. 5.29. In agreement with numerical predictions (B10, H1, H6), no boundary layer separation is predicted when there are no gradients of surface tension at the surface (H8).

Treatment of liquid drops is considerably more complex than bubbles, since the internal motion must be considered and internal boundary layers are difficult to handle. Early attempts to deal with boundary layers on liquid drops were made by Conkie and Savic (C8), McDonald (M9), and Chao (C4, W7). More useful results have been obtained by Harper and Moore (H10) and Parlange (P1). The unperturbed internal flow field is given by Hill's spherical vortex (H13) which, coupled with irrotational flow of the external fluid, leads to a first estimate of drag for a spherical droplet for $\mathrm{Re} \gg 1$ and $\mathrm{Re}_p \gg 1$. The internal flow field is then modified to account for convection of vorticity by the internal fluid to the front of the drop from the rear. The drag coefficient,

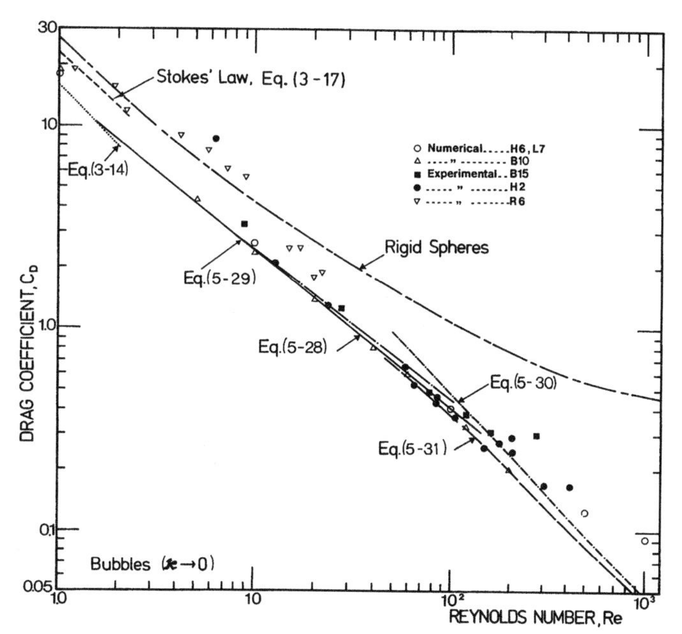

FIG. 5.29 Drag coefficients for bubbles in pure systems: predictions of numerical, Galerkin, and boundary layer theories compared with selected experimental results.

to terms of order $\mathrm{Re}^{-3/2}$, may then be written as

$$C_D = \frac{48}{\mathrm{Re}}\left[1 + \frac{3\kappa}{2} + \frac{(2 + 3\kappa)^2}{\mathrm{Re}^{1/2}}(B_1 + B_2 \ln \mathrm{Re})\right], \tag{5-32}$$

where B_1 and B_2 are functions of $\kappa\gamma$ with specific values presented in Table 5.7. In the limit as $\kappa \to 0$ and $\gamma \to 0$, Eq. (5-32) reduces to Eq. (5-31). The internal circulation relative to that for an unperturbed Hill's spherical vortex is approximately

$$\Gamma = 1 - \frac{5(2 + 3\kappa)}{2\mathrm{Re}^{1/2}}\left[\frac{1 + 2\sqrt{\kappa\gamma}}{2 + 3\sqrt{\kappa\gamma}}\right]. \tag{5-33}$$

TABLE 5.7

Values of B_1 and B_2 for Eq. (5-32)[a]

$(\kappa\gamma)$	25	4.0	1.0	0.25	0.04	0
B_1	−0.608	−0.652	−0.660	−0.642	−0.622	−0.553
B_2	0.00286	0.00877	0.0142	0.0160	0.0119	0

[a] Calculated from Table 3 of Harper and Moore (H10).

Equations (5-32) and (5-33) are only expected to be valid at relatively low κ (S14), typically $\kappa \dot{<} 2$, and for Re $\dot{>}$ 50 (H8). They should not be used when Γ predicted by Eq. (5-33) is less than 0.5, or when C_D from Eq. (5-32) exceeds the value from the standard drag curve for rigid spheres at the same Re. In these cases, the true drag will be close to the rigid sphere value, provided that the drop is nearly spherical.

4. *Comparison of the Theoretical Predictions with Experiment*

All the work discussed in the preceding sections is subject to the assumptions that the fluid particles remain perfectly spherical and that surfactants play a negligible role. Deformation from a spherical shape tends to increase the drag on a bubble or drop (see Chapter 7). Likewise, any retardation at the interface leads to an increase in drag as discussed in Chapter 3. Hence the theories presented above provide lower limits for the drag and upper limits for the internal circulation of fluid particles at intermediate and high Re, just as the Hadamard–Rybzcynski solution does at low Re.

In practice few systems approach the drag coefficient values predicted by the theoretical treatments. Since the theories provide lower limits on drag, it is reasonable to compare their predictions with the lowest available experimental values. From the restrictions noted, these will be systems of (i) low Morton number ($M \dot{<} 10^{-8}$) and (ii) low surface pressure (i.e., free of surfactants). Figure 5.29 compares selected C_D data on bubbles in very pure systems with theoretical predictions. The different theoretical approaches are in good agreement with each other and drag is predicted to be less than for rigid spheres. There is reasonable agreement with the experimental results. For drops, agreement with the boundary layer and Galerkin treatments is generally less favorable, although some of the results of Winnikow and Chao (W7) fall within 10% of the predictions of Eq. (5-32) (H8, H10). Excellent agreement has been obtained between numerical predictions and experimental results for raindrops in air (L9), where κ is sufficiently high that internal circulation does not influence C_D even in the absence of surface contaminants, and for water drops in cyclohexanol and in *n*-butyl lactate (A1).

Unfortunately there is little quantitative data, e.g., concerning internal and external velocity profiles, with which to test other aspects of the theories. On the other hand, the theories are supported by the agreement between the numerical and boundary layer approaches in their common ranges and by such qualitative features as secondary internal vortices (P6), forward displacement of the internal stagnation ring (H15, P6), delayed boundary layer separation with increasing system purity (E2, W7), and increasing dimensionless internal fluid velocities with increasing Re (G4, L9, P6).

5. *Effect of Surfactants*

Since the Schmidt number Sc tends to be much greater than unity for surfactants in solution, Re > 1 generally implies high Peclet numbers. This case has been considered by only a few investigators (D8, H9, L14, L16). The difference between the drag coefficients for rigid and fluid spheres becomes considerably wider as Re increases (see Fig. 5.29). Hence the influence of surfactants can be even more marked than at low Re. Unfortunately, accurate experimental data with known surfactant concentrations do not appear to be available. Thus theories cannot be tested except by fitting the contaminant concentration to match the data. Moreover, the conditions which must be satisfied for the theories to hold are so stringent that theories are of little practical importance (H9).

C. Heat and Mass Transfer

1. *External Resistance*

The external resistance has been evaluated under steady-state conditions using the assumption of a thin concentration boundary layer on the outer surface of a fluid sphere. Surface velocities calculated by each of the three methods described in Section B above have been used in conjunction with Eq. (3-51).

An asymptotic formula for Re $\to \infty$ is easily derived by substitution of the potential flow surface velocity,

$$(u_\theta/U_y)_{r=a} = \tfrac{3}{2}\sin\theta, \tag{5-34}$$

into Eq. (3-51) to yield

$$\mathrm{Sh} = (2/\sqrt{\pi})\,\mathrm{Pe}^{1/2}. \tag{5-35}$$

A first-order correction for finite Pe (W2) adds a constant term of 0.88 to the right-hand side of Eq. (5-35). This constant term is nearly the same for potential flow as for creeping flow [cf. Eq. (3-48)], and this fact has already been used in designing the mass transfer correlations for rigid spheres. Modifying the constant slightly to unity, we write

$$(\mathrm{Sh} - 1)/\mathrm{Pe}^{1/2} = 2/\sqrt{\pi} = 1.13 \tag{5-36}$$

as an approximate limiting condition for large Re. The expression on the left is now in a convenient form for bringing together numerical results for finite Re, both for Sc $\to \infty$ (L7) and for finite Sc (O1). The results are shown by the solid curves in Fig. 5.30.

The thin concentration boundary layer approximation, Eq. (3-51), has also been solved for bubbles ($\kappa = 0$) using surface velocities from the Galerkin method (B3) and from boundary layer theory (L15, W8). The Galerkin method agrees with the numerical calculations only over a small range of Re (L7). Boundary layer theory yields

$$\mathrm{Sh} = \frac{2}{\sqrt{\pi}}\left(1 - \frac{2.89}{\mathrm{Re}^{1/2}}\right)^{1/2} \mathrm{Pe}^{1/2}. \tag{5-37}$$

This result is within 7% of the numerical solution shown in Fig. 5.30 for Re > 70.

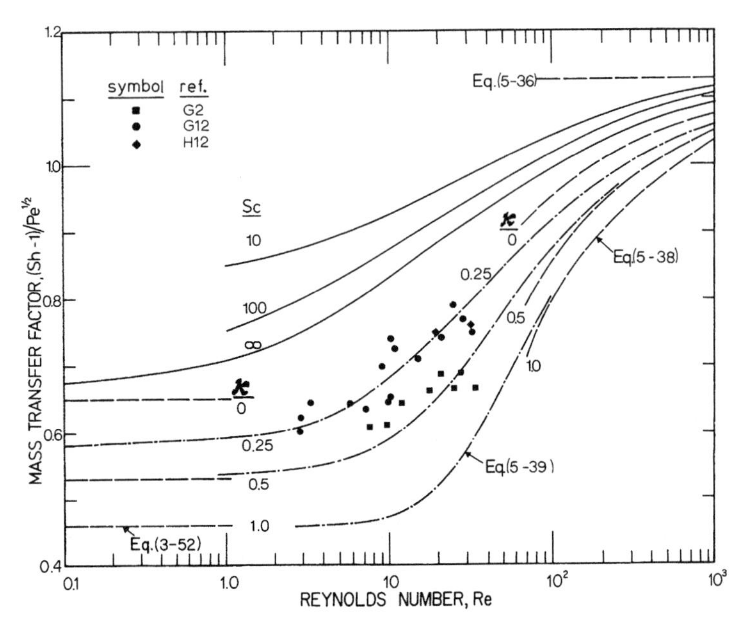

FIG. 5.30 Mass transfer factor as a function of Reynolds number for spherical fluid particles: —— numerical solutions for $\kappa = 0$ (L7, O1); --- asymptotic solutions (Sc, Pe $\rightarrow \infty$); ----- Eq. (5-39). Data for transfer of water to isobutanol ($\kappa = 0.39$, $\gamma = 1.2$, Sc $\doteq$ 12,000) from (G2, G12, H12).

For liquid drops Eq. (3-51) has been solved (W3) using the boundary layer velocities of Harper and Moore (H10). The resulting solution is valid for $\kappa \mathrel{\dot{<}} 2$. The Sherwood number was only weakly dependent upon γ with the results well approximated by

$$\mathrm{Sh} = \frac{2}{\sqrt{\pi}}\left[1 - \frac{1}{\mathrm{Re}^{1/2}}(2.89 + 2.15\kappa^{0.64})\right]^{1/2} \mathrm{Pe}^{1/2}. \tag{5-38}$$

Based on the result for bubbles, this should be accurate for Re > 70.

The surface velocities of Abdel-Alim and Hamielec (A1) can be used to obtain Sherwood numbers at intermediate κ and Re. An equation which fits these calculated values, the numerical results for Sc $\rightarrow \infty$, and the asymptotic solutions for $\kappa \mathrel{\dot{<}} 2$ is

$$\frac{\mathrm{Sh}}{\mathrm{Pe}^{1/2}} = \frac{2}{\sqrt{\pi}}\left[1 - \frac{\dfrac{2 + 3\kappa}{3(1 + \kappa)}}{\left\{1 + \left[\dfrac{(2 + 3\kappa)\,\mathrm{Re}^{1/2}}{(1 + \kappa)(8.67 + 6.45\kappa^{0.64})}\right]^{n}\right\}^{1/n}}\right]^{1/2} \tag{5-39}$$

where

$$n = \tfrac{4}{3} + 3\kappa. \tag{5-40}$$

The predictions of Eq. (5-39) are shown in Fig. 5.30 for $\kappa = 0.25$, 0.5, and 1.0.

Experimental data for mass transfer from freely circulating fluid spheres are difficult to obtain because of deformation and because of the presence of surface-active agents which reduce circulation. Shown in Fig. 5.30 are data from three studies on water droplets in isobutanol where the droplets were nearly spherical and were observed to be circulating. The data are in fair agreement with each other and with Eq. (5-39). The effects of shape changes and surface-active agents are discussed in Chapter 7.

The case of a fluid sphere moving at constant velocity and suddenly exposed to a step change in the composition of the continuous phase has been treated by solving Eq. (3-56), with Eqs. (3-40), (3-41), (3-42), and (3-57) as boundary conditions for potential flow (R14). The transient external resistance is given within 3% by

$$\mathrm{Sh}/\mathrm{Pe}^{1/2} = [1.829 + (2/\sqrt{\pi\tau\,\mathrm{Pe}})^5]^{1/5}. \tag{5-41}$$

2. *Transfer with Variable Particle Concentration*

The only situation with variable concentration inside the particle and finite internal resistance for which a theoretical treatment is available is for $\mathrm{Pe}_p \to \infty$. In this case the diffusion time is long compared to the time for circulation of the fluid within the sphere. Thus the concentration contours are identical to the streamlines of the Hill's spherical vortex except in thin boundary layers near the particle surface. As $\mathrm{Re} \to \infty$ the rate of diffusion normal to the streamlines in the bulk of the drop determines the rate of mass transfer (B12). Since the streamlines of Hill's spherical vortex are identical in form to the Hadamard–Rybczynski solution in creeping flow, the rate of extraction is identical to that shown in Fig. 3.22. This conclusion has been supported by experimental studies [e.g., (B13, K1)] which have shown that the Kronig–Brink solution gives a good prediction of mass transfer at Reynolds numbers well above those corresponding to creeping flow. For negligible resistance within the particle, a situation which occurs for gas bubbles, it has been shown (D1) that a quasi-steady treatment, i.e., substitution of Eq. (5-35) in Eq. (3-89), is valid.

REFERENCES

A1. Abdel-Alim, A. H., and Hamielec, A. E., *Ind. Eng. Chem., Fundam.* **14**, 308–312 (1975).
A2. Abraham, F. F., *Phys. Fluids* **13**, 2194–2195 (1970).
A3. Achenbach, E., *J. Fluid Mech.* **54**, 565–575 (1972).
A4. Achenbach, E., *J. Fluid Mech* **62**, 209–221 (1974).
A5. Allen, M. S., *Philos. Mag.* **50**, 323–338, 519–534 (1900).
A6. Al-Taha, T. R., Ph.D. Thesis, Imperial Coll., London, 1969.
A7. Arnold, H. D., *Philos. Mag.* **22**, 755–775 (1911).

B1. Bailey, A. B., and Hiatt, J., *AIAA J.* **10**, 1436–1440 (1972).
B2. Bailey, A. B., *J. Fluid Mech.* **65**, 401–410 (1974).
B3. Baird, M. H. I., and Hamielec, A. E., *Can. J. Chem. Eng.* **40**, 119–121 (1962).
B4. Batchelor, G. K., "An Introduction to Fluid Dynamics." Cambridge Univ. Press, London and New York, 1967.
B5. Beard, K. V., and Pruppacher, H. R., *J. Atmos. Sci.* **26**, 1066–1072 (1969).
B6. Beard, K. V., and Pruppacher, H. R., *J. Atmos. Sci.* **28**, 1455–1464 (1971).
B7. Beard, K. V., Ph.D. Thesis, Univ. of California, Los Angeles 1971.
B8. Bilham, E. G., and Relf, E. F., *Q. J. R. Meteorol. Soc.* **63**, 149–162 (1937).
B9. Boulos, M. I., and Pei, D. C. T., *Int. J. Heat Mass Transfer* **17**, 767–783 (1974).
B10. Brabston, D. C., and Keller, H. B., *J. Fluid Mech.* **69**, 179–189 (1975).
B11. Brauer, H., and Mewes, D., *Chem.-Ing.-Tech.* **44**, 865–868 (1972).
B12. Brignell, A. S., *Int. J. Heat Mass Transfer* **18**, 61–68 (1975).
B13. Brounshtein, B. I., Zheleznyak, A. S., and Fishbein, G. A., *Int. J. Heat Mass Transfer* **13**, 963–973 (1970).
B14. Brown, W. S., Pitts, C. C., and Leppert, G., *J. Heat Transfer* **84**, 133–140 (1962).
B15. Bryn, T., *David Taylor Model Basin Transl.* No. 132 (1949).
C1. Calderbank, P. H., and Korchinski, I. J. O., *Chem. Eng. Sci.* **6**, 65–78 (1956).
C2. Calderbank, P. H., and Moo-Young, M. B., *Chem. Eng. Sci.* **16**, 39–54 (1961).
C3. Calvert, J. R., *Aeronaut. J.* **76**, 248–250 (1972).
C4. Chao, B. T., *Phys. Fluids* **5**, 69–79 (1962).
C5. Christiansen, E. B., and Barker, D. H., *AIChE J.* **11**, 145–151 (1965).
C6. Clift, R., and Gauvin, W. H., *Proc. Chemeca '70*, **1**, 14–28 (1970).
C7. Cometta, C., Tech. Rep. WT-21. Div. Eng., Brown Univ., Providence, Rhode Island (1957).
C8. Conkie, W. R., and Savic, P., *Natl. Res. Counc. Can., Rep.* No. MT-23 (1953).
D1. Dang, V. D., Ruckenstein, E., and Gill, W. N., *Chem. Eng. (London)* **241**, 248–251 (1970).
D2. Davies, C. N., *Proc. Phys. Soc., London* **57**, 259–270 (1945).
D3. Dennis, S. C. R., and Walker, J. D. A., *J. Fluid Mech.* **48**, 771–789 (1971).
D4. Dennis, S. C. R., and Walker, J. D. A., *Phys. Fluids* **15**, 517–525 (1972).
D5. Dennis, S. C. R., Walker, J. D. A., and Hudson, J. D., *J. Fluid Mech.* **60**, 273–283 (1973).
D6. Dimopoulos, H. G., and Hanratty, T. J., *J. Fluid Mech.* **33**, 303–319 (1968).
D7. Dryden, H. L., Schubauer, G. B., Mock, W. G., and Skramstad, H. K., *NACA Rep.* No. 581 (1937).
D8. Dukhin, S. S., and Deryagin, B. V., *Russ. J. Phys. Chem.* **35**, 611–616 (1961).
E1. Eastop, T. D., Ph.D. Thesis, C.N.A.A., London, 1971.
E2. Elzinga, E. R., and Banchero, J. T., *AIChE J.* **7**, 394–399 (1961).
F1. Fage, A., *Aeronaut. Res. Counc. Rep. Memo.* No. 1766 (1937).
F2. Foch, A., and Chartier, C., *C. R. Acad. Sci.* **200**, 1178–1181 (1935).
F3. Frössling, N., *Gerlands Beitr. Geophys.* **52**, 170–216 (1938).
F4. Frössling, N., *Lunds Univ. Arsskr., Avd. 2* **36**, No. 4 (1940). [Engl. transl., *NACA Tech. Memo.* **NACA TM 1432** (1958).]
G1. Galloway, T. R., and Sage, B. H., *AIChE J.* **18**, 287–293 (1972).
G2. Garner, F. H., Foord, A., and Tayeban, M., *J. Appl. Chem.* **9**, 315–323 (1959).
G3. Garner, F. H., and Grafton, R. W., *Proc. R. Soc., Ser. A* **224**, 64–82 (1954).
G4. Garner, F. H., and Lane, J. J., *Trans. Inst. Chem. Eng.* **37**, 162–172 (1959).
G5. Gibert, H., Ph.D. Thesis, Univ. Paul Sabatier, Toulouse, 1972.
G6. Gibert, H., and Angelino, H., *Chem. Eng. Sci.* **28**, 855–867 (1973).
G7. Gilbert, M., Davies, L., and Altman, D., *Jet Propul.* **25**, 26–30 (1955).
G8. Gluckman, M. J., Weinbaum, S., and Pfeffer, R., *Annu. AIChE Meet., 64th, San Francisco* (1971).
G9. Goin, K. L., and Lawrence, W. R., *AIAA J.* **6**, 961–963 (1968).

G10. Goldburg, A., and Florsheim, B. H., *Phys. Fluids* **9**, 45–50 (1966).
G11. Govier, G. W., and Aziz, K., "The Flow of Complex Mixtures in Pipes." Van Nostrand-Reinhold, New York, 1972.
G12. Griffith, R. M., *Chem. Eng. Sci.* **12**, 198–213 (1960).
G13. Griffith, R. M., Ph.D. Thesis, Univ. of Wisconsin, Madison, 1958.
G14. Gunn, R., and Kinzer, G. D., *J. Meteorol.* **6**, 243–248 (1949).
H1. Haas, U., Schmidt-Traub, H., and Brauer, H., *Chem.-Ing.-Tech.* **44**, 1060–1068 (1972).
H2. Haberman, W. L., and Morton, R. K., *David Taylor Model Basin Rep.* No. 802 (1953).
H3. Hamielec, A. E., Ph.D. Thesis, Univ. of Toronto, 1961.
H4. Hamielec, A. E., Hoffman, T. W., and Ross, L. L., *AIChE J.* **13**, 212–219 (1967).
H5. Hamielec, A. E., and Johnson, A. I., *Can. J. Chem. Eng.* **40**, 41–45 (1962).
H6. Hamielec, A. E., Johnson, A. I., and Houghton, W. T., *AIChE J.* **13**, 220–224 (1967).
H7. Hamielec, A. E., Storey, S. H., and Whitehead, J. H., *Can. J. Chem. Eng.* **41**, 246–251 (1963).
H8. Harper, J. F., *Adv. Appl. Mech.* **12**, 59–129 (1972).
H9. Harper, J. F., *Q. J. Mech. Appl. Math.* **27**, 87–100 (1974).
H10. Harper, J. F., and Moore, D. W., *J. Fluid Mech.* **32**, 367–391 (1968).
H11. Hatim, B. M. H., Ph.D. Thesis, Imperial Coll., London, 1975.
H12. Heertjes, P. M., Holve, W. A., and Talsma, H., *Chem. Eng. Sci.* **3**, 122–142 (1954).
H13. Hill, M. J. M., *Philos. Trans. R. Soc. London, Ser. A* **185**, 213–245 (1894).
H14. Hoerner, S., *NACA Tech. Memo.*, **NACA TM 777** (1935).
H15. Horton, T. J., Fritsch, T. R., and Kintner, R. C., *Can. J. Chem. Eng.* **43**, 143–146 (1965).
I1. Ihme, F., Schmidt-Traub, H., and Brauer, H., *Chem.-Ing.-Tech.* **44**, 306–313 (1972).
J1. Jenson, V. G., *Proc. R. Soc. Ser. A* **249**, 346–366 (1959).
K1. Kadenskaya, N. I., Zheleznyak, A. S., and Brounshtein, B. I. *Zh. Prikl. Khim.* **38**, 1156–1159 (1965).
K2. Kalra, T. R., and Uhlherr, P. H. T., *Aust. Conf. Hydraul. Fluid Mech., 4th, Melbourne*, 1971.
K3. Kashko, A. V., *Fluid Dyn. (USSR)* **5**, 514–516 (1970).
K4. Kawaguti, M., and Jain, P., *J. Phys. Soc. Jpn.* **21**, 2055–2062 (1966).
K5. Kawaguti, M., *J. Phys. Soc. Jpn.* **10**, 694–699 (1955).
K6. Kendall, J. M., reported by Küchemann, D., *J. Fluid Mech.* **21**, 1–20 (1965).
K7. Kramers, H., *Physica (Utrecht)* **12**, 61–80 (1946).
K8. Kürten, H., Raasch, J., and Rumpf, H., *Chem.-Ing.-Tech.* **38**, 941–948 (1966).
L1. Landau, L. D., and Lifshitz, E. M., "Fluid Mechanics." Pergamon, Oxford, 1959.
L2. Langmuir, I., and Blodgett, K. B., *U.S. Army Air Force Tech. Rep.* No. 5418 (1948).
L3. Lapple, C. E., "Particle Dynamics." Eng. Res. Lab., E. I. DuPont de Nemours and Co., Wilmington, Delaware (1951).
L4. Lapple, C. E., and Shepherd, C. B., *Ind. Eng. Chem.* **32**, 605–617 (1940).
L5. LeClair, B. P., Ph.D. Thesis, McMaster Univ., Hamilton, Ontario, 1970.
L6. LeClair, B. P., and Hamielec, A. E., *Fluid Dyn. Symp., McMaster Univ., Hamilton, Ont.*, 1970.
L7. LeClair, B. P., and Hamielec, A. E., *Can. J. Chem. Eng.* **49**, 713–720 (1971).
L8. LeClair, B. P., Hamielec, A. E., and Pruppacher, H. R., *J. Atmos. Sci.* **27**, 308–315 (1970).
L9. LeClair, B. P., Hamielec, A. E., Pruppacher, H. R., and Hall, W. D., *J. Atmos. Sci.* **29**, 728–740 (1972).
L10. Lee, K., and Barrow, M., *Int. J. Heat Mass Transfer* **11**, 1013–1026 (1968).
L11. Levich, V. G., "Physicochemical Hydrodynamics." Prentice-Hall, New York, 1962.
L12. Liebster, H., *Ann. Phys., (Leipzig)* **82**, 541–562 (1927).
L13. Linton, M., and Sutherland, K. L., *Chem. Eng. Sci.* **12**, 214–229 (1960).
L14. Lochiel, A. C., *Can. J. Chem. Eng.* **43**, 40–44 (1965).
L15. Lochiel, A. C., and Calderbank, P. H., *Chem. Eng. Sci.* **19**, 471–484 (1964).

L16. Lyman, G. J., personal communication (1974).
M1. MacCready, P. B., and Jex, H. R., *NASA Tech. Memo.* **NASA TMX-53089** (1964).
M2. Masliyah, J. H., Ph.D. Thesis, Univ. of British Columbia, Vancouver, 1970.
M3. Masliyah, J. H., *Int. J. Heat Mass Transfer* **14**, 2164–2165 (1971).
M4. Masliyah, J. H., *Phys. Fluids* **15**, 1144–1146 (1972).
M5. Masliyah, J. H., and Epstein, N., *Phys. Fluids* **14**, 750–751 (1971).
M6. Masliyah, J. H., and Epstein, N., *Prog. Heat Mass Transfer* **6**, 613–632 (1972).
M7. Maxworthy, T., *J. Fluid Mech.* **23**, 369–372 (1965).
M8. Maxworthy, T., *J. Appl. Mech.* **91**, 598–607 (1969).
M9. McDonald, J. E., *J. Meteorol.* **11**, 478–494 (1954).
M10. Millikan, C. B., and Klein, A. L., *Aircr. Eng.* **5**, 169–174 (1933).
M11. Möller, W., *Phys. Z.* **39**, 57–80 (1938).
M12. Moore, D. W., *J. Fluid Mech.* **16**, 161–176 (1963).
M13. Mujumdar, A. S., and Douglas, W. J. M., *Int. J. Heat Mass Transfer* **13**, 1627–1629 (1970).
M14. Murrow, H. N., and Henry, R. M., *J. Appl. Meteorol.* **4**, 131–138 (1965).
N1. Nakamura, I., *Phys. Fluids* **19**, 5–8 (1976).
N2. Nakano, Y., and Tien, C., *Can. J. Chem. Eng.* **45**, 135–140 (1967).
N3. Newman, L. B., Sparrow, E. M., and Eckert, E. R. G., *J. Heat Transfer* **94**, 7–15 (1972).
N4. Newton, I., "Principia Mathematica," Book II, Proposition XXXVIII, 1687.
N5. Newton, I., "Principia Mathematica," Book II, Scholium to Proposition XL, 1687.
N6. Nguyen, T. H., Tech. Pap. Dep. Chem. Eng., McGill Univ., Montreal (1973).
N7. Nisi, H., and Porter, A. W., *Philos. Mag.* **46**, 754–768 (1923).
O1. Oellrich, L., Schmidt-Traub, H., and Brauer, H., *Chem. Eng. Sci.* **28**, 711–721 (1973).
P1. Parlange, J. Y., *Acta Mech.* **9**, 323–328 (1970).
P2. Pei, D. C. T., *Int. J. Heat Mass Transfer* **12**, 1707–1709 (1969).
P3. Perry, J. H., ed., "Chemical Engineers Handbook," 4th Ed. McGraw-Hill, New York, 1963.
P4. Pettyjohn, E. A., and Christiansen, E. B., *Chem. Eng. Prog.* **44**, 157–172 (1948).
P5. Preukschat, A. W., Aeronaut. Eng. Thesis, Calif. Inst. Technol., Pasadena, 1962.
P6. Pruppacher, H. R., and Beard, K. V., *Q. J. R. Meteorol. Soc.* **96**, 247–256 (1970).
P7. Pruppacher, H. R., LeClair, B. P., and Hamielec, A. E., *J. Fluid Mech.* **44**, 781–790 (1970).
P8. Pruppacher, H. R., and Steinberger, E. H., *J. Appl. Phys.* **39**, 4129–4132 (1968).
R1. Rafique, K., Ph.D. Thesis, Imperial Coll., London, 1971.
R2. Raithby, G. D., and Eckert, E. R. G., *Int. J. Heat Mass Transfer* **11**, 1233–1252 (1968).
R3. Raithby, G. D., and Eckert, E. R. G., *Wärme-Stoffübertrag.* **1**, 87–94 (1968).
R4. Ranger, A. A., and Nicholls, J. A., *AIAA J.* **7**, 285–290 (1969).
R5. Ranz, W. E., and Marshall, W. R., *Chem. Eng. Prog.* **48**, 141–146, 173–180 (1952).
R6. Redfield, J. A., and Houghton, G., *Chem. Eng. Sci.* **20**, 131–139 (1965).
R7. Rhodes, J. M., Ph.D. Thesis, Univ. of Tennessee, Knoxville, 1967.
R8. Rimon, Y., and Cheng, S. I., *Phys. Fluids* **12**, 949–959 (1969).
R9. Rivkind, V. Y., Ryskin, G. M., and Fishbein, G. A., *Fluid Mech.—Sov. Res.* **1**, 142–151 (1972).
R10. Roos, F. W., and Willmarth, W. W., *AIAA J.* **9**, 285–291 (1971).
R11. Roshko, A., *J. Fluid Mech.* **10**, 345–356 (1961).
R12. Roshko, A., *Phys. Fluids* **10**, 5181–5183 (1967).
R13. Rowe, P. N., Claxton, K. T., and Lewis, J. B., *Trans. Inst. Chem. Eng.* **43**, 14–31 (1965).
R14. Ruckenstein, E., *Int. J. Heat Mass Transfer* **10**, 1785–1792 (1967).
S1. Schiller, L., and Nauman, A. Z., *Ver. Deut. Ing* **77**, 318–320 (1933).
S1a. Schlichting, H., "Boundary Layer Theory," 6th ed. McGraw-Hill, New York, 1968.
S2. Schmiedel, J., *Phys. Z.* **29**, 593–610 (1928).
S3. Schuepp, P. H., and List, R., *J. Appl. Meteorol.* **8**, 254–263 (1969).
S4. Scoggins, J. R., *J. Geophys. Res.* **69**, 591–598 (1964).
S5. Scoggins, J. R., *NASA Tech. Note* **NASA TN D-3994** (1967).

S6. Seeley, L. E., Ph.D. Thesis, Univ. of Toronto, 1972.
S7. Seeley, L. E., Hummel, R. L., and Smith, J. W., *J. Fluid Mech.* **68**, 591–608 (1975).
S8. Shafrir, U., and Gal-Chen, T., *J. Atmos. Sci.* **28**, 741–751 (1971).
S9. Shakespear, G. A., *Philos. Mag.* **28**, 728–734 (1914).
S10. Sheth, R. B., M. S. Thesis, Brigham Young Univ., Provo, Utah, 1970.
S11. Snyder, L. J., Spriggs, T. W., and Stewart, W. E., *AIChE J.* **10**, 535–540 (1964).
S12. Son, J. S., and Hanratty, T. J., *J. Fluid Mech.* **35**, 353–368 (1969).
S13. Sumner, B. S., and Moore, F. K., *NASA Contract. Rep.* **NASA CR-1669** (1970).
S14. Sumner, B. S., and Moore, F. K., *NASA Contract. Rep.* **NASA CR-1362** (1969).
T1. Tanaka, Z., and Iinoya, K., *J. Chem. Eng. Jpn.* **3**, 261–262 (1970).
T2. Taneda, S., *J. Phys. Soc. Jpn.* **11**, 1104–1108 (1956).
T3. Tomotika, S., *Aeronaut. Res. Counc. Rep. Memo.* No. 1678 (1935).
V1. Venezian, E., Crespo, M. J., and Sage, B. H., *AIChE J.* **8**, 383–388 (1962).
V2. Viets, H., and Lee, D. A., *AIAA J.* **9**, 2038–2042 (1971).
V3. Vlajinac, M., and Covert, E. E., *J. Fluid Mech.* **54**, 385–392 (1972).
W1. Wadsworth, J., *Natl. Res. Counc. Can., Rep.* No. MT-39 (1958).
W2. Watts, R. G., *J. Heat Transfer* **94**, 1–6 (1972).
W3. Weber, M. E., *Ind. Eng. Chem., Fundam.* **14**, 365–366 (1975).
W4. Wieselsberger, C., *Phys. Z.* **23**, 219–224 (1922).
W5. Willis, J. T., Browning, K. A., and Atlas, D., *J. Atmos. Sci.* **21**, 103–108 (1964).
W6. Willmarth, W. W., and Enlow, R. L., *J. Fluid Mech.* **36**, 417–432 (1969).
W7. Winnikow, S., and Chao, B. T., *Phys. Fluids* **9**, 50–61 (1966).
W8. Winnikow, S., *Chem. Eng. Sci.* **22**, 477 (1967).
W9. Woo, S.-W., Ph.D. Thesis, McMaster Univ., Hamilton, Ontario, 1971.
X1. Xenakis, G., Amerman, A. E., and Michelson, R. W., Tech. Rep. 53–117. Wright Air Dev. Cent., Dayton (1953).
Y1. Yuge, T., *Rep. Inst. High Speed Mech., Tohoku Univ.* **11**, 209–230 (1959–1960).

Chapter 6

Nonspherical Rigid Particles at Higher Reynolds Numbers

I. INTRODUCTION

Nonspherical particles are more difficult to treat than spheres because of the influence of particle orientation and the lack of a single unambiguous dimension upon which to base dimensionless groups. In this chapter we treat rigid nonspherical particles at higher Reynolds numbers than were covered in Chapter 4. We begin by reviewing spheroids, disks, and finite cylinders,† shapes for which considerable work has been reported. General correlations for arbitrary shapes are discussed in Section IV. The fall of other specific shapes or specific types of particles is covered very briefly in Section V. There are no data nor numerical calculations for heat or mass transfer with variable particle concentration and finite resistance in each phase corresponding to the nonspherical particles considered. Hence, only the external resistance is treated in this chapter.

It is convenient to distinguish two regimes for freely falling nonspherical bodies. In the *intermediate regime*, particles adopt preferred orientations and C_D varies with Re although less strongly than at low Re. Particles usually align themselves with their maximum cross section normal to the direction of relative motion (K7, K10). In this regime there is no appreciable secondary motion so that results for flow past fixed objects of the same shape can be used if the orientation corresponds to a preferred orientation. In the *Newton's law regime*, on the other hand, C_D is insensitive to Re and secondary motion occurs,

† Two-dimensional flow past infinite cylinders is not treated in detail since such bodies do not meet our definition of a particle (see Chapter 1).

generally associated with wake shedding. In this regime the density ratio γ plays an important role in determining the type of motion, the mean terminal velocity and the transfer rate. Freely falling isometric particles generally begin to show pitching motion for Re (based on d_e) in the range $70 \lessdot \mathrm{Re} \lessdot 300$ (P4).

II. SPHEROIDS AND DISKS

Spheroids are of special interest, since they represent the shape of such naturally occurring particles as large hailstones (C2, L2, R4) and water-worn gravel or pebbles. The shape is also described in a relatively simple coordinate system. A number of workers have therefore examined rigid spheroids. Disks are obtained in the limit for oblate spheroids as $E \to 0$. The sphere is a special case where $E = 1$. Throughout the following discussion, Re is based on the equatorial diameter $d = 2a$ (Fig. 4.2).

A. Axisymmetric Motion

1. *Flow Patterns*

As shown in Chapters 3 and 4, creeping flow analyses have little value for $\mathrm{Re} \gtrdot 1$. A number of workers (M4, M7, M11, P5, R3) have obtained numerical solutions for intermediate Reynolds numbers with motion parallel to the axis of a spheroid. The most reliable results are those of Masliyah and Epstein (M4, M7) and Pitter *et al.* (P5). Flow visualization has been reported for disks (K2, W5) and oblate spheroids (M5).

At intermediate Re, phenomena are similar to those described for spheres in Chapter 5. Figure 6.1 shows streamlines calculated by Masliyah (M4) for steady flow past spheroids at $\mathrm{Re} = 100$. As the body becomes more "streamlined" (i.e., as E increases), the wake volume decreases. Figure 6.2 shows predicted and observed wake lengths. The Reynolds number at which separation first occurs decreases with aspect ratio to less than 2 for a disk. The calculations do not show clearly whether separation first occurs at the edge of a disk, but separation is certainly at the edge for $\mathrm{Re} \geq 10$ (R3). For spheroids with $E \geq 0.2$, separation is still aft of the equator for $\mathrm{Re} = 100$ (M4). Flow visualization generally confirms these predictions (M5), although numerical calculations tend to overpredict the wake length as for spheres due to difficulty in defining precisely the wake "tail." Disk wakes start to oscillate at $\mathrm{Re} \doteq 100$ (W5), while spheroids with $E \geq 0.2$ have steady wakes to higher Re (M5). At high Re, flow patterns continue to be qualitatively similar to those around a sphere (S5, W5), except that disks show nothing equivalent to the critical transition because the separation circle is fixed by the body shape. For spheroids of finite aspect ratio, the critical Reynolds number decreases slightly with increasing E (L4), and the drop in C_D at the critical transition becomes more marked (R1).

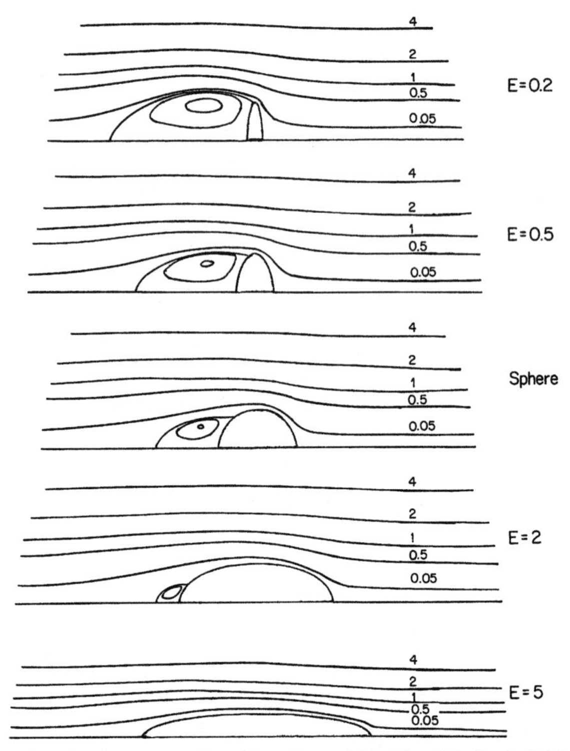

FIG. 6.1 Streamlines for flow past spheroids at Re = 100. After Masliyah (M4). Flow from right to left. Values of ψ/a^2U indicated.

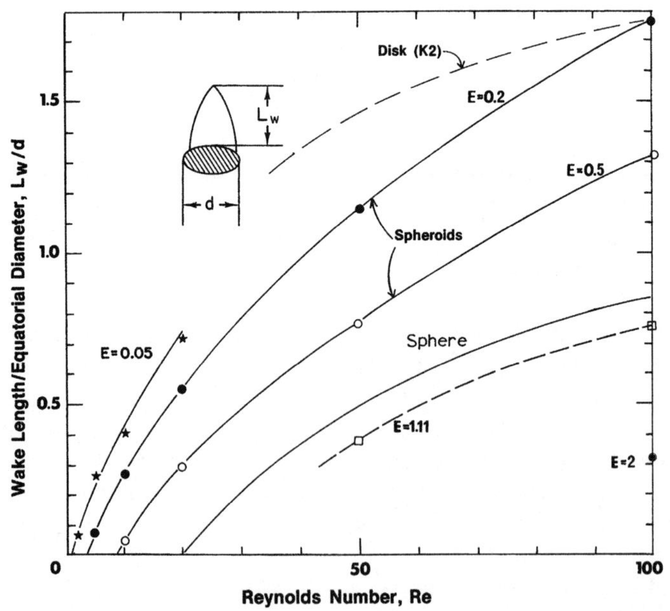

FIG. 6.2 Wake lengths for spheroids and disks. Numerical predictions for spheroids (M4, P5); flow visualization for disks (K2).

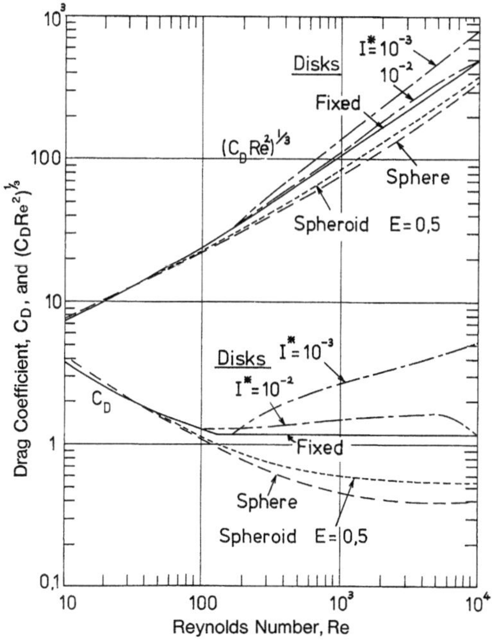

FIG. 6.3 Drag coefficient and values of $(C_D\mathrm{Re}^2)^{1/3}$ for spheroids and disks.

2. *Drag*

A number of authors have measured the drag on disks (J1, K1, L5, P5, R5, S2, S5, S8, W4, W5). For supported disks with steady motion parallel to the axis, numerical and experimental results at low and intermediate Re are well correlated (P5) by:

$$C_D = (64/\pi\,\mathrm{Re})[1 + 10^x] \qquad (0.01 < \mathrm{Re} \le 1.5), \tag{6-1}$$

where

$$x = -0.883 + 0.906\log_{10}\mathrm{Re} - 0.025(\log_{10}\mathrm{Re})^2 \tag{6-2}$$

and[†]

$$C_D = (64/\pi\,\mathrm{Re})(1 + 0.138\,\mathrm{Re}^{0.792}) \qquad (1.5 < \mathrm{Re} \le 133). \tag{6-3}$$

At lower Re, the Oseen result can be used (see Chapter 4):

$$C_D = (64/\pi\,\mathrm{Re})[1 + (\mathrm{Re}/2\pi)] \qquad (\mathrm{Re} \le 0.01). \tag{6-4}$$

Once wake shedding occurs, C_D is insensitive to Re, and is constant at 1.17 for Re $\gtrsim$ 1000 (H5). There is some indication that C_D passes through a minimum of about 1.03 for Re $\simeq$ 400 (L5, W5), but most data are correlated within 10% by Eq. (6-3), with $C_D = 1.17$ for Re > 133. Figure 6.3 compares the drag curve

[†] Pitter *et al.* (P5) applied Eq. (6-3) only to Re = 100. However, their correlation applies to freely falling disks for Re > 100.

for disks with the corresponding curve for spheres. Some authors have shown C_D for disks passing through a maximum at Re ≈ 300, but this is almost certainly a misinterpretation (R5).

Data are scant for spheroids other than disks and spheres. Experimental results for axisymmetric flow outside the Stokes range appear to be limited to

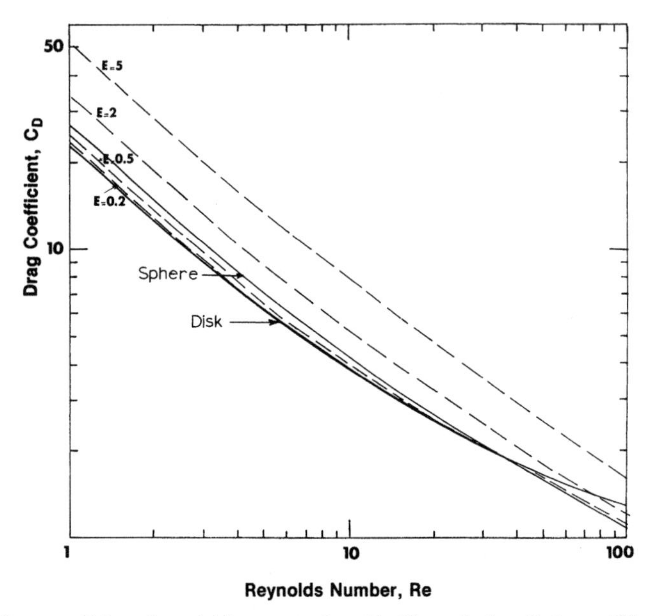

FIG. 6.4 Drag coefficients for axial flow past spheroids. Numerical predictions of Masliyah (M4).

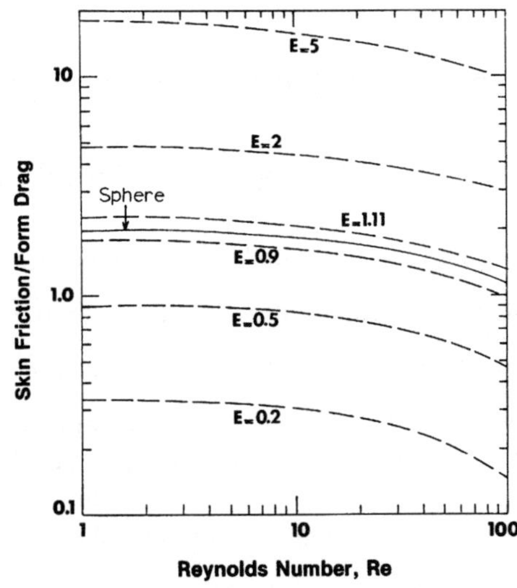

FIG. 6.5 Ratio of skin friction to form drag for spheroids in axial flow. Numerical predictions of Masliyah (M4).

oblate spheriods for which this is the preferred orientation (L3, L4, P5, S8). For Re up to 100, numerical results of Masliyah and Epstein (M4, M7) agree closely with experimental data (see Fig. 4.6). Figure 6.4 shows the dependence of C_D on Re predicted by Masliyah, and Fig. 6.5 shows the ratio of skin friction to form drag. For Re < 10, this ratio is only weakly dependent on Re, and drag on a spheroid can be estimated closely by multiplying the sphere drag by the drag ratio for Stokes flow, Δ_a (see Chapter 4). At higher Re, the dependence of C_D on E is more complex; for Re > 37, spheroids with aspect ratio close to unity have less drag than either very oblate or very prolate shapes. In this range, the drag on a spheroid with $E = 0.2$ is almost indistinguishable from that on a disk.

For $100 < \mathrm{Re} < 10^3$, the only data on spheroids appear to be those of Stringham *et al.* (S8) for $E = 0.5$.† Figure 6.3 shows an equation fitted to these results:

$$\log_{10} C_D = 2.0351 - 1.660w + 0.3958w^2 - 0.0306w^3 \quad (E = 0.5;\ 40 < \mathrm{Re} < 10^4), \tag{6-5}$$

where $w = \log_{10} \mathrm{Re}$.

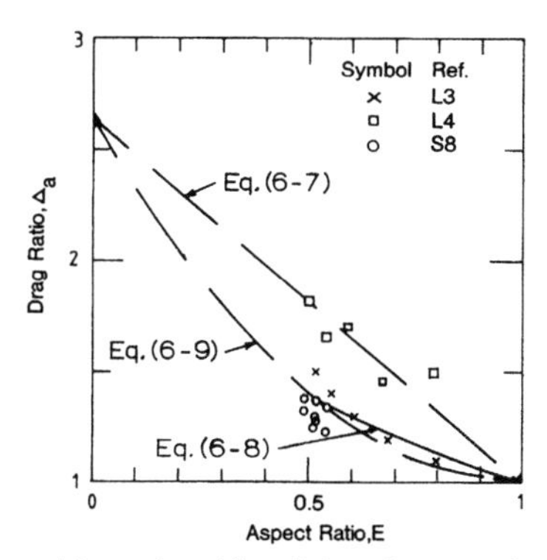

FIG. 6.6 Ratio Δ_a of drag on oblate spheroid or disk to drag on sphere of same equatorial radius.

For $\mathrm{Re} \gtrsim 10^3$, C_D is essentially constant at its "Newton's law" value. Available data are shown in Fig. 6.6. In view of the scatter in the data, it is reasonable to use the "Newton's law" value $C_D = 0.445$ for spheres (see Chapter 5), i.e.,

$$C_D = 0.445\Delta_a. \tag{6-6}$$

Hence $\Delta_a = 2.63$ for a disk. The results of List and Dussault (L3) are interpolated from wind-tunnel measurements on approximately spheroidal hailstone models (L2) while those of List *et al.* (L4) are for true spheroids in a wind

† Wall effects are significant for these results, and have been corrected using the correlations in Chapter 9.

tunnel. If all the data are reliable, the difference between these sets of results can only be ascribed to an effect of Re. For near-critical conditions, the results of List *et al.* (L4) suggest that

$$\Delta_a = 1 + 1.63(1 - E) \qquad (\mathrm{Re} > 4 \times 10^4). \tag{6-7}$$

For lower Re, Charlton and List (C2) suggest

$$\Delta_a = \frac{E}{(0.8E + 0.2)^2} \qquad (0.5 \leq E \leq 1). \tag{6-8}$$

The data are approximated equally well (see Fig. 6.6) by

$$\Delta_a = 1 + 1.63(1 - E)^2 \qquad (980 < \mathrm{Re} < 10^4), \tag{6-9}$$

which has the advantage of giving the correct limit at $E = 0$, but its reliability for $0 < E < 0.5$ is untested.

B. Free Fall

1. *Disks*

As for spheres, it is convenient to express drag in terms of $C_D \mathrm{Re}_T^2$, which contains a dimension or dimensions of the particle, but not the velocity. For a disk of thickness δ at its terminal velocity,

$$C_D \mathrm{Re}_T^2 = 2\rho\, \Delta\rho\, g\delta d^2/\mu^2 = 2\Delta\rho g E d^3/\rho\nu^2, \tag{6-10}$$

where $E = \delta/d$. For $0.1 \lessdot \mathrm{Re}_T \lessdot 100$ [i.e., $1.3 \lessdot (C_D \mathrm{Re}_T^2)^{1/3} \lessdot 23.4$] a disk in free motion moves steadily with its axis vertical (M9) and the drag is identical to that on a fixed disk at the same relative velocity. The terminal Re can then be calculated from the relationship between C_D and Re given by Eqs. (6-1) to (6-3). Figure 6.3 gives the resulting relationship between $(C_D \mathrm{Re}_T^2)^{1/3}$ and Re_T together with the curve for spheres (see Chapter 5). For $(C_D \mathrm{Re}_T^2)^{1/3} \gtrdot 2$, i.e., $\mathrm{Re}_T \gtrdot 0.5$, C_D and the terminal velocity for given $C_D \mathrm{Re}_T^2$ are independent of E (J1).

Willmarth *et al.* (W5) showed that secondary motion of a freely falling disk depends on a dimensionless moment of inertia,

$$I^* = \pi\gamma\delta/64d = \pi\gamma E/64. \tag{6-11}$$

The upper bound of the region of stable steady motion is shown in Fig. 6.7 as a function of $(C_D \mathrm{Re}_T^2)^{1/3}$ and I^*. For large I^*, secondary motion starts at $\mathrm{Re}_T \doteq 100$, i.e., $(C_D \mathrm{Re}_T^2)^{1/3} = 23.4$. At lower I^*, steady motion persists to higher Re_T; the boundary shows a maximum at $\mathrm{Re}_T = 172$, $(C_D \mathrm{Re}_T^2)^{1/3} = 32.6$ for $I^* = 8 \times 10^{-4}$. Three kinds of secondary motion have been observed (S8), although the distinctions between them are not sharp. Immediately above the transition to unsteady motion, a disk shows regular *oscillations* about a diameter: the amplitude of oscillation and of the associated horizontal motion increases with

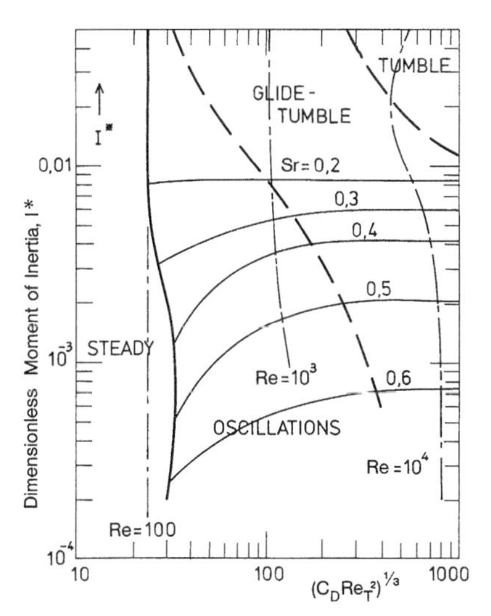

FIG. 6.7 Regimes of motion for disks in free fall or rise. Contours of constant Strouhal number Sr and constant Reynolds number are also shown.

$(C_D\,Re_T{}^2)^{1/3}$ and decreases with I^*. At higher $(C_D\,Re_T{}^2)^{1/3}$, the amplitude of the oscillation increases so much that the disk "flies" in a succession of curved arcs: at the end of each arc, the axis of the disk is inclined at a large angle to the vertical, and a vortex is shed from the wake. Stringham *et al.* (S8) termed this regime *glide-tumble*. At higher I^* and $(C_D\,Re_T{}^2)^{1/3}$, a disk shows a *tumbling* motion, rotating continually about a diameter and following a trajectory which is approximately rectilinear, but not vertical. Figure 6.7 shows approximate boundaries between these regimes. The Strouhal number of oscillation, $Sr = fd/U_T$, decreases with I^*, and increases with $(C_D\,Re_T{}^2)^{1/3}$ close to the boundary of unsteady motion. Figure 6.7 shows Sr contours, from Willmarth *et al.* (W5). For $I^* > 0.01$, the data of Stringham *et al.* (S8) show $Sr < 0.3$ but are too scattered for contours to be drawn. Once free fall motion becomes unsteady, the mean drag can differ significantly from that on a fixed disk with steady relative velocity. Generally, a disk with low I^* experiences higher drag and correspondingly lower mean vertical velocity. However, the data of Willmarth *et al.* (W5) and Stringham *et al.* (S8) indicate that drag is significantly lower near transition from glide-tumble to tumbling. Figure 6.3 shows curves for two values of I^*, and contours of terminal Re are indicated in Figure 6.7. The data on which these curves are based show considerable scatter. Apparently the velocity of a given particle may even vary between experiments (S8). Hence the curves must be interpreted as approximate.[†]

[†] Jayaweera and Cottis (J1) and Pitter *et al.* (P5) have given C_D and $(C_D\,Re_T{}^2)^{1/3}$ as functions of Re_T for $100 < Re_T < 600$. However, their curve neglects the effect of I^*, which varied over a wide range in the original experiments, and its general validity is therefore uncertain.

2. *Oblate Spheroids*

For an oblate spheroid moving at its terminal velocity,

$$C_D \mathrm{Re_T}^2 = 4\rho \Delta\rho g E d^3/3\mu^2. \tag{6-12}$$

For spheroids with $E \doteq 0.5$, Stringham *et al.* (S8) showed that steady motion with the axis vertical persists over a much wider range of $\mathrm{Re_T}$ than for thin disks. Secondary motion started at $\mathrm{Re_T} \doteq 4 \times 10^4$, $(C_D \mathrm{Re_T}^2)^{1/3} = 10^3$. On increasing γ, steady motion persisted to higher $\mathrm{Re_T}$ but the data are too scant to show whether the transition can be correlated by a dimensionless moment of inertia. The limit of steady motion must decrease on reducing E, but quantitative data are lacking. Two types of secondary motion have been observed for oblate spheroids (K6, K9, S8): *oscillation* with the minor axis rotating to trace out a cone, and continuous rotation or *tumbling* about a horizontal axis. List *et al.* (K8, K9, L4) explained this behavior qualitatively, based on measurements for spheroids at steady inclination, but it is not possible to predict which kind of motion will occur.

Figure 6.3 shows the relationship between $(C_D \mathrm{Re_T}^2)^{1/3}$ and $\mathrm{Re_T}$ for $E = 0.5$, fitted to the data of Stringham *et al.* (S8):

$$\begin{gathered}\log_{10} \mathrm{Re_T} = -1.7239 + 3.8068W - 0.9477W^2 + 0.1277W^3 \\ (E = 0.5;\ 15 < (C_D \mathrm{Re_T}^2)^{1/3} < 400;\ 40 < \mathrm{Re_T} < 10^4),\end{gathered} \tag{6-13}$$

where $W = \log_{10}[(C_D \mathrm{Re_T}^2)^{1/3}]$. For other aspect ratios and $10^3 < \mathrm{Re_T} < 10^4$, C_D may be estimated from Eqs. (6-6) and (6-9), giving:

$$U_T \doteq 1.73\left\{\frac{\Delta\rho g E d}{\rho[1 + 1.63(1 - E)^2]}\right\}^{1/2}. \tag{6-14}$$

C. Heat and Mass Transfer

The mechanism of mass transfer to the external flow is essentially the same as for spheres in Chapter 5. Figure 6.8 shows numerically computed streamlines and concentration contours with $\mathrm{Sc} = 0.7$ for axisymmetric flow past an oblate spheroid ($E = 0.2$) and a prolate spheroid ($E = 5$) at $\mathrm{Re} = 100$. Local Sherwood numbers are shown for these conditions in Figs. 6.9 and 6.10. Figure 6.9 shows that the minimum transfer rate occurs aft of separation as for a sphere. Transfer rates are highest at the edge of the oblate ellipsoid and at the front stagnation point of the prolate ellipsoid.

A number of computations of average Sherwood number have been made (A3, M6) for $\mathrm{Re} < 100$, $0.2 \leq E \leq 5$, and $0.7 < \mathrm{Sc} < 2.4$. Some values are also available at $E = 0.05$ and $\mathrm{Sc} = 0.7$ for $\mathrm{Re} \leq 20$ (P5) and for higher Sc with

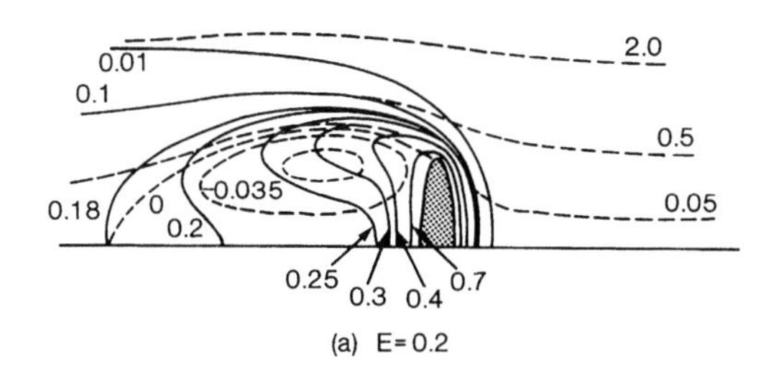

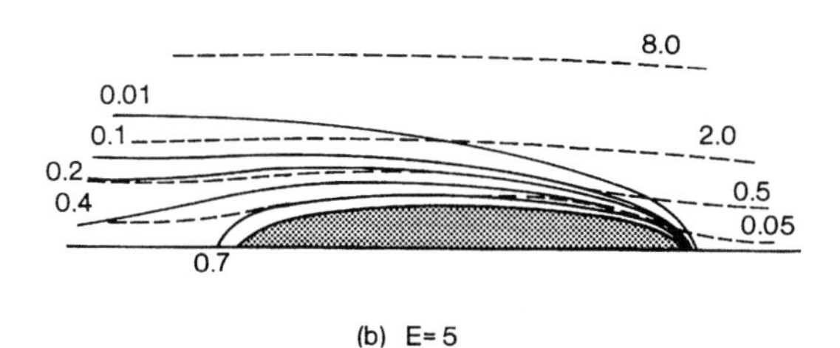

FIG. 6.8 Concentration contours for flow past spheroids at Re = 100 and Sc = 0.7. Flow from right to left. Dashed lines are streamlines as in Fig. 6.1 with values of ψ/a^2U indicated. Dimensionless concentration values are marked on the solid lines which trace lines of constant concentration (M6).

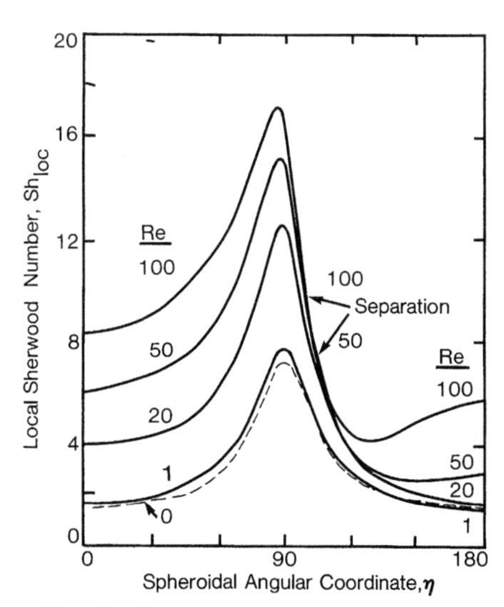

FIG. 6.9 Local Sherwood number for an oblate ($E = 0.2$) spheroid with Sc = 0.7. After Masliyah and Epstein (M6). Axial flow.

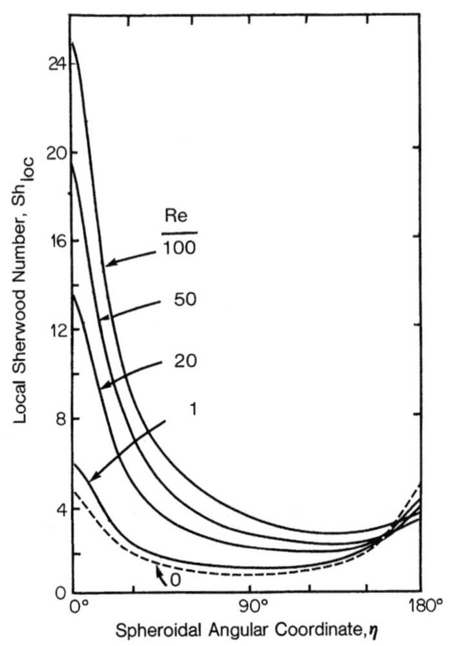

FIG. 6.10 Local Sherwood number for a prolate ($E = 5$) spheroid with Sc = 0.7. After Masliyah and Epstein (M6). Axial flow.

$0.4 \leq E \leq 1$ and Re < 10 (A3). The calculated values are correlated within 5% by

$$\frac{\text{Sh} - \text{Sh}_0/2}{\text{Sh}_{\text{sphere}} - 1} = \frac{1.25}{1 + 0.25E^{0.9}} \qquad (0.2 \leq E \leq 5, \quad 1 \leq \text{Re} \leq 100). \qquad (6\text{-}15)$$

where Sh and Sh_0 are based on the equatorial diameter, Sh_0 is given in Table 4.2, and $\text{Sh}_{\text{sphere}}$ is for a sphere at the same Re given by Eq. (5-25).

For axisymmetric flow at higher Re the most reliable data are those of Beg for the sublimation of oblate naphthalene spheroids (B4) ($0.25 < E < 1$) and disks (B3). His correlations are in terms of the characteristic length L' defined in Eq. (4-67). For spheroids

$$\text{Sh}' = 0.62(\text{Re}')^{0.50}\,\text{Sc}^{1/3} \qquad (200 < \text{Re}' \leq 2000), \qquad (6\text{-}16)$$

$$\text{Sh}' = 0.26(\text{Re}')^{0.61}\,\text{Sc}^{1/3} \qquad (2000 < \text{Re}' \leq 3.2 \times 10^4), \qquad (6\text{-}17)$$

while for disks

$$\text{Sh}' = 0.266(\text{Re}')^{0.60}\,\text{Sc}^{1/3} \qquad (270 < \text{Re}' \leq 3.5 \times 10^4). \qquad (6\text{-}18)$$

These equations are shown in Fig. 6.11. The correlations overlap with numerical calculations only for spheres ($E = 1$) and show a reasonable match with these

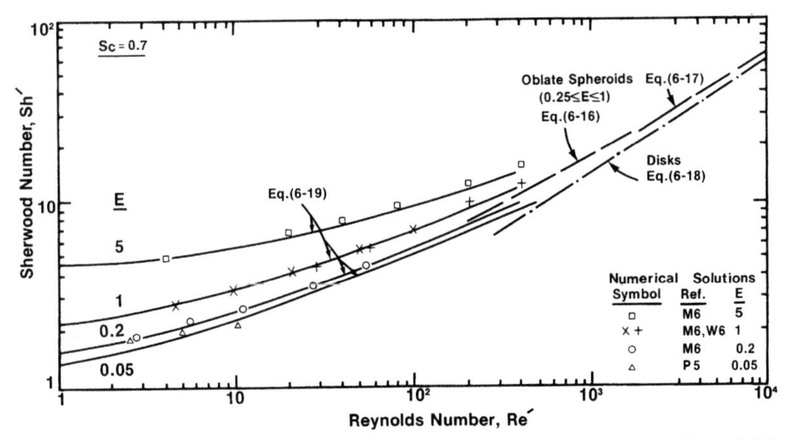

FIG. 6.11 Correlations and numerical calculations for heat transfer to spheroids and disks with $\mathrm{Pr} = 0.7$.

values. At lower Re′, calculated values of Sh′ are well correlated by an expression similar to that for spheres:

$$\frac{\mathrm{Sh}' - \mathrm{Sh_0}'/2}{\mathrm{Sc}^{1/3}} = \left[1 + \frac{(K')^3 - 1}{(\mathrm{Re}')^{1/8}} + \frac{(\mathrm{Sh_0}'/2)^3}{\mathrm{Re}'\,\mathrm{Sc}}\right]^{1/3} (\mathrm{Re}')^{0.41} \qquad (1 \leq \mathrm{Re}' < 400). \tag{6-19}$$

Equation (6-19) matches Eq. (4-70) at $\mathrm{Re}' = 1$ and is plotted in Fig. 6.11. Note the decreasing dependence of Sh′ on E as Re′ increases. There is reasonable matching between Eqs. (6-16) and (6-19) in their common range of application.

No data are available for heat and mass transfer to or from disks or spheroids in free fall. When there is no secondary motion the correlations given above should apply to oblate spheroids and disks. For larger Re where secondary motion occurs, the equations given below for particles of arbitrary shape in free fall are recommended.

III. CYLINDERS

A. MOTION IN FREE FALL OR RISE

1. *Steady Motion*

In the following discussion, cylinders are characterized by the length/diameter ratio E and Re is based on the cylinder diameter.† As noted in Section II, drag on a disk in steady free motion is relatively insensitive to its thickness; cylinders

† Other definitions are often used. For example, Stringham *et al.* (S8) based Re on the area-equivalent diameter d_A, while Christiansen and Barker (C3) used cylinder length.

with $E < 1$ can therefore be treated as disks. For $\mathrm{Re} \dot{>} 0.01$, a cylinder with $E > 1$ falls with its axis horizontal. Steady motion with this orientation persists up to Re of order 100 (J2, M3, S8). The upper bound of steady motion increases with decreasing γ, and increases sharply for $E \dot{>} 20$ (J2), but data are too scant to enable reliable prediction of the onset of secondary motion.

Steady flow normal to the axis of a long cylinder has been investigated even more thoroughly than flow past a sphere [e.g., see (A4, J2, K4, P8, T1)]. Qualitatively, the flow pattern shows features similar to those described for spheres in Chapter 5. Separation occurs for $\mathrm{Re} \dot{>} 5$ (D2, U1); wake oscillation is apparent for $\mathrm{Re} \dot{>} 30$, and wake shedding for $\mathrm{Re} \dot{>} 40$ (H6, R6, T1). Shedding from a cylinder gives a regular succession of vortices, termed the "von Karman vortex street," recognizable over the range $70 \dot{<} \mathrm{Re} \dot{<} 2.5 \times 10^3$. Above $\mathrm{Re} = 10^5$, the critical region is entered, with flow transitions similar to those described for a sphere in Chapter 5 (A1, R7, S6). For cylinders of finite length, flow past the ends sets up a three-dimensional circulation pattern, and the wake adopts a pyramidal shape (J2).

Figure 6.12 shows a curve fitted by Pruppacher *et al.* (P8) to the many determinations of C_D for steady crossflow past long cylinders in the Re range applicable to free motion. We have approximated this curve by the following expressions:

$$C_D = C_D'(1 + 0.147\,\mathrm{Re}^{0.82}) \qquad (0.1 < \mathrm{Re} \le 5), \tag{6-20}$$

$$C_D = C_D'(1 + 0.227\,\mathrm{Re}^{0.55}) \qquad (5 < \mathrm{Re} \le 40), \tag{6-21}$$

$$C_D = C_D'(1 + 0.0838\,\mathrm{Re}^{0.82}) \qquad (40 < \mathrm{Re} \le 400), \tag{6-22}$$

where

$$C_D' = 9.689\,\mathrm{Re}^{-0.78}. \tag{6-23}$$

The junctions between these expressions correspond to changes in flow pattern. For lower Re, see Chapter 4.

For a cylinder with $E > 1$ in free motion,

$$C_D\,\mathrm{Re}_T^{\,2} = \pi g \rho\, \Delta\rho\, d^3/2\mu^2, \tag{6-24}$$

where C_D is based on the area projected normal to the axis. In the range where motion is steady with the axis horizontal, Eqs. (6-20) to (6-22) can be used to obtain relationships between Re_T and $(C_D\,\mathrm{Re}_T^{\,2})^{1/3}$ for a long cylinder: the resulting curve is shown in Fig. 6.12. Jayaweera and Cottis (J1) have given similar curves for cylinders of finite length† based on data of Jayaweera and Mason (J2). Expressions fitted to these curves are given in Table 6.1. Corresponding

† Their curve for a long cylinder corresponds to drag coefficients 10–20% lower than those given by Pruppacher *et al.* (P8). The Pruppacher values are preferred, since they are based on a more extensive data compilation.

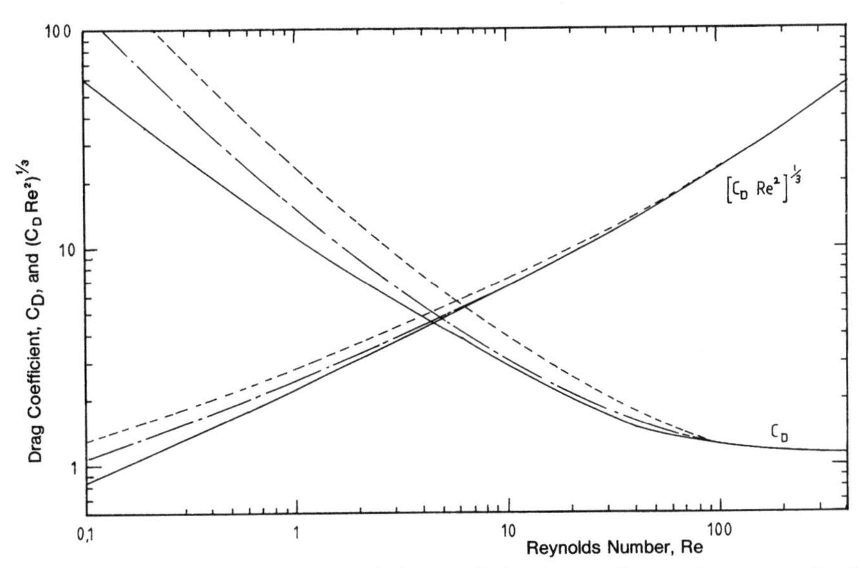

FIG. 6.12 Drag coefficient and $(C_D\mathrm{Re}^2)^{1/3}$ for cylinders in crossflow: —$E = \infty$; --- $E = 2$; --- $E = 1$.

TABLE 6.1

Correlations for Terminal Reynolds Number of Cylindrical Particles

$$C_D\,\mathrm{Re_T}^2 = \frac{\pi g \rho\,\Delta\rho\,d^3}{2\mu^2};\ w = \log_{10}(C_D\,\mathrm{Re_T}^2)^{1/3}$$

$$\log_{10}\mathrm{Re_T} = a_0 + a_1 w + a_2 w^2 + a_3 w^3$$

where

$$a_0 = -0.81824 - 0.55689/E$$

$$a_2 = 2.41277 + 1.54674/E - 0.53872/E^2$$

$$a_3 = -0.20560 - 1.34714/E + 0.65696/E^2$$

$$a_4 = 0.82343 - \frac{27}{64}a_0 - \frac{9a_1}{16} - \frac{3a_2}{4};\ \text{as } E \to \infty,\ a_4 = -0.03436$$

values for $E = 1$ and $E = 2$ appear in Fig. 6.12. For a long cylinder ($E \to \infty$) these expressions agree within 3.5% with Eqs. (6-20) to (6-22).

2. *Motion at Higher Reynolds Numbers*

As noted above, for $\mathrm{Re_T}$ greater than a value of order 100, a cylinder in free motion has a secondary oscillatory motion superimposed on its steady fall or rise. For cylinders with $E > 1$, the axis oscillates in a vertical plane about the mean (horizontal) orientation, and the trajectory oscillates about the mean path in the same plane as the cylinder "sideslips" when its axis is not horizontal

(C3, I2, J2, M3, S8). The amplitude of angular oscillations decreases as E increases, and a very long cylinder falls steadily to high Re_T(I2, J2). If $Re_T \gtrsim 3500$ (S8), motion also occurs in a horizontal plane. For relatively low γ, the cylinder oscillates about a vertical axis (I2, S8), while for dense particles in liquids or particles in gases the cylinder rotates continuously about a vertical axis (C3, I2). A cylinder with $E \doteq 1$ follows a trajectory inclined to the vertical, and "tumbles" in the direction of horizontal travel (I2). For $E < 1$, the axis oscillates and rotates about a vertical line, so that the secondary motion resembles the final stages of motion of a coin spinning on a flat surface (I2).

As for disks and spheroids, the terminal velocity in this regime depends upon γ as well as on particle shape. Table 6.2 summarizes correlations (I2) which may be used for γ typical of particles in liquids. The correlations do not extrapolate to the high γ-values typical of particles in gases, and comparison with available data (C3) shows that predicted terminal velocities are 25 to 35% too high.[†] We therefore propose that terminal velocities for cylinders in gases in the "Newton's law" range be estimated for $E > 1$ by multiplying the values given by Table 6.2 by 0.77. For $E < 1$, particles are best treated as disks.

TABLE 6.2

Drag Coefficients and Terminal Velocities for Cylinders with Secondary Motion[a]

	$E > 1$	$E < 1$
C_D	$0.99\gamma^{-0.12}E^{-0.08}$	$1.25\gamma^{-0.05}E^{-0.18}$
U_T	$1.26\gamma^{0.06}E^{0.04}\sqrt{dg\,\Delta\rho/\rho}$	$1.265\gamma^{0.025}E^{0.59}\sqrt{dg\,\Delta\rho/\rho}$

[a] After Isaacs and Thodos (I2). The area used in defining C_D is d^2E for $E > 1$ and $\pi d^2/4$ for $E < 1$.

Marchildon *et al.* (M3) related oscillation of a falling cylinder to movement of the front stagnation point, and obtained an expression for the frequency:

$$f = \frac{0.11U_T}{d}\left[\frac{3E}{\gamma(3+4E^2)}\right]^{1/2}. \tag{6-25}$$

This result agrees closely with their own data and those of Stringham *et al.* (S8). However, its validity for particles in gases appears to be untested.

B. Heat and Mass Transfer

Mass transfer rates in steady two-dimensional flow normal to the axis of a long cylinder have been computed numerically over a range of Re (D3, M8, W6).

[†] Christiansen and Barker (C3) correlated the drag on cylinders falling through gases. However, they indicate anomalously high dependence on γ and E. Moreover, we have been unable to interpret these correlations in a way which is consistent with either their own data or that of Isaacs and Thodos (I2) and Marchildon *et al.* (M3).

These results, which are expected to be reliable for Re $\dot{<}$ 40, are correlated within 10% by the expression

$$\frac{\mathrm{Sh}}{\mathrm{Sc}^{1/3}} = 0.68\left(1 + \frac{1}{\mathrm{ReSc}}\right)^{1/3}\mathrm{Re}^{0.46} \qquad (\mathrm{Re} > 0.1). \tag{6-26}$$

Figure 6.13 compares Eq. (6-26) with available numerical solutions and experimental data of Hilpert (F1, H4) for heat transfer to air. Agreement is good even for Re as high as 10^3. The review of Morgan (M12) should be consulted for additional data and discussion on transfer to cylinders in crossflow.

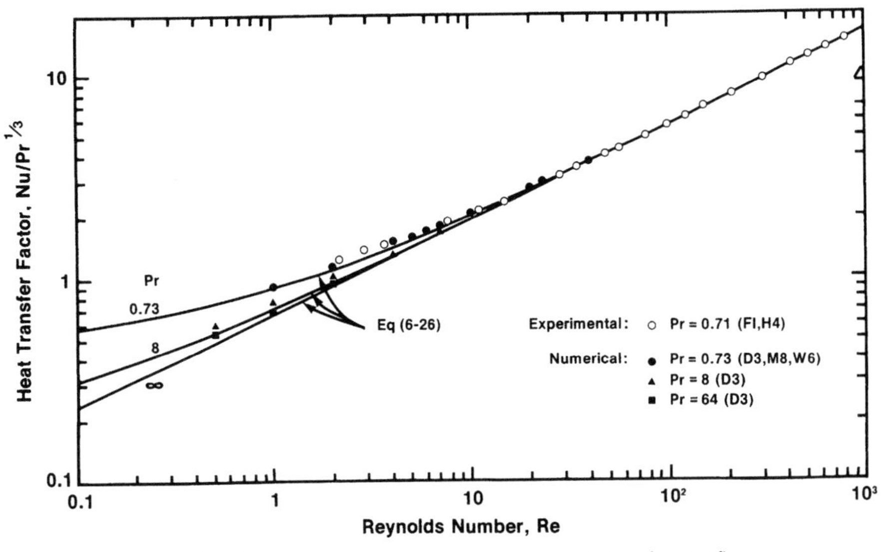

FIG. 6.13 Heat transfer factors for long cylinders in steady crossflow.

Although there are no data for cylinders in free fall, the following suggestions are offered for calculating transfer rates. For cylinders with $E > 1$ falling with axis horizontal and without secondary motion, Eq. (6-26) should be used with the transfer coefficient over the flat ends of the cylinder taken as equal to that over the curved surface. Cylinders with $E < 1$ falling without secondary motion can be treated as oblate spheroids of the same E. For higher Re, the recommendations given below for particles of arbitrary shape in free fall at high Re should be followed.

IV. PARTICLES OF ARBITRARY SHAPE

As shown in Chapter 4, the terminal velocity of a particle of arbitrary shape cannot be predicted with complete confidence, even at low Re. In this chapter, we have shown that the behavior of particles with well-characterized shapes is not well understood at higher Re, especially when secondary motion occurs. In view of these factors and the difficulties noted in Chapter 2 in characterizing the shape of irregular particles, it is not surprising that there are no fully successful methods for predicting the behavior of particles of arbitrary shape. Torobin and Gauvin (T2) reviewed various correlations which have been proposed.

A. Free Fall at Intermediate Reynolds Numbers

For calculating terminal velocities, it is convenient to use groups like those defined in Chapter 5:

$$N_D^{1/3} = [4\Delta\rho\, g/3\rho v^2]^{1/3} \times \text{(equivalent diameter)}, \tag{6-27}$$

$$N_U^{1/3} = [3\rho/4\Delta\rho\, gv]^{1/3} U_T, \tag{6-28}$$

where the equivalent diameter to be used depends on the correlation to be applied. The velocity correction factor for an arbitrary particle is defined as:

$$K = U_T/U_{\text{sphere}}, \tag{6-29}$$

where U_{sphere}, the terminal velocity of a spherical particle of equivalent diameter, can be found from the Re_T or $N_U^{1/3}$ vs. $N_D^{1/3}$ correlations in Chapter 5 and the Appendices.

1. *Sphericity*

Wadell (W1, W2) proposed that the sphericity ψ, defined in Chapter 2, could be used to correlate drag on irregular particles. The appropriate dimension for definition of Re and $N_D^{1/3}$ is then d_e, the diameter of the sphere with the same volume as the particle. Figure 6.14 shows velocity correction factors calculated on this basis (G5). This approach has found widespread acceptance, although there is experimental evidence that terminal velocity does not correlate well with sphericity (B8, S8).

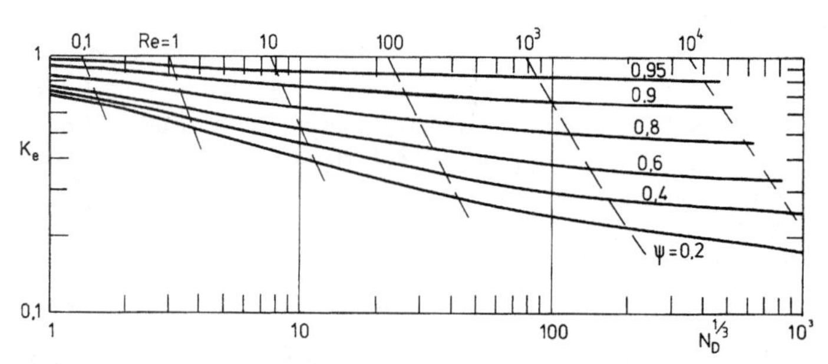

Fig. 6.14 Ratio K_e of terminal velocity of particle of arbitrary shape to that of sphere having the same volume. Based on Wadell (W2) and Govier and Aziz (G5).

Some indication of the validity of this correlation can be obtained by comparing it with results for specific shapes. For oblate spheroids with an aspect ratio of 0.5, predicted values of $N_U^{1/3}$ agree with available data within 10%. For more oblate spheroids, the Wadell correlation predicts terminal velocities as much as 20% too low. For cylinders, agreement is even worse, with Wadell's correlation underpredicting U_T by up to 40%, except for aspect ratios of order 10. Combining these results with the inherent difficulty in measuring ψ for an irregular particle, we conclude that sphericity is not a good basis for predicting terminal velocities, even in the "intermediate" range, except for oblate shapes with ψ approaching unity.

2. *Heywood's Correlation*

Heywood's "volumetric shape factor" k, defined in Chapter 2, can be estimated rapidly, even for irregular particles, using Eq. (2-2). Table 6.3 gives values for regular shapes and some natural particles. Heywood (H2, H3) suggested that k be employed to correlate drag and terminal velocity, using d_A and the projected

TABLE 6.3

Values for Heywood's Volumetric Shape Factor

Regular shapes:	
Sphere	0.524
Cube	0.696
Tetrahedron	0.328
Cylinder with $E = 1$:	
viewed along axis	0.785
viewed normal to axis	0.547
Spheroids: $E = 0.5$	0.262
$E = 2$	0.370
Approximate values for isometric irregular shapes, k_e(H2):	
Rounded	0.56
Subangular	0.51
Angular	
tending to prismoidal	0.47
tending to a tetrahedron	0.38
Selected natural particles (D1):	
Sand	0.26
Bituminous coal	0.23
Blast furnace slag	0.19
Limestone	0.16
Talc	0.16
Gypsum	0.13
Flake graphite	0.023
Mica	0.003

area to define Re and C_D, respectively. There are justifications for this approach because many natural particles have an oblate shape, with one dimension much smaller than the other two. Over large Re_T ranges in the "intermediate" regime, such particles present their maximum area to the direction of motion, and this is the area characterized by d_A. There is also evidence that the shape of this projected area, which does not influence k, has little effect on drag; for example, Jayaweera and Cottis (J1) found essentially the same C_D vs. Re relationship for hexagonal and circular disks.

Heywood gave drag curves for various values of k (H3), and tabulated the velocity correction factor K_A (H2). Figure 6.15 shows K_A plotted from Heywood's table. There is empirical evidence for the validity of this approach (D1). As with sphericity, comparison for specific shapes is informative. For oblate spheroids (for which d_A is the equatorial diameter) and $Re_T \lessdot 100$,

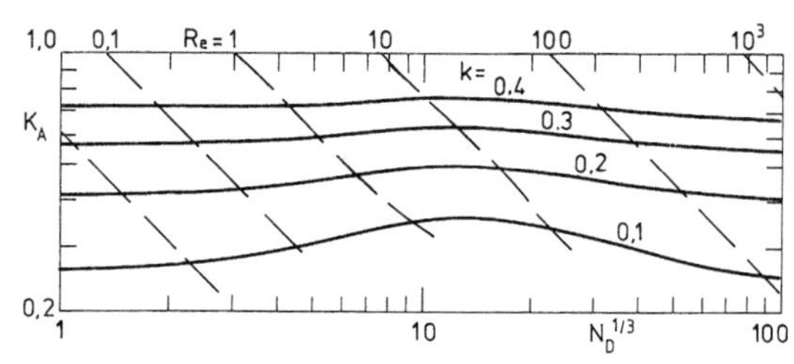

FIG. 6.15 Ratio K_A of terminal velocity of particle of arbitrary shape to that of a sphere having the same projected area. After Heywood (H2).

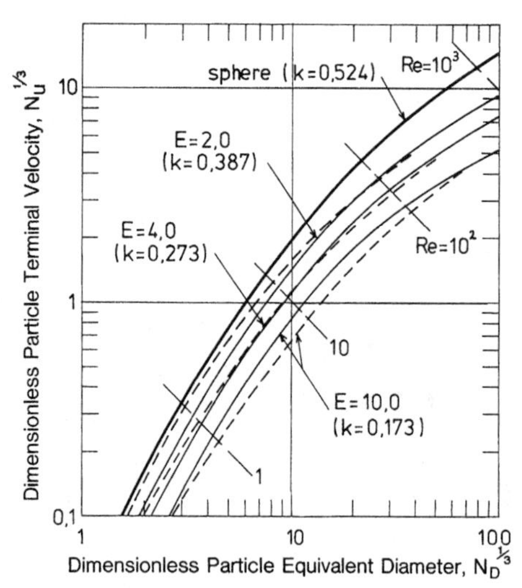

FIG. 6.16 Comparison of Heywood correlation with experimental results for cylinders of $E = 2$, 4 and 10. $N_D^{1/3}$ and Re are based on d_A. Dashed lines are calculated from the correlations of experimental data given in Table 6.1. Light solid lines are the corresponding Heywood predictions.

predicted values of $N_U^{1/3}$ are within 10% of available data. At higher $\mathrm{Re_T}$ where Eq. (6-13), limited to $E = 0.5$, is the only available result, Heywood's correlation underpredicts the terminal velocity by up to 20%. For cylinders, the comparison with the correlation in Table 6.1 is shown in Fig. 6.16. For $E = 2$, agreement is within 10% over the range of $\mathrm{Re_T}$, based on d_A, from 10 to 200, but at lower $\mathrm{Re_T}$ the Heywood correlation underpredicts the terminal velocity by up to 25%. For $E = 4$, agreement is within 10% down to $\mathrm{Re_T} = 0.5$, while for $E = 10$, U_T is overpredicted over the whole range.

Since most irregular particles of practical concern tend to be oblate, lenticular, or rod-like with moderate aspect ratio, these comparisons generally support Heywood's approach. Combining this observation with the fact that the volumetric shape factor is more readily determined than sphericity, we conclude that Heywood's approach is preferred for the "intermediate" range. For convenience in estimating U_T, Table 6.4 gives correlations, fitted to Heywood's values, for $0.1 \le k \le 0.4$ at specific values of $N_D^{1/3}$. Since K_A is relatively insensitive to $N_D^{1/3}$, interpolation for K_A at other values of $N_D^{1/3}$ is straightforward. In common with Heywood's tabulated values, the correlations in Table 6.4 do not extrapolate to $K_A = 1$ for a sphere ($k = 0.524$).

TABLE 6.4

Correlations for Velocity Correction Factor
($0.1 \le k \le 0.4$)

$N_D^{1/3}$	K_A = Velocity correction factor
1	$0.104 + 1.538k$
$10^{1/2}$	$0.127 + 1.526k - 0.1k^2$
10	$0.1975 + 1.575k - 0.45k^2$
$10^{3/2}$	$0.166 + 1.496k - 0.3k^2$
100	$0.0665 + 1.907k - 1.05k^2$

B. Free Fall in the Newton's Law Regime

Since the motion of particles of simple shapes in free fall or rise is poorly understood when secondary motion occurs, it is not surprising that the behavior of particles with more complex or irregular shapes in this range cannot be predicted with certainty. As for the regular shapes, C_D is only weakly dependent on Re, but depends on γ (T2). The correlation developed by Wadell (Fig. 6.14) is not recommended since it shows dependence on Re but not on γ and has already been shown to be unreliable in the intermediate range.

Pettyjohn and Christiansen (P4) reported extensive data for isometric particles. Heywood's volumetric shape factor was not a good basis for correlation in the "Newton's law" range, but sphericity was found suitable. Subsequently,

Barker (B1, T2) showed that the data included a significant density effect, and gave a modified correlation,

$$C_D = \gamma^{-1/18}(5.96 - 5.51\psi) \qquad (1.1 < \gamma < 8.6), \tag{6-30}$$

where C_D is based on the cross-sectional area of the volume-equivalent sphere, $\pi d_e^2/4$. The terminal velocity is then

$$U_T = 0.49\gamma^{1/36}\,[g\,\Delta\rho\, d_e/\rho(1.08 - \psi)]^{1/2}. \tag{6-31}$$

The dependence on γ is compatible with results for short cylinders (see Table 6.2). Data are too scant for these equations to be recast in a form analogous to Eqs. (5-17) and (5-18) and to extrapolate to high γ.

For other particles, Christiansen and Barker (C3) proposed that shapes be classified according to the ratio of maximum to minimum lengths on sections through the centroid. If this ratio is less than 1.7, the particle should be treated as isometric, and Eqs. (6-30) and (6-31) can then be applied. Otherwise, the particle should be classified as rod-like or disk-like, and results given for these shapes in earlier sections should be applied. For irregular particles which do not approximate a shape for which data are available, there are no very satisfactory methods available. The situation is complicated by the fact that particles with high γ can maintain nonpreferred orientations over large distances (M2). The state of affairs is such that if one wishes to estimate the terminal velocities of irregular particles in the Newton's law range one should measure them whenever possible.

C. Heat and Mass Transfer

For particles at high Re, the total heat or mass transfer is made up of a contribution from the front part of the body forward of separation and a contribution from the wake region aft of separation. The two regions should be treated differently to correlate transfer rates.† Over a broad Re range and for nonstreamlined shapes, separation can be considered, to a first approximation, to occur at the locus of the maximum perimeter normal to flow. Figure 6.17 shows mass- and heat-transfer data for the aft portion of a number of different shapes, supported rigidly from the rear. Since transfer in the separated region is particularly sensitive to wall effects and turbulence (I1, P3), data have been included only when tunnel area blockage and turbulence intensity were less than 10% and 3%, respectively. The characteristic length for both the ordinate and abscissa in Fig. 6.17 is

$$L'_{aft} = \frac{\text{surface area aft of maximum perimeter}}{\text{maximum perimeter normal to the flow}}. \tag{6-32}$$

† This idea was first used by Van der Hegge Zijnen (V1) and Douglas and Churchill (D4) in developing correlations for cylinders in crossflow.

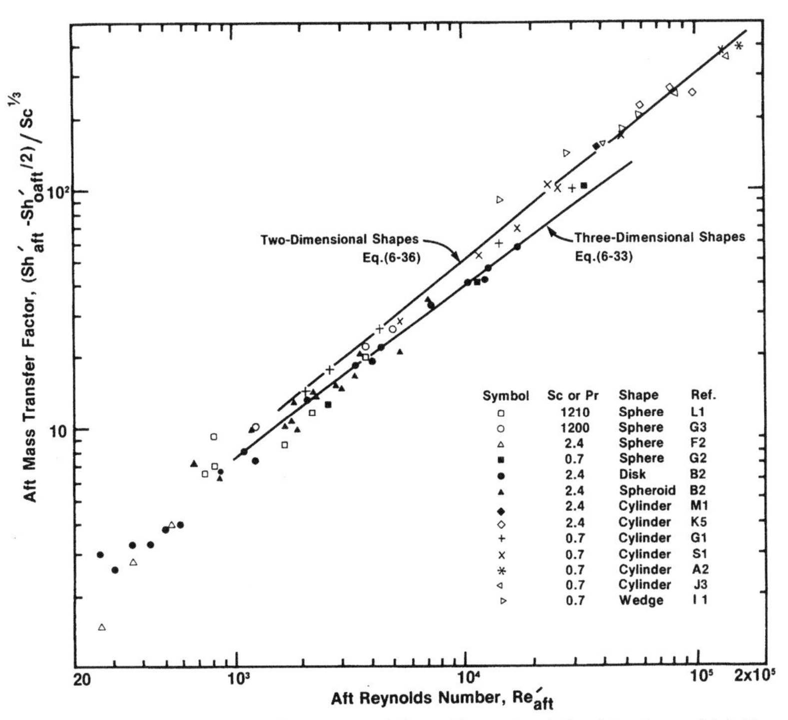

FIG. 6.17 Wake region transfer for two- and three-dimensional fixed bodies at high Reynolds number.

Different correlations are required for three-dimensional bodies (spheres, disks, and spheroids) than for the two-dimensional shapes (cylinders and wedges). For three-dimensional shapes transfer in the aft region is correlated by

$$\frac{\mathrm{Sh}'_{\mathrm{aft}} - \mathrm{Sh}'_{0\,\mathrm{aft}}/2}{\mathrm{Sc}^{1/3}} = 0.056(\mathrm{Re}'_{\mathrm{aft}})^{0.71}, \tag{6-33}$$

where $\mathrm{Sh}'_{0\,\mathrm{aft}}$ is the Sherwood number based on L'_{aft} for diffusion to a stagnant medium. For the forward portion, we assume transfer to be proportional to the square root of the Reynolds number and adjust the proportionality constant to fit the sphere data presented in Chapter 5. The overall correlation for three-dimensional shapes is then

$$\frac{\mathrm{Sh}' - \mathrm{Sh}_0'/2}{\mathrm{Sc}^{1/3}} = 0.62(1-\alpha)^{1/2}(\mathrm{Re}')^{1/2} + 0.056(\alpha\,\mathrm{Re}')^{0.71} \qquad (10^3 \lessdot \mathrm{Re}' \lessdot 10^5), \tag{6-34}$$

where α is the fraction of the total particle area which is aft of the maximum perimeter and Sh', Sh_0', and Re' are all based on L', the characteristic length introduced by Pasternak and Gauvin (P1) [see Eq. (4-67)]. If only part of the surface is active, the active area and maximum perimeter of the active area should be used. For bodies with fore-and-aft symmetry, $\alpha = 0.5$ and Eq. (6-34) becomes:

$$\frac{\mathrm{Sh}' - \mathrm{Sh}_0'/2}{\mathrm{Sc}^{1/3}} = 0.44(\mathrm{Re}')^{1/2} + 0.034(\mathrm{Re}')^{0.71}. \qquad (6\text{-}35)$$

Equation (6-35) appeared already for spheres in Table 5.4.

For two-dimensional shapes, the equations corresponding to Eqs. (6-33) to (6-35) are

$$\mathrm{Sh}'_{\mathrm{aft}}/\mathrm{Sc}^{1/3} = 0.038(\mathrm{Re}'_{\mathrm{aft}})^{0.78}, \qquad (6\text{-}36)$$

$$\mathrm{Sh}'/\mathrm{Sc}^{1/3} = 0.62(1-\alpha)^{1/2}\mathrm{Re}^{1/2} + 0.038(\alpha\,\mathrm{Re}')^{0.78}, \qquad (6\text{-}37)$$

$$\mathrm{Sh}'/\mathrm{Sc}^{1/3} = 0.44(\mathrm{Re}')^{1/2} + 0.022(\mathrm{Re}')^{0.78} \quad (\alpha = 0.5). \qquad (6\text{-}38)$$

The exponents on Re′ for the wake contribution terms in Eqs. (6-33) to (6-38) fall within the range $\frac{2}{3}$ to 1 proposed by earlier investigators (D4, R2, V1). The Sh_0' term does not appear in Eqs. (6-36) to (6-38) since $\mathrm{Sh}_0' = 0$ for a truly two-dimensional body.

Although there are few data to compare with these equations, they are in accord with the results of Pasternak and Gauvin (P1) who found that use of L' as characteristic length brought together their data for spheres, cylinders, cubes, prisms, and hemispheres for $500 < \mathrm{Re}' < 5000$. Unfortunately, the turbulence intensity in their wind tunnel was too high for their data to be used to test the equations directly. When compared with heat transfer data from cylinders in crossflow, the errors were within 10% for circular cylinders (F1, H4), 15% for square and hexagonal cylinders (H4), and 25% for elliptic cylinders with 2:1 axis ratios (K3). In using Eq. (6-37) for the square and hexagonal cylinders, the aft area was taken to include the area of sides parallel to the flow.

For particles of arbitrary shape held in a flow, Eqs. (6-34) and (6-37) should be used for $\mathrm{Re}' \gtrsim 1000$. For particles in free fall the only data available (P2) show that the transfer is little affected by particle rotation with rotational velocities less than 50% of the particle velocity. The correlation for fixed particles was adequate provided that the equivalent diameter d_e was used in place of L'. For particles of arbitrary shape falling in the Newton's law regime, Eq. (6-35) should be used with d_e replacing L' and Sh_0' taken as 2.

For particles with rough surfaces, e.g., with roughness elements of height less than 20% of d_e, the mass transfer coefficient is usually larger than predicted here (A5, J4, S3, S4), but at most by about 50%. Roughness is treated in more detail in Chapter 10. For a particle made up of a small number of particles in a cluster, the use of d_e in Eq. (6-35) gives good results (S4).

In the intermediate regime it is recommended that the particle be treated as an oblate spheroid with major and minor axes determined from the particle volume and the projected area of the particle lying in its orientation of maximum stability on a horizontal surface; both parameters are required in any case for the determination of the volumetric shape factor k. The aspect ratio E is then

$$E = (6/\pi)k, \tag{6-39}$$

and the equatorial diameter d is given by

$$d = 2a = (6V/\pi E)^{1/3}. \tag{6-40}$$

If $k > 0.524$, E should be taken as unity.

V. FREE FALL OF OTHER SPECIFIC SHAPES OR TYPES OF PARTICLE

Excellent reviews of work on other specific shapes have been prepared by Hoerner (H5) and Torobin and Gauvin (T2). Some more recent references are listed in Table 6.5. Particles of special shapes often show interesting preferred orientations in the "intermediate regime." For example, a freely falling cone of uniform density falls with its base horizontal for Re_T (defined with basal diameter as characteristic length) $\dot{>}$ 0.05; the apex points up if the apex angle is less than 45° and down if the angle is greater than 45° (J2). Cubes and isometric tetrahedra adopt an orientation with a flat face perpendicular to the direction of motion for $Re_T \dot{>} 10$, while octahedra show similar behavior for $Re_T \dot{>} 20$ (P4).

Work has also been done on specific types of particles encountered in agricultural, meteorological, and other applications. Some relevant references are listed in Table 6.5. When data for the specific shape or the specific type of

TABLE 6.5

Sources of Data on Drag and Free Fall Behavior of Some Nonspherical Particles

Shape or type of particle	References
Non circular plates	(L5, P6, P7)
Cones	(C1, G4, J2, L5, W3)
Prisms	(C3, C4, M2)
Straws and stems	(B6, B7, M10)
Grain, seeds, kernels	(B5, B6, B7, H1)
Soybeans	(H1)
Blueberries	(S7)
Walnuts	(M13)
Hail, ice crystals	(L2, L5, P6, R4)
Snow crystals	(L5, N1)
Sand	(B8)

particles of interest are available, these data often provide a more accurate basis for predicting free fall behavior than the general and approximate methods outlined previously in Section IV.

REFERENCES

A1. Achenbach, E., *J. Fluid Mech.* **34**, 625–639 (1968).
A2. Achenbach, E., *Heat Transfer 1974* **2**, 229–233 (1974).
A3. Al-Taha, T. R., Ph.D. Thesis, Imperial Coll., London, 1969.
A4. Apelt, C. J., *Aeronaut. Res. Counc., Rep. Memo.* No. 3175 (1961).
A5. Aufdermaur, A. N., and Joss, J., *Z. Angew. Math. Phys.* **18**, 852–866 (1967); **19**, 377 (1968).
B1. Barker, D. H., Ph.D. Thesis, Univ. of Utah, Salt Lake City, 1951.
B2. Beg, S. A., Ph.D. Thesis, Imperial Coll., London, 1966.
B3. Beg, S. A., *Wärme-Stoffübertrag.* **6**, 45–51 (1973).
B4. Beg, S. A., *Wärme-Stoffübertrag.* **8**, 127–135 (1975).
B5. Bilanski, W. K., Collins, S. H., and Chu, P., *Agric. Eng.* **43**, 216–219 (1962).
B6. Bilanski, W. K., and Lal, R., *Trans. ASAE* **8**, 411–416 (1965).
B7. Bilanski, W. K., and Lal, R., *Meet. Am. Soc. Agric. Eng.*, Pap. No. 66–610 (1966).
B8. Briggs, L. I., McCulloch, D. S., and Moser, F., *J. Sediment. Petrol.* **32**, 645–656 (1962).
C1. Calvert, J. R., *J. Fluid Mech.* **27**, 273–289 (1967).
C2. Charlton, R. B., and List, R., *J. Rech. Atmos.* **6**, 55–62 (1972).
C3. Christiansen, E. B., and Barker, D. H., *AIChE J.* **11**, 145–151 (1965).
C4. Clift, R., and Gauvin, W. H., *Can. J. Chem. Eng.* **49**, 439–448 (1971).
D1. Davies, C. N., *Part.-Fluid Interact., Harold Heywood Memorial Symp., Loughborough Univ., 1973.*
D2. Dennis, S. C. R., and Chang, G. Z., *J. Fluid Mech.* **42**, 471–489 (1970).
D3. Dennis, S. C. R., Hudson, J. D., and Smith, N., *Phys. Fluids* **11**, 933–940 (1968).
D4. Douglas, W. J. M., and Churchill, S. W., *Chem. Eng. Prog., Symp. Ser.* **52**, No. 18, 23–28 (1956).
F1. Fand, R., and Keswani, K. K., *J. Heat Transfer* **95**, 224–226 (1973).
F2. Frössling, N., *Gerlands Beitr. Geophys.* **52**, 170–216 (1938).
G1. Galloway, T. R., and Sage, B. H., *AIChE. J.* **13**, 563–570 (1967).
G2. Galloway, T. R., and Sage, B. H., *AIChE J.* **18**, 287–293 (1972).
G3. Gibert, H., and Angelino, H., *Chem. Eng. Sci.* **28**, 855–867 (1973).
G4. Goldburg, A., and Florsheim, B. H., *Phys. Fluids* **9**, 45–50 (1966).
G5. Govier, G. W., and Aziz, K., "The Flow of Complex Mixtures in Pipes." Van Nostrand-Reinhold, New York, 1972.
H1. Hawk, A. L., Brooker, D. B. and Cassidy, J. J., *Trans. ASAE* **9**, 48–51 (1966).
H2. Heywood, H., *Symp. Interact. Fluids Part., Inst. Chem. Eng., London* pp. 1–8 (1962).
H3. Heywood, H., *Proc. Inst. Mech. Eng.* **140**, 257–347 (1938).
H4. Hilpert, R., *Forsch. Geb. Ingenieurwes.* **4**, 215–224 (1933).
H5. Hoerner, S. F., "Fluid Dynamic Drag." Published by the author, Midland Park, New Jersey, 1958.
H6. Homann, F., *Z. Angew. Math. Mech.* **16**, 153–164 (1936).
I1. Igarashi, T., and Hirata, M., *Heat Transfer 1974* **2**, 300–304 (1974).
I2. Isaacs, J. L., and Thodos, G., *Can. J. Chem. Eng.* **45**, 150–155 (1967).
J1. Jayaweera, K. O. L. F., and Cottis, R. E., *Q. J. R. Meteorol. Soc.* **95**, 703–709 (1969).
J2. Jayaweera, K. O. L. F., and Mason, B. J., *J. Fluid Mech.* **22**, 709–720 (1965).
J3. Johnson, T. R., and Joubert, P. N., *J. Heat Transfer* **91**, 91–99 (1969).
J4. Joss, J., and Aufdermaur, A. N., *Int. J. Heat Mass Transfer* **13**, 213–215 (1970).
K1. Kajikawa, M., *J. Meteorol. Soc. Jpn.* **49**, 367–375 (1971).

K2. Kalra, T. R., and Uhlherr, P. H. T., *Can. J. Chem. Eng.* **51**, 655–658 (1973).
K3. Katinas, W. I., Ziugzda, I. I., and Zukauskas, A. A., *Heat Transfer—Sov. Res.* **3**(6), 10–33 (1971).
K4. Keller, H. B., and Takami, H., *Proc. Adv. Symp. Numerical Solutions Non-Linear Differential Equations* pp. 115–140 (1966).
K5. Kestin, J., and Wood, R. T., *J. Heat Transfer* **93**, 321–326 (1971).
K6. Knight, C. A., and Knight, N. C., *J. Atmos. Sci.* **27**, 672–681 (1970).
K7. Krumbein, W. C., *Trans. Am. Geophys. Union* **23**, 621–633 (1942).
K8. Kry, P. R., and List, R., *Phys. Fluids* **17**, 1087–1092 (1974).
K9. Kry, P. R., and List, R., *Phys. Fluids* **17**, 1093–1102 (1974).
K10. Kunkel, W. B., *J. Appl. Phys.* **19**, 1056–1058 (1948).
L1. Linton, M., and Sutherland, K. L., *Chem. Eng. Sci.* **12**, 214–229 (1960).
L2. List, R., *Z. Angew. Math. Phys.* **10**, 143–159 (1959).
L3. List, R., and Dussault, J.-G., *J. Atmos. Sci.* **24**, 522–529 (1967).
L4. List, R., Rentsch, U. W., Byram, A. C., and Lozowski, E. P., *J. Atmos. Sci.* **30**, 653–661 (1973).
L5. List, R., and Schemenauer, R. S., *J. Atmos. Sci.* **28**, 110–115 (1971).
M1. Mabuchi, I, Hiwada, M., and Kumada, M., *Heat Transfer 1974* **2**, 315–319 (1974).
M2. Marchildon, E. K., Ph.D. Thesis, McGill Univ., Montreal, 1965.
M3. Marchildon, E. K., Clamen, A., and Gauvin, W. H., *Can. J. Chem. Eng.* **42**, 178–182 (1964).
M4. Masliyah, J. H., Ph.D. Thesis, Univ. of British Columbia, Vancouver, 1970.
M5. Masliyah, J. H., *Phys. Fluids* **15**, 1144–1146 (1972).
M6. Masliyah, J. H., and Epstein, N., *Prog. Heat Mass Transfer* **6**, 613–632 (1972).
M7. Masliyah, J. H., and Epstein, N., *J. Fluid Mech.* **44**, 493–512 (1970).
M8. Masliyah, J. H., and Epstein, N., *Ind. Eng. Chem., Fundam.* **12**, 317–323 (1973).
M9. McNown, J. S., and Malaika, J., *Trans. Am. Geophys. Union* **31**, 74–82 (1950).
M10. Menzies, D., and Bilanski, W. K., *Trans. ASAE* **11**, 829–831 (1968).
M11. Michael, P., *Phys. Fluids* **9**, 466–471 (1966).
M12. Morgan, V. T., *Adv. Heat Transfer* **11**, 199–264 (1975).
M13. Mueller, R. A., Brooker, D. B., and Cassidy, J. J., *Trans ASAE* **10**, 57–61 (1967).
N1. Nakaya, U., and Terada, T., *J. Fac. Sci., Hokkaido Imp. Univ.* **1**, 191–201 (1935).
P1. Pasternak, I. S., and Gauvin, W. H., *Can. J. Chem. Eng.* **38**, 35–42 (1960).
P2. Pasternak, I. S., and Gauvin, W. H., *AIChE J.* **7**, 254–260 (1961).
P3. Petrie, A. M., and Simpson, H. C., *Int. J. Heat Mass Transfer* **15**, 1497–1513 (1972).
P4. Pettyjohn, E. A., and Christiansen, E. B., *Chem. Eng. Prog.* **44**, 157–172 (1948).
P5. Pitter, R. L., Pruppacher, H. R., and Hamielec, A. E., *J. Atmos. Sci.* **30**, 125–134 (1973).
P6. Podzimek, J., *Proc. Int. Conf. Cloud Phys., Tokyo Sapporo* pp. 224–230 (1965).
P7. Podzimek, J., *Proc. Int. Conf. Cloud Phys., Toronto* pp. 295–299 (1968).
P8. Pruppacher, H. R., LeClair, B. P., and Hamielec, A. E., *J. Fluid Mech.* **44**, 781–790 (1970).
R1. Riabouchinsky, D. P., *NACA Tech. Note* No. 44(1) (1921).
R2. Richardson, P. D., *Chem. Eng. Sci.* **18**, 149–155 (1963).
R3. Rimon, Y., and Lugt, H. J., *Phys. Fluids* **12**, 2465–2472 (1969).
R4. Roos, D. S., *J. Appl. Meteorol.* **11**, 1008–1011 (1972).
R5. Roos, F. W., and Willmarth, W. W., *AIAA J.* **9**, 285–291 (1971).
R6. Roshko, A., *NACA Rep.* No. 1191 (1954).
R7. Roshko, A., *J. Fluid Mech.* **10**, 345–356 (1961).
S1. Schmidt, E., and Wenner, K., *NACA Tech. Memo.* No. 1050 (1943).
S2. Schmiedel, J., *Phys. Z.* **29**, 593–610 (1928).
S3. Schuepp, P. H., and List, R., *J. Appl. Meteorol.* **8**, 743–746 (1969).
S4. Schuepp, P. H., and List, R., *J. Appl. Meteorol.* **8**, 254–263 (1969).
S5. Simmons, L. F. G., and Dewey, N. S., *Aeronaut. Res. Counc., Rep. Memo.* No. 1334 (1930).
S6. Son, J. S., and Hanratty, T. J., *J. Fluid Mech.* **35**, 353–368 (1969).

S7. Soule, H. M., *Trans. ASAE* **13**, 114–117 (1970).
S8. Stringham, G. E., Simons, D. B., and Guy, H. P., *U.S., Geol. Surv., Prof. Pap.* **562-C** (1969).
T1. Taneda, S., *J. Phys. Soc. Jpn.* **11**, 302–307 (1956).
T2. Torobin, L. B., and Gauvin, W. H., *Can. J. Chem. Eng.* **38** 142–153 (1960).
U1. Underwood, R. L., *J. Fluid Mech.* **37**, 95–114 (1969).
V1. Van der Hegge Zijnen, B. G., *Appl. Sci. Res., Sect. A* **6**, 129–140 (1956).
W1. Wadell, H., *J. Geol.* **41**, 310–331 (1933).
W2. Wadell, H., *J. Franklin Inst.* **217**, 459–490 (1934).
W3. Wehrman, O. H., *Phys. Fluids* **9**, 2284–2285 (1966).
W4. Wieselsberger, C., *Phys. Z.* **23**, 219–224 (1922).
W5. Willmarth, W. W., Hawk, N. E., and Harvey, R. L., *Phys. Fluids* **7**, 197–208 (1964).
W6. Woo, S. W., Ph.D. Thesis, McMaster Univ., Hamilton, Ontario, 1971.

Chapter 7

Ellipsoidal Fluid Particles

I. INTRODUCTION

The conditions under which fluid particles adopt an ellipsoidal shape are outlined in Chapter 2 (see Fig. 2.5). In most systems, bubbles and drops in the intermediate size range (d_e typically between 1 and 15 mm) lie in this regime. However, bubbles and drops in systems of high Morton number are never ellipsoidal. Ellipsoidal fluid particles can often be approximated as oblate spheroids with vertical axes of symmetry, but this approximation is not always reliable. Bubbles and drops in this regime often lack fore-and-aft symmetry, and show shape oscillations.

II. FLUID DYNAMICS

A. AIR/WATER SYSTEMS

Because of their practical importance, water drops in air and air bubbles in water have received more attention than other systems. The properties of water drops and air bubbles illustrate many of the important features of the ellipsoidal regime.

1. *Water Drops in Air*

Numerous determinations of the terminal velocities of water drops have been reported. The most careful measurements are those by Gunn and Kinzer (G13) and Beard and Pruppacher (B4). Figure 7.1, derived from these results,[†] shows

[†] Results for Re $\lesssim$ 300 were included in the data used to derive the "standard drag curve" in Chapter 5. Numerical results for spherical raindrops (valid for Re $\lesssim$ 200) are also discussed in Chapter 5.

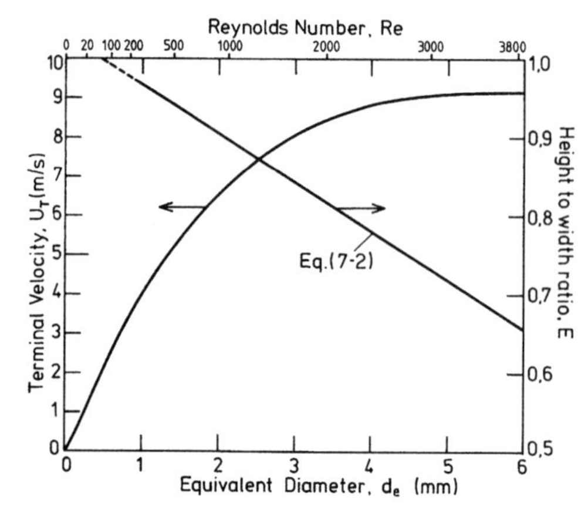

FIG. 7.1 Terminal velocity and aspect ratio of water drops falling freely in air at 20°C and 1 bar (B4, G13, P5).

terminal velocity and corresponding Reynolds number as a function of volume-equivalent diameter for water drops in air. Berry and Pranger (B7) and Beard (B3) give empirical polynomials describing the terminal velocity of drops in air, with Beard's equations covering a wider range of atmospheric conditions than the others. For water drops in air under normal atmospheric conditions at sea level, the simplest fit (B7), accurate within about 3%, gives

$$\mathrm{Re} = \exp[-3.126 + 1.013 \ln N_D - 0.01912(\ln N_D)^2]$$
$$(2.4 < N_D < 10^7;\ 0.1 < \mathrm{Re} < 3550) \qquad (7\text{-}1)$$

where N_D is defined by Eq. (5-15). For $d_e \lessdot 1$ mm (Re $\lessdot$ 300), deviations from a spherical shape and internal circulation are so small that the correlations for rigid spheres in Chapter 5 may be used to predict terminal velocities. For $d_e \lessdot$ 20 μm (B3), correction for noncontinuum effects must be made (see Chapter 10). Pitter and Pruppacher (P4) studied the motion of 200 to 350 μm water drops undergoing freezing.

Drops larger than about 1 mm in diameter are significantly nonspherical; the mean height to width ratio is approximated (P5) by:

$$E = 1.030 - 0.062 d_e \qquad (1 < d_e < 9 \text{ mm}), \qquad (7\text{-}2)$$

with d_e in mm. This ratio is plotted in Fig. 7.1. Figure 7.2 shows that deformation increases the drag coefficient above the value for a rigid sphere if C_D and Re are based on the volume-equivalent diameter d_e. The flattening of water drops at the front (lower) surface results from the increased hydrodynamic pressure there, while the rear has a more uniform hydrodynamic pressure and is therefore more rounded (M6). Blanchard (B9) discusses the popular misconception that raindrops fall with a teardrop shape. Figure 2.4(a) shows a photograph of a water drop in air. Shapes are discussed in detail in Section D.

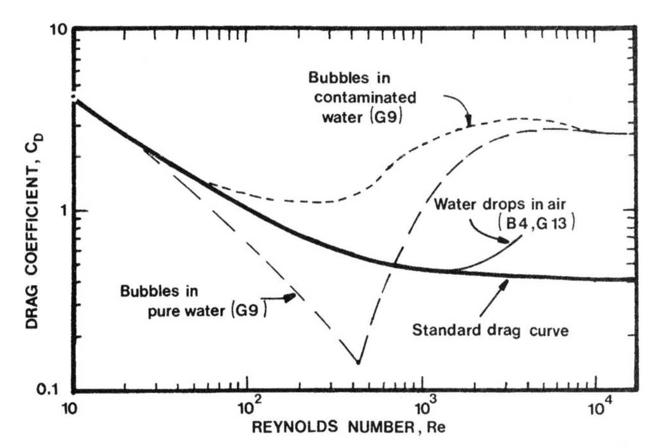

FIG. 7.2 Drag coefficient as function of Reynolds number for water drops in air and air bubbles in water, compared with standard drag curve for rigid spheres.

Water drops become unstable and tend to break up before they reach 1 cm in diameter (see Chapter 12). Drops approaching this size show periodic shape fluctuations of relatively low amplitude (J3, M4).

As for other types of fluid particle, the internal circulation of water drops in air depends on the accumulation of surface-active impurities at the interface (H9). Observed internal velocities are of order 1% of the terminal velocity (G4, P5), too small to affect drag detectably. Ryan (R6) examined the effect of surface tension reduction by surface-active agents on falling water drops.

2. *Air Bubbles in Water*

Experimental terminal velocities for air bubbles rising in water are presented in Fig. 7.3 for the ellipsoidal regime and adjacent parts of the spherical and spherical-cap regimes. Some of the spread in the data results from experimental scatter, but the greatest cause is surface contamination. For water drops in air, described in the previous section, surfactants have negligible effect on drag since κ is so high that internal circulation is small even in pure systems. For air bubbles in water, κ is so small that there is little viscous resistance to internal circulation, and hence the drag and terminal velocity are sensitive to the presence of surfactants.

The two curves in Fig. 7.3 are based on those given by Gaudin (G9) for distilled water and for water with surfactant added. The curves converge for small (spherical) bubbles, since even distilled water tends to contain sufficient surfactant to prevent circulation in this range (see Chapter 3), and for large (spherical-cap) bubbles, where surface tension forces cease to be important. Surface-active contaminants affect the rise velocity most strongly in the ellipsoidal range. Drag coefficients corresponding to these two curves appear in Fig. 7.2, and show that C_D for bubbles lies below the rigid sphere curve when

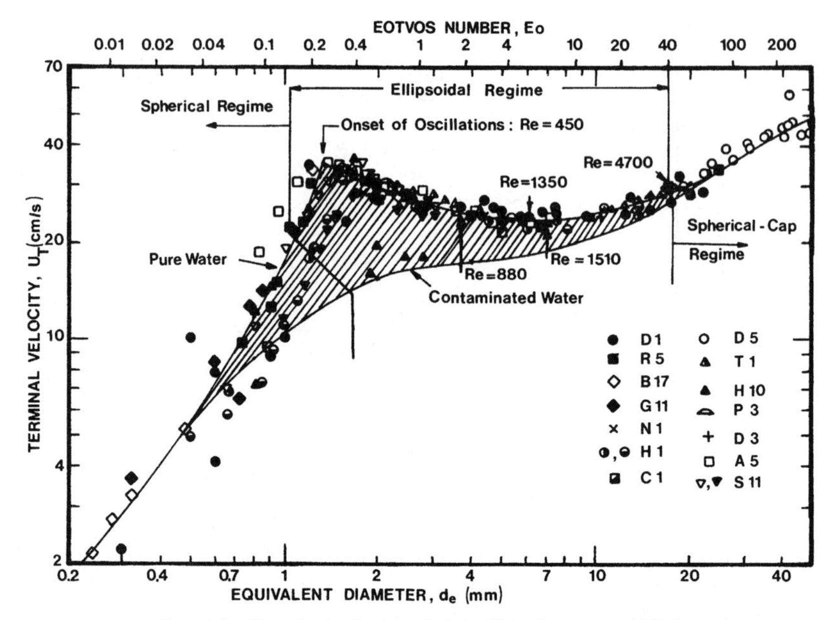

FIG. 7.3 Terminal velocity of air bubbles in water at 20°C.

internal circulation is present, but above if there is no internal circulation and the drag is dominated by deformation. For $d_e > 1.3$ mm, the uppermost (pure system) curve in Fig. 7.3 is approximated closely by

$$U_T = [(2.14\sigma/\rho d_e) + 0.505 g d_e]^{1/2}, \tag{7-3}$$

which is of the form suggested by a wave analogy (C2, M7).

Aybers and Tapucu (A4, A5) measured trajectories of air bubbles in water. When surface-active agents continue to accumulate during rise, the terminal velocity may never reach steady state (A4, B1) and may pass through a maximum (W4). Five types of motion were observed, listed in Table 7.1 with Re based on the maximum instantaneous velocity. Secondary motion of fluid par-

TABLE 7.1

Motion of Intermediate Size Air Bubbles Through Water at 28.5°C[a]

d_e (mm)	Re	$\bar{E}$	Path
<1.3	<565	>0.8	Rectilinear
1.3 to 2.0	565 to 880	0.8 to 0.5	Helical
2.0 to 3.6	880 to 1350	0.5 to 0.36	Plane (zig-zag) then helical
3.6 to 4.2	1350 to 1510	0.36 to 0.28	Plane (zig-zag)
4.2 to 17	1510 to 4700	0.28 to 0.23	Rectilinear but with rocking

[a] After Aybers and Tapucu (A5).

ticles, associated with wake phenomena, is discussed in greater detail in Section F. Lindt and De Groot (L8) give values of the Strouhal number for helical vortex shedding behind air bubbles in water.

B. Terminal Velocities of Drops and Bubbles in Liquids

The generalized graphical correlation presented in Fig. 2.5 gives one method of estimating terminal velocities of drops and bubbles in infinite liquid media. For more accurate predictions, it is useful to have terminal velocities correlated explicitly in terms of system variables. To obtain such a correlation is especially difficult for the ellipsoidal regime where surface-active contaminants are important and where secondary motion can be marked.

1. *Effect of Viscosity Ratio* κ

It is general practice to ignore the effect of the viscosity of the internal fluid in correlations of terminal velocities. We recall from Chapter 3 that decreasing μ_p, all other factors remaining fixed, can at most cause a 50% change in U_T at low Re, and this change is seldom realized in practice due to the effect of surfactants. Hamielec (H2) showed that varying κ over a tenfold range had a small but noticeable effect for cyclohexanol drops in water with Re up to about 10. Figure 7.4 shows Re (Eo) for eight systems, all having virtually the same Morton

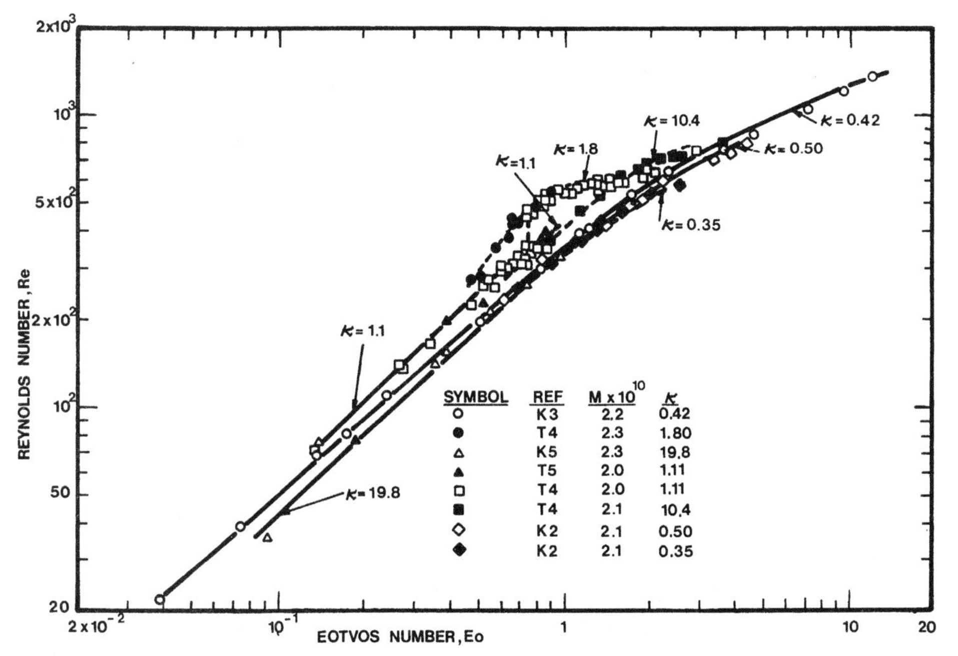

Fig. 7.4 Reynolds number as a function of Eotvos number for systems with essentially identical M studied by different workers.

Number (2.0 to 2.3 × 10^{-10}), but widely different values of κ (0.35 to 20). While the data exhibit some scatter, the observed dimensionless terminal velocities do not vary systematically with κ, but appear to reflect differences in system purity. Thorsen and coworkers (T4, T5) took greater care to purify their systems than the other authors and this is reflected in higher velocities. The internal fluid viscosity can be considered to be of secondary importance for systems in which no particular care has been taken to eliminate surfactants.

2. *Effect of Surface-Active Contaminants*

We may illustrate the effect of surfactants by comparing terminal velocities measured by different workers using the same system. Results for air bubbles in water have already been shown in Figs. 7.2 and 7.3. Results from six different studies on carbon tetrachloride drops falling through water are plotted in Fig. 7.5. The measured terminal velocities differ widely among different investigators, and one can only attribute these differences to differences in system purity. A number of workers have noted a strong influence of system purity on the drag or terminal velocity of ellipsoidal fluid particles [e.g. (E3, R1, S9, T1, T4, Z1)]. In a very careful study, Edge and Grant (E3) examined the effects of low concentrations of a surfactant, sodium lauryl sulphate, on the motion of dichloroethane drops descending through water. At very low surfactant concentrations (10^{-5} gm/liter or less) there was no observable effect. As the concentration was increased, a marked decrease in terminal velocity was observed for drops of equivalent diameter between 2 and 6 mm and this was usually accompanied by earlier boundary layer separation and irregular drop

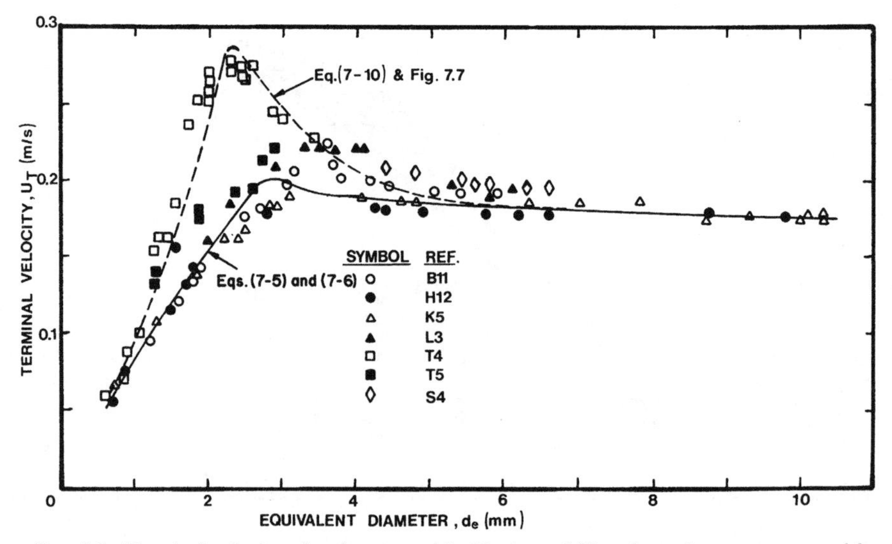

FIG. 7.5 Terminal velocity of carbon tetrachloride drops falling through water, measured by different workers, with varying system purity.

oscillations. At relatively high surfactant concentrations (10^{-2} gm/liter or greater) the systems were said to be "grossly contaminated." The drop terminal velocities again became independent of surfactant concentration while the interface remained rigid, and oscillations became more regular.

Very few workers have succeeded in eliminating all surface-active contaminants from their systems. Moreover, the type and concentration of contaminants present have seldom been characterized. Based on the available evidence, one may draw the following conclusions.

(i) Surfactants tend to damp out internal motion by rendering the interface rigid as discussed in Chapter 3. The influence of surfactants is most significant for low values of κ, since at large κ the viscous resistance of the internal fluid limits internal motion even for pure systems.

(ii) Surfactants have the greatest influence on terminal velocity near the point of transition from rectilinear to oscillating motion. This is presumably because internal circulation can drastically alter the wake structure of a fluid particle (see below) leading to delayed boundary layer separation, smaller wakes, and delayed vortex shedding.

(iii) Surfactants play a particularly important role in high σ systems (e.g., air/water) since surface tension reductions are largest for these systems (see Chapter 3).

(iv) Most of the experimental results in the literature are for "grossly contaminated" bubbles and drops. Since it is so difficult to eliminate surface-active contaminants in systems of practical importance, this is not a serious limitation.

3. *Correlation for Contaminated Drops and Bubbles*

There is a substantial body of data in the literature on the terminal velocities of bubbles and drops. In view of the influence of system purity discussed above, a separation of this data has been made. Cases where there is evidence that considerable care was taken to eliminate surfactants and where a sharp peak in the U_T vs. d_e curve at low M and κ is apparent (as for the pure systems in Figs. 7.3 and 7.5) are discussed in Section 4.

Grace *et al.* (G12) applied three types of correlation to a large body of experimental data: the form proposed by Klee and Treybal (K3); that proposed by Hu and Kintner (H12) and its extension by Johnson and Braida (J2); and a wave analogy suggested for bubbles by Mendelson (M7) and extended to drops by Marrucci *et al.* (M5). The second of these forms gave smaller residuals, especially as M is increased. Even so, it was necessary to eliminate high M systems from the resulting correlation. Cases where wall effects were too significant were also eliminated from the data treated. The criteria which the data had to meet were then Eqs. (9-33) or (9-34) and

$$M < 10^{-3}, \qquad \mathrm{Eo} < 40, \qquad \mathrm{Re} > 0.1. \tag{7-4}$$

The indices in the original Johnson and Braida correlation and the point of intersection between the two linear regions were adjusted to improve the agreement with all the data meeting the above criteria. The resulting correlation (G12) is:

$$J = 0.94H^{0.757} \qquad (2 < H \leq 59.3) \qquad (7\text{-}5)$$

and

$$J = 3.42H^{0.441} \qquad (H > 59.3), \qquad (7\text{-}6)$$

where

$$H = \tfrac{4}{3}\,\mathrm{Eo}\,M^{-0.149}(\mu/\mu_w)^{-0.14}, \qquad (7\text{-}7)$$

$$J = \mathrm{Re}\,M^{0.149} + 0.857, \qquad (7\text{-}8)$$

and μ_w is the viscosity of water in Braida's experiments, which may be taken as 0.0009 kg/ms (0.9 cp).

A plot of every fourth data point and the lines given by Eqs. (7-5) and (7-6) appears in Fig. 7.6. The gradient discontinuity corresponds approximately to the transition between nonoscillating and oscillating bubbles and drops. In the above correlation, the terminal velocity appears only in the dimensionless group J, and may be expressed explicitly as:

$$U_T = \frac{\mu}{\rho d_e} M^{-0.149}(J - 0.857). \qquad (7\text{-}9)$$

The r.m.s. deviation between measured and predicted terminal velocities is about 15% for the 774 points with $H \leq 59.3$ and 11% for the 709 points with $H > 59.3$. This correlation is recommended for calculations of bubble and drop terminal velocities when the criteria outlined above are satisfied and where some surface-active contamination is inevitable. The predictions from this correlation for carbon tetrachloride drops in water are shown on Fig. 7.5.

Many other correlations for calculating the terminal velocity of bubbles and drops are available [e.g. (H12, J2, K3, T1, V1, W2)]. None covers such a broad range of data as Eqs. (7-5) and (7-6). Moreover, a number of the earlier correlations require that values be read from graphs or that iterative procedures be used to determine U_T.

4. *Correlation for Pure Systems*

In view of the limited data available for pure systems, Grace *et al.* (G12) modified the correlation given in the previous section rather than proposing an entirely different correlation. A correction of somewhat similar form to that suggested in Chapter 3 for low Re is employed; i.e.,

$$(U_T)_{\mathrm{pure}} = U_T[1 + \Gamma/(1 + \kappa)], \qquad (7\text{-}10)$$

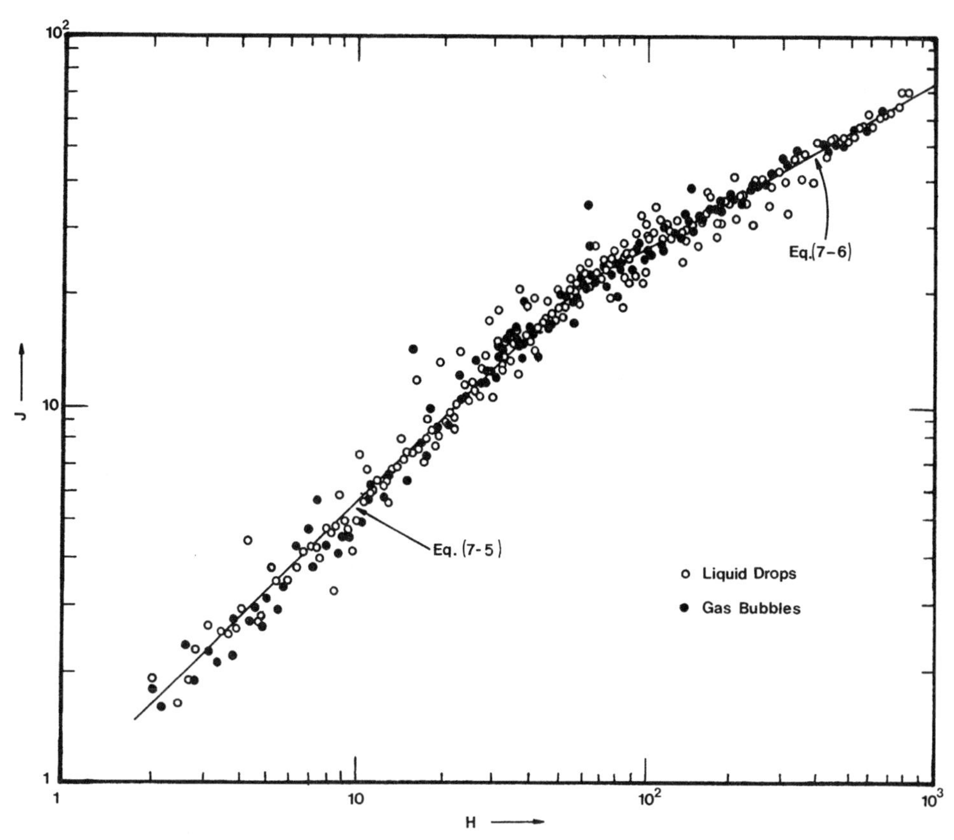

FIG. 7.6 Data (showing one point in four) used to obtain general correlation for terminal velocity of drops and bubbles in contaminated liquids, compared with Eqs. (7-5) and (7-6) (B11, B17, D2, E3, E4, G9, G10, G14, H1, H12, J4, K2, K3, K5, L3, L11, P3, T1, W3, Y4).

where Γ is to be obtained experimentally and U_T is predicted using Eqs. (7-5) to (7-9). Since the continuous fluid was water for all the pure systems for which data are available, μ and M cover very restricted ranges. Experimental values of Γ are plotted in Fig. 7.7 as a function of $\mathrm{Eo}(1 + 0.15\kappa)/(1 + \kappa)$, where the function of κ was chosen to reduce the spread in the resulting points.

Careful purification of a system has little effect for small and large drops and bubbles. Hence Γ reaches a maximum for a particular value of the abscissa and decreases to zero at large and small values of the abscissa. An envelope has been drawn to provide an estimate of the maximum increase in terminal velocity for bubbles and drops in pure systems over that for contaminated systems. This envelope, together with Eq. (7-10) and the correlation of the previous section, have been used to obtain the upper curve in Fig. 7.5 for carbon tetrachloride drops in water. The curve gives a good representation of the higher velocities observed for carefully purified systems.

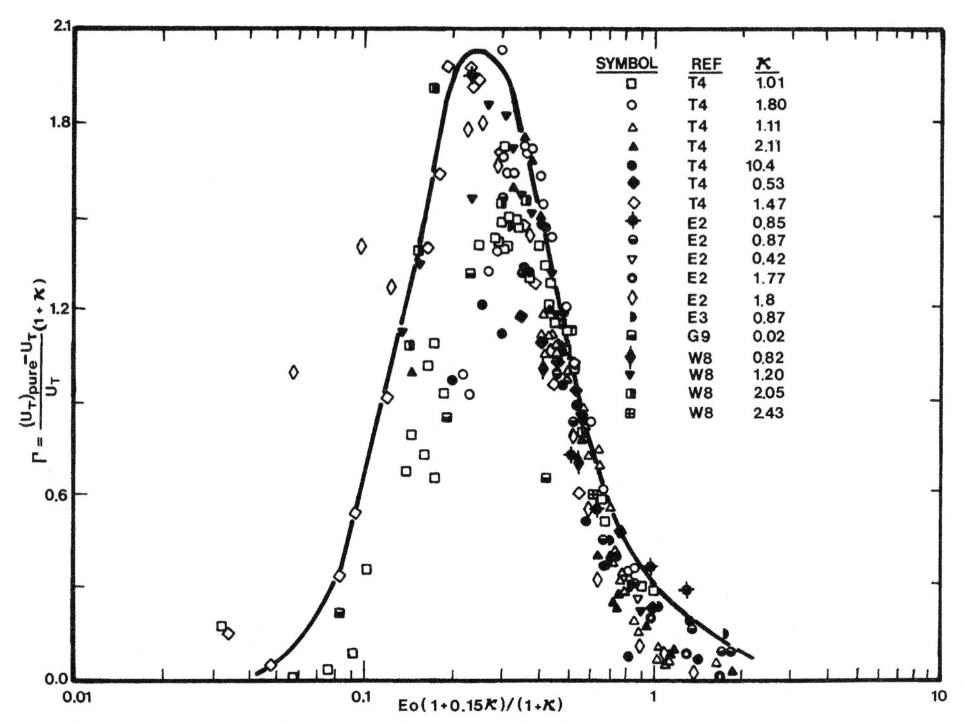

FIG. 7.7 Correction factor, Γ, relating terminal velocity in pure systems to value in corresponding contaminated systems (E2, E3, G9, T4).

The envelope in Fig. 7.5 is for the maximum increase in terminal velocity obtainable by eliminating surface-active contaminants. For systems of intermediate purity, Γ may be assigned a value between zero and that given by the envelope. Since the envelope has been derived solely from experiments for aqueous systems, it should be used with caution for nonaqueous systems.

C. Terminal Velocities of Liquid Drops in Air

As indicated in Chapter 2, liquid drops falling through gases have such extreme values of γ and κ that they must be treated separately from bubbles and drops in liquids. Few systems have been investigated aside from water drops in air, discussed above, and what data are available for other systems (F1, G5, L5, V2) show wide scatter. Rarely have gases other than air been used, and some data for these cases [e.g. (L5, N2)] cannot be interpreted easily because of evaporation and combustion effects. Results for drops in air at other than room temperature (S8) differ so radically from results of other workers that they cannot be used with confidence.

For Eo $\lesssim$ 0.15, drops are closely spherical and terminal velocities may be calculated using correlations given in Chapter 5 for rigid spheres. For larger drops, data of Finlay (F1) and Van der Leeden *et al.* (V2) can be correlated with the best data for water drops in air (B4, G13) by the equations

$$\mathrm{Re} = 1.62\,\mathrm{Eo}^{0.755} M^{-0.25} \qquad (0.5 \le \mathrm{Eo} \le 1.84), \tag{7-11}$$

$$\mathrm{Re} = 1.83\,\mathrm{Eo}^{0.555} M^{-0.25} \qquad (1.84 \le \mathrm{Eo} \le 5.0), \tag{7-12}$$

$$\mathrm{Re} = 2.0\,\mathrm{Eo}^{0.5} M^{-0.25} \qquad (\mathrm{Eo} \ge 5.0). \tag{7-13}$$

Equation (7-13) predicts that the terminal velocity approaches an upper limit,

$$U_{\mathrm{T}} = 2.0(\Delta\rho\, g\sigma/\rho^2)^{1/4}, \tag{7-14}$$

independent of the drop size and the viscosity of the gas.

An alternative correlation given by Garner and Lihou (G5) based on data for different liquids in air may be written:

$$\mathrm{Re} = 0.776\,\mathrm{Eo}^{0.66} M^{-0.28} \qquad (\mathrm{Eo} \le 164 M^{1/6}), \tag{7-15}$$

$$\mathrm{Re} = 1.37\,\mathrm{Eo}^{0.55} M^{-0.26} \qquad (\mathrm{Eo} > 164 M^{1/6}). \tag{7-16}$$

This form of correlation was used by Beard (B3) to suggest a correlation for water drops in air under different atmospheric conditions. It should be used with caution for gases with properties widely different from air under atmospheric conditions, but the range of liquid properties covered is broad.

It is an open question whether small quantities of surfactants, too small to influence the gross properties, affect the terminal velocity of liquid drops in air. This appears unlikely in view of the large values of κ, but Buzzard and Nedderman (B18) have claimed such an influence. Acceleration may have contributed to this observation. Quantities of surfactant large enough to lower σ appreciably can lead to significantly increased deformation and hence to an increase in drag and a reduction in terminal velocity (R6).

D. Shapes of Ellipsoidal Fluid Particles

1. *Theory*

General criteria for determining the shape regimes of bubbles and drops are presented in Chapter 2, where it is noted that the boundaries between the different regions are not sharp and that the term "ellipsoidal" covers a variety of shapes, many of which are far from true ellipsoids. Many bubbles and drops in this regime undergo marked shape oscillations, considered in Section F. Where oscillations do occur, we consider a shape averaged over a small number of cycles.

As noted in Chapters 2 and 3, deformation of fluid particles is due to inertia effects. For low Re and small deformations, Taylor and Acrivos (T3) used a matched asymptotic expansion to obtain, to terms of order $\mathrm{We}^2/\mathrm{Re}$,

$$\frac{r(\theta)}{a_0} = 1 - \lambda\,\mathrm{We}\,P_2(\cos\theta) - \frac{3\lambda(11\kappa + 10)}{70(1+\kappa)}\frac{\mathrm{We}^2}{\mathrm{Re}} P_3(\cos\theta), \tag{7-17}$$

where

$$\lambda = \frac{1}{32(1+\kappa)^3}\left\{(3 + 10.3\kappa + 11.4\kappa^2 + 4.05\kappa^3) - \frac{(1+\kappa)(\gamma - 1)}{3}\right\}. \quad (7\text{-}18)$$

P_2 and P_3 are the second- and third-order Legendre polynomials. For small We, the deformed bubble or drop is predicted to be exactly spheroidal. In principle, the spheroid may be either prolate or oblate, but for cases of physical significance oblate shapes are predicted. If $\mathrm{Re}_p \gg 1$, droplet shapes are predicted to differ only slightly from the case where both Re and Re_p are small (P1). Brignell (B14) extended the series expansion to terms of order $\mathrm{We}\,\mathrm{Re}^2$. Since deformation at low Re is only observable for high M systems, this approach is of little practical value.

At larger Re and for more marked deformation, theoretical approaches have had limited success. There have been no numerical solutions to the full Navier–Stokes equation for steady flow problems in which the shape, as well as the flow, has been an unknown. Savic (S3) suggested a procedure whereby the shape of a drop is determined by a balance of normal stresses at the interface. This approach has been extended by Pruppacher and Pitter (P6) for water drops falling through air and by Wairegi (W1) for drops and bubbles in liquids. The drop or bubble adopts a shape where surface tension pressure increments, hydrostatic pressures, and hydrodynamic pressures are in balance at every point. Thus

$$\Delta\rho\, gy + \sigma[(1/R_1) + (1/R_2) - (2/R_0)] + p_{\mathrm{HD}} - (p_{\mathrm{HD}})_p = 0, \quad (7\text{-}19)$$

where y is measured vertically upwards from the lowest point, 0, of the drop; R_1 and R_2 are the principal radii of curvature at a general point on the surface ($R_1 = R_2 = R_0$ at 0); and p_{HD} and $(p_{\mathrm{HD}})_p$ are the pressures due to the external and internal fluid motions, respectively, less the stagnation pressures. It is usual practice to assume that $(p_{\mathrm{HD}})_p \ll p_{\mathrm{HD}}$, although this has been criticized by Foote (F2). With this assumption, drop shapes can be determined if the distribution of p_{HD} is known. Savic assumed that the pressure distribution was the same as that about a rigid sphere at the same value of Re; Pruppacher and Pitter used the same approach, with more recent and reliable pressure data. Deformations were assumed small and the shape represented by a cosine series (P6, S3) or by Legendre polynomials (W1). The general procedure is the reverse of that employed by McDonald (M6) to calculate surface dynamic pressure distributions from observed drop shapes. The predictions become less realistic with increasing particle size and deformation because of increasing error in the assumed pressure distribution.

A reasonable approximation to the observed profile of many drops and bubbles is a combination of two half oblate spheroids with a common major axis and different minor axes (B8, F1). This observation has been used (W1) to propose a model from which bubble and drop shapes can be estimated at high Re. The pressure distribution over the front surface is assumed to be the potential flow pressure distribution over a complete spheroid of the same eccentricity, while the dynamic pressure over the rear is assumed uniform so that the rear deforms like a sessile drop or bubble of the same size in the same system.† The theory correctly predicts that drops in air deform most at the front, while some systems (e.g., bubbles in water) begin by flattening more at the front, then deforming more at the rear with increasing d_e.

2. *Experimental Results for Bubbles and Drops in Liquids*

It is possible to prepare a generalized plot of mean aspect ratio $\bar{E}$, where E = maximum vertical dimension/maximum horizontal dimension. In the literature, both We and Eo are commonly used as independent variables for correlating shape parameters for fluid particles. The Eotvos number gives a better overall representation (G12). As in Section B, it is necessary to separate data for liquid drops falling through air (see Section 4) and for very pure systems (see Section 3). The generalized graphical correlation for bubbles and drops in contaminated liquid media is given in Fig. 7.8. Wall effects have been eliminated using the same criteria as for terminal velocities, i.e., Eqs. (9-33) and (9-34).

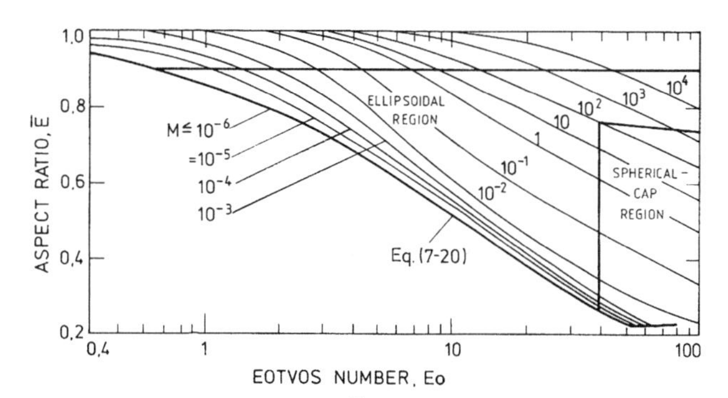

FIG. 7.8 Correlation for mean aspect ratio $\bar{E}$ of drops and bubbles in contaminated systems (B10, H7, K2, K3, K4, T6, W1, W6, Y4).

From Fig. 7.8 it is clear that deformation depends not only on Eo but also on M, higher M giving rise to less deformation at the same Eo. For low M, it is reasonable to correlate the data by a single line:

$$\bar{E} = 1/(1 + 0.163\,\mathrm{Eo}^{0.757}) \qquad (\mathrm{Eo} < 40,\ M \leq 10^{-6}), \qquad (7\text{-}20)$$

given by Wellek *et al.* (W6). For higher M, Fig. 7.8 can be used to estimate the height-to-width ratio of bubbles and drops in liquids. An alternative correlation

† Previous workers have also made use of potential flow pressure distributions about spheroids, but no allowance was made for lack of fore-and-aft symmetry, while the constant pressure condition was satisfied only near the front stagnation point (S1) or at the equator and poles (H6, M11).

obtained by Tadaki and Maeda (T1) for air bubbles with $M < 10^{-3}$ expresses $d_e/2a = \bar{E}^{1/3}$ (where $2a$ is the maximum horizontal dimension) as a function of a dimensionless group $\mathrm{Ta} = \mathrm{Re}\, M^{0.23}$. Vakrushev and Efremov (V1) extended this approach to give:

$$\bar{E} = 1 \qquad (\mathrm{Ta} \leq 1), \qquad (7\text{-}21)$$

$$\bar{E} = [0.81 + 0.206 \tanh\{2(0.8 - \log_{10} \mathrm{Ta})\}]^3, \quad (1 \leq \mathrm{Ta} \leq 39.8), \qquad (7\text{-}22)$$

$$\bar{E} = 0.24 \qquad (\mathrm{Ta} \geq 39.8) \qquad (7\text{-}23)$$

Equation (7-23) implies a spherical-cap shape with an included angle of about 50° (see Chapter 8).

3. *Experimental Results for Pure Systems*

Drops and bubbles in highly purified systems are significantly more deformed than corresponding fluid particles in contaminated systems. Increased flattening of fluid particles in pure systems results from increased inertia forces related to the increased terminal velocities discussed above. Some experimental results for drops and bubbles in water (low M systems) are shown in Fig. 7.9. The

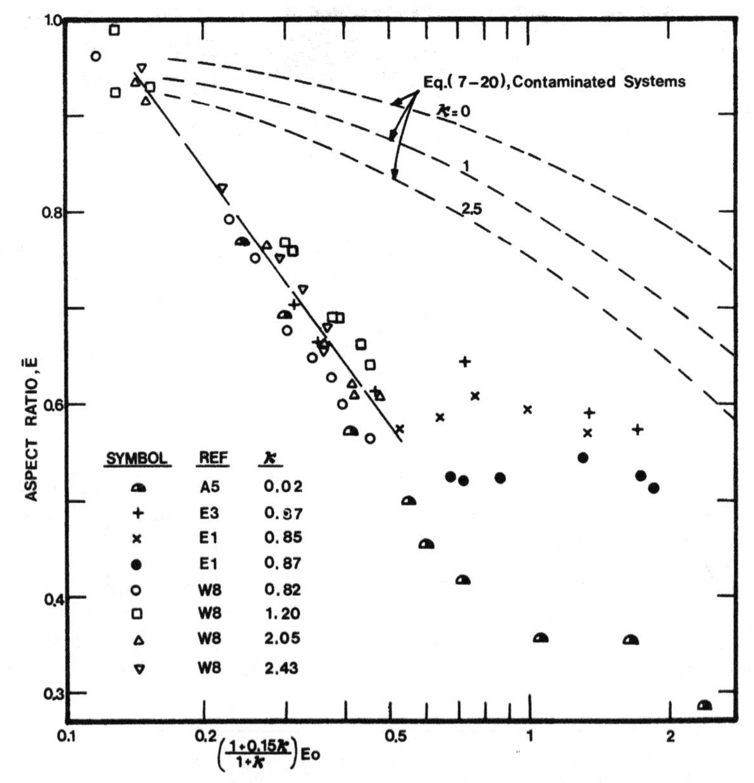

FIG. 7.9 Deformation of drops and bubbles in pure water.

aspect ratios lie significantly below the predictions of Eq. (7-20). The abscissa used in Fig. 7.7 brings the data together for $[\mathrm{Eo}(1 + 0.15\kappa)/(1 + \kappa)] \dot{<} 0.5$, but there is considerable scatter beyond this point. Once again the greatest effect of system purity is in the ellipsoidal regime, small bubbles and drops being spherical ($E = 1$) and large ones approaching $E = 0.24$ no matter how pure the system. In addition, system purity has the greatest effect at low κ.

4. *Experimental Results for Drops in Air*

The shapes of liquid drops falling through air can be conveniently represented by two oblate semispheroids with a common semimajor axis a and minor semiaxes b_1 and b_2 (B8, F1). Several workers have reported measurements of the aspect ratio, $(b_1 + b_2)/2a$, and these are shown as a function of Eo in Fig. 7.10. The data can be represented by the relationships

$$\frac{b_1 + b_2}{2a} \approx 1.0 \qquad (\mathrm{Eo} \leq 0.4), \tag{7-24}$$

$$\frac{b_1 + b_2}{2a} = \frac{1.0}{1.0 + 0.18(\mathrm{Eo} - 0.4)^{0.8}} \qquad (0.4 < \mathrm{Eo} < 8). \tag{7-25}$$

The shape factor, $b_1/(b_1 + b_2)$, is also plotted in Fig. 7.10 based on data given by Finlay (F1). The relationships

$$\frac{b_1}{b_1 + b_2} = 0.5 \qquad (\mathrm{Eo} \leq 0.5), \tag{7-26}$$

$$\frac{b_1}{b_1 + b_2} = \frac{0.5}{1.0 + 0.12(\mathrm{Eo} - 0.5)^{0.8}} \qquad (0.5 < \mathrm{Eo} < 8) \tag{7-27}$$

give an adequate fit to the data. Equations (7-25) and (7-27) are plotted in Fig. 7.10.

A good approximation to the shape of deformed drops in air may therefore be obtained from knowledge of the system properties and drop size. The ratios $(b_1 + b_2)/2a$ and $b_1/(b_1 + b_2)$ are calculated from Eo using Eqs. (7-25) and (7-27). From geometric considerations

$$d_e/a = 2\left[\frac{b_1 + b_2}{2a}\right]^{1/3}, \tag{7-28}$$

so that the semiminor axes can then be calculated. The surface area may be estimated by again assuming that the drop is composed of two half spheroids, i.e.,

$$A = 2\pi a^2 + \frac{\pi}{2}\left\{\frac{b_1^{\,2}}{e_1}\ln\left(\frac{1 + e_1}{1 - e_1}\right) + \frac{b_2^{\,2}}{e_2}\ln\left(\frac{1 + e_2}{1 - e_2}\right)\right\}, \tag{7-29}$$

where $e_1 = (1 - b_1^{\,2}/a^2)^{1/2}$ and $e_2 = (1 - b_2^{\,2}/a^2)^{1/2}$ are the eccentricities of the front and rear sections.

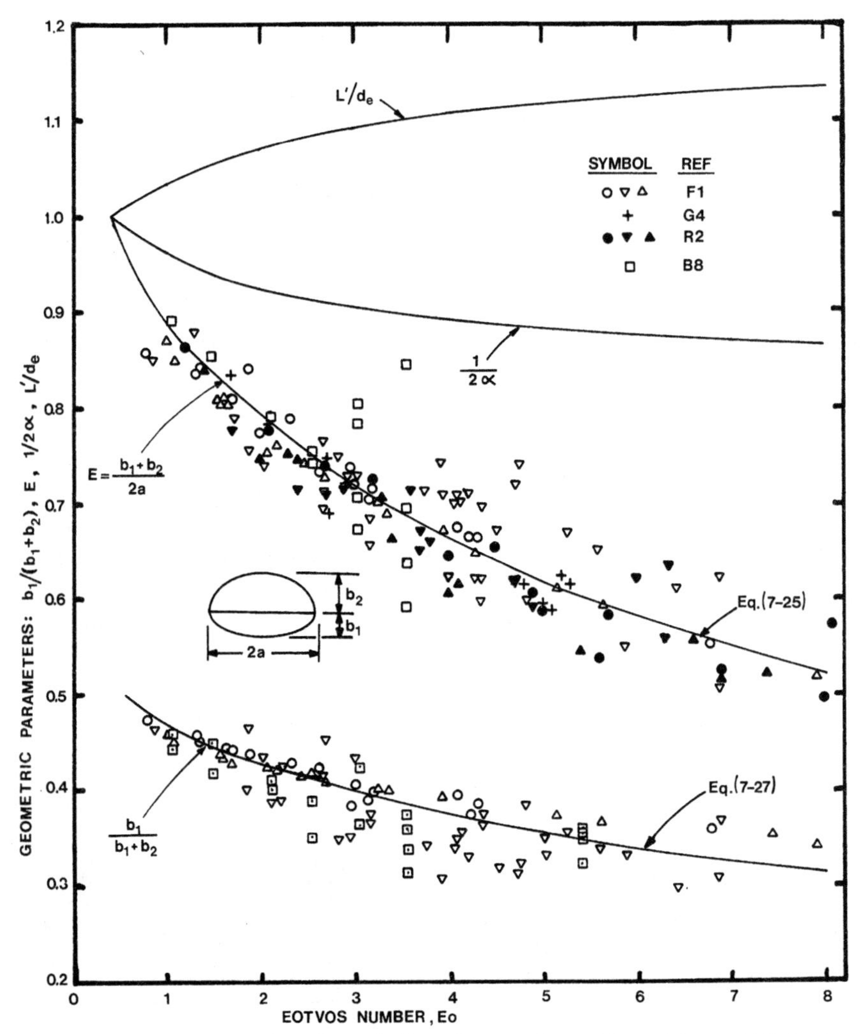

FIG. 7.10 Aspect ratio E, shape factor $b_1/(b_1 + b_2)$, area factor $1/2\alpha$, and factor L'/d_e for drops falling in air.

E. Wakes of Deformed Bubbles and Drops

The formation of an attached wake and the subsequent onset of wake shedding tend to be promoted by increasing oblateness (see Chapter 6) and by the tendency of surface-active contaminants to damp out internal circulation (see Chapter 5). Experiments have been conducted with dyes added to enable attached wakes and shedding phenomena to be visualized (H8, M1, M2, S2) and wake volumes to be measured (H8, Y4) for drops and bubbles. Since dyes tend to be surface active, the results of these experiments are probably relevant to grossly contaminated systems. Other tracers have also been used in wake visualization studies (L8, L9). The appearance of an attached wake for impure systems and the onset of wake shedding occur at Re values of about 20 and 200, respectively, as for rigid spheres, or somewhat lower values (e.g., 5 and 100) if significant deformation has already taken place before these values of Re are achieved (G6, G8, H8, S2, W6, Y4). Magarvey and co-workers (M1, M2, M3) have given an excellent series of photographs showing wakes of slightly deformed drops in liquid–liquid systems and have identified six different classes of wake.

For carefully purified systems, interfacial mobility can significantly delay both the formation of an attached eddy and wake shedding, especially for fluid particles of low κ. For example, wake shedding which began at Re $\doteq$ 200 for a contaminated system was delayed to Re $\doteq$ 800 for a carefully purified system of virtually identical properties (W8). Moreover, at a given Re, the wake volume is smaller for pure systems (E3, E4, W8). Winnikow and Chao (W8) distinguished two main classes of wake for purified drops: (a) steady vortex threads (accompanied by an attached toroidal vortex for larger κ); (b) wakes which periodically discharge vorticity, typically with convoluted geometry, initially axisymmetric but eventually becoming unsteady and asymmetric with the onset of a turbulent wake. The latter type is closely associated with shape oscillations as noted in the next section. Some photographs given by Winnikow and Chao are reproduced in Fig. 7.11.

Few observations have been reported on wakes of ellipsoidal bubbles and drops at Re > 1000. Yeheskel and Kehat (Y4) characterized shedding in this case as random. However, Lindt (L7, L8) studied air bubbles in water and distinguished a regular periodic component of drag associated with an open helical vortex wake structure. Strouhal numbers (defined as $2af/U_T$, where f is the frequency and $2a$ is the maximum horizontal dimension) increase with Re, to level off at about 0.3 as bubbles approach the transition between the ellipsoidal and spherical-cap regimes.

F. Secondary Motion

Bubbles and drops of intermediate size show two types of secondary motion:

(i) "Rigid body" type, e.g., rocking from side to side, or following a zig-zag or spiral trajectory (cf. spheres and disks in Chapters 5 and 6).
(ii) Shape dilations, usually referred to as "oscillations."

These two types of motion are often superimposed, so that the motion of intermediate size fluid particles can be particularly complex.

While other explanations have been proposed [e.g. (B6, E1, H6)], secondary motions are most plausibly related to wake shedding. The onset of oscillations coincides with the onset of vortex shedding from the wake (E1, E2, S5, W8). For high κ or contaminated drops and bubbles, the onset of oscillations

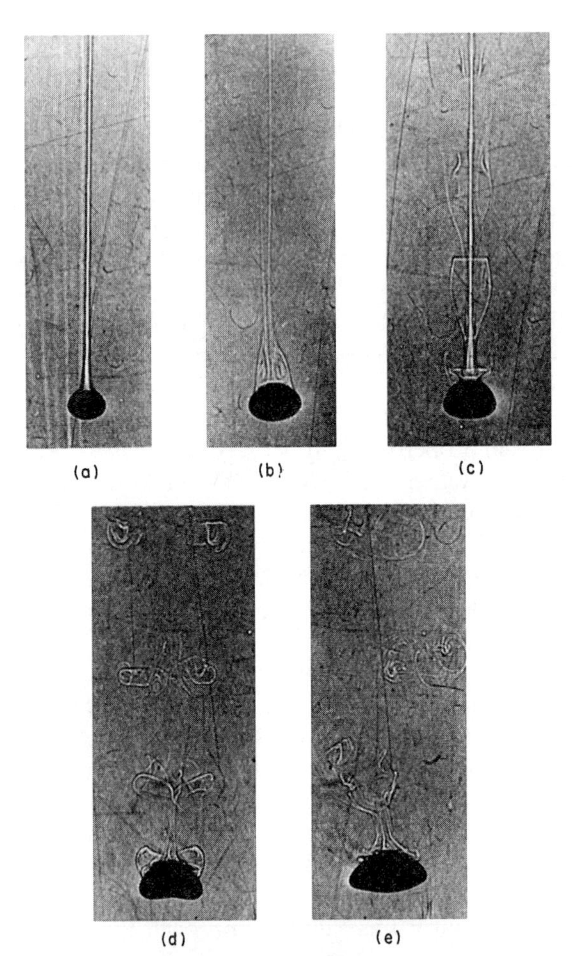

FIG. 7.11 Wake configurations for drops in water (highly purified systems), reproduced from Winnikow and Chao (W8) with permission. (a) nonoscillating nitrobenzene drop: $d_e = 0.280$ cm, Re = 515; steady thread-like laminar wake; (b) nonoscillating *m*-nitrotoluene drop: $d_e = 0.380$ cm, Re = 688; steady thread accompanied by attached toroidal vortex wake; (c) oscillating nitrobenzene drop: $d_e = 0.380$ cm, Re = 686; central thread plus axisymmetric outer vortex sheet rolled inward to give inverted bottle shape of wake; (d) oscillating nitrobenzene drop: $d_e = 0.454$ cm, Re = 775; vortex sheet in c has broken down to form vortex rings; (e) oscillating nitrobenzene drop: $d_e = 0.490$ cm, Re = 804; vortex rings in d now shed asymmetrically and the drop exhibits a rocking motion.

therefore occurs at a Reynolds number of about 200 (G8, H6, S5), while for pure systems at relatively low κ, the onset of oscillations is delayed (H6, W8), but seldom beyond Re = 1000. In viscous liquids where Re never reaches 200 over the range of practical interest (see Fig. 2.5), no oscillations occur (K4, T2). While a critical Weber number has often been suggested for the onset of oscillations in pure, low κ systems, no agreement has been reached on what the critical value should be (E2, H6, W8), and the value of Re and purity of the system appear to be better indicators of the likelihood of secondary motion.

While wake shedding appears to provide the excitation for shape oscillations, the frequency of the two phenomena may differ. For example, Winnikow and Chao (W8) measured oscillation frequencies between about 60 amd 80% of wake shedding frequencies for nitrobenzene drops in water, while Edge *et al.* (E1) found the two frequencies to be identical. To obtain a simple physical understanding of shape oscillations, consider forced vibration of a single-degree-of-freedom damped system [see, e.g., Anderson (A2)]. Suppose that the wake shedding provides a harmonic excitation of frequency f_W, while the natural frequency of the drop is given (L1) by

$$f_N = \sqrt{\frac{48\sigma}{\pi^2 d_e^{\ 3}\rho(2 + 3\gamma)}}. \tag{7-30}$$

If we define

$$\bar{f} = (f_W + f_N)/2 \tag{7-31}$$

and

$$\Delta f = (f_W - f_N)/2, \tag{7-32}$$

and if $\Delta f \ll \bar{f}$, then the motion is approximately

$$E - \bar{E} \propto \frac{\bar{f}}{\Delta f} \sin(2\pi\,\Delta f t)\cos(2\pi \bar{f} t). \tag{7-33}$$

As illustrated in Fig. 7.12, the drop then oscillates at frequency $\bar{f}$ with the amplitude modulated at frequency Δf.

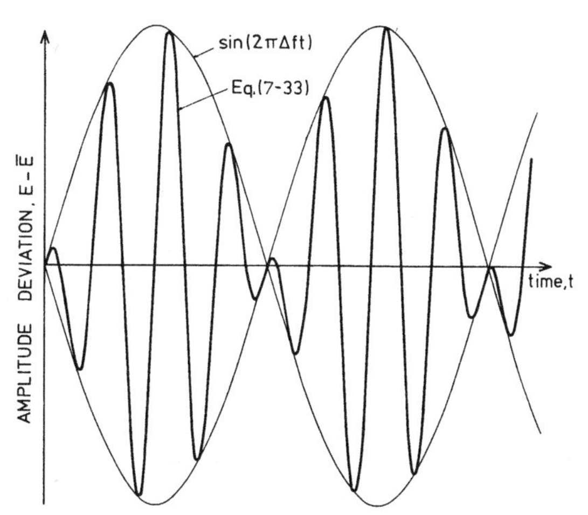

FIG. 7.12 Simple model to show nature of shape oscillations for bubbles and drops in free motion.

In practice, this model is oversimplified since the exciting wake shedding is by no means harmonic and is itself coupled with the shape oscillations and since Eq. (7-30) is strictly valid only for small oscillations and stationary fluid particles. However, this simple model provides a conceptual basis to explain certain features of the oscillatory motion. For example, the period of oscillation, after an initial transient (E1), becomes quite regular while the amplitude is highly irregular (E3, S4, S5). "Beats" have also been observed in drop oscillations (D4). If f_W and f_N are of equal magnitude, one would expect resonance to occur, and this is one proposed mechanism for breakage of drops and bubbles (Chapter 12).

Equation (7-30) gives the natural frequency of the fundamental mode for stationary fluid particles undergoing small oscillations with viscous forces neglected. It has been modified to account for viscous effects (L4, M10, S10), surface impurities (M10), finite amplitudes (S5, Y1), and translation (S10). Observed oscillation frequencies are generally less than those given by Eq. (7-30), typically by 10–20% for drops in free motion in impure systems (S4) and by 20–40% for pure systems (E1, E3, W8, Y1). The amplitude tends to be larger for pure systems (E3) and this explains the reduction in frequency.

In general, oscillations may be oblate-prolate (H8, S5), oblate-spherical, or oblate-less oblate (E2, F1, H8, R3, R4, S5). Correlations of the amplitude of fluctuation have been given (R3, S5), but these are at best approximate since the amplitude varies erratically as noted above. For low M systems, secondary motion may become marked, leading to what has been described as "random wobbling" (E2, S4, W1). There appears to have been little systematic work on oscillations of liquid drops in gases. Such oscillations have been observed (F1, M4) and undoubtedly influence drag as noted earlier in this chapter. Measurements (Y3) for 3–6 mm water drops in air show that the amplitude of oscillation increases with d_e, while the frequency is initially close to the Lamb value (Eq. 7-30) but decays with distance of fall.

Oscillating bubbles and drops may travel along zig-zag or spiral (helical) paths. Some authors have observed only one of these modes while others have observed both. There is some evidence that the type of secondary motion is affected by the mode of release (M8). Saffman (S1) performed a careful series of experiments on air bubbles in water. Rectilinear motion was found to become unstable, and gave rise to zig-zag motion which in turn gave way to spiral motion for larger bubbles.[†] The paths followed by fluid particles undergoing secondary motion are no doubt associated with the type of wake. Details of the paths, orientation, and periods of spiralling and zig-zagging drops and bubbles are presented by Mercier and Rocha (M9) and Tsuge and Hibino (T6).

Secondary motion plays an important role in increasing drag (L7) and in promoting heat and mass transfer from bubbles or drops. The onset of oscillations corresponds approximately to the maximum in $U_T(d_e)$ and minimum in

[†] See also Table 7.1.

C_D(Re) curves for drops and bubbles (B11, E1, E2, T4). The influence of oscillations on heat and mass transfer is discussed in Section III.

G. Internal Circulation

Surface-active contaminants play an important role in damping out internal circulation in deformed bubbles and drops, as in spherical fluid particles (see Chapters 3 and 5). No systematic visualization of internal motion in ellipsoidal bubbles and drops has been reported. However, there are indications that deformations tend to decrease internal circulation velocities significantly (M12), while shape oscillations tend to disrupt the internal circulation pattern of droplets and promote rapid mixing (R3). No secondary vortex of opposite sense to the prime internal vortex has been observed, even when the external boundary layer was found to separate (S11).

H. Theoretical Solutions for Deformed Bubbles and Drops

Attempts to obtain theoretical solutions for deformed bubbles and drops are limited, while no numerical solutions have been reported. A simplifying assumption adopted is that the bubble or drop is perfectly spheroidal. Saffman (S1) considered flow at the front of a spheroidal bubble in spiral or zig-zag motion. Results are in fair agreement with experiment. Harper (H4) tabulated energy dissipation values for potential flow past a true spheroid. Moore (M11) applied a boundary layer approach to a spheroidal bubble analogous to that for spherical bubbles described in Chapter 5. The interface is again assumed to be completely free of contaminants. The drag is given by

$$C_D = \frac{48}{\mathrm{Re}} f_1(E)\left[1 + \frac{f_2(E)}{\mathrm{Re}^{1/2}}\right], \tag{7-34}$$

where the first term results from the viscous energy dissipation for irrotational flow past an oblate spheroid, and the second arises from dissipation in the boundary layer and wake. Harper (H5) tabulated values of $f_1(E)$ and $f_2(E)$ and plotted drag curves for four values of M. The curves show minima and are in qualitative agreement with observed C_D(Re) curves for bubbles. No attempt has been reported to extend this treatment to deformed drops of low κ.

III. HEAT AND MASS TRANSFER

A. Regimes of Motion and Transfer

The flow and shape transitions for small and intermediate size bubbles and drops are summarized in Fig. 7.13. In pure systems, bubbles and drops circulate freely, with internal velocity decreasing with increasing κ. With increasing size they deform to ellipsoids, finally oscillating in shape when Re exceeds a value of order 10^3. In contaminated systems spherical and nonoscillating ellipsoidal

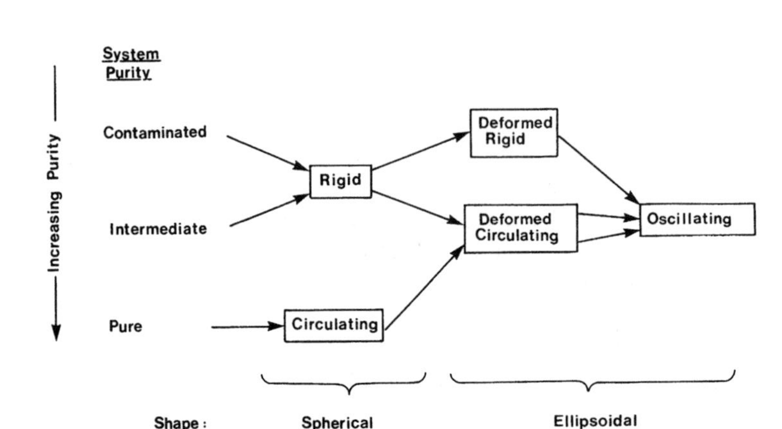

FIG. 7.13 Flow transitions for bubbles and drops in liquids (schematic).

bubbles and drops are effectively rigid but, for Re $\dot{>}$ 200, wake shedding and shape oscillations occur with associated motion of the internal fluid. In systems of intermediate purity, small bubbles and drops are rigid but, with increasing size, they become deformed and partially circulating. Circulation increases with increasing size, and shape oscillations occur at Re $\dot{>}$ 200. The Reynolds number marking the transition from rigid to circulating behavior depends on system purity.

These flow transitions lead to a complex dependence of transfer rate on Re and system purity. Deliberate addition of surface-active material to a system with low to moderate κ causes several different transitions. If Re $\dot{<}$ 200, addition of surfactant slows internal circulation and reduces transfer rates to those for rigid particles, generally a reduction by a factor of 2–4 (S6). If Re $\dot{>}$ 200 and the drop is not oscillating, addition of surfactant to a pure system decreases internal circulation and reduces transfer rates. Further additions reduce circulation to such an extent that shape oscillations occur and transfer rates are increased. Addition of yet more surfactant may reduce the amplitude of the oscillation and reduce the transfer rates again. Although these transitions have been observed (G7, S6, T5), additional data on the effect of surface active materials are needed.

The internal resistance is always decreased substantially when a bubble or drop oscillates, but the external resistance may be unaffected if the Reynolds number is high enough. A rough criterion can be obtained from Eq. (11-63) for vibration of a particle in an axial stream. Oscillation has negligible effect on the external resistance if

$$\frac{\mathrm{Re_V}}{\mathrm{Re}}\left(\frac{2a'}{d}\right)^{-0.45} < 0.2, \tag{7-35}$$

where a' is the amplitude of the oscillation, $\mathrm{Re}_v = 4a'fd/\nu$ is the vibration Reynolds number, and f is the frequency. Rearranging Eq. (7-35) yields:

$$\frac{a'f}{U_T} < 0.05\left(\frac{2a'}{d}\right)^{0.45}. \tag{7-36}$$

Assuming spherical-oblate oscillations with amplitude $2a' = (1 - \bar{E})d_e$, taking $\bar{E} = 0.5$ as a rough approximation and replacing d by d_e and f by f_N from Eq. (7-30), we find no effect of oscillation on the external resistance if

$$d_e f_N/U_T \lessdot 0.15. \tag{7-37}$$

For liquid drops in gases the terminal velocity is so large that the inequality is obeyed and oscillation has essentially no effect on transfer. For drops and bubbles in liquids, the effect of oscillation on transfer is significant.

B. Definitions

Mass transfer rates from drops are obtained by measuring the concentration change in either or both of the phases after passage of one or more drops through a reservoir of the continuous phase. This method yields the average transfer rate over the time of drop rise or fall, but not instantaneous values. For measurements of the resistance external to the drop this is no drawback, because this resistance is nearly constant, but the resistance within the drop frequently varies with time. The fractional approach to equilibrium, F, is calculated from the compositions and is then related to the product of the overall mass transfer coefficient and the surface area:

$$(\overline{KA})_p = -(\pi d_e^3/6t)\ln(1 - F), \tag{7-38}$$

where t is the time of free rise or fall and $(\overline{KA})_p$ is the time-average coefficient-area product based on dispersed phase concentrations. If the resistance in each phase may be added,

$$\frac{1}{(\overline{KA})_p} = \frac{H}{\overline{kA}} + \frac{1}{(\overline{KA})_p}. \tag{7-39}$$

If the resistance external to the drop is negligible,

$$(\overline{kA})_p = -(\pi d_e^3/6t)\ln(1 - F). \tag{7-40}$$

Many investigators base mass transfer coefficients upon the area of the volume-equivalent sphere, especially for oscillating drops:

$$(\overline{kA})_p/A_e = -(d_e/6t)\ln(1 - F). \tag{7-41}$$

The Sherwood number based on this coefficient is

$$\mathrm{Sh}_{pe} = [(\overline{kA})_p/A_e]d_e/\mathscr{D}_p. \tag{7-42}$$

A similar definition is frequently used for the continuous phase Sherwood number

$$\mathrm{Sh_e} = [\overline{kA}/A_\mathrm{e}]d_\mathrm{e}/\mathscr{D} \tag{7-43}$$

In some studies the surface area of the particle is measured and area-free Sherwood numbers are reported

$$\mathrm{Sh_p} = \bar{k}_\mathrm{p} d_\mathrm{e}/\mathscr{D}_\mathrm{p}, \tag{7-44}$$

$$\mathrm{Sh} = \bar{k} d_\mathrm{e}/\mathscr{D}. \tag{7-45}$$

Careful reading of papers is required to determine which definition has been used. Measurements of the continuous phase resistance around bubbles frequently use photographic, volumetric, or pressure change techniques to yield instantaneous rates of mass transfer, and thus kA. Here too, both definitions of the Sherwood number, Eqs. (7-43) and (7-45), have been used.

C. External Resistance

Figure 7.14 gives area-free Sherwood numbers for organic drops in water. In the furfural–water system ($\kappa = 1.7$), the transition from circulation at low Re to circulation at high Re agrees well with the treatment of Chapter 5, i.e., deformation has little effect on the area-free Sherwood number. For this value of κ, however, it is not clear whether the drops were circulating for $\mathrm{Re} < 10$. For the diol–water system ($\kappa = 80$), circulation is so slow that Sh agrees with the result for rigid spheres up to $\mathrm{Re} \doteq 200$ where oscillation begins. At this Reynolds number, $d_\mathrm{e} f_\mathrm{N}/U_\mathrm{T} \doteq 0.6$ and oscillations are expected to affect the Sherwood number; see Eq. (7-37). The chlorobenzene/benzene drop system ($\kappa = 0.7$) shows the effect of addition of surfactant. Without surfactant, Sh departs from the line for solids at $\mathrm{Re} \doteq 20$ and deviation increases with Re as circulation becomes stronger. The data with added surfactant follow the line for solids up to $\mathrm{Re} \doteq 50$ and remain below the pure system values at higher Re. Even the system without surfactant was contaminated, since the data should lie above those for $\kappa = 1.7$. The presence of surface-active materials acts in the same way as an increase in the drop viscosity with respect to terminal velocity. Transition from a stagnant drop to a drop with circulation may occur at any Re below 200. The data for aniline drops ($\kappa = 4.4$) lie between the systems with $\kappa = 1.7$ and 80, and show reasonable agreement with Eq. (5-39). Oscillation in contaminated systems and circulation in less contaminated systems both cause Sh to rise more rapidly than $\mathrm{Re}^{1/2}$.

1. *Particles without Shape Oscillations*

For nonspherical particles the only theoretical treatment available is for potential flow around a spheroid (L10). For an oblate spheroid the area-free

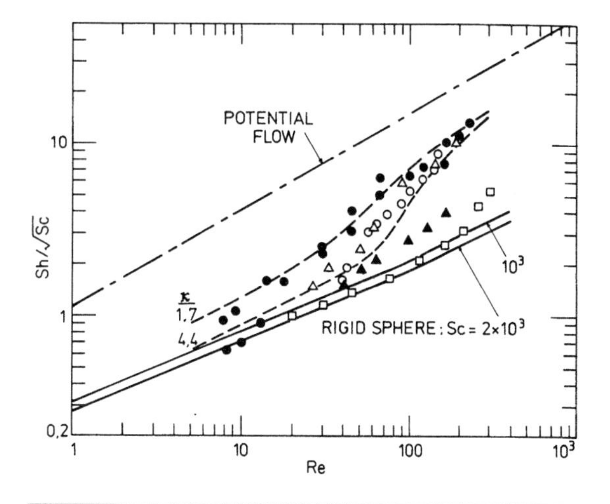

Symbol	κ	γ	Sc	Ref.
□	80	0.95	1660	(G8)
○	4.4	1.02	1100	(H11)
●	1.7	1.16	975	(G8, H11)
△	0.7	1.02	1020	(T5)
▲	same as △, but contaminated with 10^{-3} gm/liter sodium oleyl-*p*-anisidine sulfonate			
---	Eq. (5-39)			
—	Rigid sphere, Eq. (5-25), Table 5.4E			
----	Sphere in potential flow, Eq. (5-35)			

FIG. 7.14 Area-free mass transfer factors, $\mathrm{Sh}/\sqrt{\mathrm{Sc}}$, for drops.

Sherwood number is

$$\mathrm{Sh} = \frac{2}{\sqrt{\pi}} \mathrm{Pe}^{1/2} \left[\frac{8e^3 E^{1/3}}{3(\sin^{-1} e - eE)} \right]^{1/2} \Bigg/ \left[1 + \frac{E^2}{2e} \ln\left(\frac{1+e}{1-e} \right) \right], \tag{7-46}$$

where Pe is based on d_e and

$$e = (1 - E^2)^{1/2}. \tag{7-47}$$

Since the area ratio is given by

$$\frac{A}{A_e} = \frac{1}{2E^{2/3}} \left[1 + \frac{E^2}{2e} \ln\left(\frac{1+e}{1-e} \right) \right], \tag{7-48}$$

then

$$\mathrm{Sh}_e = \frac{2}{\sqrt{\pi}} \mathrm{Pe}^{1/2} \left[\frac{2e^3}{3E(\sin^{-1} e - eE)} \right]^{1/2}. \tag{7-49}$$

Comparison of these equations shows that the area-free Sherwood number is only slightly affected by eccentricity; e.g. $\mathrm{Sh}/\mathrm{Pe}^{1/2}$ for a spheroid with $E = 0.4$ is only 8.5% larger than that for the equivalent sphere while the area ratio A/A_e is 17% larger. Therefore, we expect little effect of deformation on the area-free Sherwood number for bubbles and drops at high Re. This is borne out by the agreement of the data in Fig. 7.14 with Eq. (5-39), derived for fluid spheres.

a. *Drops in Gases* For liquid drops in gases at low pressure the equations for solid particles in Chapter 6 can be used to predict heat and mass transfer rates. Figure 7.10 shows the area ratio α and the ratio L'/d_e as functions of Eo, to facilitate use of Eq. (6-34), while areas may be calculated from Eq. (7-29) or from Eq. (7-48). Surface-active materials should have little effect. For drops in high-pressure gases, oscillations may become important if $\mathrm{Re} > 200$ and the terminal velocity is small enough that $d_e f_N/U_T > 0.1$.

Near the point of drop release, transfer coefficients can be much different from those predicted, due to large amplitude oscillation and internal circulation induced by departure from the nozzle or tip (A1, G4, Y3).

b. *Drops in Liquids* For drops in pure liquid systems, the area-free Sherwood number may be taken as the larger of the values calculated from the equations for solid spheres in Chapter 5 or Eq. (5-39) for fluid spheres. This provides a transition from the lines for solids in Fig. 7.14 to the potential flow line with increasing Re. For impure systems, surface-active materials may immobilize the drop surface and reduce the coefficients to those for solid particles. The area-free Sherwood number should be equal to or above that for a solid sphere, yet below that for a fluid sphere given by Eq. (5-39). If the system is grossly contaminated, oscillations occur if $\mathrm{Re} > 200$.

c. *Bubbles in Water* Water is the only continuous fluid for which reliable mass transfer data are available at low M. Figure 7.15 presents the mass transfer factor $(\overline{kA}/A_e)/\mathscr{D}^{1/2}$ for bubbles in water including only data in which wall effects are small ($d_e/D < 0.12$) and for which the water had been degassed. Dissolved gases can transfer into the bubble and reduce the driving force appreciably (B13, L6, W5). The scatter in the figure is due to different methods of bubble release (Z2), different techniques of measuring the mass transfer rate (G1, W7), and different system purities (R1). Figure 7.15 also shows the mass transfer factor for a rigid spheroid with its aspect ratio given by Eq. (7-20), its velocity by the lower curve in Fig. 7.3, and its Sherwood number calculated from Eqs. (6-16) and (6-17) with $\mathrm{Sc} = 500$. Predictions for potential flow from Eq. (7-46) are also shown, based on the properties of water at 25°C with terminal velocity from the upper curve in Fig. 7.3. Curve 1 corresponds to pure systems, with bubble shape from Fig. 7.9, while curve 2 corresponds to the shape in a contaminated system given by Eq. (7-20).

For $d_e > 0.5$ cm, the data agree closely with the potential flow solution with the shape appropriate to a contaminated system. For $d_e < 0.5$ cm, system purity

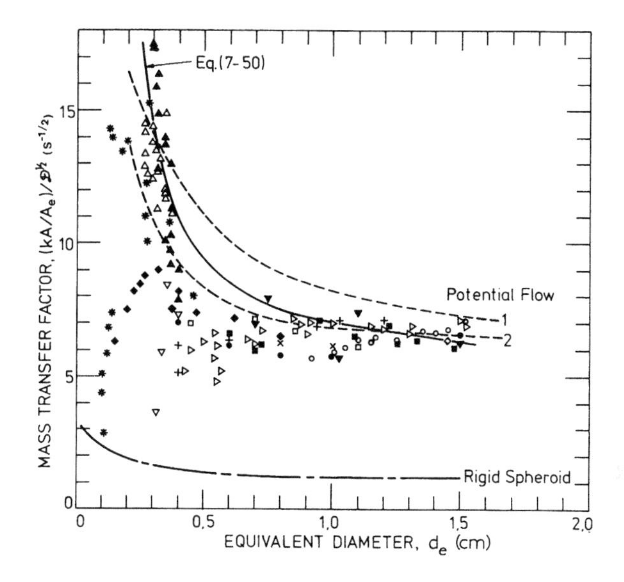

Experimental Data

Symbol	Gas	Ref.	Symbol	Gas	Ref.
+	CO_2	(J1)	△	CO_2	(Z2)
*	CO_2	(W4)	▽	CO_2	(R1)[a]
●	CO_2	(C1)	○	O_2	(V3)
◇	CO_2	(G15)	▲	C_2H_2	(L2)
▷	CO_2	(D3)	×	C_2H_4	(J1)
▼	CO_2	(B2)	◆	C_2H_4	(G3)
■	CO_2	(L6)	□	C_4H_8	(J1)

[a] As Δ, but with 1.2 ppm *n*-nonanol.

FIG. 7.15 Mass transfer factor $(kA/A_e)/\sqrt{\mathscr{D}}$ for gas bubbles in water.

has a pronounced effect, just as on terminal velocity (see Fig. 7.3). In carefully purified systems [e.g., (Z2)], the mass transfer coefficient increases sharply with decreasing d_e, but contaminated systems do not show such a sharp increase. With 1.2 ppm *n*-nonanol added, the coefficient decreases towards the value for a rigid spheroid. Garner and Hammerton (G3) and Weiner (W4) apparently used systems of intermediate purity. Weiner also found that the mass transfer coefficient and terminal velocity decreased with bubble age due to accumulation of surfactants. The data for pure systems with $d_e < 0.5$ cm are better predicted by the potential flow solution with shape given by Fig. 7.9, but the predicted mass transfer factors increase less rapidly with decreasing bubble size than the data. The failure of the prediction results from zig-zag and helical motion in the range $0.2 \text{ cm} < d_e < 0.4 \text{ cm}$ (see Table 7.1).

A reasonable upper limit on the mass transfer factor from bubbles to well-purified water at room temperature is given by:

$$\frac{\overline{kA}/A_e}{\mathscr{D}^{1/2}} = \frac{0.14}{d_e^{\,3}} + \frac{6.94}{d_e^{1/4}}, \tag{7-50}$$

with d_e in cm and the left side in $s^{-1/2}$. For contaminated systems, the data for $d_e > 0.5$ cm are well represented by taking $(\overline{kA}/A_e)/\mathscr{D}^{1/2} = 6.5\ s^{-1/2}$.

2. *Particles with Shape Oscillations*

When the shape of a particle oscillates, the surface area changes with time. This situation has been modeled by neglecting the motion adjacent to the surface due to the terminal velocity of the particle, i.e., by considering the particle to be oscillating but stationary, with material transferred by transient molecular diffusion over a time equal to the period of oscillation. For $Sc \gg 1$ the thin concentration boundary layer assumptions are invoked (see Chapter 1).

Two alternative assumptions have been made for the manner in which the area variation occurs. The more realistic postulates that all elements of the surface remain in the surface throughout an oscillation cycle. Increasing surface area stretches the surface (A3, B5) and causes a velocity normal to the surface which increases the diffusion rate. For a surface of area A_0 suddenly exposed at $t = 0$, the mass transfer product averaged over time is given by

$$\frac{\overline{kA}}{A_0} = 2\sqrt{\frac{\mathscr{D}}{\pi t}}\left[\frac{1}{t}\int_0^t \left(\frac{A}{A_0}\right)^2 dt'\right]^{1/2}, \tag{7-51}$$

where the bracketed term represents the effect of the area variation. The value of $\overline{kA}$ is proportional to the r.m.s. interfacial area, so that the transfer rate is larger when the area oscillates.

The alternate assumption is that new elements are brought to the surface as the area increases, and the oldest elements are removed from the surface when the area decreases (B16). For a surface of area A_0 exposed at $t = 0$, the time-averaged mass transfer product is then

$$\frac{\overline{kA}}{A_0} = 2\sqrt{\frac{\mathscr{D}}{\pi t}} + \sqrt{\frac{\mathscr{D}}{\pi}}\frac{1}{t}\int_0^t\left[\int_0^T \frac{1}{\sqrt{T-t'}}\frac{d(A/A_0)}{dt'}dt'\right]dT. \tag{7-52}$$

The first term on the right-hand side represents transfer to the elements of surface present over the entire time period t, while the second represents transfer to appearing or disappearing elements. The fresh surface model, Eq. (7-52), predicts larger coefficients than the surface stretch model, Eq. (7-51).

Given the time variation of the area of a fluid particle, the $\overline{kA}$ product is easily calculated. For oscillating droplets, Angelo *et al.* (A3) showed that the time variation of area is given closely by:

$$A/A_0 = 1 + \varepsilon \sin^2(\pi f t'), \tag{7-53}$$

where $1 + \varepsilon$ is the ratio of maximum area to minimum area, A_0. Assuming that the averaging time is the period of oscillation, f^{-1}, and that the oscillation is spherical-oblate, we obtain from Eqs. (7-51) and (7-53) for the surface stretch model:

$$\mathrm{Sh_e} = \frac{2}{\sqrt{\pi}} \sqrt{\frac{d_e^2 f}{\mathscr{D}}} \sqrt{1 + \varepsilon + \frac{3\varepsilon^2}{8}}, \tag{7-54}$$

while from Eqs. (7-52) and (7-53) the fresh surface model yields

$$\mathrm{Sh_e} = \frac{2}{\sqrt{\pi}} \sqrt{\frac{d_e^2 f}{\mathscr{D}}} (1 + 0.687\varepsilon). \tag{7-55}$$

These results are remarkably close to each other; e.g., for an extreme value of $\varepsilon = 0.5$ the fresh surface prediction is only 6% larger than the surface stretch prediction. The amplitude of the area oscillation, ε, has a relatively small effect since $\varepsilon \doteq 0.3$ in many systems (R3, Y1).

Mass transfer data for oscillating liquid drops have been obtained in several studies in liquids (G2, G8, Y2) and a single study in gases (L5). Comparison with Eqs. (7-54) and (7-55) is difficult due to uncertainty in predicting the frequency f, and the lack of data on the amplitude factor ε. As noted earlier, the frequency of oscillation is generally less than the natural frequency given by Eq. (7-30). The following empirical equation applies to the liquid–liquid data with an average deviation of 6%:

$$\mathrm{Sh_e} = 1.2\sqrt{d_e^2 f_N/\mathscr{D}} \qquad \text{or} \qquad \overline{kA}/A_e = 1.2\sqrt{f_N \mathscr{D}}. \tag{7-56}$$

Data for drops in gases show an average deviation of about 30% from Eq. (7-56).

D. Internal Resistance

For circulating fluid particles without shape oscillations the internal resistance varies with time in a way similar to that discussed in Chapter 5 for fluid spheres. The occurrence of oscillation, with associated internal circulation, always has a strong effect on the internal resistance. If the oscillations are sufficiently strong to promote vigorous internal mixing, the resistance within the particle becomes constant.

1. *Particles without Shape Oscillations*

Although there are no solutions for circulating ellipsoidal fluid particles similar to the Kronig–Brink model for spheres, Fig. 3.22, which includes the external resistance, should be a good approximation with $d = d_e$ and k taken to be the area-free external mass transfer coefficient. This procedure is supported by the work on freely suspended drops in gases by Garner and Lane (G4), who found that the Kronig–Brink model applied up to $\mathrm{Re} \doteq 3000$ after decay of strong initial circulation, caused during drop formation. Some of their data are shown in Fig. 7.16. The ethylene glycol and monoethanolamine drops did not oscillate. The data from these experiments, in which there was a small external resistance, agree well with the curve for Bi = 50 from Fig. 3.22. Similar agreement with the Kronig–Brink model has been found for drops in liquids (B15, K1) as noted in Chapter 5. Although their data for nonoscillating drops in liquids were in fair agreement with the Kronig—Brink model, Skelland and Wellek (S7) proposed an empirical equation which is widely used. In impure systems, where surface-active materials make the particle effectively rigid, the drop may approach equilibrium at rates given by Fig. 3.21.

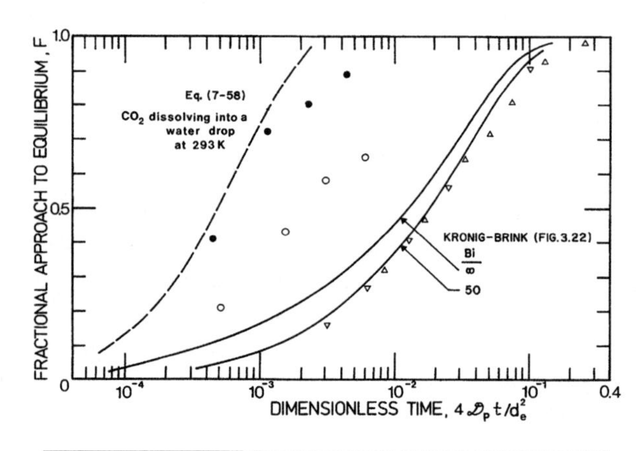

Symbol	d_e, cm	Drop	Controlling resistance	Re
●	0.58	water	internal	3500
○	0.46	dekalin	internal	3000
△	0.29	glycol	external	1500
▽	0.53	ethanolamine	external	3500

FIG. 7.16 Fractional approach to equilibrium for circulating and oscillating drops in gases. Data of Garner and Lane (G4).

2. *Particles with Shape Oscillations*

If a fluid particle oscillates violently enough to mix its contents in each oscillation cycle, the average internal resistance is constant if the driving force is based upon the mixed mean concentration within the drop. The fractional approach to equilibrium is then given by Eq. (7-40) or (7-41).

A model of transfer within an oscillating droplet was proposed by Handlos and Baron (H3). They assumed that transfer within the drop was entirely by turbulent motion, random radial movement, superimposed upon toroidal circulation streamlines. No allowance was made for the variation of shape or surface area. The results of the model are expressed in terms of a series solution for the fractional approach to equilibrium. For long times, only the first term is required, yielding a constant internal resistance:

$$\bar{k}_p = 0.00375 U_T/(1 + \kappa). \tag{7-57}$$

Calculations valid for short times and including external resistance are available (P2). Equation (7-57) gives a rough estimate of $(\overline{kA})_p/A_e$ for organic-water systems.

The assumption of transfer by a purely turbulent mechanism in the Handlos–Baron model leads to the prediction that the internal resistance is independent of molecular diffusivity. However, such independence has not been found experimentally, even for transfer in well-stirred cells or submerged turbulent jets (D4). In view of this fact and the neglect of shape and area oscillations, models based upon the surface stretch or fresh surface mechanism appear more realistic. For rapid oscillations in systems with $\mathrm{Sc} \gg 1$, mass transfer rates are described by identical equations on either side of the drop surface, so that the mass transfer results embodied in Eqs. (7-54) and (7-55) are valid for the internal resistance if $\mathscr{D}$ is replaced by $\mathscr{D}_p$. Measurements of the internal resistance of oscillating drops show that the surface stretch model predicts the internal resistance with an average error of about 20% (B16, Y1). Agreement of the data for drops in liquids with Eq. (7-56) considerably improves if the constant is increased to 1.4, i.e.,

$$(\overline{kA})_p/A_e = 1.4\sqrt{f_N \mathscr{D}_p}. \tag{7-58}$$

Figure 7.16 shows the fractional approach to equilibrium of an oscillating 5.8 mm water drop in a CO_2–air mixture, predicted from Eqs. (7-41) and (7-58). The large decrease in internal resistance with shape oscillation is readily apparent by comparison with the Kronig–Brink lines. The prediction is a good approximation of the rapid approach to equilibrium found by Garner and Lane (G4) for oscillating water droplets with negligible external resistance. Their data for dekalin are intermediate between the oscillating droplet prediction and the Kronig–Brink model, possibly because oscillation was not vigorous enough to mix the contents of the drop fully. Brunson and Wellek (B16) review other models for oscillating drops.

REFERENCES

A1. Ahmadzadeh, J. and Harker, J. H., *Trans. Inst. Chem. Eng.* **52**, 108–111 (1974).
A2. Anderson, R. A., "Fundamentals of Vibrations." Macmillan, New York, 1967.
A3. Angelo, J. B., Lightfoot, E. N. and Howard, D. W., *AIChE J.* **12**, 751–760 (1966).
A4. Aybers, N. M., and Tapucu, A., *Wärme-Stoffübertrag.* **2**, 118–128 (1969).
A5. Aybers, N. M., and Tapucu, A., *Wärme-Stoffübertrag.* **2**, 171–177 (1969).
B1. Bachhuber, C., and Sanford, C., *J. Appl. Phys.* **45**, 2567–2569 (1974).
B2. Baird, M. H. I., and Davidson, J. F., *Chem. Eng. Sci.* **17**, 87–93 (1962).

B3. Beard, K. V., *J. Atmos. Sci.* **33**, 851–864 (1976).
B4. Beard, K. V., and Pruppacher, H. R., *J. Atmos. Sci.* **26**, 1066–1072 (1969).
B5. Beek, W. J., and Kramers, H., *Chem. Eng. Sci.* **17**, 909–921 (1962).
B6. Berghmans, J., *Chem. Eng. Sci.* **28**, 2005–2011 (1973).
B7. Berry, E. X., and Pranger, M. R., *J. Appl. Meteorol.* **13**, 108–113 (1974).
B8. Best, A. C., *Meteorol. Res. Pap.* No. 277 (1946); No. 330 (1947).
B9. Blanchard, D. C., "From Raindrops to Volcanoes." Doubleday, Garden City, New York, 1967.
B10. Bonato, L. M., *Termotec. Ric.* **20**, 11–18 (1971).
B11. Braida, L., M.A.Sc. Thesis, Univ. of Toronto, 1956.
B12. Brian, P. L. T., and Hales, H. B., *AIChE J.* **15**, 419–425 (1969).
B13. Bridgwater, J., and McNab, G. S., *Chem. Eng. Sci.* **27**, 837–840 (1972).
B14. Brignell, A. S., *Q. J. Mech. Appl. Math.* **26**, 99–107 (1973).
B15. Brounshtein, B. I., Zheleznyak, A. S., and Fishbein, G. A., *Int. J. Heat Mass Transfer* **13**, 963–973 (1970).
B16. Brunson, R. J., and Wellek, R. M., *Can. J. Chem. Eng.* **48**, 267–274 (1970).
B17. Bryn, T., *David Taylor Model Basin Transl.* No. 132 (1949).
B18. Buzzard, J. F., and Nedderman, R. M., *Chem. Eng. Sci.* **22**, 1577–1586 (1967).
C1. Calderbank, P. H., Johnson, D. S. L., and Loudon, J., *Chem. Eng. Sci.* **25**, 235–256 (1970).
C2. Comolet, R., *C. R. Acad. Sci., Ser. A* **272**, 1213–1216 (1971).
D1. Datta, R. L., Napier, D. H., and Newitt, D. M., *Trans. Inst. Chem. Eng.* **28**, 14–26 (1950).
D2. Davenport, W. G., Ph.D. Thesis, Imperial College, London, 1964.
D3. Davenport, W. G., Richardson, F. D., and Bradshaw, A. V., *Chem. Eng. Sci.* **22**, 1221–1235 (1967).
D4. Davies, J. T., "Turbulence Phenomena." Academic Press, New York, 1972.
D5. Davies, R. M., and Taylor, Sir G. I., *Proc. Roy. Soc., Ser. A* **200**, 375–390 (1950).
E1. Edge, R. M., Flatman, A. T., Grant, C. D., and Kalafatoglu, I. E., *Symp. Multiphase Flow Syst., Inst. Chem. Eng., London* Pap. C3 (1974).
E2. Edge, R. M., and Grant, C. D., *Chem. Eng. Sci.* **26**, 1001–1012 (1971).
E3. Edge, R. M., and Grant, C. D., *Chem. Eng. Sci.* **27**, 1709–1721 (1972).
E4. Elzinga, E. R., and Banchero, J. T., *AIChE J.* **7**, 394–399 (1961).
F1. Finlay, B. A., Ph.D. Thesis, Univ. of Birmingham, 1957.
F2. Foote, G. B., *J. Atmos. Sci.* **26**, 179–181 (1969).
G1. Garbarini, G. R., and Tien, C., *Can. J. Chem. Eng.* **47**, 35–41 (1969).
G2. Garner, F. H., Foord, A., and Tayeban, M., *J. Appl. Chem.* **9**, 315–323 (1959).
G3. Garner, F. H., and Hammerton, D., *Trans. Inst. Chem. Eng.* **32**, 518–524 (1954).
G4. Garner, F. H., and Lane, J. J., *Trans. Inst. Chem. Eng.* **37**, 162–172 (1959).
G5. Garner, F. H., and Lihou, D. A., *DECHEMA-Monogr.* **55**, 155–178 (1965).
G6. Garner, F. H., and Skelland, A. H. P., *Chem. Eng. Sci.* **4**, 149–158 (1955).
G7. Garner, F. H., and Skelland, A. H. P., *Ind. Eng. Chem.* **48**, 51–58 (1956).
G8. Garner, F. H., and Tayeban, M., *An. Fis. Quim.* **LVI-B** 479–498 (1960).
G9. Gaudin, A. M., "Flotation," 2nd ed. McGraw-Hill, New York, 1957.
G10. Gibbons, J. H., Houghton, G., and Coull, J., *AIChE J.* **8**, 274–276 (1962).
G11. Gorodetskaya, A., *Zh. Fiz. Khim.* **23**, 71–77 (1949).
G12. Grace, J. R., Wairegi, T., and Nguyen, T. H., *Trans. Inst. Chem. Eng.* **54**, 167–173 (1976).
G13. Gunn, R., and Kinzer, G. D., *J. Meteorol.* **6**, 243–248 (1949).
G14. Guthrie, R. I. L., Ph.D. Thesis, Imperial College, London, 1967.
G15. Guthrie, R. I. L., and Bradshaw, A. V., *Chem. Eng. Sci.* **28**, 191–203 (1973).
H1. Haberman, W. L., and Morton, R. K., *David Taylor Model Basin Rep.* No. 802 (1953).
H2. Hamielec, A. E., Ph.D. Thesis, Univ. of Toronto, 1961.
H3. Handlos, A. E., and Baron, T., *AIChE J.* **3**, 127–136 (1957).
H4. Harper, J. F., *Chem. Eng. Sci.* **25**, 342–343 (1970).

H5. Harper, J. F., *Adv. Appl. Mech.* **12**, 59–129 (1972).
H6. Hartunian, R. A., and Sears, W. R., *J. Fluid Mech.* **3**, 27–47 (1957).
H7. Hayashi, S., and Matunobu, Y., *J. Phys. Soc. Jpn.* **22**, 905–910 (1967).
H8. Hendrix, C. D., Dave, S. B., and Johnson, H. F., *AIChE J.* **13**, 1072–1077 (1967).
H9. Horton, T. J., Fritsch, T. R., and Kintner, R. C., *Can J. Chem. Eng.* **43**, 143–146 (1965).
H10. Houghton, G., Ritchie, P. D., and Thomson, J. A., *Chem. Eng. Sci.* **7**, 111–112 (1957).
H11. Hozawa, M., Tadaki, T., and Maeda, S., *Kagaku Kogaku* **34**, 315–320 (1970).
H12. Hu, S., and Kintner, R. C., *AIChE J.* **1**, 42–50 (1955).
J1. Johnson, A. I., Besik, F., and Hamielec, A. E., *Can. J. Chem. Eng.* **47**, 559–564 (1969).
J2. Johnson, A. I., and Braida, L., *Can. J. Chem. Eng.* **35**, 165–172 (1957).
J3. Jones, D. M., *J. Meteorol.* **16**, 504–510 (1959).
J4. Jones, D. R. M., Ph.D. Thesis, Cambridge Univ., 1965.
K1. Kadenskaya, N. I., Zheleznyak, A. S., and Brounshtein, B. I., *Zh. Prikl. Khim.* (*Leningrad*) **38**, 1156–1159 (1965).
K2. Keith, F. W., and Hixson, A. N., *Ind. Eng. Chem.* **47**, 258–267 (1955).
K3. Klee, A. J., and Treybal, R. E., *AIChE J.* **2**, 444–447 (1956).
K4. Kojima, E., Akehata, T., and Shirai, T., *J. Chem. Eng. Jpn.* **1**, 45–50 (1968).
K5. Krishna, P. M., Venkateswarlu, D., and Narasimhamurty, G. S. R., *J. Chem. Eng. Data* **4**, 336–343 (1959).
L1. Lamb, H., "Hydrodynamics," 6th ed. Cambridge Univ. Press, London, 1932.
L2. Lessard, R. R., and Zieminski, S. A., *Ind. Eng. Chem., Fundam.* **10**, 260–269 (1971).
L3. Licht, W., and Narasimhamurty, G. S. R., *AIChE J.* **1**, 366–373 (1955).
L4. Lihou, D. A., *Trans. Inst. Chem. Eng.* **50**, 392–393 (1972).
L5. Lihou, D. A., Lowe, W. D., and Hattangady, K. S., *Trans. Inst. Chem. Eng.* **50**, 217–223 (1972).
L6. Lindt, J. T., Dissertation, Technische Hogeschool, Delft. (Bronder—Offset N. V., Rotterdam, 1971.)
L7. Lindt, J. T., *Chem. Eng. Sci.* **27**, 1775–1781 (1972).
L8. Lindt, J. T., and De Groot, R. G., *Chem. Eng. Sci.* **29**, 957–962 (1974).
L9. List, R., and Hand, M. J., *Phys. Fluids* **14**, 1648–1655 (1971).
L10. Lochiel, A. C., and Calderbank, P. H., *Chem. Eng. Sci.* **19**, 471–484 (1964).
L11. Loutaty, R., and Vignes, A., *Chem. Eng. Sci.* **25**, 201–217 (1970).
M1. Magarvey, R. H., and Bishop, R. L., *Phys. Fluids* **4**, 800–805 (1961).
M2. Magarvey, R. H., and Bishop, R. L., *Can. J. Phys.* **39**, 1418–1422 (1961).
M3. Magarvey, R. H., and Blackford, B. L., *Can. J. Phys.* **40**, 1036–1040 (1962).
M4. Magono, C., *J. Meteorol.* **11**, 77–79 (1954).
M5. Marrucci, G., Apuzzo, G., and Astarita, G., *AIChE J.* **16**, 538–541 (1970).
M6. McDonald, J. E., *J. Meteorol.* **11**, 478–494 (1954).
M7. Mendelson, H. D., *AIChE J.* **13**, 250–252 (1967).
M8. Mercier, J., and Anciaes, W., *Houille Blanche* No. 5, 421–425 (1972).
M9. Mercier, J., and Rocha, A., *Chem. Eng. Sci.* **24**, 1179–1183 (1969).
M10. Miller, C. A., and Scriven, L. E., *J. Fluid Mech.* **32**, 417–435 (1968).
M11. Moore, D. W., *J. Fluid Mech.* **23**, 749–766 (1965).
M12. Moore, F. K., *NASA Contract. Rep.* **NASA CR-1368** (1972).
N1. Napier, D. H., Newitt, D. M., and Datta, R. L., *Trans. Inst. Chem. Eng.* **28**, 14–31 (1950).
N2. Natarajan, R., *Combust. Flame* **20**, 199–209 (1973).
P1. Pan, F. Y., and Acrivos, A., *Ind. Eng. Chem., Fundam.* **7**, 227–232 (1968).
P2. Patel, J. M., and Wellek, R. M., *AIChE J.* **13**, 384–386 (1967).
P3. Peebles, F. N., and Garber, H. J., *Chem. Eng. Prog.* **49**(2), 88–97 (1953).
P4. Pitter, R. L., and Pruppacher, H. R., *Q. J. R. Meteorol. Soc.* **99**, 540–550 (1973).
P5. Pruppacher, H. R., and Beard, K. V., *Q. J. R. Meteorol. Soc.* **96**, 247–256 (1970).
P6. Pruppacher, H. R., and Pitter, R. L., *J. Atmos. Sci.* **28**, 86–94 (1971).

R1. Raymond, D. R., and Zieminski, S. A., *AIChE J.* **17**, 57–65 (1971).
R2. Reinhart, A., *Chem.-Ing.-Tech.* **36**, 740–746 (1964).
R3. Rose, P. M., Ph.D. Thesis, Illinois Inst. of Technol., Chicago, 1965.
R4. Rose, P. M., and Kintner, R. C., *AIChE J.* **12**, 530–534 (1966).
R5. Rosenberg, B., *David Taylor Model Basin Rep.* No. 727 (1950).
R6. Ryan, R. T., *J. Appl. Meteorol.* **15**, 157–165 (1976).
S1. Saffman, P. G., *J. Fluid Mech.* **1**, 249–275 (1956).
S2. Satapathy, R., and Smith, W., *J. Fluid Mech.* **10**, 561–570 (1961).
S3. Savic, P., *Natl. Res. Counc. Can., Rep.* No. MT-22 (1953).
S4. Schroeder, R. R., Ph.D. Thesis, Illinois Inst. of Technol., Chicago, 1964.
S5. Schroeder, R. R., and Kintner, R. C., *AIChE J.* **11**, 5–8 (1965).
S6. Skelland, A. H. P., and Caenepeel, C. L., *AIChE J.* **18**, 1154–1163 (1972).
S7. Skelland, A. H. P., and Wellek, R. M., *AIChE J.* **10**, 491–496 (1964).
S8. Srikrishna, M., and Narasimhamurty, G. S. R., *Indian Chem. Eng.* **13**, 4–11 (1971).
S9. Stuke, B., *Naturwissenschaften* **39**, 325–326 (1952).
S10. Subramanyam, S. V., *J. Fluid Mech.* **37**, 715–725 (1969).
S11. Sumner, B. S., and Moore, F. K., *NASA Contract. Rep.* **NASA CR-1669** (1970).
T1. Tadaki, T., and Maeda, S., *Kagaku Kogaku* **25**, 254–264 (1961).
T2. Tapucu, A., Document IGN-87, Ecole Polytechnique, Montreal, 1974.
T3. Taylor, J. D., and Acrivos, A. J., *J. Fluid Mech.* **18**, 466–476 (1964).
T4. Thorsen, G., Stordalen, R. M., and Terjesen, S. G., *Chem. Eng. Sci.* **23**, 413–426 (1968).
T5. Thorsen, G., and Terjesen, S. G., *Chem. Eng. Sci.* **17**, 137–148 (1962).
T6. Tsuge, H., and Hibino, S., *Kagaku Kogaku* **35**, 65–71 (1971).
V1. Vakhrushev, I. A., and Efremov, G. I., *Chem. Technol. Fuels Oils (USSR)* **5/6**, 376–379 (1970).
V2. Van der Leeden, P., Nio, L. D., and Suratman, P. C., *Appl. Sci. Res., Sect. A* **5**, 338–348 (1956).
V3. Vogtländer, J. G., and Meijboom, F. W., *Chem. Eng. Sci.* **29**, 799–803 (1974).
W1. Wairegi, T., Ph.D. Thesis, McGill Univ., Montreal, 1974.
W2. Wallis, G. B., *Int. J. Multiphase Flow* **1**, 491–511 (1974).
W3. Warshay, M., Bogusz, E., Johnson, M., and Kintner, R. C., *Can. J. Chem. Eng.* **37**, 29–36 (1959).
W4. Weiner, A., Ph.D. Thesis, Univ. of Pennsylvania, Philadelphia, 1974.
W5. Wellek, R. M., Andoe, W. V., and Brunson, R. J., *Can. J. Chem. Eng.* **48**, 645–655 (1970).
W6. Wellek, R. M., Agrawal, A. K., and Skelland, A. H. P., *AIChE J.* **12**, 854–862 (1966).
W7. Weller, K. R., *Can. J. Chem. Eng.* **50**, 49–58 (1972).
W8. Winnikow, S., and Chao, B. T., *Phys. Fluids* **9**, 50–61 (1966).
Y1. Yamaguchi, M., Fujimoto, T., and Katayama, T., *J. Chem. Eng. Jpn.* **8**, 361–366 (1975).
Y2. Yamaguchi, M., Watanabe, S., and Katayama, T., *J. Chem. Eng. Jpn.* **8**, 415–417 (1975).
Y3. Yao, S.-C., and Schrock, V. E., *J. Heat Transfer* **98**, 120–125 (1976).
Y4. Yeheskel, J., and Kehat, E., *Chem. Eng. Sci.* **26**, 1223–1233 (1971).
Z1. Zabel, T., Hanson, C., and Ingham, J., *Trans. Inst. Chem. Eng.* **51**, 162–164 (1973).
Z2. Zieminski, S. A., and Raymond, D. R., *Chem. Eng. Sci.* **23**, 17–28 (1968).

Chapter 8

Deformed Fluid Particles of Large Size

I. INTRODUCTION

This chapter is devoted to bubbles and drops with $\mathrm{Eo} > 40$ and $\mathrm{Re} > 1.2$ (see Chapter 2). These inequalities are generally satisfied by bubbles and drops with volumes greater than about 3 cm^3 (i.e., $d_e \gtrsim 1.8$ cm). Considerable work has been carried out for large gas bubbles, primarily in connection with underwater explosions, fluidized beds, and processing of liquid metals, and reviews have been prepared by Wegener and Parlange (W5) and Harper (H2). Relatively little attention has been devoted to large drops. Drops falling in gases almost always break up before an Eotvos number of 40 is reached (see Chapter 12) so that the present chapter is restricted to cases where the continuous phase is a liquid.

In the present chapter, we neglect wall effects and unsteady motion including splitting. These factors are considered in Chapters 9, 11, and 12, respectively. The fluid mechanics of large bubbles and drops are discussed before turning to mass transfer.

II. FLUID MECHANICS

A. Shape

Over most of the range covered by this chapter, the shape of bubbles and drops can be closely approximated as a segment of a sphere (see Fig. 2.4). Hence, most of the fluid particles under discussion are said to be "spherical-caps." For $\mathrm{Re} \gtrsim 150$, the rear or base is quite flat, though sometimes irregular, and the wake angle very nearly 50°. At lower Re, the wake angle is larger (G4), as shown in Fig. 8.1. For $\mathrm{Re} \lesssim 40$, the leading edge tends to be oblate ellipsoidal

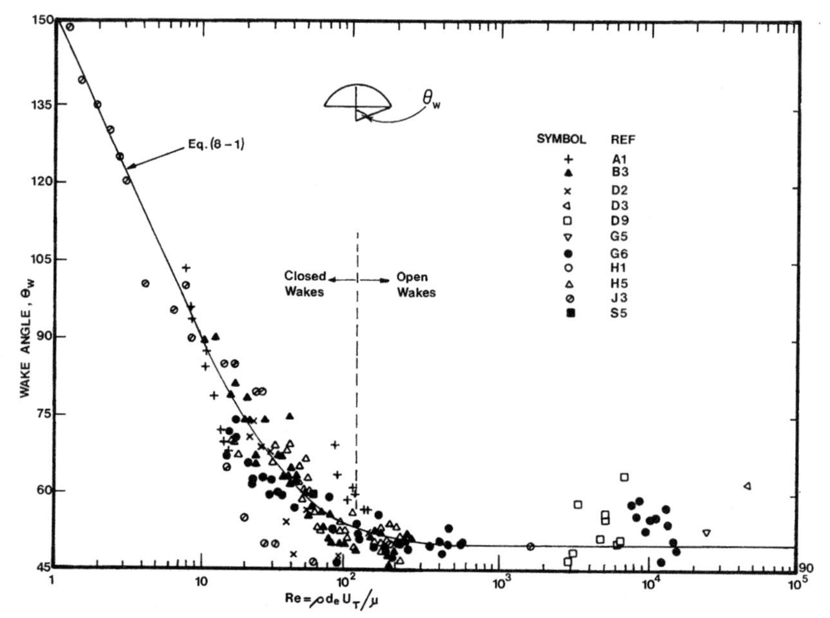

FIG. 8.1 Wake angle θ_W for spherical-cap bubbles as a function of bubble Reynolds number.

(B3, W3) while the rear is indented or dimpled. Skirt formation may also occur, as discussed in Section D. The wake angle for spherical-caps (expressed in degrees) is well represented by the empirical equation

$$\theta_W = 50 + 190\exp[-0.62\,\mathrm{Re}^{0.4}] \qquad (\mathrm{Eo} \geq 40,\ \mathrm{Re} > 1.2). \qquad (8\text{-}1)$$

This equation is shown in Fig. 8.1 together with available data. Somewhat different angles are obtained for Re $\dot{<}$ 40 if the angle is measured from the center of an enclosing ellipsoid rather than from the center of a sphere which fits the front portion of the bubble (B3). Attempts to predict wake angles theoretically for spherical-cap (C5, M3, R4) or two-dimensional circular-cap (B2) fluid particles have met with only limited success. The volume of continuous phase material, V_R, contained in the indentation at the rear of ellipsoidal- and spherical-cap bubbles has also been measured (B3, H5) by subtracting the true bubble volume from the apparent volume assuming a flat base. Results indicate that the fractional indentation volume, V_R/V, increases from zero at Re $\doteq$ 1 to about 0.35 at Re $\doteq$ 50, decreasing to essentially zero again for Re > 150.

B. TERMINAL VELOCITY

While the shape of a large fluid particle cannot be predicted accurately from first principles, the terminal velocity can be obtained from the observed shape. Interfacial tension forces are ignored. Flow is considered only in the neighborhood of the nose, where the external fluid is assumed to flow as an inviscid fluid over a complete sphere or spheroid of which the fluid particle forms the cap. The surface pressure distribution in the continuous fluid may then be calculated using Bernoulli's theorem. For a spherical-cap, this gives

$$p_s - p_0 = \pm ga\rho(1 - \cos\theta) - \tfrac{9}{8}\rho U_T{}^2 \sin^2\theta, \tag{8-2}$$

where p_0 is the pressure at the nose ($\theta = 0$) and the ($+$) and ($-$) signs apply to upward and downward moving caps, respectively. The pressure distribution at the surface in the dispersed phase is assumed to be the hydrostatic pressure distribution. This will apply if $\mathrm{Re}_p = \rho_p d_e U_T/\mu_p$ is sufficiently large, e.g., of order 100 or greater, so that there is a thin interior boundary layer across which the pressure distribution is impressed by the slow moving interior fluid (W3). For a spherical-cap, the pressure distribution is then

$$p_s - p_0 = \pm\rho_p ga(1 - \cos\theta). \tag{8-3}$$

Equating the two expressions for $(p_s - p_0)$ and solving for the terminal velocity U_T we obtain

$$U_T{}^2 = \frac{8}{9} ga \frac{\Delta\rho}{\rho}\left(\frac{1 - \cos\theta}{\sin^2\theta}\right). \tag{8-4}$$

Equation (8-4) cannot be satisfied over the entire spherical-cap surface, but if it is satisfied for $\theta \to 0$ to terms of order θ^2, the terminal velocity reduces to

$$U_T = \tfrac{2}{3}\sqrt{ga\,\Delta\rho/\rho}, \tag{8-5}$$

which is the celebrated Davies and Taylor (D9) equation.† For spheroidal-cap drops or bubbles of eccentricity e and vertical semiaxis b, an analogous procedure yields

$$U_T = f(e)\sqrt{gb\,\Delta\rho/\rho}, \tag{8-6}$$

where for oblate spheroidal-caps (W3)

$$f(e) = (1/e^3)\{\sin^{-1} e - e\sqrt{1 - e^2}\}, \tag{8-7}$$

while for prolate spheroidal-caps (G5)

$$f(e) = (\sqrt{1 - e^2}/e^3)\{e - (1 - e^2)\tanh^{-1} e\}. \tag{8-8}$$

Collins (C5) obtained a second approximation to the velocity of a large bubble using a perturbation analysis to balance the pressures along the interface. The result, in generalized form, is

$$U_T = 0.652\sqrt{g\bar{a}\,\Delta\rho/\rho}, \tag{8-9}$$

where $\bar{a}$ is the average radius of curvature over the surface from $\theta = 0$ to $\theta = 37.5°$. Experimental results obtained by various workers are shown in Fig. 8.2.

† This result also applies to the rise under gravity of a large mass of hot gas in a colder gas (S2, T1).

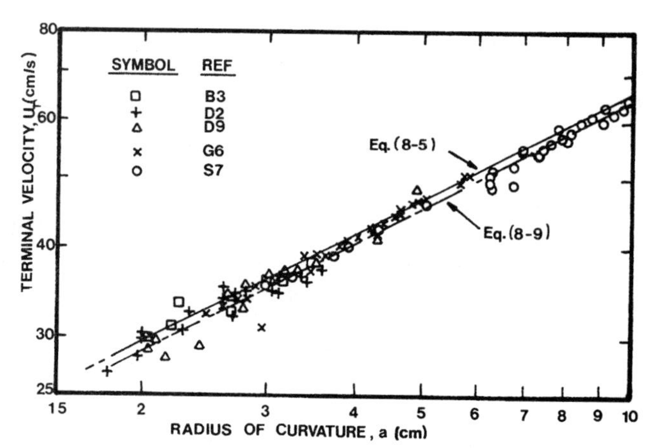

FIG. 8.2 Rise velocity of spherical-cap bubbles as a function of the radius of curvature of the leading surface.

It is clear that there is no reason to prefer Eq. (8-9) over Eq. (8-5). Therefore Eq. (8-5) is usually adopted for simplicity and is recommended for Re $\gtrsim$ 40. Agreement with Eq. (8-6) has also been found to be good (B3, G5, W3) and Eqs. (8-6) and (8-7) are recommended for $1.2 < \mathrm{Re} < 40$.

As shown in Fig. 8.1, spherical-cap fluid particles are geometrically similar with a wake angle θ_W of approximately 50° once Re is greater than about 150. The radius of curvature may then be related directly to either V or d_e yielding

$$U_T = 0.792 g^{1/2} V^{1/6} \sqrt{\Delta\rho/\rho} \qquad (\mathrm{Re} \gtrsim 150, \mathrm{Eo} \geq 40), \tag{8-10}$$

$$U_T = 0.711 \sqrt{g d_e \Delta\rho/\rho} \qquad (\mathrm{Re} \gtrsim 150, \mathrm{Eo} \geq 40), \tag{8-11}$$

or

$$C_D = \frac{4}{3}\frac{g d_e}{U_T^2}\frac{\Delta\rho}{\rho} = \frac{8}{3} \qquad (\mathrm{Re} \gtrsim 150, \mathrm{Eo} \geq 40). \tag{8-12}$$

Equations (8-10) to (8-12) have been confirmed many times [e.g. (D4, W7)]. For $M \gtrsim 10^2$, bubbles and drops change directly from spherical to spherical-cap, as noted in Chapter 2. The drag coefficient is then closely approximated by

$$C_D = \frac{8}{\mathrm{Re}}\frac{(2 + 3\kappa)}{(1 + \kappa)} + \frac{8}{3} \qquad (\mathrm{M} > 10^2, \text{all Re}), \tag{8-13}$$

the generalized form of an equation suggested by Darton and Harrison (D1). To solve for U_T over the entire range of Re, it is more convenient to rewrite Eq. (8-13) as a quadratic equation in Re:

$$2\mathrm{Re}^2 + 6\mathrm{Re}[(2 + 3\kappa)/(1 + \kappa)] - \mathrm{Ar} = 0, \tag{8-14}$$

where $\mathrm{Ar} = \mathrm{Eo}^{3/2} M^{-1/2} = g\rho \Delta\rho\, d_e^3/\mu^2$ is an Archimedes number analogous to $N_D = C_D \mathrm{Re}^2$ introduced in Chapter 5.

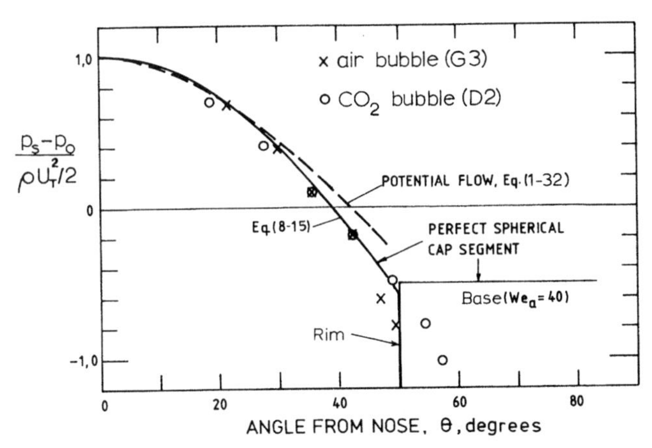

FIG. 8.3 Surface modified pressure distribution for spherical-caps at high Re, derived using the method of McDonald (M2). Experimental profiles obtained from photographs of bubbles in water.

Large "two-dimensional" or plane bubbles give results analogous to those presented above, and have been considered by Collins (C4), Grace and Harrison (G5), and Hills (H4).

C. SURFACE PRESSURE DISTRIBUTION

The method developed by McDonald (M2) to calculate surface dynamic pressure distributions for falling drops (see Chapter 7) may also be applied to large fluid particles. Equation (7-19) may therefore be applied. For a perfect spherical-cap whose terminal velocity U_T is given by Eq. (8-5), the modified pressure over the leading surface is given by

$$(p_s - p_0)/(\tfrac{1}{2}\rho U_T^2) = 4.5 \cos\theta - 3.5 \qquad (0 \le \theta < \theta_W). \tag{8-15}$$

For $\theta_W = 50°$ and a flat rear, the pressure on the rear surface is

$$(p_s - p_0)/(\tfrac{1}{2}\rho U_T^2) = -0.61 + (4/\mathrm{We_a}) \qquad (\text{rear surface, } \theta_W = 50°), \tag{8-16}$$

where $\mathrm{We_a} = \rho U_T^2 a/\sigma$. Since $\mathrm{We_a}$ is generally greater than 20 for the large fluid particles considered in this section, σ plays a relatively minor role except at the rim where the spherical-cap surface intersects the base.†

Actual shapes of fluid particles deviate from the idealized shape which leads to Eqs. (8-15) and (8-16). Surface pressure distributions derived from observed shapes (W2) are shown in Fig. 8.3 for spherical-cap bubbles at high Re. It is seen that the pressure variation is well described by Eq. (8-15) for $0 \le \theta < \theta_W$ while the potential flow pressure distribution, Eq. (1-32), gives good agreement up to about 30° from the nose.

† The rim cannot be sharp, as it often appears, or the capillary pressure component would be infinite.

D. Fluid Skirts

Large bubbles and drops in high M systems at Re of order 10 to 50 commonly trail thin annular films of the dispersed fluid as shown in Fig. 2.4g and 2.4h. These thin films are usually referred to as "skirts." Systems in which skirts are observed have Morton numbers greater than about 0.1; hence the continuous fluid must have a rather high viscosity, generally greater than 1 poise. Approximate boundaries for the occurrence of skirts are shown in the Re vs. Eo diagram, Fig. 2.5. Experimentally, it has been found that trailing skirts have a negligible influence on the terminal velocity of bubbles and drops (B3, W1), although they affect the nature of the wake as discussed below. This lends further support to the theory above, where the terminal velocity of large bubbles and drops is derived considering only the shape and motion near the nose.

For systems in which skirt formation can occur and d_e is slightly less than required for skirt formation, large bubbles or drops tend to be indented at the rear. Skirt formation occurs when viscous forces acting at the rim or corner of the dimpled bubble or drop are strong enough to overcome interfacial tension forces and pull the rim out into a thin sheet (B3, H5, W1, W5). The onset of skirts is dependent both on the ratio $\mathrm{We}/\mathrm{Re} = \mu U_T/\sigma$, sometimes called a capillary or skirt number, and on Re. Figure 8.4 shows data for the transition from unskirted to skirted bubbles or drops. For bubbles, skirts exist for $\mathrm{Re} \dot{>} 9$ and

$$\mathrm{We}/\mathrm{Re} > 2.32 + [11/(\mathrm{Re} - 9)^{0.7}]. \tag{8-17}$$

For drops, skirt formation occurs for $\mathrm{We}/\mathrm{Re} \dot{>} 2.3$ and $\mathrm{Re} \dot{>} 4$. It is possible that some upper bound on Re exists above which skirts are no longer observed, but this has not been determined precisely. The highest Re for which skirts have been reported is 500 (H5).

Once skirts are formed, they may be steady and axisymmetric, growing with time, or asymmetric with finite amplitude waves traveling towards the rear (W3). Wairegi (W2) classified skirt configurations into:

(i) smooth skirts curled inwards;
(ii) straight skirts perpendicular to the base of the bubble or drop;
(iii) wavy skirts;
(iv) exfoliating skirts;
(v) fluttering skirts.

Transitions between these categories are not abrupt. Bhaga (B3) delineated the conditions under which wavy skirts are found.

The skirt thickness Δ may be predicted from an approach suggested by Guthrie and Bradshaw (G8) which in extended form (W2) yields

$$\Delta = (6\mu_p U_T/g\,\Delta\rho)^{1/2}. \tag{8-18}$$

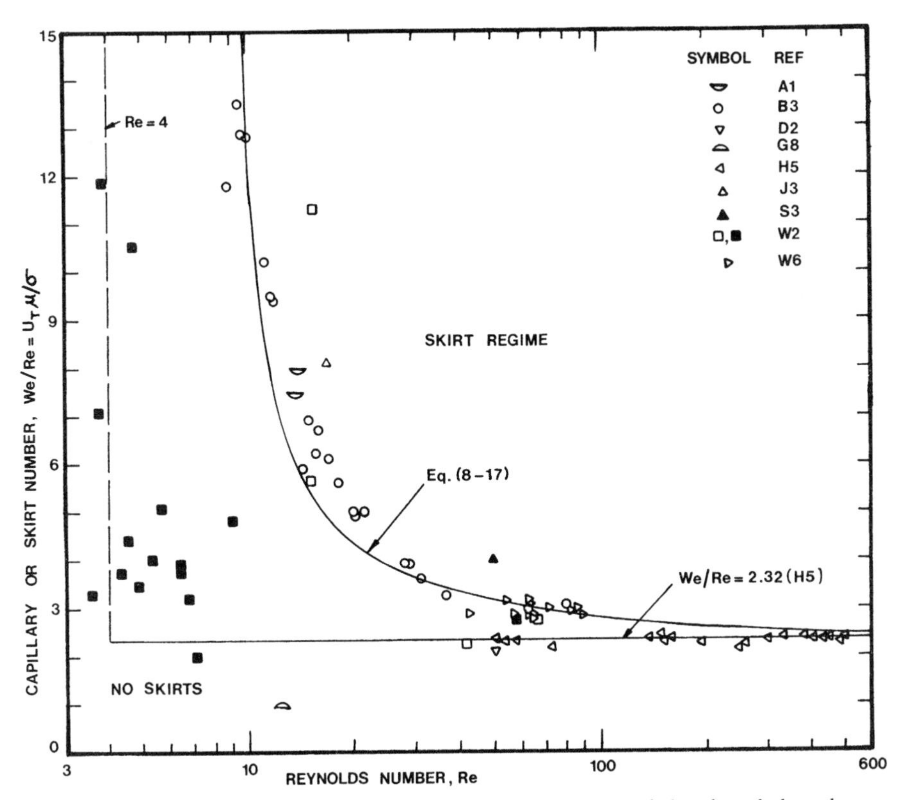

FIG. 8.4 Onset of skirt formation: open symbols refer to bubbles and closed symbols to drops.

Experimental measurements of skirt thickness (B3, B5, G8, W2) show reasonable agreement with Eq. (8-18). In practice, skirts become thinner with increasing distance from the rear of the bubble or drop (B3, H5). Skirts behind bubbles are of order 50 μm thick, while the thickness of liquid skirts behind drops is of order 1 mm.

Steady skirt lengths increase with Re (B3, H5, W2). Wairegi (W2) and Bhaga (B3) also reported skirt lengths which increased with time. The length of steady skirts is controlled by a balance of viscous and capillary forces at the rim of the skirt (B3), whereas the length of wavy skirts appears to be determined by growth of Helmholtz instability waves (H5).

E. Internal Circulation

The dispersed phase fluid must circulate for large fluid particles in qualitatively the same manner as for small fluid particles. Because of the large values of Eo, surface-active contaminants are not expected to damp out internal fluid motion entirely (see Chapter 3), although interfacial motion may be impeded over part of the leading surface of a spherical-cap away from the nose (W4).

Internal circulation measurements are very difficult to obtain for gas bubbles (D8). Some results have been obtained for large liquid skirted drops using tracer particles (W2), and provide a qualitative picture of the internal motion as shown in Fig. 8.5. It is not clear whether there is a reverse vortex motion in the interior of a large fluid particle (as indicated by the dotted lines). Such a secondary vortex would appear to be necessary to satisfy velocity and stress continuity, but experimental evidence is inconclusive.

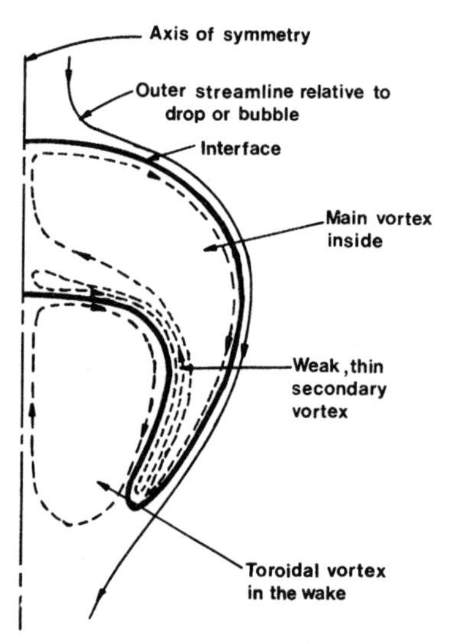

FIG. 8.5 Schematic diagram of internal and external flow patterns for a skirted bubble or drop.

F. WAKES

At low Re, wakes behind large bubbles and drops are closed (B3, H5, S5, W2, W6), whereas at high Re open turbulent wakes are formed (H5, M1, W6). The value of Re for transition between these two types of wake has been determined as 110 ± 2 (B3) for skirtless bubbles. There is some evidence (H5) that the transition Reynolds number may be increased if skirts are present.

Closed wakes have been modeled as completing the sphere or spheroid of which the particle forms the cap [e.g. (C5, P2)]. However, the wake is smaller than that required to complete a spheroid for Re $\dot{<} 5$ and greater for larger Re (B3). The wake becomes more nearly spherical as Re $\rightarrow 100$, but is still somewhat "egg-shaped" (B3, H5). Wake volumes, normalized with respect to the volume of the fluid particle, are shown in Fig. 8.6 for Re up to 110. Note the close agreement with results (K1) for solid spherical caps of the same aspect ratio. This is not surprising since separation necessarily occurs at the rim of the

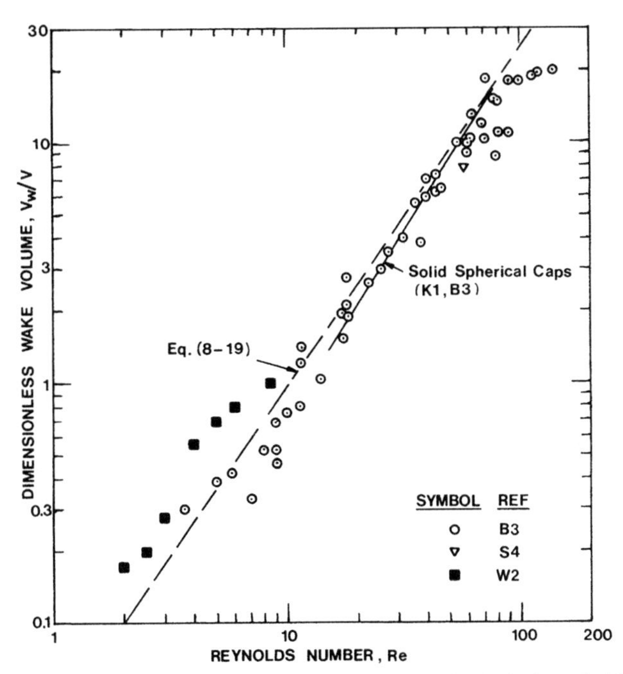

FIG. 8.6 Dimensionless wake volumes for ellipsoidal-cap and spherical-cap bubbles and drops, compared with solid spherical-caps.

spherical-cap whether it is rigid or circulating, even if the boundary layers over the curved portion of the cap differ in the two cases. The wake volume is well represented by

$$V_W/V = 0.037\,\mathrm{Re}^{1.4} \qquad (3 \leq \mathrm{Re} \leq 110). \tag{8-19}$$

Bhaga (B3) determined the fluid motion in wakes using hydrogen bubble tracers. Closed wakes were shown to contain a toroidal vortex with its core in the horizontal plane where the wake has its widest cross section. The core diameter is about 70% of the maximum wake diameter, similar to a Hill's spherical vortex. When the base of the fluid particle is indented, the toroidal motion extends into the indentation. Liquid within the closed wake moves considerably more slowly relative to the drop or bubble than the terminal velocity U_T. If a skirt forms, the basic toroidal motion in the wake is still present (see Fig. 8.5), but the strength of the vortex is reduced. Momentum considerations require that there be a velocity defect behind closed wakes and this accounts for the "tail" observed by some workers (S5). Crabtree and Bridgwater (C8) and Bhaga (B3) measured the velocity decay and drift in the far wake region.

There has been considerably less work on open turbulent wakes, although some excellent photographs have been published (B3, M1, W5, W6). Wake shedding appears to be responsible for the wobbling motion often shown by spherical-cap bubbles with Re $\gtrsim$ 150, and for erratic motion of trailing satellite bubbles. Wakes for large two-dimensional bubbles have received some attention (C3, C7, L1, L2).

G. External Flow Field

Bhaga (B3) determined streamlines relative to rising ellipsoidal and spherical-cap bubbles in high M systems using hydrogen bubbles as tracers. Results for Re $= 29$ and 82 are shown in Fig. 8.7. As the external liquid moves past the bubble and wake boundary, its velocity decreases, especially at low Re. Thus the motion deviates from the potential flow field used in deriving U_T (see above), but the deviation decreases as Re increases and is very small at distances from the bubble of the order of the radius of curvature. The value of u_t/U_T at the equator increases from about 0.56 to 0.81 as Re increases from 2.5 to 42, whereas a value of 1.5 is expected for potential flow; see Eq. (1-31). At higher Re, the modification to Hill's spherical vortex proposed by Harper and Moore (H3), applied to the spherical region approximately containing the spherical-cap and its closed wake, gives a reasonable description of the flow field both outside and inside this region.

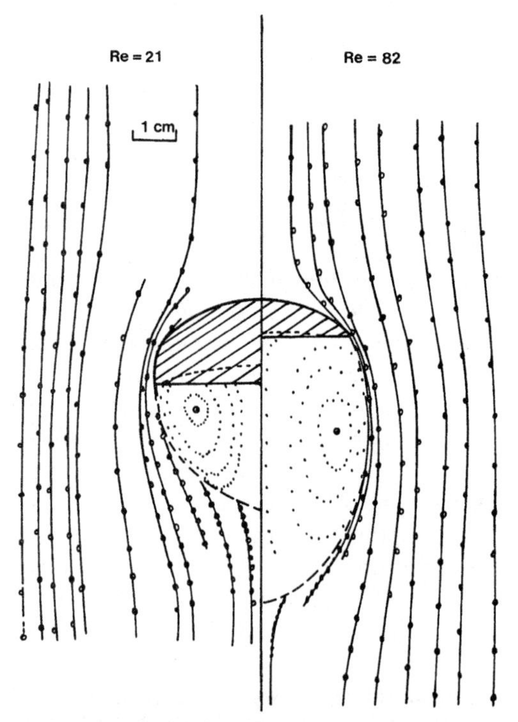

FIG. 8.7 Streamlines in the outer fluid relative to an ellipsoidal-cap and a spherical-cap bubble, after Bhaga (B3). Points on streamlines show positions at intervals of 0.03 sec.

III. MASS TRANSFER IN THE CONTINUOUS PHASE

Transfer from large bubbles and drops may be estimated by assuming that the front surface is a segment of a sphere with the surrounding fluid in potential flow. Although bubbles are oblate ellipsoidal for Re $\dot{<}$ 40, less error should result from assumption of a spherical shape than from the assumption of potential flow.

Transfer from a spherical segment in potential flow is described (B1, B4, J2, L4) by

$$kA = [8\pi a^3 U_T \mathscr{D} g(\theta_W)]^{1/2}, \tag{8-20}$$

where a is the radius of the spherical-cap, θ_W is the maximum angle of the segment measured from the stagnation point, and

$$g(\theta_W) = 2 - 3\cos\theta_W + \cos^3\theta_W. \tag{8-21}$$

A. High Reynolds Number

For a spherical cap with a flat base:

$$\frac{a}{d_e} = [2g(\theta_W)]^{-1/3}, \tag{8-22}$$

and Eq. (8-20) for transfer over the forward portion can be rewritten

$$\frac{(kA)_F}{A_e} = \frac{2}{\sqrt{\pi}}\left(\frac{U_T \mathscr{D}}{d_e}\right)^{1/2}, \tag{8-23}$$

or

$$[(kA)_F/A_e]d_e/\mathscr{D} = 2(\mathrm{Pe}/\pi)^{1/2}, \tag{8-24}$$

regardless of the wake angle. Equation (8-24) is the so-called Higbie equation, Eq. (5-35), written in terms of the equivalent diameter, and represents transfer from the entire particle if there is negligible transfer through the base. Many investigations have shown that Eq. (8-24) provides a fair estimate of the continuous phase resistance regardless of bubble size as long as Re is large [e.g., (C1)].

At high Re the transfer through the base is not negligible. Weber (W4) showed that basal transfer may be estimated using the penetration theory, assuming complete renewal each time vortices are shed. He obtained

$$\frac{(\overline{kA})_R}{A_e} = \left(\frac{\mathrm{Sr}\sin^3\theta_W}{4\pi}\right)^{1/2}\left(\frac{2a}{d_e}\right)^{3/2}\left(\frac{U_T \mathscr{D}}{d_e}\right)^{1/2}, \tag{8-25}$$

where $\mathrm{Sr} = f_W w/U_T$ is the Strouhal number for eddy shedding based on the maximum width of the bubble, w. For Re $\dot{>}$ 150, $\theta_W = 50°$ as discussed earlier. For Sr = 0.3 (L3)

$$(\overline{kA})_R/A_e = 0.357(U_T \mathscr{D}/d_e)^{1/2}. \tag{8-26}$$

Assuming that base and frontal transfer are independent, we obtain

$$\frac{(\overline{kA})}{A_e} = \frac{(kA)_F + (\overline{kA})_R}{A_e} = 1.49\left(\frac{U_T\mathscr{D}}{d_e}\right)^{1/2}. \tag{8-27}$$

Use of Eq. (8-11) leads to the final recommended relationship:

$$(\overline{kA})/A_e = 1.25(g\,\Delta\rho/\rho)^{1/4}\mathscr{D}^{1/2}d_e^{-1/4}. \tag{8-28}$$

Equation (8-28) gives good agreement for spherical-cap bubbles in liquids at Re > 100 (W4) as shown in Fig. 8.8. No data are available for large drops. For bubbles in liquids, $\Delta\rho/\rho \doteq 1$ and Eq. (8-28) becomes

$$[(\overline{kA})/A_e]/\mathscr{D}^{1/2} = 6.94d_e^{-1/4}, \tag{8-29}$$

with d_e in cm and the left side in $s^{-1/2}$. This form was used for bubbles in water in Eq. (7-50). Surfactants reduce transfer rates from spherical-cap bubbles in low-viscosity liquids (B1), and this effect has been analyzed by Weber (W4).

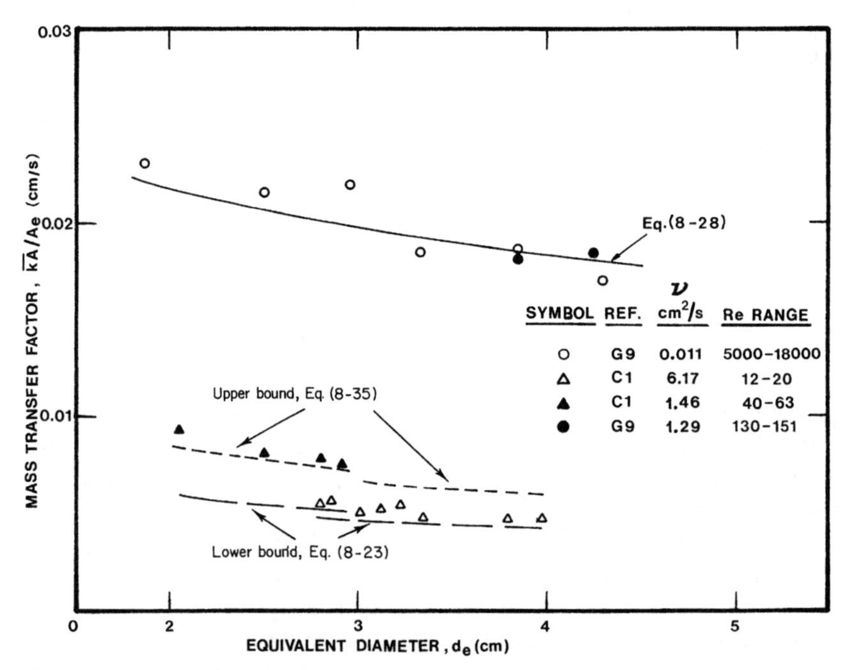

FIG. 8.8 Mass transfer factor kA/A_e for dissolution of CO_2 bubbles in aqueous solutions.

B. Low Reynolds Number

For Re < 110 the wake is closed and laminar as discussed above. Transfer over the front portion of the cap is again described by Eq. (8-20). Transfer from the base occurs by diffusion into the wake fluid as it moves along the bubble base, producing a concentration boundary layer. The solute in this boundary layer then diffuses both into the interior of the wake and into the continuous phase as the wake fluid circulates (B4).

Figure 8.9 shows a model of a large indented or dimpled fluid particle at low Re composed of two spherical segments. The front surface has radius a and angle θ_W, while the rear surface or base has radius a_R and angle θ_R. Brignell (B4) showed that if θ_W is small, the wake large, and $a = a_R$, the transfer coefficient from the rear segment is equal to that from the front segment, both being given by Eq. (8-20). In reality these assumptions are not valid. The velocity in the wake is less than the velocity over the forward segment. These considerations suggest that transfer rates may be bounded. The lower bound is given by neglecting transfer from the rear surface.

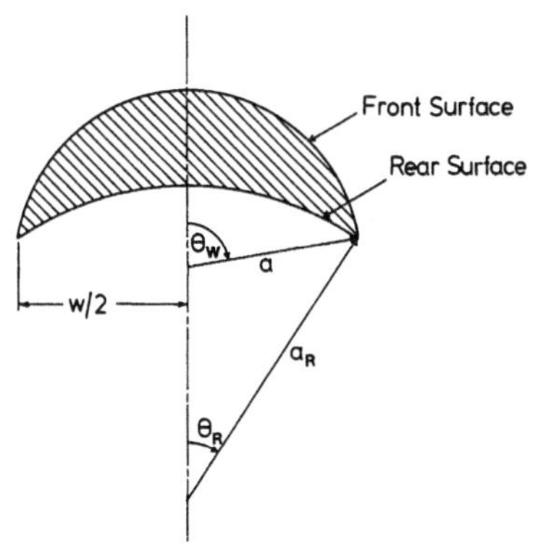

FIG. 8.9 Two spherical-segment model of an indented fluid particle.

For the geometry of Fig. 8.9, the volume of the fluid particle is

$$V = (\pi a^3/3)g(\theta_W) - V_R, \tag{8-30}$$

where

$$V_R = (\pi a_R{}^3/3)g(\theta_R) \tag{8-31}$$

is the volume of the rear spherical-cap and $g(\theta)$ is given by Eq. (8-21). The frontal radius is given by:

$$a/d_e = [2g(\theta_W)/(1 + (V_R/V))]^{-1/3}. \tag{8-32}$$

This value of a with θ_W from Eq. (8-1) is used with Eqs. (8-20) and (8-23) to obtain the lower bound.

The upper bound is found by adding to the lower bound the transfer from the rear segment given by Eq. (8-20) written with a_R and θ_R in place of a and θ_W. Application of Eq. (8-20) to the front and rear surfaces gives

$$\frac{(kA)_F}{(kA)_R} = \left(\frac{a}{a_R}\right)^{3/2}\left[\frac{g(\theta_W)}{g(\theta_R)}\right]^{1/2}. \tag{8-33}$$

Combination of Eqs. (8-30), (8-31), and (8-33) yields

$$\frac{(kA)_F}{(kA)_R} = \left(1 + \frac{V}{V_R}\right)^{1/2}. \tag{8-34}$$

If the front and rear transfer rates are independent,

$$\frac{kA}{A_e} = \frac{(kA)_F}{A_e}\left[1 + \frac{(kA)_R}{(kA)_F}\right]. \tag{8-35}$$

The upper bound is then calculated from Eqs. (8-35), (8-34), and (8-20).

Although V_R/V, the fractional indentation, varies with Re for Re > 2 as discussed above, a reasonable average is $V_R/V = 0.2$. Using this approximation and measured bubble terminal velocities, we have calculated the upper and lower bounds shown in Fig. 8.8. In all cases agreement with the experimental data is good.

Mass transfer rates for skirted bubbles in polyvinyl alcohol solutions have been measured by Guthrie and Bradshaw (G9) and Davenport *et al.* (D4). When a skirt is present the transfer rate increases, but not in proportion to the increase in surface area. Davenport attributes this to the accumulation on the surface of the skirt of surface-active impurities which immobilize the interface and reduce the transfer rate. Presumably transfer rates from skirted bubbles or drops in very pure liquids would be appreciably higher than from fluid particles without skirts.

IV. SPECIAL SYSTEMS

Gas bubbles in liquid metals and in fluidized beds have been the subject of special studies because of their practical importance and because of the experimental difficulties associated with studying bubble properties in opaque media. Much of the work has been carried out in so-called "two-dimensional" columns, where a sheet of liquid or fluidized particles, typically 1 cm thick, is confined between two parallel transparent walls. Bubbles span the gap between the front and rear faces and can be observed with backlighting.

A. Bubbles in Liquid Metals

There is considerable evidence (D3, G7, P1, P4, S1) that bubbles in liquid metals show the behavior expected from studies in more conventional liquids. Because of the large surface tension forces for liquid metals, Morton numbers tend to be low (typically of order 10^{-12}) and these systems are prone to contamination by surface-active impurities. Figure 8.10a shows a two-dimensional nitrogen bubble in liquid mercury. For experimental convenience, the bubbles studied have generally been rather large, so that there are few data available for spherical or slightly deformed ellipsoidal bubbles in liquid metals. Data

(a)

(b)

Fig. 8.10 (a) "Two-dimensional" nitrogen bubble in liquid mercury [Paneni and Davenport (P1), *Trans. Metall. Soc. AIME*, copyrighted by the American Institute of Mining, Metallurgical and Petroleum Engineers, Inc]. (b) X-ray photograph of bubbles in a fluidized bed. Reference grid spacing is 2 cm in vertical direction. Fluidized particles are 79 μm silicon carbide particles. (Reproduced with permission of Prof. P. N. Rowe).

tend to be subject to abnormally high scatter due to experimental difficulties associated with opaque media and high temperatures. For reviews of the properties of bubbles in liquid metals, see (D5, R1, R2).

B. Bubbles in Fluidized Beds

Fluidized beds are beds of solid particles supported by upward flow of a gas or liquid. Because of their temperature uniformity, excellent heat transfer characteristics, and solids handling possibilities, fluidized beds have found wide application for physical and chemical processes.

Gas fluidized beds are inherently unstable to the growth of voidage disturbances and this is believed to be the origin of bubbles in fluidized beds (J1). Rowe (R5) has reviewed the properties of these bubbles and experimental techniques used in their study. As a first approximation, the particulate phase (particles and interstitial gas) is usually treated as a Newtonian liquid of zero surface tension and of kinematic viscosity of order 5 cm^2/s (G4). There is a strong analogy between bubbles in liquids and in fluidized beds (D7). In view of the negligible surface tension forces and the fact that bubbles are usually at least a centimeter in diameter, bubbles are generally in the spherical-cap regime. Reynolds numbers tend to be of order 10 to 100 with the result that bubbles have large values of θ_W (see Fig. 8.1) and closed laminar wakes. A photograph taken with the aid of x-rays is reproduced in Fig. 8.10b. The terminal velocity of bubbles in fluidized beds is usually estimated using Eq. (8-10) or (8-11) with $\Delta\rho/\rho = 1$ (D6), while the influence of bubble size on shape for bubbles from 1 to 16 cm in diameter has been represented (R7) by

$$V = (\pi/6) d_B{}^3 e^{-0.057 d_B}, \tag{8-36}$$

where d_B is the maximum bubble width in centimeters in a plane normal to the direction of motion. There is evidence that the particulate phase is significantly non-Newtonian. Slip surfaces (P3) give evidence of yield stresses, and this had led some workers [e.g. (G1)] to treat the particulate phase as a Bingham plastic.

Internal circulation for bubbles in fluidized beds is an aspect in which the analogy between liquids and fluidized beds ceases to apply since the bubble/particulate phase interface is permeable. There is a net upward gas flow through a bubble (G2). If the bubble rises more quickly than gas which is percolating through the particle interstices in the remote particulate phase, gas recirculation occurs in an annular shell called a "cloud" surrounding each bubble (R6). Cloud formation has considerable importance with regard to efficient utilization or treatment of gases in fluidized beds. Transfer between bubble and particulate phase results both from diffusion and from convection by the gas "throughflow." The overall transfer rate is commonly estimated by treating these components as additive (D6), although they probably interact strongly (C2, H6).

Three-phase (solid/liquid/gas) fluidized systems are also of some practical importance. There is again a strong analogy between the rise of gas bubbles in normal liquids and in liquid fluidized beds (D1, R3), although there is evidence of solid/liquid segregation in wakes (R3, S6) which has no parallel for two-phase systems.

REFERENCES

A1. Angelino, H., *Chem. Eng. Sci.* **21**, 541–550 (1966).
B1. Baird, M. H. I., and Davidson, J. F., *Chem. Eng. Sci.* **17**, 87–93 (1962).
B2. Berghmans, J., *Chem. Eng. Sci.* **29**, 1645–1650 (1974).
B3. Bhaga, D., Ph.D. Thesis, McGill Univ., Montreal, 1976.
B4. Brignell, A. S., *Chem. Eng. Sci.* **29**, 135–147 (1974).
B5. Brophy, J., Course 302–495B Final Rep., McGill Univ., Montreal, 1973.
C1. Calderbank, P. H., Johnson, D. S. L., and Loudon, J., *Chem. Eng. Sci.* **25**, 235–256 (1970).
C2. Chavarie, C., and Grace, J. R., *Chem. Eng. Sci.* **31**, 741–749 (1976).
C3. Collins, R., *Chem. Eng. Sci.* **20**, 851–855 (1965).
C4. Collins, R., *J. Fluid Mech.* **22**, 763–771 (1965).
C5. Collins, R., *J. Fluid Mech.* **25**, 469–480 (1966).
C6. Collins, R., *Chem. Eng. Sci.* **22**, 89–97 (1967).
C7. Crabtree, J. R., and Bridgwater, J., *Chem. Eng. Sci.* **22**, 1517–1520 (1967).
C8. Crabtree, J. R., and Bridgwater, J., *Chem. Eng. Sci.* **26**, 838–851 (1971).
D1. Darton, R. C., and Harrison, D., *Trans. Inst. Chem. Eng.* **52**, 301–306 (1974).
D2. Davenport, W. G., Ph.D. Thesis, Imperial College, London, 1964.
D3. Davenport, W. G., Bradshaw, A. V., and Richardson, F. D., *J. Iron Steel Inst., London* **205**, 1034–1042 (1967).
D4. Davenport, W. G., Richardson, F. D., and Bradshaw, A. V., *Chem. Eng. Sci.* **22**, 1221–1235 (1967).
D5. Davenport, W. G., Wakelin, D. H., and Bradshaw, A. V., *in* "Heat and Mass Transfer in Process Metallurgy" (A. W. D. Hills, ed.), pp. 207–240. Inst. Min. Metall., London, 1967.
D6. Davidson, J. F., and Harrison, D., "Fluidised Particles." Cambridge Univ. Press, London and New York, 1963.
D7. Davidson, J. F., Harrison, D., and Guedes de Carvalho, J. R. F., *Annu. Rev. Fluid Mech.* **9**, 55–86 (1977).
D8. Davidson, J. F., and Kirk, F. A., *Chem. Eng. Sci.* **24**, 1529–1530 (1969).
D9. Davies, R. M., and Taylor, Sir G. I., *Proc. R. Soc., Ser. A* **200**, 375–390 (1950).
G1. Gabor, J. D., *Chem. Eng. J.* **4**, 118–126 (1972).
G2. Garcia, A., Grace, J. R., and Clift, R., *AIChE J.* **19**, 369–370 (1973).
G3. Grace, J. R., Ph.D. Thesis, Cambridge Univ. 1968.
G4. Grace, J. R., *Can. J. Chem. Eng.* **48**, 30–33 (1970).
G5. Grace, J. R., and Harrison, D., *Chem. Eng. Sci.* **22**, 1337–1347 (1967).
G6. Guthrie, R. I. L., Ph.D. Thesis, Imperial College, London, 1967.
G7. Guthrie, R. I. L., and Bradshaw, A. V., *Trans. Metall. Soc. AIME* **245**, 2285–2292 (1969).
G8. Guthrie, R. I. L., and Bradshaw, A. V., *Chem. Eng. Sci.* **24**, 913–917 (1969).
G9. Guthrie, R. I. L., and Bradshaw, A. V., *Chem. Eng. Sci.* **28**, 191–203 (1973).
H1. Haberman, W. L., and Morton, R. K., *David Taylor Model Basin Rep.* No. 802 (1953).
H2. Harper, J. F., *Adv. Appl. Mech.* **12**, 59–129 (1972).
H3. Harper, J. F., and Moore, D. W., *J. Fluid Mech.* **32**, 367–391 (1968).
H4. Hills, J. H., *J. Fluid Mech.* **68**, 503–512 (1975).
H5. Hnat, J. G., and Buckmaster, J. D., *Phys. Fluids* **19**, 182–194, 611 (1976).
H6. Hovmand, S., and Davidson, J. F., *Trans. Inst. Chem. Eng.* **46**, 190–203 (1968).

J1. Jackson, R., *in* "Fluidization" (J. F. Davidson and D. Harrison, eds.), pp. 65–119. Academic Press, New York, 1971.

J2. Johnson, A. I., Besik, F., and Hamielec, A. E., *Can. J. Chem. Eng.* **47**, 559–564 (1969).

J3. Jones, D. R. M., Ph.D. Thesis, Cambridge Univ., 1965.

K1. Kalra, T. R., and Uhlherr, P. H. T., *Aust. Conf. Hydraul. Fluid Mech., 4th, Melbourne*, 1971.

L1. Lazarek, G. M., and Littman, H., *J. Fluid Mech.* **66**, 673–687 (1974).

L2. Lindt, J. T., *Chem. Eng. Sci.* **26**, 1776–1777 (1971).

L3. Lindt, J. T., and DeGroot, R. G., *Chem. Eng. Sci.* **29**, 957–962 (1974).

L4. Lochiel, A. C., and Calderbank, P. H., *Chem. Eng. Sci.* **19**, 471–484 (1964).

M1. Maxworthy, T., *J. Fluid Mech.* **27**, 367–368 (1967).

M2. McDonald, J. E., *J. Meteorol.* **11**, 478–494 (1954).

M3. Moore, D. W., *J. Fluid Mech.* **6**, 113–130 (1959).

P1. Paneni, M., and Davenport, W. G., *Trans. Metall. Soc. AIME* **245**, 735–738 (1969).

P2. Parlange, J. Y., *J. Fluid Mech.* **37**, 257–263 (1969).

P3. Partridge, B. A., Lyall, E., and Crooks, H. E., *Powder Technol.* **2**, 301–305 (1969).

P4. Patel, P., *Arch. Eisenhuttenwes.* **44**, 435–441 (1973).

R1. Richardson, F. D., *Metall. Trans.* **2**, 2747–2756 (1971).

R2. Richardson, F. D., "Physical Chemistry of Melts in Metallurgy," Vol. 2. Academic Press, New York, 1974.

R3. Rigby, G. R., and Capes, C. E., *Can. J. Chem. Eng.* **8**, 343–348 (1970).

R4. Rippin, D. W. T., and Davidson, J. F., *Chem. Eng. Sci.* **22**, 217–228 (1967).

R5. Rowe, P. N., *in* "Fluidization" (J. F. Davidson and D. Harrison, eds.), pp. 121–191. Academic Press, New York, 1971.

R6. Rowe, P. N., Partridge, B. A., and Lyall, E., *Chem. Eng. Sci.* **19**, 973–985 (1964).

R7. Rowe, P. N., and Widmer, A. J., *Chem. Eng. Sci.* **28**, 980–981 (1973).

S1. Schwerdtfeger, K., *Chem. Eng. Sci.* **23**, 937–938 (1968).

S2. Scorer, R. S., "Natural Aerodynamics." Pergamon, Oxford, 1958.

S3. Shoemaker, P. D., and de Chazal, L. E. M., *Chem. Eng. Sci.* **24**, 795–797 (1969).

S4. Slaughter, I., Ph.D. Thesis, Univ. of Newcastle, 1967.

S5. Slaughter, I., and Wraith, A. E., *Chem. Eng. Sci.* **23**, 932 (1968).

S6. Stewart, P. S. B., and Davidson, J. F., *Chem. Eng. Sci.* **19**, 319–321 (1964).

S7. Sundell, R. E., Ph.D. Thesis, Yale Univ., New Haven, 1971.

T1. Taylor, Sir G. I., *Proc. R. Soc., Ser. A* **201**, 175–186 (1950).

W1. Wairegi, T., M. Eng. Thesis, McGill Univ., Montreal, 1972.

W2. Wairegi, T., Ph.D. Thesis, McGill Univ., Montreal, 1974.

W3. Wairegi, T., and Grace, J. R., *Int. J. Multiphase Flow* **3**, 67–77 (1976).

W4. Weber, M. E., *Chem. Eng. Sci.* **30**, 1507–1510 (1975).

W5. Wegener, P. P., and Parlange, J. Y., *Annu. Rev. Fluid Mech.* **5**, 79–100 (1973).

W6. Wegener, P. P., Sundell, R. E., and Parlange, J. Y., *Z. Flugwiss.* **19**, 347–352 (1971).

W7. Wu, B. J. C., Deluca, R. T., and Wegener, P. P., *Chem. Eng. Sci.* **29**, 1307–1309 (1974).

Chapter 9

Wall Effects

I. INTRODUCTION

In terms of the analytic solutions for flow around rigid and circulating particles, the effect of containing walls is to change the boundary conditions for the equations of motion and continuity of the continuous phase. In place of the condition of uniform flow remote from the particle, containing walls impose conditions which must be satisfied at definite boundaries.

Consider the example shown schematically in Fig. 9.1; a sphere of diameter d is moving parallel to the axis of a cylindrical tube of radius R through which a

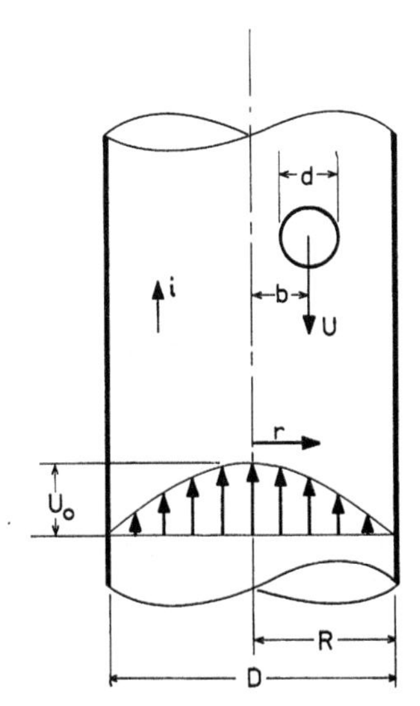

FIG. 9.1 Sphere falling through a fluid in laminar flow: schematic.

fluid is passing in Poiseuille flow with centerline velocity U_0. Taking a reference frame moving with the particle, the new boundary conditions are:

$$\mathbf{u} = U\mathbf{i} \qquad \text{at the walls,} \tag{9-1}$$

where $\mathbf{i}$ is the unit vector directed vertically upwards and U is the absolute downward velocity of the particle, and

$$\mathbf{u} = \mathbf{i}[U + U_0(1 - r^2/R^2)] \tag{9-2}$$

at large distances upstream and downstream from the particle. These new boundary conditions cause changes in the drag force and transfer rate. For fluid particles there is the additional effect of the container walls on the particle shape.

Here we concentrate on cylindrical containing walls, although there is some work on particles near plane boundaries and surfaces of arbitrary shape. Most of the work on rigid particles refers to spheres, and it is then convenient to use the diameter ratio

$$\lambda = d/D. \tag{9-3}$$

For fluid particles, the volume-equivalent diameter is used in defining λ.

II. RIGID PARTICLES

A. Flow Patterns

Little work has been reported on the motion of bounded fluids past rigid particles, except for the creeping flow range. Coutanceau (C8) reported visualization of the flow around a sphere moving along the axis of a tube containing an otherwise stationary fluid. The walls were found to delay formation of the attached recirculatory wake, and the onset of separation was given for $\lambda < 0.8$ by

$$\mathrm{Re}_s = 20(1 - \lambda)^{-0.56}. \tag{9-4}$$

Taking detectable departure from fore-and-aft symmetry as the upper limit of Stokes flow, Coutanceau found that increasing λ increased the range of validity of the creeping flow approximation.† The upper limit of Stokes flow was proposed as:

$$\mathrm{Re}_L = (\mathrm{Re}_s/7) - 2.75 \qquad (\lambda < 0.8). \tag{9-5}$$

Johansson (J1) reported numerical calculations of the flow around a sphere fixed on the axis of a Poiseuille flow (Fig. 9.1 with $b = 0$, $U = 0$). Only solutions for $\lambda = 0.1$ were considered, and wake formation was predicted for $\mathrm{Re} = 20.4$ based on the centerline velocity U_0.

† As noted in Chapter 3, the inconsistency in Stokes' solution occurs in the outer flow field.

At the other extreme of Re, Achenbach (A1) investigated flow around a sphere fixed on the axis of a cylindrical wind tunnel in the critical range. Wall effects can increase the supercritical drag coefficient well above the value of 0.3 arbitrarily used to define Re_c in an unbounded fluid (see Chapter 5). If Re_c is based on the mean approach velocity† and corresponds to C_D midway between the sub- and super-critical values, the critical Reynolds number decreases from 3.65×10^5 in an unbounded fluid to 1.05×10^5 for $\lambda = 0.916$.

B. Drag and Terminal Velocity

There are three useful measures of the effect of bounding walls on drag. A drag factor can be defined, based on the same particle at the same fluid velocity:

$$K_F = \frac{\text{drag in bounded fluid}}{\text{drag in infinite fluid}} = \frac{F_D}{F_{D\infty}}. \tag{9-6}$$

Alternatively, a velocity ratio can be defined, based on constant particle dimensions (i.e., constant N_D):

$$K_U = \frac{\text{terminal velocity in infinite fluid}}{\text{terminal velocity in bounded fluid}} = \frac{U_{T\infty}}{U_T}. \tag{9-7}$$

For falling sphere viscometry, it is most convenient to define a viscosity ratio based on constant particle dimensions and terminal velocity (S7):

$$K_\mu = \mu_s/\mu, \tag{9-8}$$

where μ is the actual fluid viscosity and μ_s is the viscosity of the unbounded fluid which would give the observed U_T if Stokes' law were to apply, i.e.,

$$\mu_s = d^2 g(\rho_p - \rho)/18U_T. \tag{9-9}$$

The term K_μ includes the effect of departures from the creeping flow approximations. In creeping flow, $K_F = K_U = K_\mu \equiv K$, but at higher Re the relationships are more complex. In general, any of these ratios is a function of λ and one other group (such as Re) chosen to suit the problem at hand.

1. *Low Reynolds Numbers*

For a complete review on low Re motion in bounded fluids, see Happel and Brenner (H3). Some general results are of immediate interest. For a particle moving through an otherwise undisturbed fluid, without rotation and with velocity U parallel to a principal axis both of the body and the container,

$$K = [1 - CF_{D\infty}/6\pi\mu U l + O(c/l)^3]^{-1}, \tag{9-10}$$

where c is the maximum particle dimension, l the shortest distance from the center of the particle to the wall, and C depends on the nature of the boundary

† Achenbach based Re on flow conditions in the smallest cross section between sphere and tube. With this definition, wall effects increase Re_c.

TABLE 9.1

Wall Correction Coefficient C in Eq. (9-10) for Rigid Boundaries

Boundary	Location of particle	Direction of motion	C
Circular cylinder	Axis	Axial	2.10444
	Eccentric	Axial	Fig. 9.2
Parallel plane walls	Midplane	Parallel to walls	1.004
	$\frac{1}{4}$distance across channel	Parallel to walls	0.6526
Single plane wall		Parallel to wall	$\frac{9}{16}$
		Normal to wall	$\frac{9}{8}$
Spherical	Center		$\frac{9}{4}$

but not on the shape of the particle.† Thus correction factors determined for one particle shape can be applied to another shape, provided that c/l is sufficiently small for the higher-order terms to be neglected. Table 9.1 gives C for various rigid boundaries; different values apply for free surfaces. For a particle settling eccentrically in a cylinder, C depends upon distance from the axis as shown in Fig. 9.2. For small b/R,

$$C = 2.10444 - 0.6977(b/R)^2 + O(b/R)^4. \tag{9-11}$$

Note that K is insensitive to position provided that $b/R \dot{<} 0.6$.

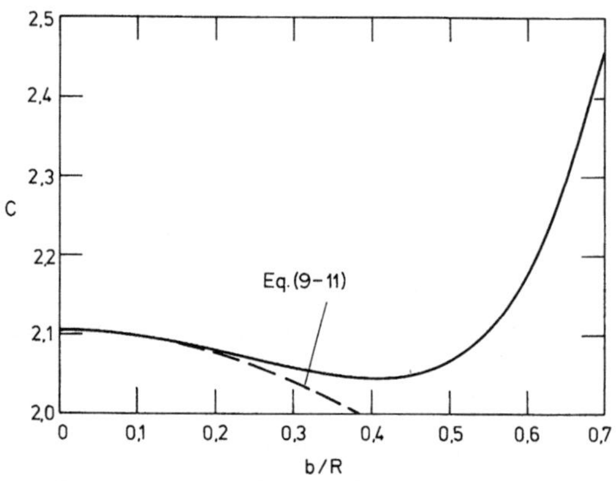

FIG. 9.2 Wall correction coefficient [C in Eq. (9-10)] for a rigid particle settling eccentrically in a circular cylinder.

To obtain higher-order approximations, it is necessary to consider specific shapes. Various correction factors have been proposed for rigid spheres moving through an otherwise undisturbed fluid. The most widely used are summarized in Table 9.2. Experimental determinations of K reported by Fidleris and

† This result also applies to a circulating particle provided that the appropriate $F_{D\infty}$ is used.

TABLE 9.2

Wall Correction Factor K for a Rigid Spherical Particle Moving on the Axis of a Cylindrical Tube in Creeping Flow

Author	Expression for K
Ladenburg (L1)	$1 + 2.105\lambda$
Faxén (F1)	$(1 - 2.104\lambda + 2.09\lambda^3 - 0.95\lambda^5)^{-1}$
Haberman and Sayre (H1)	$\dfrac{1 - 0.75857\lambda^5}{1 - 2.1050\lambda + 2.0865\lambda^3 - 1.7068\lambda^5 + 0.72603\lambda^6}$
Francis (F6) (empirical)	$[(1 - 0.475\lambda)/(1 - \lambda)]^4$

Whitmore (F4) for $\lambda \leq 0.6$ lie between the expressions of Haberman and Sayre and of Francis. The Haberman and Sayre result shows about 1% less deviation, but the equation due to Francis has the virtue of simplicity. Experimental results due to Sutterby (S7) for $\lambda \leq 0.13$ with $\mathrm{Re} \rightarrow 0$ agree with the Faxén, Haberman, and Francis curves which are virtually indistinguishable in this range. The Ladenburg result is only accurate for $\lambda \dot{<} 0.05$.

The results in Table 9.2 apply when no end effects are present. Sutterby (S7) determined simultaneous wall and end correction factors for the creeping flow range. His correlations are shown in Fig. 9.3 where the cylindrical column has closed ends a distance L_c apart and the center of the spherical particle is distance Z from one end of the tube. The curve for $D/L_c = 1.0$ and $Z/L_c = 1/2$ is

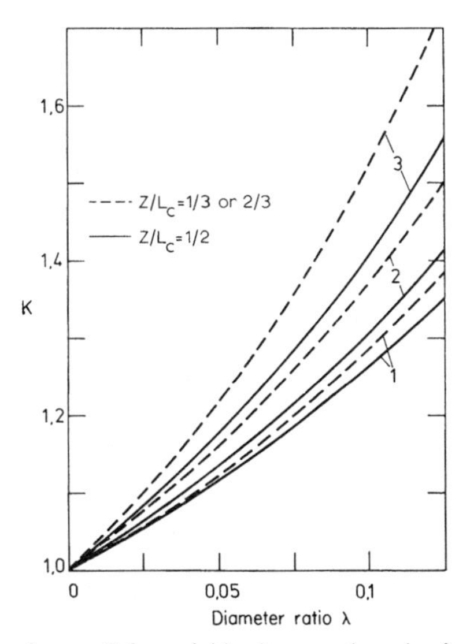

FIG. 9.3 Wall correction factors K for a rigid sphere on the axis of a cylinder of finite length in creeping flow (S7): (1) $L_c/D = 1$; (2) $L_c/D = \frac{2}{3}$; (3) $L_c/D = \frac{1}{2}$.

indistinguishable from the Haberman and Sayre result for a long tube (Table 9.2), so that the relative displacement of the other curves indicates the magnitude of end effects. These are least when the sphere is at the midpoint of the column. For a column with $L_c/D = 2.0$, end effects are negligible if $1/3 < Z/L_c < 2/3$. A theoretical treatment of simultaneous wall and end effects by Tanner (T2) gives values of K which are asymptotic to the curves in Fig. 9.3 for small λ, but underpredicts K otherwise.

For a rigid sphere on the axis of a tube through which a fluid moves in laminar flow (Fig. 9.1 with $b = 0$), Haberman and Sayre (H1) showed that the magnitude of the drag force is

$$F_D = -3\pi\mu d K(U - K'U_0), \tag{9-12}$$

where

$$K' = [1 - (2\lambda^2/3) - 0.20217\lambda^5]/[1 - 0.75857\lambda^5]. \tag{9-13}$$

2. *Higher Reynolds Numbers*

As for particles in infinite fluids, analytic solutions have not been successfully extended beyond the creeping flow range. Faxén (F1) applied the Oseen linearization to a sphere moving axially in a tube, but the resulting drag predictions are no more reliable than for an unbounded fluid (F4, H3, S7). However, reliable experimental results are available for freely settling spheres (F4, M4, S7), spheres fixed in a fluid flow (A1, M5), and spheres freely suspended in an upward-flowing liquid (R2). The results of these investigations are in remarkably good agreement. For particles in ducts of noncircular section, it is usual to define D as the conventional "hydraulic diameter," but the accuracy of this approximation does not appear to have been seriously assessed.

Figure 9.4 shows curves for the drag coefficient (based on the velocity for a freely settling sphere and the mean approach velocity for a fixed or suspended sphere) and for the fractional increase in drag caused by wall effects, $(K_F - 1)$. Up to Re of order 50, the results are approximated closely by an equation proposed by Fayon and Happel (F2):

$$C_D = C_{D\infty} + (24/\text{Re})(K - 1), \tag{9-14}$$

i.e.,

$$K_F = 1 + (24/\text{Re}C_{D\infty})(K - 1), \tag{9-15}$$

where K is given by Table 9.2. For Re in the range from roughly 100 to 10^4, available data indicate that K_F is independent of Re, and given within 6% by:

$$K_F = 1/(1 - 1.6\lambda^{1.6}) \qquad (\lambda \leq 0.6). \tag{9-16}$$

For Re $> 10^5$, Achenbach's result (A1) may be used:

$$K_F = (1 + 1.45\lambda^{4.5})/(1 - \lambda^2)^2 \qquad (\lambda \leq 0.92). \tag{9-17}$$

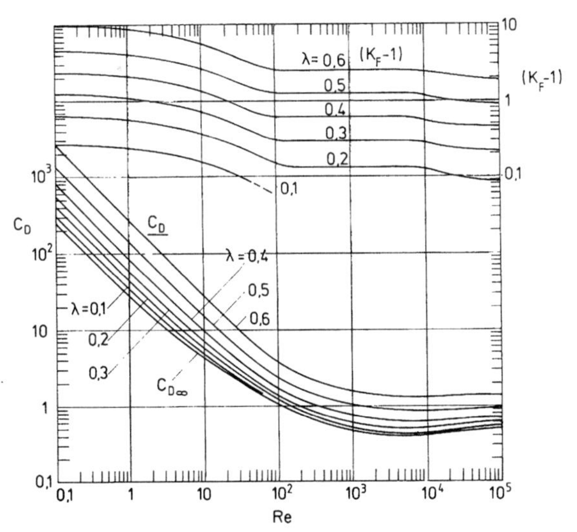

FIG. 9.4 Drag coefficient C_D and fractional drag increase $(K_F - 1)$ for rigid spheres on the axis of circular ducts.

For treatment of terminal settling velocities, it is more convenient to work in terms of N_D defined in Eq. (5-15). Figure 9.5 shows the terminal Reynolds number and $(K_U - 1)$ as functions of $N_D^{1/3}$. For $N_D^{1/3} > 10^3$, K_U is approximated closely by $\sqrt{K_F}$, with K_F given by Eq. (9-17).

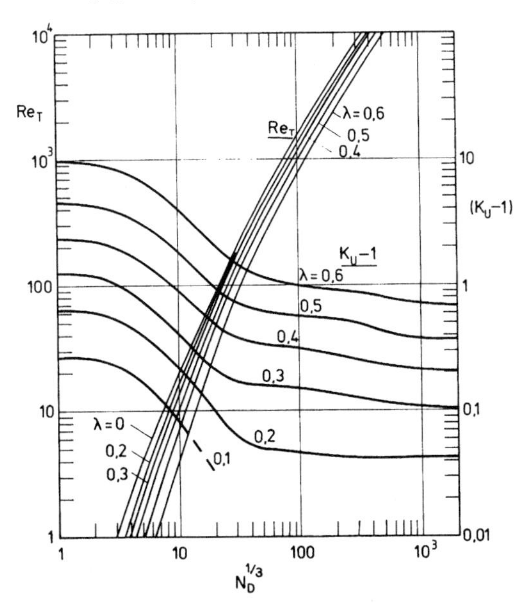

FIG. 9.5 Terminal Reynolds number and velocity correction factor for rigid spheres on the axis of circular ducts.

Sutterby (S7) gave a useful tabulation of the viscosity ratio K_μ, defined in Eq. (9-8), for relatively low Re and λ. These values, intended primarily to correct for departures from Stokes' law in falling sphere viscometry, are shown in Fig. 9.6. Reynolds number is defined using the measured U_T and μ_s defined in Eq. (9-9). The curve for $\lambda = 0$ accounts for departures from the creeping flow approximations in an unbounded fluid, and the relative displacement of the other curves indicates the wall effect.

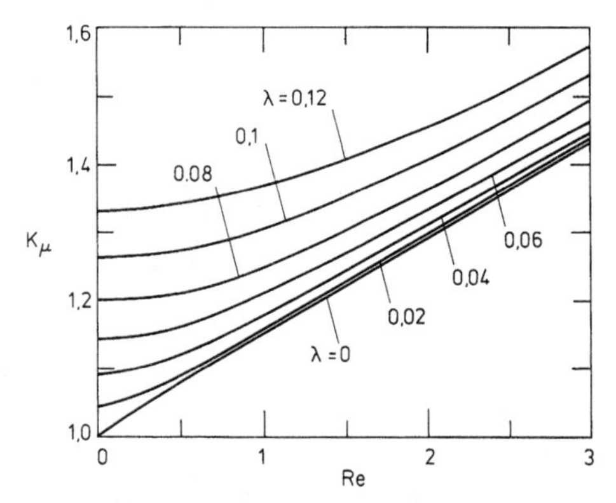

FIG. 9.6 Viscosity correction factor K_μ for rigid spheres settling axially in circular columns with $\mathrm{Re} = U_T \rho d/\mu_s$ (S7).

C. Pressure Drop

In addition to the effect of the walls on the drag on the particle, the particle alters the shear on the duct. Consider a particle settling through a quiescent fluid (Fig. 9.1 with $U_0 = 0$). Brenner (B3) showed that, for low particle Re with the particle small by comparison with the distance between particle and wall (i.e., $\lambda \ll 1 - \beta$, where $\beta = b/R$), there is an excess pressure drop, ΔP^+, between points far below and far above the particle given by

$$\Delta P^+ A/F_D = 2(1 - \beta^2) - \tfrac{4}{3}\lambda^2 + O(\lambda^3) \qquad (\lambda \ll 1 - \beta), \tag{9-18}$$

where F_D is the force on the particle and A is the cross-sectional area of the duct. Bungay and Brenner (B7) carried out a complementary analysis which predicts ΔP^+ for particles close to the wall. For a small particle settling on the axis

$$\Delta P^+ A = 2F_D, \tag{9-19}$$

so that the walls exert a total force F_D downwards on the fluid. Surprisingly in view of the assumptions, Eq. (9-19) appears to apply up to quite high particle Re, for a variety of different particle shapes (F3, L3), but then $\Delta P^+ A/F_D$ falls sharply from two to unity. This transition corresponds roughly to

$$2RU_T/\nu = 2300, \tag{9-20}$$

where U_T is the terminal velocity of the particle. The coincidence with the Reynolds number characterizing laminar/turbulent transition in pipe flow remains a curiosity (L3).

D. Particle Migration

It is noted in Chapter 10 that, if inertial terms are neglected, a freely rotating particle suspended in a sheared fluid experiences no lift. However, in a classic series of experiments, Segré and Silberberg (S3) demonstrated that neutrally buoyant rigid spheres suspended in a Poiseuille flow migrate to a position given roughly by $\beta = 0.6$. This effect has been confirmed many times (B3, G4, H2, L4, T1). If $\gamma \neq 1$, a sphere in a Poiseuille flow migrates towards the wall if its velocity exceeds the local undisturbed fluid velocity, but towards the center line if its velocity lags the fluid. Both neutrally buoyant and sedimenting spheres in a Couette flow migrate to the central plane (V2). The reason for this migration has been widely debated. It must result from a lift force, but this cannot be explained by particle rotation [see (L4)]. The only explanation is the presence of inertial effects. Ho and Leal (H5) and Vasseur (V2) have confirmed this by applying the method of matched asymptotic expansions (C9). The migration velocity depends on the particle size and position, and on the duct and particle Reynolds numbers. Resulting trajectory predictions agree closely with observation. Eichorn and Small (E2) measured the lift on a solid sphere fixed in a Poiseuille flow with $80 < \mathrm{Re} < 250$.

E. Heat and Mass Transfer

There are two useful measures of the effect of bounding walls on the heat- or mass-transfer rate. A mass transfer factor can be defined based on the same relative velocity between the particle and the fluid:

$$K_{\mathrm{MU}} = \frac{\text{Sherwood number in bounded fluid}}{\text{Sherwood number in infinite fluid}} = \left(\frac{\mathrm{Sh}}{\mathrm{Sh}_\infty}\right)_{\mathrm{U}} \tag{9-21}$$

where the subscript U denotes the fact that the relative velocity is the same in the bounded and unbounded fluids. Alternatively, a mass transfer factor can be based upon constant particle dimensions

$$K_{\mathrm{MD}} = \frac{\text{Sherwood number in infinite fluid}}{\text{Sherwood number in bounded fluid}} = \left(\frac{\mathrm{Sh}_\infty}{\mathrm{Sh}}\right)_{\mathrm{D}} \tag{9-22}$$

where the subscript D denotes the fact that the particle dimensions, i.e., $N_{\mathrm{D}}^{1/3}$, are the same in the bounded and unbounded fluids.

1. *Low Reynolds Numbers*

The stream function expressions of Haberman and Sayre (H1) for creeping flow permit the calculation of the effect of cylindrical containing walls on the

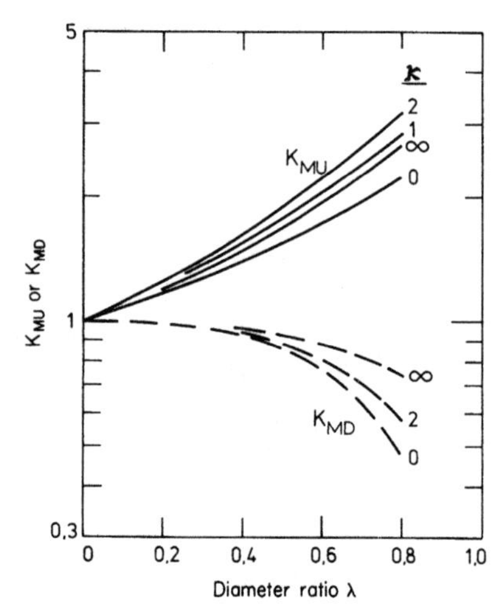

FIG. 9.7 Mass transfer correction factors K_{MU} and K_{MD} for a sphere on the axis of a cylinder in creeping flow.

Sherwood number under the thin concentration boundary layer assumption through Eq. (3-46). The results are plotted in terms of K_{MU} and K_{MD} in Fig. 9.7. For a rigid sphere in creeping flow, the relationship between these quantities and the velocity ratio K is

$$K_{MD} = K_{MU}/K^{1/3}. \tag{9-23}$$

For a rigid sphere ($\kappa = \infty$) on the axis of a cylindrical tube, the Sherwood number is larger than in an unbounded fluid with the same particle/fluid velocity. The ratio of Sherwood numbers is approximated within 3% for $\lambda \leq 0.6$ by

$$K_{MU} = (1 - 0.35\lambda)/(1 - \lambda). \tag{9-24}$$

The presence of container walls has a much smaller effect on Sherwood number than on drag since the mass transfer coefficient is only proportional to the one-third power of the surface vorticity. For a sphere with given $N_D^{1/3}$ settling on the axis of a cylindrical container, the Sherwood number decreases with λ, but it is still within 8% of the Sherwood number in an infinite fluid for $\lambda = 0.5$. No data are available to test these predictions.

2. *Higher Reynolds Numbers*

For Re $\gtrsim 10^3$ there are a number of studies of the effect of walls on heat and mass transfer from solid particles in wind and water tunnels. In these studies it was customary to define a velocity ratio $K_{\chi e}$ based on the same Sherwood number in bounded and infinite fluids:

$$K_{Me} = U_e/U_T \tag{9-25}$$

where U_T is the true relative velocity between the particle and the fluid and U_e, the effective velocity, is the velocity required in an unbounded fluid to give the same Sherwood number. The large number of relationships proposed for K_{Me} have been reviewed by Pei (P1) and Morgan (M6). The expression proposed by Leppert and coworkers (P2, V3) gives good agreement with data for rigid spheres located on the axis of cylindrical ducts:

$$K_{Me} = 1/(1 - \tfrac{2}{3}\lambda^2). \tag{9-26}$$

Equation (9-26) can be used with the Sherwood number equations for solid spheres in Chapter 5 to determine the increase in Sh due to container walls. For a settling sphere, a more useful velocity ratio is $U_e/U_{T\infty}$, the ratio of the effective velocity to the terminal velocity of the sphere in an infinite fluid:

$$U_e/U_{T\infty} = 1/(1 - \tfrac{2}{3}\lambda^2)K_U. \tag{9-27}$$

Here K_U is obtained from Fig. 9.5. Equation (9-27) and the equations of Chapter 5 can be used to determine the decrease in Sh for a rigid sphere with fixed $N_D^{1/3}$ settling on the axis of a cylindrical tube. For example, for a settling sphere with $\lambda = 0.4$ and $N_D^{1/3} = 200$, $U_T/U_{T\infty} = 0.76$ and $U_e/U_T = 0.85$. Since the Sherwood number is roughly proportional to the square root of Re, the Sherwood number for the settling particle is reduced only 8%, while its terminal velocity is reduced 24%. As in creeping flow, the effect of container walls on mass and heat transfer is much smaller than on terminal velocity.

III. BUBBLES AND DROPS

It is convenient to divide the discussion of wall effects for bubbles and drops into two parts. Section A covers cases where the diameter ratio, $\lambda = d_e/D$, is less than about 0.6. At low λ, the walls cause little deformation beyond that which may be present for the fluid particle in an infinite medium, so that the discussion of wall effects for rigid particles forms a good starting point. Section B treats the case of slug flow ($\lambda > 0.6$) where the container walls have a dominant effect on the shape of the bubble or drop.

A. WALL CORRECTIONS FOR $\lambda \leq 0.6$

1. *Low Reynolds Numbers*

We recall from Chapters 2 and 3 that fluid particles at low Re in infinite media tend to be spherical and that the interface is usually stagnant due to surface-active contaminants or large values of $\kappa = \mu_p/\mu$. If λ is less than about 0.3, deformation due to the container walls tends to be minor and the corrections given above for rigid spheres at low Re may be used.

For an interface free of surface-active contaminants, Haberman and Sayre (H1) obtained approximate solutions for a circulating sphere traveling in steady motion along the axis of a cylindrical tube. As with rigid particles, the drag force on a circulating particle tends to increase as λ increases. However, the effect is less than for a rigid particle under corresponding conditions. The drag force on a circulating sphere is given by

$$F_D = \pi\mu dU\, K(2 + 3\kappa)/(1 + \kappa), \tag{9-28}$$

where the correction factor, K is given by

$$K = \frac{1 + 2.2757\lambda^5\left(\dfrac{1-\kappa}{2+3\kappa}\right)}{1 - 0.7017\left(\dfrac{2+3\kappa}{1+\kappa}\right)\lambda + 2.0865\left(\dfrac{\kappa}{1+\kappa}\right)\lambda^3 + 0.5689\left(\dfrac{2-3\kappa}{1+\kappa}\right)\lambda^5 - 0.72603\left(\dfrac{1-\kappa}{1+\kappa}\right)\lambda^6}, \tag{9-29}$$

which reduces to the result in Table 9.2 as $\kappa \to \infty$. This correction factor was found to give good agreement with experimental results for relatively large aqueous glycerine or silicone oil drops falling on the axes of cylindrical tubes through castor oil (H1). As λ increased, the presence of the walls caused droplet deformation, elongation occurring in the vertical direction to yield approximately prolate ellipsoid shapes. The theory gave an accurate prediction of the wall correction for λ up to about 0.5, although significant droplet deformation had occurred.

The analysis was extended to apply to circulating particles on the axis of cylinders where there is a parabolic (laminar) velocity profile well upstream and downstream of the particle (Fig. 9.1 with $b = 0$). The drag force is given by

$$F_D = -\pi\mu d[(2 + 3\kappa)/(1 + \kappa)](KU - K'U_0), \tag{9-30}$$

where K is given by Eq. (9-29) and

$$K' = \frac{1 - \left(\dfrac{2\kappa}{2+3\kappa}\right)\lambda^2 + 0.60651\left(\dfrac{1-\kappa}{2+3\kappa}\right)\lambda^5}{1 + 2.2757\left(\dfrac{1-\kappa}{2+3\kappa}\right)\lambda^5} \tag{9-31}$$

The above results give good predictions for bubbles and drops that would normally be spherical, provided that λ is less than about 0.5, Re less than unity, and the fluid particle near the axis of the tube.

2. *Intermediate Size Drops and Bubbles* (Eo < 40)

All studies of drops and bubbles have been carried out in containers of finite dimensions; hence wall effects have always been present to a greater or lesser extent. However, few workers have set out to determine wall effects directly using a series of different columns of varying diameter. Where studies have been carried out, the sole aim has usually been to determine the influence of λ on the terminal velocity. While it is known that the containing walls tend to cause elongation of fluid particles in the vertical direction, suppress secondary motion, and alter the wake structure, there is insufficient experimental evidence on these factors to allow useful quantitative generalizations to be drawn.

Previous correlations of the influence of λ on terminal velocities (E1, H4, M1, S1, S6, T3, U1) are limited to specific systems, fail to recognize the different regimes of fluid particles (see Chapter 2), or are difficult to apply. In the present section we consider both bubbles and drops, but confine our attention to those of intermediate size (see Chapter 7) where $\mathrm{Eo} < 40$ and $\mathrm{Re} > 1$. Only the data of Uno and Kintner (U1), Strom and Kintner (S6) and Salami *et al.* (S1) are used since other workers either failed to use a range of column sizes for the same fluid–fluid systems, or it was impossible to obtain accurate values of the original data. This effectively limits the Reynolds number range to $\mathrm{Re} > 10$ for the low M systems studied.

A plot of all the data as $U_T/U_{T\infty} = K_U^{-1}$ (where $U_{T\infty}$ is the terminal velocity which the drop or bubble would have in an infinite container, as taken or extrapolated from the authors' own data) versus Re shows that, as for rigid spheres in cylindrical columns, the terminal velocity ratio deviates further from unity as λ increases and as Re or N_D decreases. In fact, the curves in Fig. 9.5 may be taken over and used directly for the prediction of $U_T/U_{T\infty}$ for $\lambda \leq 0.6$. This may appear surprising, but it should be remembered that unbounded drops and bubbles in this range tend to be flattened in the vertical direction, while the containing walls tend to cause elongation. Hence the resulting shape may not deviate greatly from a sphere. For wall effects to have negligible influence (less than about 2%) on terminal velocities, the following conditions should apply:

$$\mathrm{Re} \leq 0.1 \qquad \lambda \leq 0.06, \tag{9-32}$$

$$0.1 < \mathrm{Re} < 100 \qquad \lambda \leq 0.08 + 0.02 \log_{10} \mathrm{Re}, \tag{9-33}$$

$$\mathrm{Re} \geq 100 \qquad \lambda \leq 0.12. \tag{9-34}$$

These empirical relationships have been used in Chapters 7 and 8 to eliminate experimental results subject to significant wall effects.

For Re greater than about 200, the effect of Re on $U_T/U_{T\infty}$ is relatively small (see Fig. 9.5). It is therefore possible to represent the results by a unique relationship between $U_T/U_{T\infty}$ and λ. The experimental results are shown in Fig. 9.8 together with the equation

$$U_T/U_{T\infty} = [1 - \lambda^2]^{3/2}, \tag{9-35}$$

which gives an excellent fit for λ up to about 0.6.† There appears to be a systematic and inexplicable difference between the Salami *et al.* (S1) data and the other data at λ values greater than about 0.5. Equation (9-35) is recommended for bubbles and drops for $\mathrm{Eo} < 40$, $\mathrm{Re} > 200$ and $\lambda \leq 0.6$. When the first and third of these conditions apply but $1 \leq \mathrm{Re} \leq 200$, Fig. 9.5 should be used.

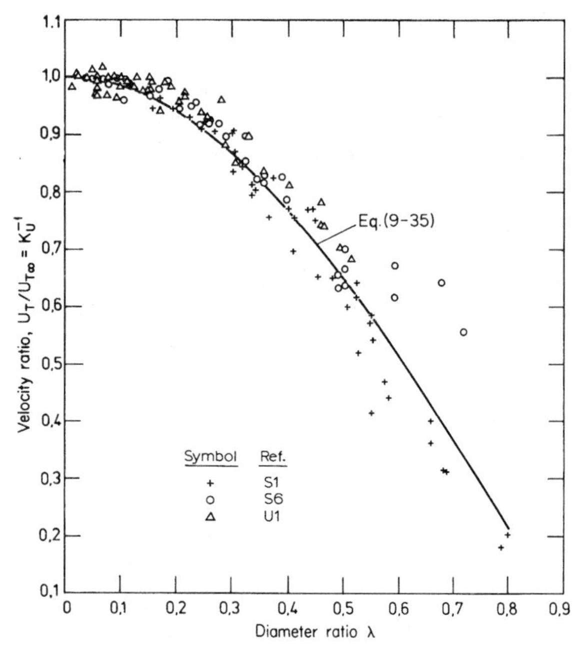

FIG. 9.8 Retarding effect of column walls on the terminal velocity of drops and bubbles of intermediate size.

3. *Large Bubbles and Drops* (Eo $\geq$ 40)

Collins (C5, C6) carried out a thorough study of the influence of containing walls on the velocity of spherical-cap bubbles. The work was extended to lower Re by Bhaga (B1) who also investigated the influence of wall proximity on wake size, external flow fields, bubble shape, and skirt behavior. Generally speaking, increasing λ for a given large fluid particle in a system of fixed fluid properties was found to cause bubble elongation, a decrease in terminal velocity, a marked reduction in the wake volume and the rate of fluid circulation within the wake, and a delay in the onset and waviness of skirts. Excellent photographs of bubbles subject to wall effects have been published (B1, C1). Tracings showing the effect of increasing λ at constant bubble volume on bubble shape are shown in Fig. 9.9. Some data illustrating the strong dependence of wake volume on λ appear in Table 9.3.

Experimental results show that wall effects are negligible for λ up to about 0.125 for spherical-caps in low M systems. Collins suggested semiempirical equations for $U_T/U_{T\infty}$. A simpler equation proposed by Wallis (W1) which agrees well with the results of Collins is

$$U_T/U_{T\infty} = 1.13e^{-\lambda} \qquad (0.125 \leq \lambda \leq 0.6). \tag{9-36}$$

† This result also fits the values for rigid particles at $N_D^{1/3} \doteq 100$ and $0.25 < \lambda < 0.6$.

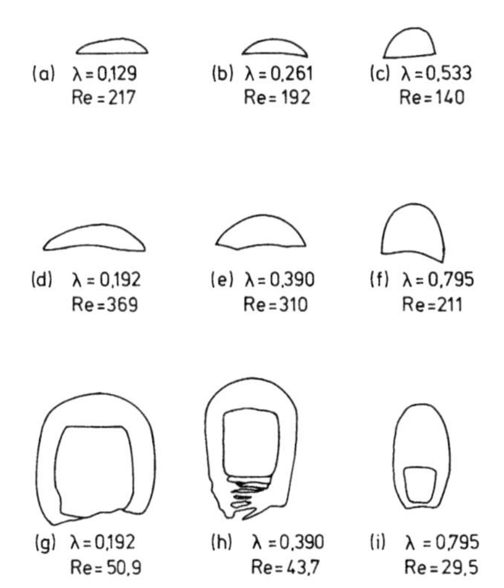

FIG. 9.9 Bubble shapes traced from photographs (B1) showing the influence of λ on the shape of large bubbles: (a, b, c) $V = 27.8\ \text{cm}^3$, $M = 1.64 \times 10^{-3}$; (d, e, f) $V = 92.6\ \text{cm}^3$, $M = 1.64 \times 10^{-3}$; (g, h, i) $V = 92.6\ \text{cm}^3$, $M = 4.2$.

TABLE 9.3

Wall Effect on Wake Volume for Large Bubbles in Viscous Liquids (B1)[a]

D (cm)	λ	U (cm/s)	Re	Wake volume (cm^3)
(a) V = 18.8 cm^3, Eo = 184:				
29.2	0.112	38.2	58.3	187.1
14.4	0.228	35.6	54.4	129.2
7.1	0.468	27.1	41.5	28.2
(b) V = 9.3 cm^3, Eo = 116:				
29.2	0.089	33.1	40.2	55.6
14.4	0.181	32.0	38.9	46.8
7.1	0.369	26.5	32.1	20.2

[a] Air bubbles in sucrose solution ($\mu = 2.89$ poise, $\rho = 1.35\ \text{g/cm}^3$, $\sigma = 77.7$ dyne/cm, $M = 0.109$).

Lin (L6) derived a further equation which shows reasonable agreement with experiment for λ less than about 0.6. While Collins' work was restricted to low M, high Re systems, Bhaga's results show that Eq. (9-36) can be applied down to Re $\doteq$ 10 regardless of whether skirts are being trailed.

While all the data discussed in this section are for large bubbles, it is reasonable to expect the results to apply also to large liquid drops for which Eo $\geq$ 40. For drops and bubbles in columns of noncircular cross section the results derived for cylindrical columns may be used with D replaced by the conventional hydraulic diameter. This practice is expected to give reasonable results for cross sections which do not deviate radically from circular, but experimental confirmation is lacking.

B. Slug Flow ($\lambda > 0.6$)

1. *Slug Flow in Vertical Tubes of Cylindrical Cross Section*

For appreciable values of λ it is obvious that wall effects influence fluid particles differently from rigid particles since a rigid particle will block the tube if too large whereas a bubble or drop can deform and maintain a nonzero terminal velocity even for $\lambda \gg 1$. When the diameter ratio λ exceeds a value of about 0.6, the tube diameter D becomes the controlling length governing the velocity and the frontal shape of a bubble or drop. Bubbles and drops are then called slugs† (or Taylor bubbles) and tend to be bullet-shaped as shown in Figs. 2.4i and 9.9f. The slug can be considered to be composed of two parts, a rounded nose region whose shape and dimensions are independent of the overall slug length and a cylindrical section surrounded by an annular film of the continuous fluid (C3). Since the slug flow regime is of special interest for applications of boiling heat transfer (G9) and fluidized beds (S5), almost all work has been devoted to gaseous slugs. Reviews of the behavior of slugs have been given by Wallis (W1) and Govier and Aziz (G5).

The terminal velocity of slugs may be estimated quite accurately using a very useful graphical correlation presented by White and Beardmore (W2), reproduced in Fig. 9.10. Although originally derived for gaseous slugs, the correlation can be generalized to apply to liquid slugs as well (H4, R1, W1), and it is in the generalized form that it appears in Fig. 9.10. Angelino (A2) found that the correlation could be extrapolated to larger values of $\mathrm{Eo_D}$. Figure 9.10 can also be replotted in terms of any three independent dimensionless groups, e.g., as Re vs. $\mathrm{Eo_D}$ with M as parameter, analogous to Fig. 2.5 for the case where wall effects are negligible.

Providing that the length of a slug exceeds about $1.5D$, slug length has virtually no influence on slug velocity (G9, L2, R1, Z1). The terminal velocity is achieved within a distance of $2D$ from release (W2). Expressions are available for predicting the terminal velocity of slugs for the following special cases:

a. *Viscosity and Surface Tension Forces Negligible* ($M \leq 10^{-6}$ *and* $\mathrm{Eo_D} > 100$) For this case it can be shown, based on potential flow theory, that

$$\mathrm{Fr_D} = \sqrt{\rho/\Delta\rho}\,(U_\mathrm{T}/\sqrt{gd}) = \text{constant}, \tag{9-37}$$

where values of 0.33 (D1), 0.35 (D2), 0.36 (L5, N1), and 0.37 (T4) have been

† In North American usage, the word slug is often used to refer to the plugs of continuous liquid separating a series of elongated bubbles or drops. This difference can create considerable confusion. Here we use the term slug to refer to the elongated bubble or drop of dispersed phase fluid.

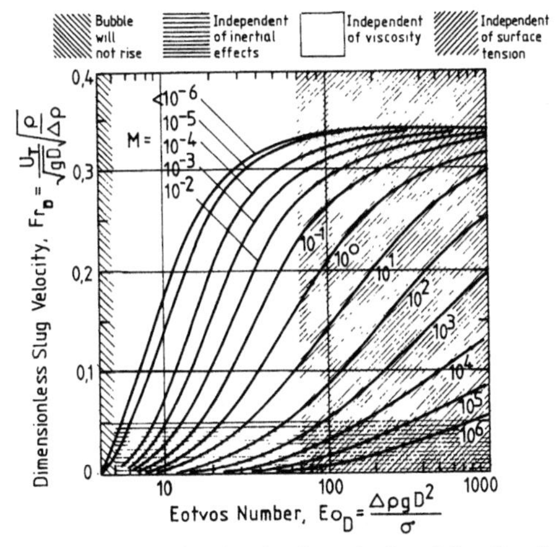

FIG. 9.10 General correlation for the rise velocity of slug flow bubbles (W2).

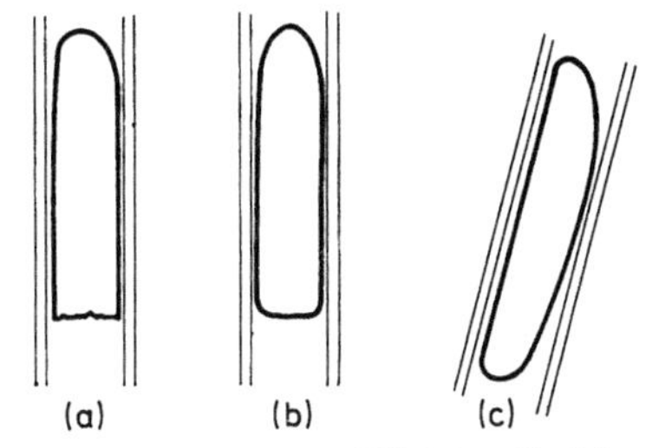

FIG. 9.11 Slug flow bubble shapes: (a) $Eo_D > 10^2$, $M < 10^{-6}$ (low viscosity liquid); (b) viscous liquid; (c) inclined tube.

derived for the constant. Experimental results (N2, S4) favor a value of 0.35.† Garabedian (G1) showed that the shape (and hence the value of Fr_D) is not uniquely determined and suggested that the shape observed is that which leads to the maximum rise velocity. Tung and Parlange (T4) applied this argument to obtain a first estimate for the lowering of U_T by surface tension. Brown (B6) extended the theory by considering the effects of viscosity in the annular film. The slug velocity was shown to depend only on the frontal radius of curvature which is influenced in a minor way by the fluid properties through their control of the thickness of the annular film. Equation (9-37) is for single slugs rising along the axis of a vertical tube as shown in Fig. 9.11a. If a slug adheres to the

† The mean value for fluidized beds has been found to be 0.36 (O1, S5). Given the scatter in the experimental data, the difference between this value and the value of 0.35 for slugs in low viscosity liquids is not significant.

wall of the tube, the terminal velocity is increased by a factor of approximately $\sqrt{2}$ (S5).

b. *Surface Tension Dominant* ($\mathrm{Eo_D} < 3.4$) In this case, the slug remains motionless with its shape determined by a balance between hydrostatic and capillary forces (B4, W2).

c. *Viscosity Dominant* ($\mathrm{Eo_D} > 70$, $\mathrm{Fr_D} < 0.05$) The terminal velocity for these conditions is given by

$$U_T = gD^2\,\Delta\rho/102\mu, \tag{9-38}$$

where the numerical constant suggested here is a mean of values given in the literature (W1, W2). The theory for this case has been presented and verified experimentally by Goldsmith and Mason (G3). The front of the slug was found to be prolate spheroidal while the rear was oblate spheroidal (see Fig. 9.11b).

When none of the sets of conditions given in (a), (b), or (c) apply, Fig. 9.10 should be used to predict the slug velocity.

2. *Slug Flow in Vertical Tubes of Noncylindrical Cross Section*

For the special case (a) above where surface tension and viscous effects are negligible, the terminal velocity of a slug in a column of rectangular cross section ($D_1 \times D_2$) is given by

$$\mathrm{Fr}_{D_1} = \frac{U_T}{\sqrt{gD_1}}\sqrt{\frac{\rho}{\Delta\rho}} = 0.23 + \frac{0.13D_2}{D_1}, \tag{9-39}$$

where $D_2 \le D_1$ (G8, W1). In the limit $D_2 \ll D_1$, this relationship reduces to the theoretical result (B2, C4, G1) for a plane slug. For a concentric annulus with inner diameter D_i and outer diameter D_o and for the special case where inertia effects are dominant, the data of Griffith (G8) can be fitted by the simple relationship

$$\mathrm{Fr_D} = 0.35 + 0.06D_i/D_o. \tag{9-40}$$

Extensive data for slugs rising in annular sections have recently been obtained (R1) in connection with blowouts in oil drilling operations. Bubbles were shown to assume the shape of "hot dog buns" with the fractions of the annular cross section occupied by downflowing liquid increasing with increasing viscosity. Eccentricity of the central tube, vibrations, changes in slug length, and surfactants were all found to have little influence on the terminal velocity of annular slugs.

For the more general case when surface tension and viscous effects are appreciable, there are few data available. Grigorev and Krokhin (G10) presented some results for the rise of bubbles in thin rectangular slits and wedge-shaped channels, while Schad and Bishop (S2) investigated bubble rise in thin annular and planar gaps.

3. *Slug Flow in Inclined Tubes*

Slug flow has been investigated for inclined cylindrical tubes (M3, R3, W2, Z1) and inclined rectangular tubes (G2, G10, M2). Slugs in inclined tubes tend to cling to one wall and the shape is altered as indicated in Fig. 9.11c. The terminal velocity tends to increase as the tube is inclined away from the vertical reaching a maximum at an orientation of about 45° (W1, Z1). Some experimental results for air slugs in water are given in Fig. 9.12. Since

$$U_T/U_{T(\text{vertical})} = \text{Fr}_D/\text{Fr}_{D(\text{vertical})} = f(\text{Eo}_D, M, \theta), \tag{9-41}$$

it is not possible to give a simple two-dimensional representation of the experimental results valid for all systems. Wallis (R3, W1) published a family of curves like those plotted in Fig. 9.12 for different ranges of Eo_D. For liquid draining from a horizontal cylindrical tube ($\theta = 90°$) and viscous and surface tension effects negligible, experimental results (G2, Z1) appear to support the experimental prediction of Brooke Benjamin (B5) giving $\text{Fr}_D = 0.54$ for the advancing slug free surface. Results for the case where the tube is rotating are given by Collins and Hoath (C7).

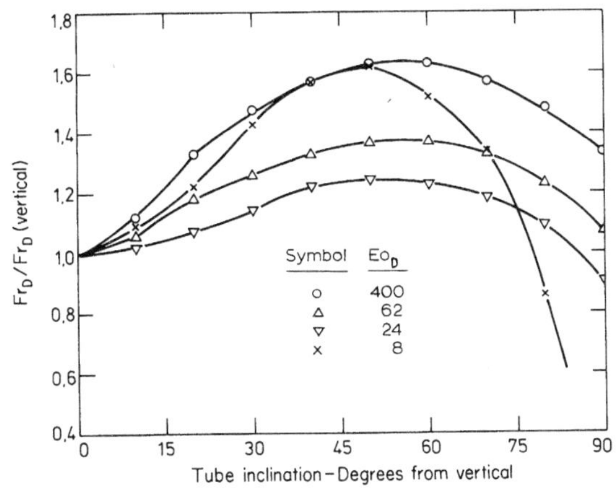

FIG. 9.12 Effect of tube inclination on the rise velocity of slug flow air bubbles in water (Z1).

C. Bubbles and Drops Enclosing Vertical Tubes or Rods

For large bubbles where inertia effects are dominant, enclosed vertical tubes lead to bubble elongation and increased terminal velocities (G7). The bubble shape tends towards that of a prolate spheroid and the terminal velocity may be predicted using the Davies and Taylor assumptions discussed in Chapter 8, but with the shape at the nose ellipsoidal rather than spherical. The maximum increase in terminal velocity is about 16% for the case where λ is small (G6) and 25% for a bubble confined between parallel plates (G6, G7) and occurs for the enclosed tube relatively close to the bubble axis.

Griffith (G8) obtained some data for gaseous slugs rising in a vertical column of diameter 5 cm containing seven vertical tubes, one concentric with the main tube and the others on a circle of diameter $0.59D$. The experimental results, reproduced by Wallis (W1), show that the terminal velocity increases with increasing enclosed tube diameter reaching a value more than 70% greater than the empty tube slug velocity for a tube to column diameter ratio of 0.2.

The results referred to in this section refer primarily to gaseous slugs and to large values of Eo or Eo_D where inertia effects tend to be dominant. Experimental results for liquid drops and for smaller bubbles and columns are lacking.

D. Heat and Mass Transfer

1. *Low Reynolds Numbers*

The surface velocities of Haberman and Sayre (H1), when used in the thin concentration boundary layer equation for circulating spheres, Eq. (3-51), yield the mass transfer factors K_{MU} and K_{MD} shown in Fig. 9.7 for $\kappa \leq 2$. For a fluid sphere in creeping flow the relationship between the mass transfer factors is

$$K_{MD} = K_{MU}/K^{1/2}, \tag{9-42}$$

where K is given by Eq. (9-29). For a sphere of given size, the wall effect on mass transfer is larger than for a solid sphere at the same λ, but it is still less than 15% at $\lambda = 0.5$ for $\kappa = 0$. Although deformation was not taken into account, the experimental results of Haberman and Sayre (H1) suggest that it is sufficiently small for $\lambda < 0.5$ that both K_{MU} and K_{MD} can be taken to be ratios of the mass transfer rates.

2. *Intermediate and Large Drops and Bubbles*

The influence of λ on the mass transfer rate has not been determined. Consideration of the tracings shown in Fig. 9.9 and of data of Bhaga (B1) suggests several different effects. Increasing λ elongates the bubble, at first making it more spherical and reducing its surface area; however, as λ approaches unity, the bubble becomes more cylindrical and the surface area increases again. At the same time, increasing λ decreases the rise velocity causing a reduction in the mass transfer coefficient. Increasing λ also decreases the fraction of the surface in contact with the wake which tends to increase the overall transfer rate. Comparison of the data of Calderbank *et al.* (C2) in a 10-cm-diameter column and of Guthrie and Bradshaw (G11) in a 45-cm-diameter column indicates that λ has little effect on the rate of mass transfer for spherical-cap bubbles when $\lambda < 0.5$, suggesting that the effects of λ cited above are compensatory. This conclusion must be considered tentative, however, because these studies used different techniques to obtain the mass transfer rate. Measurements of the rate of transfer in a series of columns using a single experimental method are needed.

3. *Slug Flow*

Mass transfer to the liquid phase around a slug can be treated with the thin concentration boundary layer assumption through Eq. (1-63). Van Heuven and Beek (V1) completed these calculations for a slug with viscous and surface tension forces negligible ($Eo_D > 100$, $M \leq 10^{-6}$). The results can be represented by

$$\overline{kA}/\pi DL = C_S \mathscr{D}^{1/2}(g/L)^{1/4}, \tag{9-43}$$

where $\overline{kA}$ is the mass transfer coefficient × surface area product and the values of C_S are given in Table 9.4 as a function of L/D, the ratio of the slug length to the tube diameter. For slugs with $L/D > 1$ the coefficient is essentially constant. Van Heuven and Beek's experimental data are in fair agreement with Eq. (9-43).

The resistance to mass transfer within a slug in a liquid of low viscosity has been measured by Filla *et al.* (F5), who found that $(\overline{kA})_p$ was approximately proportional to the square root of the diffusivity within the bubble, $\mathscr{D}_p$, as predicted by the thin concentration boundary layer approximation. In addition, $(\overline{kA})_p/A_e$ was independent of slug length for $1 \leq L/D \leq 2.5$.

TABLE 9.4

Slug Flow Mass Transfer Constants C_s in Eq. (9-43)

L/D	C_s	L/D	C_s
0.3	0.73	3	0.98
0.5	0.80	4	0.99
1	0.88	5	1.00
2	0.95	∞	1.09

REFERENCES

A1. Achenbach, E., *J. Fluid Mech.* **65**, 113–125 (1974).
A2. Angelino, H., *Chem. Eng. Sci.* **21**, 541–550 (1966).
B1. Bhaga, D., Ph.D. Thesis, McGill Univ., Montreal, 1976.
B2. Birkhoff, G., and Carter, D., *J. Math. Mech.* **6**, 769–779 (1957).
B3. Brenner, H., *Adv. Chem. Eng.* **6**, 287–438 (1966).
B4. Bretherton, F. P., *J. Fluid Mech.* **10**, 166–188 (1961).
B5. Brooke Benjamin, T., *J. Fluid Mech.* **31**, 209–248 (1968).
B6. Brown, R. A. S., *Can. J. Chem. Eng.* **43**, 217–223 (1965).
B7. Bungay, P. M., and Brenner, H., *J. Fluid Mech.* **60**, 81–96 (1973).
C1. Calderbank, P. H., *Chem. Eng.* (*London*) No. 212, 209–233 (1967).
C2. Calderbank, P. H., Johnson, D. S. L., and Loudon, J., *Chem. Eng. Sci.* **25**, 235–256 (1970).
C3. Clift, R., Grace, J. R., and Sollazzo, V., *J. Heat Transfer* **96**, 371–376 (1974).
C4. Collins, R., *J. Fluid Mech.* **22**, 763–771 (1965).
C5. Collins, R., *U.K. At. Energy Res. Establ., Rep.* **AERE-R 5402** (1967).

C6. Collins, R., *J. Fluid Mech.* **28**, 97–112 (1967).
C7. Collins, R., and Hoath, M. T., *J. Fluid Mech.* **67**, 763–768 (1975).
C8. Coutanceau, M., *C. R. Acad. Sci., Ser. A* **274**, 853–856 (1972).
C9. Cox, R. G., and Brenner, H., *Chem. Eng. Sci.* **23**, 147–173 (1968).
D1. Davies, R. M., and Taylor, Sir G. I., *Proc. R. Soc., Ser. A* **200**, 375–390 (1950).
D2. Dumitrescu, D. T., *Z. Angew. Math. Mech.* **23**, 139–149 (1943).
E1. Eaton, L. R., and Hoffer, T. E., *J. Appl. Meteorol.* **9**, 269–275 (1970).
E2. Eichhorn, R., and Small, S., *J. Fluid Mech.* **20**, 513–527 (1964).
F1. Faxén, H., *Ark. Mat., Astron. Fys.* **17**(27), 1–28 (1923).
F2. Fayon, A. M., and Happel, J., *AIChE J.* **6**, 55–58 (1960).
F3. Feldman, G. A., and Brenner, H., *J. Fluid Mech.* **32**, 705–720 (1968).
F4. Fidleris, V., and Whitmore, R. L., *Br. J. Appl. Phys.* **12**, 490–494 (1961).
F5. Filla, M., Davidson, J. F., Bates, J. F., and Eccles, M. A., *Chem. Eng. Sci.* **31**, 359–367 (1976).
F6. Francis, A. W., *Physics* **4**, 403–406 (1933).
G1. Garabedian, P. R., *Proc. R. Soc., Ser. A* **241**, 423–431 (1957).
G2. Gardner, G. C., and Crow, I. G., *J. Fluid Mech.* **43**, 247–255 (1970).
G3. Goldsmith, H. L., and Mason, S. G., *J. Fluid Mech.* **14**, 42–58 (1962).
G4. Goldsmith, H. L., and Mason, S. G., *in* "Rheology: Theory and Applications" (F. R. Eirich, ed.), Vol. 4, pp. 85–250. Academic Press, New York, 1967.
G5. Govier, G. W., and Aziz, K., "The Flow of Complex Mixtures in Pipes." Van Nostrand-Reinhold, New York, 1972.
G6. Grace, J. R., Ph.D. Thesis, Cambridge Univ., 1968.
G7. Grace, J. R., and Harrison, D., *Chem. Eng. Sci.* **22**, 1337–1347 (1967).
G8. Griffith, P., *Am. Soc. Mech. Eng., Pap.* No. 63-HT-20 (1963).
G9. Griffith, P., and Wallis, G. B., *J. Heat Transfer* **83**, 317–320 (1961).
G10. Grigorev, V. A., and Krokhin, Y. I., *High Temp. (USSR)* **9**, 1138–1142 (1971).
G11. Guthrie, R. I. L., and Bradshaw, A. V., *Chem. Eng. Sci.* **28**, 191–203 (1973).
H1. Haberman, W. L., and Sayre, R. M., *David Taylor Model Basin Rep.* No. 1143 (1958).
H2. Halow, J. S., and Wills, G. B., *Ind. Eng. Chem., Fundam.* **9**, 603–607 (1970).
H3. Happel, J., and Brenner, H., "Low Reynolds Number Hydrodynamics." 2nd ed. Noordhoff, Leyden, Netherlands, 1973.
H4. Harmathy, T. Z., *AIChE J.* **6**, 281–288 (1960).
H5. Ho, B. P., and Leal, L. G., *J. Fluid Mech.* **65**, 365–400 (1974).
J1. Johansson, H., *Chem. Eng. Commun.* **1**, 271–280 (1974).
L1. Ladenburg, R., *Ann. Phys. (Leipzig)* **23**, 447–458 (1907).
L2. Laird, A. D. K., and Chisholm, D., *Ind. Eng. Chem.* **48**, 1361–1364 (1956).
L3. Langins, J., Weber, M. E., and Pliskin, I., *Chem. Eng. Sci.* **26**, 693–696 (1971).
L4. Lawler, M. T., and Lu, P.-C., *in* "Advances in Solid-Liquid Flow in Pipes and its Application" (I. Zandi, ed.), pp. 39–57. Pergamon, Oxford, 1971.
L5. Layser, D., *Astrophys. J.* **122**, 1–12 (1955).
L6. Lin, S. P., *Phys. Fluids* **10**, 2283–2285 (1967).
M1. Maneri, C. C., and Mendelson, H. D., *AIChE J.* **14**, 295–300 (1968).
M2. Maneri, C. C., and Zuber, N., *Int. J. Multiphase Flow* **1**, 623–645 (1974).
M3. Mattar, L., and Gregory, G. A., *J. Can. Pet. Technol.* **13**, 69–76 (1974).
M4. McNown, J. S., Lee, H. M., McPherson, M. B., and Engez, S. M., *Proc. Int. Congr. Appl. Mech., 7th* **2**, 17–29 (1948).
M5. McNown, J. S., and Newlin, J. T., *Proc. U.S. Natl. Congr. Appl. Mech., 1st*, pp. 801–806 (1951).
M6. Morgan, V. T., *Adv. Heat Transfer* **11**, 199–264 (1975).
N1. Nicklin, D. J., Ph.D. Thesis, Cambridge Univ., 1961.
N2. Nicolitsas, A. J., and Murgatroyd, W., *Chem. Eng. Sci.* **23**, 934–936 (1968).

O1. Ormiston, R. M., Mitchell, F. R. G., and Davidson, J. F., *Trans. Inst. Chem. Eng.* **43**, 209–216 (1965).
P1. Pei, D. C. T., *Int. J. Heat Mass Transfer* **12**, 1707–1709 (1969).
P2. Perkins, H. C., and Leppert, G., *Int. J. Heat Mass Transfer* **7**, 143–158 (1964).
R1. Rader, D. W., Bourgoyne, A. T., and Ward, R. H., *J. Pet. Technol.* **27**, 571–584 (1975).
R2. Round, G. F., Latto, B., and Anzenavs, R., *Proc. Hydrotransp.* 2, *Br. Hydromech. Res. Assoc.* Pap. F1 (1972).
R3. Runge, D. E., and Wallis, G. B., *USAEC* Rep. No. NYO-3114-8 (1965).
S1. Salami, E., Vignes, A., and Le Goff, P., *Génie Chim.* **94**(3), 67–77 (1965).
S2. Schad, H. O., and Bishop, A. A., *AIChE Symp., Fundam. Fluid-Part. Syst., New York, 1972.* (Abstr.)
S3. Segré, G., and Silberberg, A., *J. Fluid Mech.* **14**, 115–157 (1962).
S4. Stewart, P. S. B., Ph.D. Thesis, Cambridge Univ., 1965.
S5. Stewart, P. S. B., and Davidson, J. F., *Powder Technol.* **1**, 61–80 (1967).
S6. Strom, J. R., and Kintner, R. C., *AIChE J.* **4**, 153–156 (1958).
S7. Sutterby, J. L., *Trans. Soc. Rheol.* **17**, 559–585 (1973).
T1. Tachibara, M., *Rheol. Acta* **12**, 58–69 (1973).
T2. Tanner, R. I., *J. Fluid Mech.* **17**, 161–170 (1963).
T3. Tsuge, H., and Hibino, S., *Int. Chem. Eng.* **15**, 186–192 (1975).
T4. Tung, K. W., and Parlange, J. Y., *Acta Mechanica.* **24**, 313–317 (1976).
U1. Uno, S., and Kintner, R. C., *AIChE J.* **2**, 420–425 (1956).
V1. Van Heuven, J. W., and Beek, J. W., *Chem. Eng. Sci.* **18**, 377–390 (1963).
V2. Vasseur, P., Ph.D. Thesis, McGill Univ., Montreal, 1973.
V3. Vliet, C. C., and Leppert, G., *J. Heat Transfer* **83**, 163–175 (1962).
W1. Wallis, G. B., "One-Dimensional Two-Phase Flow." McGraw-Hill, New York, 1969.
W2. White, E. T., and Beardmore, R. H., *Chem. Eng. Sci.* **17**, 351–361 (1962).
Z1. Zukoski, E. E., *J. Fluid Mech.* **25**, 821–837 (1966).

Chapter 10

Surface Effects, Field Gradients, and Other Influences

I. INTRODUCTION

In previous chapters, we have considered smooth particles moving steadily under the action of gravity in uniform fluids. In this chapter, we consider factors which commonly complicate the motion and transfer processes for solid and fluid particles. Surface roughness for rigid particles and interfacial effects for fluid particles are treated first. Natural convection resulting from density gradients associated with heat or mass transfer is the subject of the next section. We then give a brief review of the effects of shear and particle rotation. Free-stream turbulence can greatly influence particle motion and transfer processes, and this is treated next. Finally, we give a brief review of the effects of compressibility and noncontinuum flow on particle motion and heat transfer.

II. SURFACE ROUGHNESS (RIGID PARTICLES)

A. Effect on Drag

Roughness on the surface of a solid particle is normally characterized by the "relative roughness," ε/d, the ratio of the effective roughness height to the mean outside diameter of the particle. This is analogous to the relative roughness employed to characterize flow through pipes. Spherical elements on the surface have an "effective roughness height" equal to 55% of the sphere diameter (A2). The Reynolds number for a rough sphere is conventionally based on the diameter of the circumscribing sphere.

The most significant effects of surface roughness on flow past a particle occur in the critical range (see Chapter 5). Achenbach (A2) investigated this range in detail. Roughness induces earlier transition to turbulence in the attached boundary layer, as in Prandtl's classic experiment in which a small wire hoop was attached to the surface of a sphere [see (G14)]. The sudden rearward shift in the final separation point occurs at Re values which decrease with increasing surface roughness down to 8×10^4 for a relative roughness of 0.007 (cf. $\mathrm{Re_c} = 3.65 \times 10^5$ for a smooth sphere). The resulting drop in C_D occurs at lower Re,† while the minimum value of C_D increases with increasing roughness. In the transcritical range, well above transition, C_D for roughened spheres is independent of Re and constant at 0.38 for relative roughness greater than about 10^{-3}. There is some evidence (S20) that C_D may be higher for a very rough sphere of low γ in free flight. Similar phenomena have been observed for flow around cylinders (A1, A3, F1) and are likely for any body lacking edges which fix the separation position. In the supercritical range, smooth spheres in free motion are subject to random variations in flow which give rise to an erratic trajectory (see Chapter 5). Scoggins (S19) showed that roughening the surface reduces these fluctuations, causing particles to follow a much more regular path. This effect has been applied in the design of balloon wind sensors (S20).

Just below the critical range, roughness has little effect on drag (A2), but for Re of order 10^3 roughness can increase C_D substantially (S23, S28). The increase in C_D appears to depend on the ratio of roughness height to boundary layer thickness. For Re < 500, large-scale roughness reduces the drag (S28). The general result of Hill and Power (see Chapter 4) suggests that this should extend into the creeping flow range, since the volume of a roughened particle is less than that of the circumscribing sphere.

B. Effect on Transfer

As for drag, the most dramatic effects of surface roughness on heat or mass transfer occur near the critical range. The earlier transition to turbulence in the attached boundary layer yields a maximum in the local transfer rate located 40–60° from the front stagnation point for spheres (J3, S15) and cylinders (A3). Near the point of final separation on the rearward portion of the body, the local transfer rate exhibits a minimum. The overall transfer rate is increased to a maximum of 2 to 3 times the rate for a smooth particle. At Reynolds numbers above the critical value, an increase in relative roughness causes the overall transfer rate to increase to a maximum and then decrease for relative roughness greater than 0.1 (S17). The few data available at lower Re indicate that roughness has little effect for transfer in gases (Sc, Pr $\approx$ 1) when $\varepsilon/d < 10/\mathrm{Re}^{3/4}$ (A8, J3). However, for large Pr or Sc, roughness increases the transfer rate down to

† An interesting result of this effect occurs for falling ice spheres. When the surface melts, and therefore becomes smooth, flow can pass from supercritical to subcritical and the terminal velocity is suddenly reduced (W6).

lower Re because of the thinness of the thermal or concentration boundary layer (S16, S17).

III. INTERFACIAL EFFECTS (FLUID PARTICLES)

The effects of surface-active agents on the motion of and transfer from bubbles and drops have been discussed in earlier chapters. The main effect is to reduce the mobility of all or part of the interface. In this section we consider briefly two other interfacial phenomena: interfacial convection during mass transfer and interfacial barriers to mass transfer.

A. Interfacial Convection: The Marangoni Effect

Movements in the plane of the interface result from local variations of interfacial tension during the course of mass transfer. These variations may be produced by local variations of any quantity which affects the interfacial tension. Interfacial motions have been ascribed to variations in interfacial concentration (H6, P6, S33), temperature (A9, P6), and electrical properties (A10, B19). In ternary systems, variations in concentration are the major factor causing interfacial motion; in partially miscible binary systems, interfacial temperature variations due to heat of solution effects are usually the cause.

On the interface between quiescent fluids, interfacial motions may take the form of ripples (E4, O2) or of ordered cells (B5, L5, O2, S22). Slowly growing cells may exist for long periods of time (B5, O2), or the cells may oscillate and drift over the surface (L6, L7). When the phases are in relative motion, interfacial disturbances usually take the form of localized eruptions, often called "interfacial turbulence" (M3). This form of disturbance can also be observed at the interface of a drop (S8). A thorough review of interfacial phenomena, including a number of striking photographs, has been presented by Sawistowski (S7).

The shape of a drop moving under the influence of gravity may be affected by interfacial motions; the drop may also wobble and move sideways (S27, W3). In one system (S22) the terminal velocity was reduced yielding a drag coefficient nearly equal to that of a solid particle. Interfacial convection tends to increase the rate of mass transfer above that which would occur in the absence of interfacial motion. The interaction between mass transfer and interfacial convection has been treated by Sawistowski (S7) and Davies (D4, D5).

1. *Cellular Interfacial Motions*

The factors determining the appearance of ordered cell-like motions were first investigated by Sternling and Scriven (S33) who considered the two-dimensional stability of a plane interface separating two immiscible semi-infinite fluid phases with mass transfer occurring between the phases. This system was shown to be unstable for mass transfer in one direction, but stable for transfer in the opposite direction. For an interfacial tension-lowering solute, instability was predicted for transfer out of the phase with lower diffusivity or out of the phase with higher kinematic viscosity. A similar analysis by Brian and co-workers (B16, B17, B18) for a gas–liquid system included the adsorption and subsequent movement of the solute at the interface. The inclusion of interfacial convection and diffusion reduced interfacial concentration variations, making the system more stable.

Cellular interfacial motions are generally observed in quiescent systems when the mass transfer driving forces and interfacial tension gradients are small, and when natural or buoyancy-driven convection is suppressed. Under these conditions, the occurrence of cell-like motions is in agreement with the Sternling–Scriven theory. The presence of these cells enhances the rate of mass transfer (B4), since fresh fluid is brought to the interface. The maximum increase in the rate of mass transfer has been predicted (B5) by assuming that cells of depth δ on either side of the interface are in equilibrium as they grow. If there is no solute in the continuous phase, the amount of mass transferred per unit area, m, is

$$m = [\delta/(1 + H)]c_{p\infty}, \tag{10-1}$$

where $c_{p\infty}$ is the bulk concentration of solute. In the absence of interfacial motion and assuming diffusivities in each phase to be equal, the penetration theory gives

$$m^* = (2/\pi)[\delta_{pen}/(1 + H)]c_{p\infty}, \tag{10-2}$$

where δ_{pen}, the penetration depth, is $\sqrt{\pi \mathscr{D} t}$. Bakker *et al.* (B5) found experimentally that

$$1 < \delta/\delta_{pen} < 2. \tag{10-3}$$

Hence, the maximum increase in mass transfer duc to interfacial convection is

$$\pi/2 < m/m^* < \pi. \tag{10-4}$$

These limits are in good agreement with data on plane interfaces (B4, M3).

2. *Interfacial Turbulence*

Disordered interfacial motion can occur when there is mass transfer in either direction. When an eddy of solute-rich fluid reaches the interface, the interfacial tension is reduced locally at the point of impingement. Small regions of reduced interfacial tension formed in this manner tend to spread. As spreading occurs, bulk fluid of lower solute concentration is brought to the interface causing a local increase in interfacial tension which retards and eventually stops the spreading. The interfacial motion then reverses toward the original point of impingement. This reversed flow, if sufficiently strong, produces two jets of fluid, one ejected into each phase. This ejection is seen as an eruption from the interface (S8, T9).

The original eddy motion which sets up the chain of events leading to eruptions may be caused by forced flow of the bulk phases, density differences due to concentration or temperature gradients (B12), or earlier eruptions. Strong eruptions occur when a critical concentration driving force or a critical interfacial tension depression is exceeded (O3, S8, S9). At lower concentration differences ripples may result (E4), eruptions may occur only over part of the interface (S8) with the jets taking some time to form (T9), or no interfacial motion at all may occur. Attempts to correlate the minimum driving force required for spontaneous interfacial motions have met with little success.

Mass transfer rates are increased in the presence of eruptions because the interfacial fluid is transported away from the interface by the jets. For mass transfer from drops with the controlling resistance in the continuous phase, the maximum increase in the transfer rate is of the order of three to four times (S8), not greatly different from the estimate of Eq. (10-4) for cellular convection. This may indicate that equilibrium is attained in thin layers adjacent to the interface during the spreading and contraction. When the dispersed-phase resistance controls, on the other hand, interfacial turbulence may increase the mass transfer rate by more than an order of magnitude above the expected value. This is almost certainly due to vigorous mixing caused by eruptions within the drop.

The maximum effect of interfacial turbulence on the mass transfer coefficient can be estimated using the correlation of Davies and Rideal (D6) for the initial spreading velocity, U_s, of a surface tension-lowering material spreading at the interface between two fluid phases:

$$U_s = \Delta\sigma \times 10^{-2}/(\mu + \mu_p), \tag{10-5}$$

where $\Delta\sigma$ is the surface tension depression causing spreading. This velocity is then used in the Handlos–Baron (H1) expression, Eq. (7-57) to give

$$k_p = 3.75 \times 10^{-5}\,\Delta\sigma/\mu(1 + \kappa)^2. \tag{10-6}$$

In this equation, $\Delta\sigma$ is taken as the maximum possible surface tension lowering. Hence for a solute-free continuous phase, $\Delta\sigma$ is the difference between the interfacial tension for the solvent-free system and the equilibrium interfacial tension corresponding to the solute concentration in the dispersed phase. Equation (10-6) indicates a strong effect of the viscosity ratio κ on the mass transfer coefficient as found experimentally (L11). For the few systems in which measurements are reported (B11, L11, O4), estimates from Eq. (10-6) have an average error of about 30% for the first 5–10 seconds of transfer when interfacial turbulence is strongest.

B. Interfacial Barriers to Mass Transfer

The existence of interfacial barriers to mass transfer caused by films of surface-active materials has long been recognized (L1). When surfactants are added to a system undergoing mass transfer, they reduce the interfacial tension and make it less sensitive to variations in solute concentration. In addition, surfactants cause a resistance (surface viscosity) to motion at the interface. Interfacial motion, whether caused by forced flow or by the Marangoni effect, is thereby reduced. The reduction in the rate of transfer caused by addition of surfactant to a system undergoing interfacial turbulence can be very striking (S10). In certain systems the surfactant itself provides a significant resistance to mass transfer. This is sometimes called the barrier effect. Interfacial resistances to transfer between quiescent liquid phases in the presence of surface-active materials have been determined by several workers (D7, M14). The effects appear to be specific to the solute–surfactant combination. Similar results have been obtained for gas–liquid systems (G15, P11, S1). Whether there are interfacial resistances to transfer in surfactant-free systems is still hotly debated [e.g., see (B14, C2, H15, T14)].

IV. NATURAL CONVECTION AND MIXED FLOW

"Natural" or "free" convective flows are generated by density gradients resulting from heat or mass transfer. Gradients of temperature and/or concentration cause body forces to be nonuniform throughout the flow field, and these forces generate the "natural" motion. Because the density depends on composition or temperature, the momentum and continuity equations are coupled to the species continuity or energy equations. Since these equations are extremely difficult to solve, a set of simplifying assumptions, called the Boussinesq approximation, is widely used.

A. The Boussinesq Approximation

Analyses of time-steady free convection usually assume that:

1. density is constant in the continuity and momentum equations, except in the body force term;
2. density variations are caused only by temperature and composition gradients;
3. all other properties are constant.

Under these assumptions Eq. (1-1), the Navier–Stokes equation, becomes

$$\mathbf{u} \cdot \nabla \mathbf{u} = (\rho/\rho_\infty)\mathbf{g} - (\nabla p/\rho_\infty) + \nu \nabla^2 \mathbf{u}. \tag{10-7}$$

The variable density is expanded in a Taylor series about the density of the fluid far from the body, ρ_∞:

$$\rho = \rho_\infty + (T - T_\infty)\left(\frac{\partial \rho}{\partial T}\right)_{p_\infty, w_\infty} + (w - w_\infty)\left(\frac{\partial \rho}{\partial w}\right)_{p_\infty, T_\infty}, \tag{10-8}$$

where w is the mass fraction. The density has been considered only a function of temperature and concentration of one component. Substitution of Eq. (10-8) into Eq. (10-7) yields

$$\mathbf{u} \cdot \nabla \mathbf{u} = -\frac{\nabla p}{\rho_\infty} + \nu \nabla^2 \mathbf{u} + \frac{1}{\rho_\infty} \frac{\partial \rho}{\partial T}(T - T_\infty)\mathbf{g} + \frac{1}{\rho_\infty} \frac{\partial \rho}{\partial w}(w - w_\infty)\mathbf{g}, \quad (10\text{-}9)$$

where ∇p is the hydrostatic pressure gradient far from the body. It is convenient to introduce compressibility coefficients

$$\beta_t = -\frac{1}{\rho_\infty}\left(\frac{\partial \rho}{\partial T}\right)_{p_\infty, w_\infty}, \quad (10\text{-}10)$$

and

$$\beta_w = -\frac{1}{\rho_\infty}\left(\frac{\partial \rho}{\partial w}\right)_{p_\infty, w_\infty}. \quad (10\text{-}11)$$

Making Eq. (10-9) dimensionless through reference quantities as in Chapter 1 yields

$$\mathbf{u}' \cdot \nabla' \mathbf{u}' = -\nabla' p' + \frac{1}{\mathrm{Re}} (\nabla')^2 \mathbf{u}' - \frac{\mathrm{Gr}_t}{\mathrm{Re}^2} T' \frac{\mathbf{g}}{g} - \frac{\mathrm{Gr}_w}{\mathrm{Re}^2} w' \frac{\mathbf{g}}{g}. \quad (10\text{-}12)$$

The other governing equations—the overall continuity equation, the species continuity equation, and the energy equation—are identical to the dimensionless forms presented in Chapter 1. Two new dimensionless groups, a thermal Grashof number

$$\mathrm{Gr}_t = L^3 \beta_t (T_s - T_\infty) g / \nu^2 \quad (10\text{-}13)$$

and a composition Grashof number,

$$\mathrm{Gr}_w = L^3 \beta_w (w_s - w_\infty) g / \nu^2 \quad (10\text{-}14)$$

appear. These are algebraic quantities and may be negative. When natural convection coexists with forced convection (termed mixed convection) the relative effect of natural to forced convection is indicated by $\mathrm{Gr}/\mathrm{Re}^2$ where $\mathrm{Gr} = \mathrm{Gr}_t$ or Gr_w.

For isothermal mass transfer ($\mathrm{Gr}_t = 0$),

$$\mathrm{Sh} = f(\mathrm{Gr}_w, \mathrm{Re}, \mathrm{Sc}), \quad (10\text{-}15)$$

while for heat transfer under uniform composition conditions,

$$\mathrm{Nu} = f(\mathrm{Gr}_t, \mathrm{Re}, \mathrm{Pr}). \quad (10\text{-}16)$$

The functional forms for these two equations are identical for equivalent boundary conditions when $\mathrm{Pr}, \mathrm{Sc} \gg 1$.

The adequacy of the Boussinesq approximations has been tested for natural convection from a vertical plate (S31) and for mixed convection from a horizontal plate (R4). The approximations give an adequate representation of velocity and temperature profiles except near the point where buoyancy causes flow separation. Fluid properties are best evaluated at the reference conditions given by Sparrow and Gregg (S31) although the film condition is adequate for the calculation of average Nusselt and Sherwood numbers.

B. NATURAL CONVECTION

We consider either isothermal mass transfer ($Gr_t = 0$) or uniform composition heat transfer ($Gr_w = 0$) from a particle with constant surface composition or temperature. The Rayleigh number Ra is used for both $Gr_t Pr$ and $Gr_w Sc$.

1. *Flow Around Spheres*

The details of natural convective flows over surfaces other than flat plates have only recently been studied experimentally (A7, J1, P3, S12). We consider a heated sphere in an infinite, stagnant medium. Flow is directed toward the surface over the bottom hemisphere and away from the surface over the top hemisphere with a stagnation point at each pole (P3, S12). The lower pole is considered the forward stagnation point.

The buoyancy force can be resolved into components parallel and normal to the surface. The parallel component acts in the direction of increasing θ, measured from the forward stagnation point. Over the lower hemisphere the normal component of the buoyancy force is directed toward the surface, while over the upper hemisphere the normal component is directed away from the surface. Therefore, the flow over the lower hemisphere is similar to that over a heated, inclined plate. This flow is of boundary layer type near the leading edge, but exhibits an instability in the form of longitudinal waves triggered by two-dimensional disturbances (L10, P4). The flow over the upper hemisphere is more unstable than the flow over an inclined flat plate because the periphery available for flow and the angle of inclination decrease as θ increases. Near the rear stagnation point the normal component of the buoyancy force makes the fluid turn away from the surface and form an axisymmetric plume above the sphere (J1). Some distance above the sphere the plume becomes turbulent. As the Grashof number is increased, the point of plume instability approaches the sphere (J1) until at a sufficiently high Grashof number the flow over the rear hemisphere is disturbed (S12, W4). At higher Grashof numbers the location of velocity disturbances moves forward (S12). Even at high Grashof numbers no standing eddy occurs; the flow turns away from the sphere under the action of the normal component of buoyancy (P3).

2. *Mass or Heat Transfer from Spheres*

For $Gr = 0$, the Sherwood or Nusselt number is given by Eq. (3-44). For $Gr \to 0$, neither perturbation nor asymptotic expansion methods have proved capable of yielding solutions for Sh comparable to Eq. (3-55). At larger Gr the boundary layer approximations become appropriate (G10). For an axisymmetric body the resulting equations (for $Gr_t = 0$) are Eq. (1-55) and

$$u_x \frac{\partial u}{\partial x} + u_y \frac{\partial u}{\partial y} = g\beta_w(w - w_\infty)\sin\alpha + \nu \frac{\partial^2 u_x}{\partial y^2}, \tag{10-17}$$

$$u_x \frac{\partial w}{\partial x} + u_y \frac{\partial w}{\partial y} = \mathscr{D} \frac{\partial^2 w}{\partial y^2}, \tag{10-18}$$

where α is the angle between the outward normal to the surface and the direction of gravity. For a body with constant surface composition, w_s, the boundary conditions are

$$\begin{aligned} \text{at} \quad & y = 0 \quad && u_x = 0,\ u_y = 0,\ w = w_s, \\ & y \to \infty && u_y = 0,\ w = w_\infty, \\ & x = 0 && u_y = 0, \end{aligned} \tag{10-19}$$

where the boundary condition on u_y at the surface, $y = 0$, is correct for large Sc and for uniform composition heat transfer. A similarity solution of these equations is possible for $Sc \to \infty$ for any arbitrary body contour which does not have horizontal planes, sharp corners, or surface depressions (A5, S34). The mean Sherwood number is given by

$$\frac{Sh}{Ra^{1/4}} = 0.6705 \frac{\left[\int_0^{X_M} \mathscr{R}^{4/3}(\sin\alpha)^{1/3}\,dx'\right]^{3/4}}{\int_0^{X_M} \mathscr{R}\,dx'}, \tag{10-20}$$

where $\mathscr{R} = R/L$ and $X = x/L$; Sh and Ra are based on length L. For spheres with L taken as d, Eq. (10-20) yields

$$Sh = 0.589 Ra^{1/4}. \tag{10-21}$$

For finite Sc similarity solutions are not possible for most shapes and approximate methods have been used [e.g., see (L4)].

Experimental local Sherwood numbers on a sphere are shown in Fig. 10.1 as a function of angle from the front stagnation point. The curves for $Sc = 0.72$ and 10 are approximate boundary layer solutions (L4), while the curve for infinite Sc is the asymptotic similarity solution (A5). Except for high Ra, the data are in good agreement with the results of the boundary layer calculations. The large increase in mass transfer rate beyond 120° at the largest Ra results from the instability in the flow discussed earlier. The minimum transfer rate occurs before the flow instability (S12) since instability and longitudinal waves bring freestream fluid to the surface, thus increasing the transfer rate.

Experimental mean Sherwood numbers are shown in Fig. 10.2. The asymptotic solution, Eq. (10-21), gives a good representation of the data for large Sc when $Ra \gtrsim 10^5$. For extremely large Ra, a turbulent range is expected where

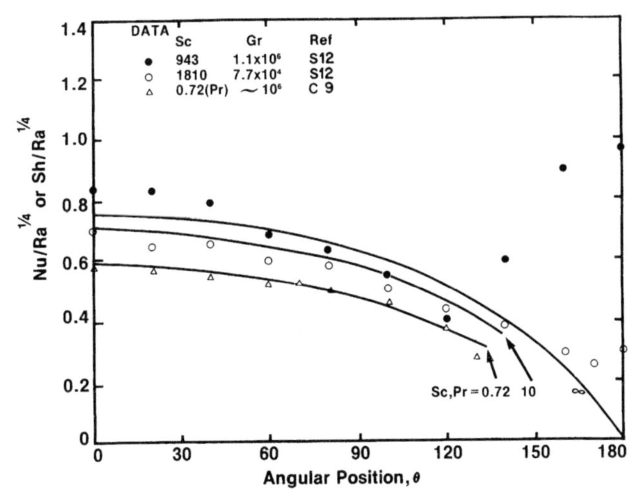

FIG. 10.1 Local values of dimensionless transfer rate from spheres as a function of angular position. The lines correspond to boundary layer solutions (L4) for the Sc or Pr indicated.

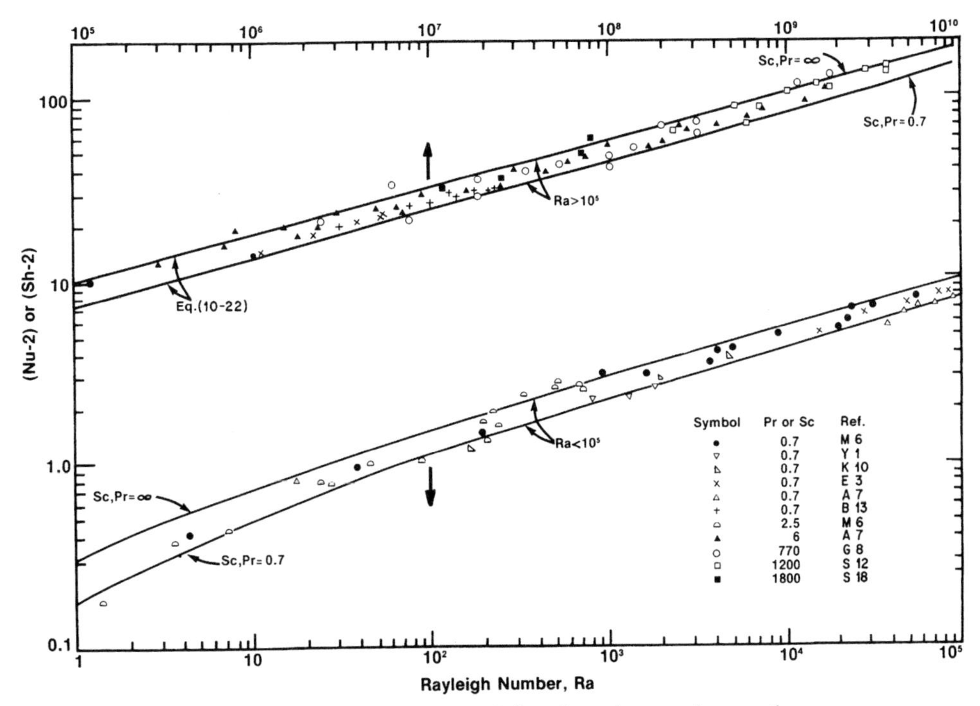

FIG. 10.2 Average (Sh-2) or (Nu-2) for spheres in natural convection.

$Sh \propto Ra^{1/3}$. Additional data are needed to confirm the existence of such a regime. For $Ra < 10^4$ the asymptotic solution is inadequate because the boundary layer thickness becomes large compared to the sphere radius. Since no solutions have yet been obtained in this region, the data must be fitted empirically. There are few solutions available for finite Sc, but the function of Churchill and Churchill (C5) fits the available solutions and data for flat plates and horizontal cylinders. The following correlation is proposed for $1 < Ra < 10^{10}$:

$$Sh = 1.7 + 0.3\{1 + 14.86[f(Sc)]Ra\}^{1/4}, \tag{10-22}$$

with

$$f(Sc) = [1 + (0.5/Sc)^{9/16}]^{-16/9}. \tag{10-23}$$

Equation (10-22) becomes essentially identical to the asymptotic solution, Eq. (10-21), for $Sc \gg 1$ and $Ra > 10^5$. In Fig. 10.2, lines are drawn corresponding to Eq. (10-22) for $Sc = 0.7$ and $Sc = \infty$.

3. *Mass or Heat Transfer from Arbitrary Shapes*

The boundary layer equations for an axisymmetric body, Eqs. (1-55), (10-17), and (10-18) have been solved approximately for arbitrary Sc (L4). For $Sc \to \infty$ the mean value of Sh can be computed from Eq. (10-20). Solutions have also been obtained for $Sc \to \infty$ for some shapes without axial symmetry, e.g., inclined cylinders (S34). Data for nonspherical shapes are shown in Fig. 10.3 for large Rayleigh number. The characteristic length in Sh′ and Ra′ is analogous to that used in Chapters 4 and 6:

$$l' = \frac{\text{area for mass transfer}}{\text{maximum perimeter projected on a plane normal to the direction of gravity}}. \tag{10-24}$$

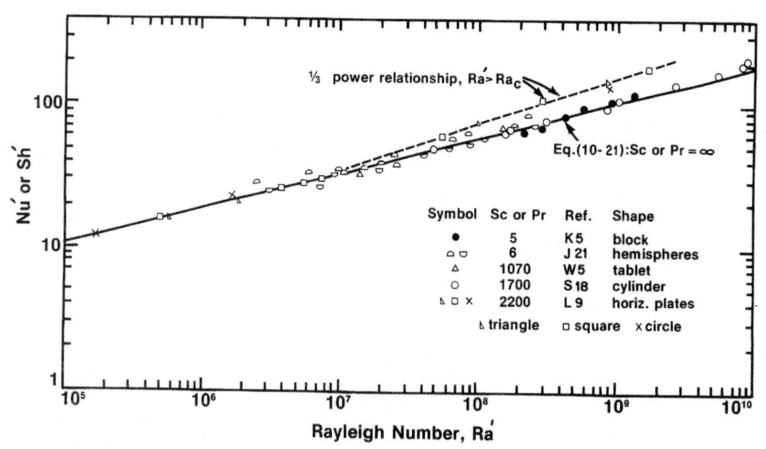

FIG. 10.3 Average Sh′ or Nu′ for nonspherical particles at high Ra.

For shapes which have an appreciable fraction of area over which the normal component of the buoyancy force is directed away from the surface (e.g., a heated horizontal plate facing upward) the turbulent regime where $Sh' \propto (Ra')^{1/3}$ occurs at low Ra'. For example, the horizontal plates of Fig. 10.3 have only one side exposed, the side for which the normal component of buoyancy is directed away from the plate. Here the $\frac{1}{3}$-power relation applies at $Ra' \gtrsim 10^7$. For horizontal cylinders, on the other hand, such a relationship is exhibited for $Ra' \gtrsim 5 \times 10^9$. For spheres in Fig. 10.2, there is no indication of this transition even at $Ra' \doteq 10^{10}$. The critical Rayleigh number, Ra_c', above which the $\frac{1}{3}$-power relationship applies is correlated by

$$Ra_c' = 10^7/f_A^4, \tag{10-25}$$

where f_A is the fraction of the total surface area over which the normal component of the buoyancy force is directed away from the surface and the angle between the outward normal to the surface and the vertical is less than 45°. The solid line in Fig. 10.3 is the asymptotic ($Sc \to \infty$) solution, Eq. (10-21) with the Sherwood and Rayleigh numbers based on the l' of Eq. (10-24). For $Ra' > Ra_c'$ the $\frac{1}{3}$-power relationship should be used as shown by the dashed line for horizontal plates in Fig. 10.3.

Based on Eq. (10-22) the following relationship is recommended for arbitrary shapes at any $Ra' < Ra_c'$:

$$Sh' = 0.85\,Sh_0' + 0.15\,Sh_0'\left\{1 + \left(\frac{0.589}{0.15\,Sh_0'}\right)^4 [f(Sc)]Ra'\right\}^{1/4}, \tag{10-26}$$

where Sh_0' is the Sherwood number for diffusion into a stagnant medium discussed in Chapter 4 and $f(Sc)$ is given by Eq. (10-23). Equation (10-26) agrees well with the only set of data available at low Ra' (G13).

4. *Simultaneous Heat and Mass Transfer*

When heat and mass are transferred simultaneously, the two processes interact through the Gr_w and Gr_t terms in Eq. (10-12) and the energy and diffusion equations. Although solutions to the governing equations are not available for spheres, results should be qualitatively similar to those for flat plates (T4), where for aiding flows ($Gr_w/Gr_t > 0$) the transfer rate and surface shear stress are increased, and for opposing flows ($Gr_w/Gr_t < 0$) the surface shear stress is predicted to drop to zero yielding an unstable flow.

Solutions to the boundary layer form of Eq. (10-12) have been obtained for spheres in aiding flow with $Sc \geq Pr$ and $Sc \to \infty$, a situation relevant to a sphere in a liquid (T4). These results are approximated (S6) within 10% by:

$$Nu = 0.589\left[1 + \frac{Gr_w/Gr_t}{(Sc/Pr)^{1/2}}\right]^{1/4} Ra^{1/4}, \tag{10-27}$$

$$Sh = 0.589\left[1 + \frac{(Sc/Pr)^{1/3}}{Gr_w/Gr_t}\right]^{1/4} Ra^{1/4}. \tag{10-28}$$

C. Mixed Free and Forced Convection

In mixed convection the orientation of the freestream velocity with respect to the gravity vector is an important variable. The three orientations which have received most attention are opposing flow, aiding flow, and crossflow. Aiding flow results when the velocity which would be induced by buoyancy acting alone is in the same direction as the forced flow, e.g., a stationary heated sphere in an upward flowing gas stream or a heated sphere falling in a stagnant gas. Opposing flow is the reverse while crossflow occurs when the freestream velocity vector and the gravity force vector are at right angles to each other.

1. *Creeping Flow*

In creeping flow the effect of aiding and opposing buoyancy has been obtained for uniform composition heat transfer by the method of matched asymptotic expansions (H9) and numerically (W7). For $Re \leq 1$, the buoyancy increases the drag coefficient in aiding flow and decreases it in opposing flow, e.g., a sphere which is hotter than a gas settles more slowly than if the sphere and gas were at the same temperature. Figure 10.4 shows the effect of temperature difference upon the terminal settling velocity at $Re \leq 1$. The parameter on the

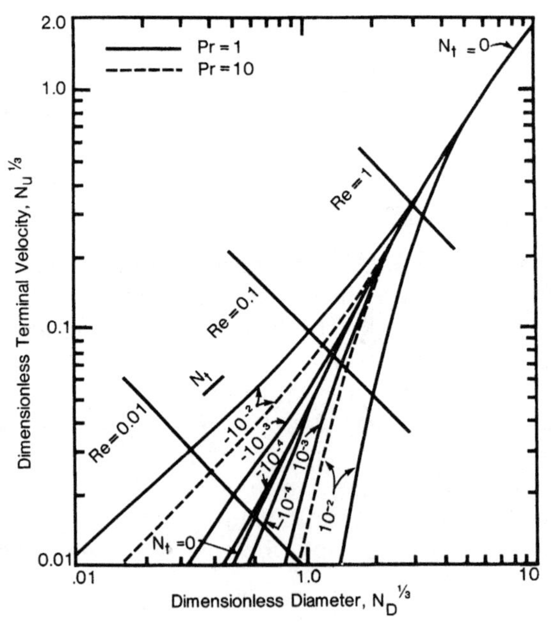

FIG. 10.4 Dimensionless terminal velocity $N_U^{1/3}$ as a function of dimensionless diameter $N_D^{1/3}$ for spheres whose temperature differs from the fluid temperature. Calculated from Woo (W7) and Hieber and Gebhart (H9).

curves is a thermal number, N_t, defined by

$$N_t = \frac{1}{C_D}\frac{\mathrm{Gr}_t}{\mathrm{Re}^2} = \frac{3}{4}\frac{\rho\beta_t(T_s - T_\infty)}{|\rho_p - \rho|} = \frac{3}{4}\frac{\rho_\infty - \rho_s}{|\rho_p - \rho_\infty|}. \tag{10-29}$$

In Fig. 10.4 the sphere diameter, terminal velocity, and temperature difference each appear in only one dimensionless group. The effect of natural convection on $N_U^{1/3}$ is smaller at Pr = 10 because the region over which the buoyancy force acts is much thinner than for Pr = 1. As Pr → ∞ the effect should disappear altogether. For Pr = 0, numerical solutions (W7) show effects about 50% larger than for Pr = 1.

The effect of natural convection on C_D is shown in Fig. 10.5 for aiding and opposing flow. The ordinate is the ratio of the drag coefficient in mixed convection to C_{D0}, the drag coefficient in pure forced convection ($\mathrm{Gr}_t = 0$). The abscissa is the parameter $\mathrm{Gr}_t\mathrm{Re}^{-1.85}\mathrm{Pr}^{-0.5}$ which brings all of the calculated values together for $0.7 \leq \mathrm{Pr} \leq 10$ and $\mathrm{Re} \leq 30$. The effect on the mean Nusselt number is appreciably less than on C_D. Nu increases less than 3% in aiding flow and decreases less than 3% in opposing flow for Re ≤ 1.

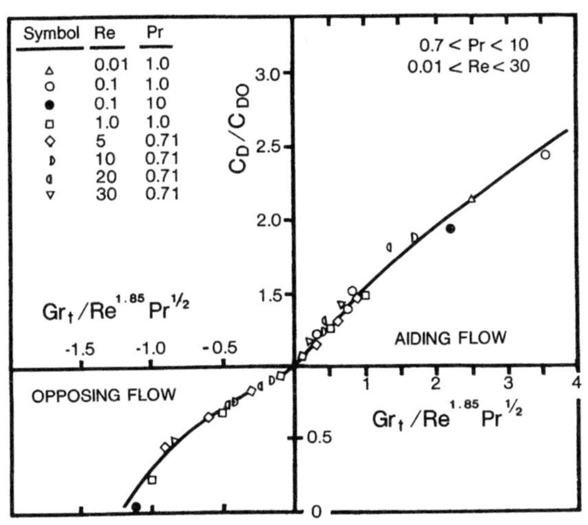

FIG. 10.5 Effect of Gr_t, Re, and Pr on drag coefficient for spheres whose temperatures differ from the fluid temperature.

2. *Higher Reynolds Numbers*

Fluid velocities have been predicted numerically (W7) for mixed convection to spheres in aiding and opposing flow at Re ≤ 30, Pr = 0.71, and $\mathrm{Gr}_t/\mathrm{Re} \leq 10$. Aiding flow delays separation, while opposing flow moves the separation point forward, e.g., at Re = 5 the flow separates at 154° with Pr = 0.71 and $\mathrm{Gr}_t = 15$

in opposing flow. As in creeping flow, C_D is increased in aiding flow. Although no computations of Nu are available, the average Nusselt number should increase in aiding flow because of the decreased size of the attached wake where local transfer rates are low. The reverse should occur for opposing flow. These expectations are borne out by data for spheres (P1, Y1) and cylinders (H5) at Re less than required for eddy shedding in forced flow. Data for heat transfer from spheres to an air stream (Y1) are shown in Fig. 10.6 on coordinates suggested by a boundary layer analysis (A4). In crossflow Nu exceeds that expected for forced flow at all Re, while for opposing flow Nu goes through a minimum and approaches the forced flow limit from below. Aiding flow data lie slightly above those for crossflow. Similar behavior occurs for cylinders (H5, O1).

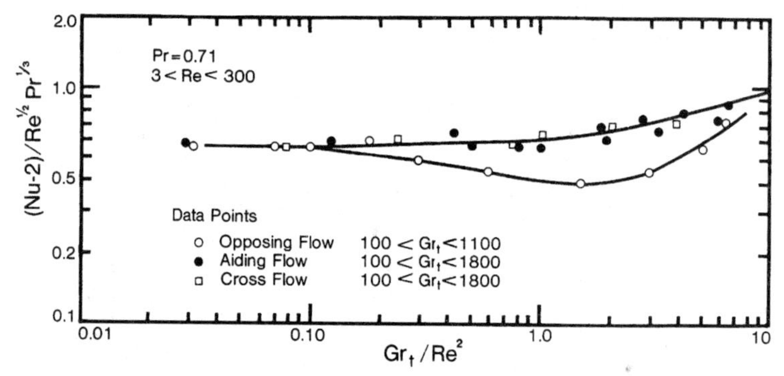

FIG. 10.6 Heat transfer in combined convection to spheres with $\mathrm{Pr} = 0.71$ and $3 < \mathrm{Re} < 300$.

At higher Re, where eddy shedding occurs in forced flow, the behavior is similar to that in Fig. 10.6 if $\mathrm{Ra} \lessdot 10^6$ (O1, P2). As Ra increases, the minimum becomes shallower and moves to higher $\mathrm{Gr}_t/\mathrm{Re}^2$, while aiding and opposing flow Nu values approach each other (F2, H5, O1). For larger Ra, the situation is reversed with opposing flow yielding larger average transfer rates than aiding flow (B13, G7, W5). This reversal is caused by two factors. First, if Ra is high enough in aiding flow, the transfer rate on the rear surface is reduced below its value in pure natural convection (G7, W5). Second, if Ra is high enough in opposing flow, there is a strong interaction between the forced flow and the opposed natural convective flow at the rear of the particle. Flow visualizations (B13, W4) indicate a complex turbulence-like flow pattern which yields higher transfer rates over the entire surface than in aiding flow (G7).

Consideration of the available data for spheres indicates that forced flow correlations are accurate to about 10% for $\mathrm{Gr}_t/\mathrm{Re}^2 \lessdot 0.2$. The analogous limit for natural convection is not so well defined, being about 10 at $\mathrm{Pr} = 0.7$ and increasing with Pr. Additional studies of mixed convection are needed to elucidate the physical phenomena and provide correlations. Simultaneous mass and heat transfer have been studied for drops evaporating into gases with $Pr \doteq 1$. The aiding flow data agree with those shown in Fig. 10.6 (N1). The Nusselt numbers are larger in opposing flow than in aiding flow, probably due to flow instability induced by the mass flow outward from the surface (N1, S5).

V. PARTICLE ROTATION AND FLUID SHEAR

It is convenient to distinguish between particle or fluid rotation about axes normal and parallel to the direction of relative motion. These two types of motion may be termed respectively "top spin" and "screw motion" (T11). Top spin is of more general importance since this corresponds to particle rotation caused by fluid shear or by collision with rigid surfaces. Workers concerned with suspension rheology and allied topics have concentrated on motion at low Re, while very high Reynolds numbers have concerned aerodynamicists. The gap between these two ranges is wide and uncharted, and we make no attempt to close it here.

A. Top Spin

1. *Low Reynolds Numbers*

Figure 10.7 shows schematically a sphere undergoing top spin in an unbounded fluid† moving with undisturbed relative velocity U_R at its center. The fluid is in uniform shear in the plane of the figure, with shear rate:

$$u_x = \mathrm{G}y, \quad u_y = 0, \quad u_z = 0. \tag{10-30}$$

It is convenient to define a shear Reynolds number:

$$\mathrm{Re_G} = \mathrm{G}d^2/\nu. \tag{10-31}$$

The angular velocity of the sphere, $\mathbf{\Omega}$, is taken as positive for rotation in the same sense as that of the fluid. The resulting lift on the particle is taken as positive in the direction $\mathbf{\Omega} \times \mathbf{U_R}$.

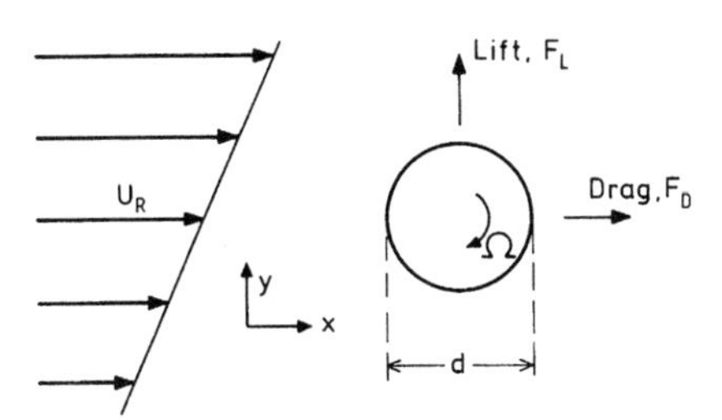

FIG. 10.7 Schematic diagram of sphere rotating in a fluid in simple shear.

† Rigid boundaries have a significant effect on lift and migration (H13) as discussed in Chapter 9.

Much of the work on particle rotation at low Re_G follows from the early work of Jeffery (J2) who considered a rigid, neutrally buoyant spheroid subject to the uniform shear field defined by Eq. (10-30). Jeffery showed that the particle center moves with the velocity which the continuous fluid would have at that point in the absence of the particle, while the axis of the spheroid undergoes rotation in one of a family of periodic orbits with angular velocities

$$\frac{d\varphi}{dt} = \frac{G}{a^2 + b^2}(a^2 \cos^2 \varphi + b^2 \sin^2 \varphi) \tag{10-32}$$

and

$$\frac{d\theta}{dt} = \frac{G(a^2 - b^2)}{(a^2 + b^2)} \sin 2\theta \sin 2\varphi, \tag{10-33}$$

where θ is the angle between the axis of symmetry of the spheroid and the z axis and φ is the angle between the yz plane and the plane which contains both the z axis and the axis of symmetry of the spheroid. Integration of these equations yields

$$\tan\theta = C_0 a/\sqrt{a^2 \cos^2 \varphi + b^2 \sin^2 \varphi} \tag{10-34}$$

and

$$\tan\varphi = (a/b)\tan[(2\pi t/\tau_r) + \varphi_0] \tag{10-35}$$

where the integration constant C_0 is the orbit constant, i.e., the eccentricity of elliptical orbits traced out by the ends of the particle; φ_0 is the initial phase angle; and τ_r is the period of rotation about the z axis given by

$$\tau_r = (2\pi/G)(E + 1/E). \tag{10-36}$$

For a sphere where $a = b$, the particle rotates with an angular velocity of $G/2$ and a period of rotation of $4\pi/G$.

Mason and co-workers (B8, F3, G11, M5, T15) have shown that Eqs. (10-32) to (10-35) can also be applied to disks and cylinders provided that one uses an apparent value of E, calculated from Eq. (10-36) and the observed τ_r. Bretherton (B15) considered more general shapes and proved that most bodies of revolution, except for some extreme shapes, show periodic rotation with no lateral migration (i.e., no lift) provided that inertia terms are neglected. In reality all these particles migrate in the direction of positive lift (see Chapter 9). For a useful extended review on particle motion in shear fields, see Goldsmith and Mason (G12).

Theoretical attempts to explain lift have concentrated on flow at small but nonzero Re, using matched asymptotic expansions in the manner of Proudman and Pearson for a nonrotating sphere (see Chapter 3). In the absence of shear, Rubinow and Keller (R6) showed that the drag is unchanged by rotation. With

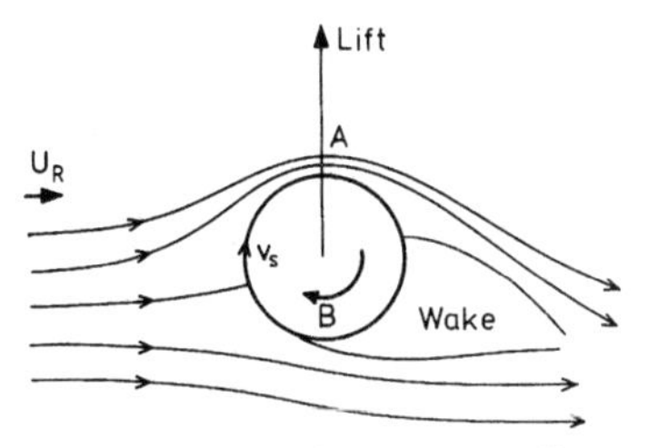

FIG. 10.8 Schematic diagram showing Magnus effect on rotating sphere.

both spin and shear, Saffman (S2) showed that the drag is slightly increased, while the lift is given by

$$F_L = 1.615 \mu d U_R \sqrt{\mathrm{Re}_G}. \tag{10-37}$$

Harper and Chang (H4) generalized the analysis for any three-dimensional body and defined a lift tensor related to the translational resistances in Stokes flow. Lin *et al.* (L3) extended Saffman's treatment to give the velocity and pressure fields around a neutrally buoyant sphere, and also calculated the first correction term for the angular velocity, obtaining

$$\Omega = (G/2)(1 - 0.0384\,\mathrm{Re}_G^{3/2}). \tag{10-38}$$

From flow visualization and angular velocity measurements, Poe and Acrivos (P12) concluded that the analysis leading to Eqs. (10-37) and (10-38) is valid only for $\mathrm{Re}_G \lessdot 0.1$, while for $\mathrm{Re}_G \gtrdot 6$ a sphere rotates unsteadily and the wake is oscillatory. Theoretical or numerical treatments appear to be lacking beyond the near-Stokesian range until much higher Reynolds numbers.

Shear fields often induce splitting of fluid particles. This is treated in Chapter 12.

2. *Higher Reynolds Numbers: The Magnus Effect*

At high Re (based on U_R), rotating cylinders and spheres experience significant lift in the absence of fluid shear. This effect is well known to players of golf and tennis, and is normally known by the name of its supposed discoverer (M2).[†] The conventional Magnus lift acts in the direction shown schematically in Fig. 10.8, i.e., in the same sense as positive lift at low Re. Lord Rayleigh (R3) gave a qualitative explanation for the lift in terms of ideal fluid theory, showing that the fluid velocity is higher and the pressure therefore lower near A than near B. The phenomenon is rather more complex and related to the formation of an asymmetric wake (see Fig. 10.8), demonstrated for spheres by Maccoll (M1) and Taneda (T3) and for cylinders by Prandtl [see (G14)]. Drag and lift on a spinning sphere must be determined experimentally.

[†] Although this effect is associated with Magnus, it was investigated for spheres much earlier by Benjamin Robins (B6).

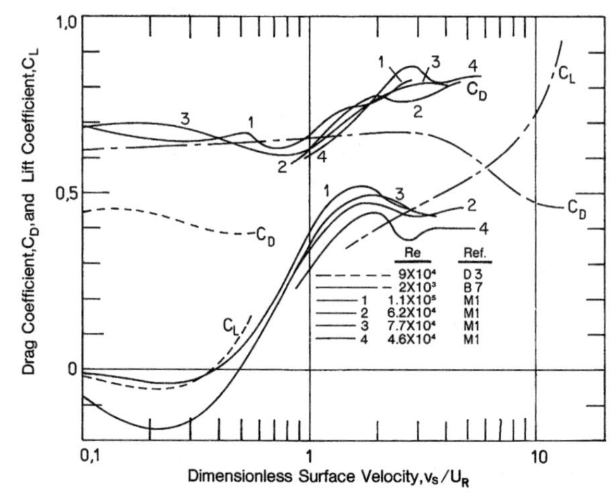

FIG. 10.9 Drag and lift coefficients for rotating spheres. All data plotted are for smooth spheres.

Figure 10.9 summarizes available measurements on smooth spheres for C_D and the lift coefficient C_L. A measure of the spin on the sphere is given by the ratio of the maximum surface speed v_s to the relative velocity U_R. Data of Maccoll (M1) and Davies (D3) for Re close to the critical transition indicate that C_D is relatively insensitive to spin. However, C_L is negative in this Re range for $v_s/U_R \dot{<} 0.6$. Taneda (T3) showed that this "negative Magnus effect" is restricted to the range $6 \times 10^4 \dot{<} \text{Re} \dot{<} 5 \times 10^5$. The cause of negative lift was shown to be earlier transition to turbulence around point B on Fig. 10.8 so that final separation on that side of the particle occurs further to the rear (see Chapter 5). In the negative lift region, the wake is distorted in the opposite sense to that shown in Fig. 10.8 with the pressure at B lower than at A. Freestream turbulence, which displaces the critical transition to lower Re (see section VI), also displaces the region of negative lift. Well below the critical range, the data of Barkla and Auchterlonie (B7) indicate a steady rise of C_L and a slight fall in C_D with increasing v_s/U_R.

Davies (D3) found that roughened spheres behave rather differently at $\text{Re} \doteq 9 \times 10^4$. Both C_D and C_L rose steadily with increasing v_s/U_R, presumably due to the effect of roughness in displacing the critical transition to lower Re (see Section II). It is therefore possible that rough spheres show negative lift at somewhat lower Re, but this has not been confirmed.

B. Screw Motion

For a sphere rotating about an axis parallel to the direction of relative motion, flow may be characterized by Re and by the ratio v_s/U_R of equatorial surface speed to the approach velocity.† As for top spin, screw rotation in

† This ratio, its inverse, and various multiples are often called the "Rossby number."

creeping flow has no effect on the drag experienced by a sphere; i.e., there is no coupling between rotation and translation (D10, G9). Results for non-spherical shapes are reviewed by Happel and Brenner (H3). The Oseen approximation leads to the following expression for the ratio of sphere drag to drag in Stokes flow (S3):

$$\frac{F_D}{F_{St}} = 1 + \frac{3}{16}\text{Re} + \left(\frac{v_s}{U_R}\right)^2\left(51.163 + \frac{60.913}{\text{Re}}\right) + O\left[\left(\frac{v_s/U_R}{\text{Re}}\right)^2\right]. \quad (10\text{-}39)$$

If v_s/U_R exceeds a critical value of order unity, reverse flow occurs near the forward stagnation point for oblate and prolate axisymmetric bodies (M8, M9), leading to formation of an "upstream separation bubble."

Luthander and Rydberg (L13) investigated the effect of screw rotation on flow patterns and drag for a sphere near critical transition. Rotation with v_s/U_R up to 1.4 had virtually no effect on transition or on C_D. Faster rotation decreased Re_c to approximately 10^5 for $v_s/U_R \doteq 3$. Again the drop in C_D was associated with increased turbulence and delayed separation of the boundary layer; very rapid rotation caused the separation point to move forward again, causing a rise in supercritical drag. Below the critical range, rotation with $v_s/U_R < 2$ had little effect on C_D. These experiments were carried out with a sphere whose location in a wind tunnel was fixed, so that the implications for bodies in free flight are not clear. It is well known that the trajectories of projectiles are stabilized by screw rotation, and this presumably results from elimination of the erratic lift forces in supercritical flow (see Chapter 5).

Somewhat similar considerations apply to a particle moving through a fluid which is rotating about an axis not necessarily passing through the center of the particle. Taylor [e.g., see (T5, T6)] did much of the early work on the subject and showed that two-dimensional cylinders tend not to be deflected by the rotation whereas three-dimensional symmetrical bodies (including spheres) are deflected. For recent work on this problem, see (M4, M7, M12, M13).

Gas bubbles in screw motion show flattening as the angular velocity is increased (R5). Coriolis forces must be considered in predicting trajectories of fluid particles and a method of doing this is given by Catton and Schwartz (C1). Slugs in rotating tubes are treated in Chapter 9.

C. Heat and Mass Transfer

The effect of rotation on transfer to a translating sphere has been studied for both screw motion (E1, F6, T2) and top spin (N3, T2) with Re > 1500. The effect of rotation on the transfer rate is less than 10% for $v_s/U_R \dot{<} 0.5$. The ratio of the Sherwood number in screw motion to that in pure translation at the same U_R is correlated within 10% by

$$\text{Sh}_r/\text{Sh} = [1 + 1.04(v_s/U_R)^2]^{1/4} \quad (10\text{-}40)$$

For top spin at higher rotation rates, Sh first decreases and then increases with increasing rotational Reynolds number, $\Omega d^2/\nu$, at constant translational Reynolds number, Re. The overall change in Sh is generally relatively small except for very rapid rotation. Analyses have been carried out (F4, P13) for spheres and cylinders with $\text{Re}_G \ll 1$ and $\text{Pe}_G = \text{Re}_G \text{ Sc} \gg 1$ in simple shear. At low Re_G there are closed streamlines around the body; at high Pe_G these streamlines are also lines of constant composition so that Sh (or Nu) becomes independent of Pe_G. For a sphere, Sh $= 8.9$ for $\text{Pe}_G \to \infty$ (P13). More complex velocity fields have also been considered at low Re_G.

VI. FREESTREAM TURBULENCE

A. Effect on Particle Motion

1. *General Considerations*

The motion of a particle in a turbulent fluid depends upon the characteristics of the particle and of the turbulent flow. Small particles show a fluctuating motion resulting from turbulent fluid motion. Generally speaking, a particle responds to turbulent fluctuations with a scale larger than the particle diameter (K9). A particle which is much larger than the scale of turbulence shows relatively little velocity fluctuation. The effect of turbulence is then to modify the flow field around the particle, so that the drag may be affected.

The range between these "small" and "large" particles is less well understood although some experimental studies have been reported (K9, U1). Similar problems arise in interpretation as with accelerated motion (see Chapter 11). Measurements are commonly correlated by a turbulence-dependent drag coefficient, which contains a number of possible acceleration-dependent components. With fundamental understanding so poorly advanced, it is impossible to say to what extent results are specific to the experimental conditions employed.

2. *Small Particles*

If the fluid turbulence is represented as a Fourier integral,

$$u' = \int_0^\infty A_\omega \omega \cos(\omega t + \phi_\omega)\, d\omega, \tag{10-41}$$

the response of the particle to individual angular frequencies ω can be examined using the results for sinusoidal fluid motion given in Chapter 11. This approach has been developed by a number of workers [e.g. (F5, H10, H12, L2, L8, S30)], with one or more drag components often neglected. The results in Table 11.2 and Fig. 11.15 can be used to estimate the particle–fluid amplitude ratio η and phase shift β for a given frequency of oscillation characterized by the dimensionless period (or Stokes number), $\tau_0 = \nu/\omega a^2$. The results are useful for evaluating techniques, e.g., flow visualization or laser-Doppler anemometry, in which suspended particles must follow the fluid closely. Figure 10.10 shows τ_0

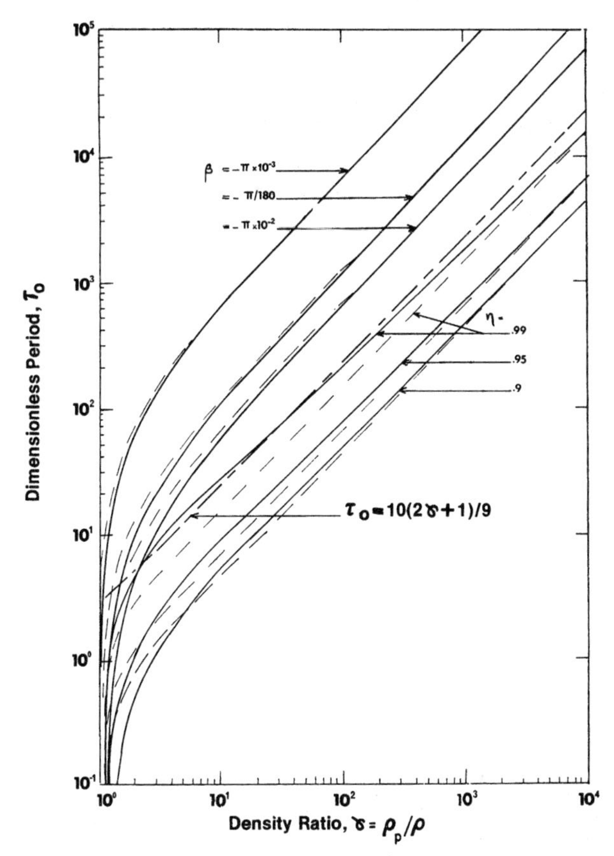

FIG. 10.10 Values of dimensionless period, $\tau_0 = \nu/\omega a^2$, for given amplitude ratio η and phase shift β for spheres. The continuous lines give the full solution, while the broken lines are for the case where the history component is neglected.

plotted as a function of the density ratio γ for constant values of η and β. A tracer particle can be considered to respond to frequencies in the range above the curve corresponding to the required accuracy, provided that the particle also meets the conditions for smallness given by Eq. (11-72) and has low Reynolds number based on the r.m.s. relative velocity.

Simplified expressions for predicting η and β are available, corresponding to neglect of various terms in the unsteady drag equation. These expressions are summarized in Table 11.2. Hjelmfelt and Mockros (H12) examined the validity of the simplifications; unfortunately, the most commonly neglected term, the Basset history term, is often significant at high frequencies.† Figure 10.10 shows

† Ahmadi and Goldschmidt (A6) showed that the history term has a negligible effect on the mean motion of a particle in a turbulent fluid. The discussion here concerns fluctuations in particle motion.

curves of constant η and β calculated neglecting the history term. Although neglect of this term is reasonable for density ratios typical of particles in gases ($\gamma > 10^3$), this approximation is not advisable for particles in liquids, especially if the amplitude is of interest.

As a rough guide, a particle follows the fluid motion faithfully if its relaxation time, $a^2(2\gamma + 1)/9\nu$, is small compared with the period of oscillation (L12), i.e., if

$$\tau_0 \gg (2\gamma + 1)/9. \tag{10-42}$$

Figure 10.10 shows the curve $\tau_0 = 10(2\gamma + 1)/9$. Equation (10-42) is a useful guide unless γ is close to unity.

The approach of representing the fluid and particle motion by their component frequencies is only valid if drag is a linear function of relative velocity and acceleration, i.e., if the particle Reynolds number is low. This is the reason for the restriction on "small" particles noted earlier. The terminal velocity of the particle relative to the fluid is superimposed on the turbulent fluctuations and is unaffected by turbulence if Re is low (see Chapter 11).

3. *Large Particles*

If the particle Re is well above the creeping flow range, mean drag may be increased or decreased by freestream turbulence. The most significant effect is on the critical Reynolds number. As noted in Chapter 5, the sharp drop in C_D at high Re results from transition to turbulence in the boundary layer and consequent rearward shift in the final separation point. Turbulence reduces Re_c, presumably by precipitating this transition.†

As in Chapter 5, it is convenient to define Re_c as the Reynolds number at which C_D falls to 0.3. Freestream turbulence may be characterized by the *relative* intensity:

$$I_R = \sqrt{\overline{u'^2}}/\bar{U}_R, \tag{10-43}$$

where $\bar{U}_R$ is the mean velocity of the particle relative to the fluid and $\sqrt{\overline{u'^2}}$ is the r.m.s. fluctuating velocity of the fluid. Since $\bar{U}_R$ is usually much smaller than the freestream velocity, I_R is generally much higher for entrained particles than for fixed particles. For fixed particles, Dryden *et al.* (D11) found that Re_c decreased as I_R increased up to 0.045. This effect has been used to estimate turbulence intensities in wind tunnels [e.g. (G14)]. A weak effect of turbulence macroscale, L_s, was also found, and Re_c correlated well with $I_R(d/L_s)^{1/5}$. This correlating group was derived by Taylor (T7), who suggested that fluctuating pressure gradients provoke boundary layer transition. Subsequently Torobin and Gauvin (T12), working with entrained spheres and $0.1 < I_R < 0.4$, found that Re_c continued to decrease, down to approximately 400 for $I_R = 0.4$ (cf.

† Seeley (S21) discounted this mechanism on the basis of flow visualization studies. However, the experiments were at $Re < Re_c$ given by Eq. (10-44), and thus appear to be in near-subcritical flow.

$\mathrm{Re_c} \doteq 3.65 \times 10^5$ for turbulence-free flow). No effect of turbulence scale was detected. The data of Torobin and Gauvin do not (C7, U1) readily extrapolate to those of Dryden *et al.* Whether the mismatch results from differences between fixed and entrained particles or from differences in turbulence characteristics is not clear. In the absence of further experimental data, $\mathrm{Re_c}$ can be estimated by modifying the empirical equations proposed by Clift and Gauvin (C7), neglecting the weak effect of d/L_s:

$$\log_{10} \mathrm{Re_c} = 5.562 - 16.4 I_R \qquad (I_R \leq 0.15), \tag{10-44}$$

$$\log_{10} \mathrm{Re_c} = 3.371 - 1.75 I_R \qquad (I_R > 0.15). \tag{10-45}$$

Equation (10-44) gives a reasonable interpolation between the lower I_R range examined by Torobin and Gauvin and the data of Dryden *et al.* (D11), but implies a stronger dependence of $\mathrm{Re_c}$ on I_R than indicated by Dryden's experiments.

Clamen and Gauvin (C6) measured C_D for entrained spheres at Re above the turbulence-induced critical transition, and C_D was found to rise for $\mathrm{Re} > \mathrm{Re_c}$ to pass through a maximum which increased with I_R. The point at which C_D again achieves 0.3 may be termed the "metacritical Reynolds number," $\mathrm{Re_M}$, and can be estimated (C7) by:

$$\log_{10} \mathrm{Re_M} = 6.878 - 23.2 I_R \qquad (I_R \leq 0.15), \tag{10-46}$$

$$\log_{10} \mathrm{Re_M} = 3.663 - 1.8 I_R \qquad (I_R > 0.15). \tag{10-47}$$

Correlations for C_D in the critical and supercritical ranges are given in Table 10.1 and plotted in Fig. 10.11. The dependence of C_D on Re in the supercritical

TABLE 10.1

Correlations for the Effect of Turbulence on Sphere Drag[a]

Range	Correlation for C_D	Source
1. Subcritical		
a) $\mathrm{Re} < 50$; $0.05 < I_R < 0.5$	$C_D = 162 I_R^{1/3} \mathrm{Re}^{-1}$	(U1)
b) $50 < \mathrm{Re} < 700$; $0.07 < I_R < 0.5$	$C_D = 0.133(1 + 150/\mathrm{Re})^{1.565} + 4 I_R$	(U1)
2. Critical		
$0.9\mathrm{Re_c} < \mathrm{Re} < \mathrm{Re_m}$	$C_D = 0.3\,(\mathrm{Re}/\mathrm{Re_c})^{-3}$	(C7)
3. Supercritical		
a) $\mathrm{Re_m} < \mathrm{Re} < \mathrm{Re_M}$	$C_D = 0.3(\mathrm{Re}/\mathrm{Re_M})^{(0.45 + 20 I_R)}$	(C7)
b) $\mathrm{Re_M} < \mathrm{Re} < 3 \times 10^4$; $I_R > 0.07$	$C_D = 3990\mathrm{Re}^{-6.10} - 4.47 \times 10^5 I_R^{-0.97} \mathrm{Re}^{-1.80}$	(C6)

[a] In the above equations, $\mathrm{Re_c}$, the critical Reynolds number, is given by Eqs. (10-44) and (10-45); $\mathrm{Re_m}$, the metacritical Reynolds number, is given by Eqs. (10-46) and (10-47), and $\mathrm{Re_m}^{(3.45 + 20 I_R)} = \mathrm{Re_c}^3 \mathrm{Re}_\chi^{(0.45 + 20 I_R)}$.

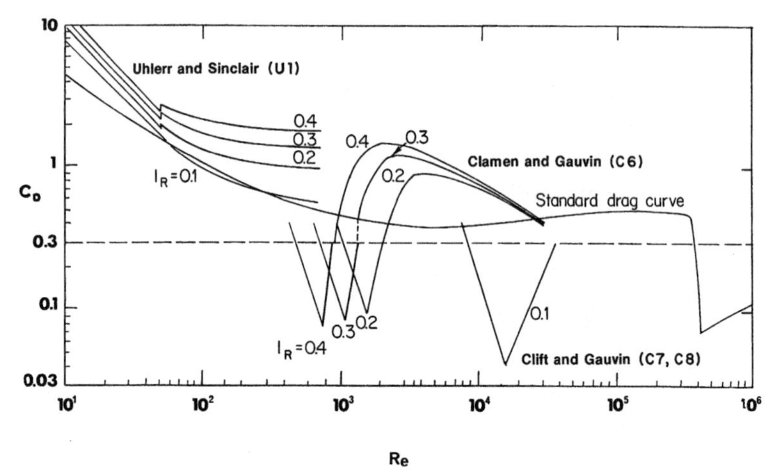

FIG. 10.11 Effect of relative turbulence intensity I_R on drag coefficient C_D.

range has not been verified and is not readily explainable, although some tentative explanations have been advanced (C6).

Several workers investigated the effect of turbulence in subcritical flow, but their results are somewhat contradictory. Uhlherr and Sinclair (U1) determined drag on entrained particles, and gave correlations which appear in Table 10.1 and Fig. 10.11. Generally, turbulence increased C_D, but Fig. 10.11 shows a region in which C_D is reduced. They report that for $50 < \text{Re} < 700$, C_D has a minimum value for $I_R \doteq 0.04$. Petrak (P7) also found a decrease in C_D in approximately the same Re range. However, Nicholls and co-workers (S28, Z1, Z2) found that turbulence levels up to $I_R = 0.08$ had negligible effect on C_D for $\text{Re} < 200$, while at higher Re, C_D increased monotonically with I_R, but less rapidly than indicated by Uhlherr and Sinclair. Neither set of experiments indicates any effect of d/L_s in the lower Re range, although Zarin (Z1) found that C_D decreased with increasing d/L_s for $600 < \text{Re} < 5000$. In addition, C_D may depend on the turbulent energy spectrum (K3). Qualitatively, turbulence increases the Reynolds number at which separation first occurs (U1), but decreases the value at which wake shedding occurs (S28, Z2). Relatively high I_R causes disturbances in the boundary layer, especially near the front stagnation point and the separation circle (S21), and reduces the width and length of the attached wake (S21, U1).

Clift and Gauvin (C7, C8) discuss application of the correlations in Table 10.1 to particles falling at their terminal velocities. High $\sqrt{\overline{u'^2}}$ can stabilize the critical regime by making it a region in which drag increases with $\bar{U}_R$. It was noted in Chapter 5 that the critical range is unstable for a particle falling through a quiescent fluid. Clamen's correlation indicates that there can be a range of particle sizes with two stable terminal velocities, the upper corresponding to critical flow and the lower to supercritical flow above $\mathrm{Re_M}$. For a given particle diameter, the terminal velocity can be increased or decreased, depending on $\sqrt{\overline{u'^2}}$.

4. *Fluid Particles*

There has been relatively little work on the motion of bubbles and drops in well-characterized turbulent flow fields. There is some evidence (B3, K7) that mean drag coefficients are not greatly altered by turbulence, although marked fluctuations in velocity (B3) and shape (K7) can occur relative to flows which are free of turbulence. The effect of turbulence on splitting of bubbles and drops is discussed in Chapter 12.

B. Effect on Heat and Mass Transfer

Experimental data are available for large particles at Re greater than that required for wake shedding. Turbulence increases the rate of transfer at all Reynolds numbers. Early experimental work on cylinders (V1) disclosed an effect of turbulence scale with a particular scale being "optimal," i.e., for a given turbulence intensity the Nusselt number achieved a maximum value for a certain ratio of scale to diameter. This led to speculation on the existence of a similar effect for spheres. However, more recent work (R1, R2) has failed to support the existence of an optimal scale for either cylinders or spheres. A weak scale effect has been found for spheres (R2) amounting to less than a 2% increase in Nusselt number as the ratio of sphere diameter to turbulence macroscale increased from zero to five. There has also been some indication (M15, S21) that the spectral distribution of the turbulence affects the transfer rate, but additional data are required to confirm this. The major variable is the intensity of turbulence. Early experimental work has been reviewed by several authors (G3, G4, K3).

In subcritical flow, increasing turbulence increases the rate of transfer only slightly over the forward portion of a sphere (G5, N2); larger increases are experienced over the separated (rear) portion. For cylinders, there appears to be a larger effect on the forward portion (G2, K4, M10) as well as a significant effect on the rear portion (M15, P8). Figure 10.12 shows local Nusselt numbers for heat transfer from a sphere to a turbulent air stream at $\mathrm{Re} = 2 \times 10^4$. The experimental results show little effect of turbulence over the forward portion of the sphere at low turbulence intensities. The highest intensity is above $I_{\mathrm{Rc}} = 0.077$, the turbulence intensity required to make $\mathrm{Re_c} = 2 \times 10^4$ as calculated from Eq. (10-44). The large Nu on the forward hemisphere and the local maximum on the rear hemisphere for $I_{\mathrm{R}} > 0.077$ indicate a turbulent boundary layer and supercritical flow conditions.

The average Nusselt number for a sphere also increases with turbulence intensity. For spheres in air at $\mathrm{Re} \lessdot 1.5 \times 10^4$ there is a rather rapid rise of

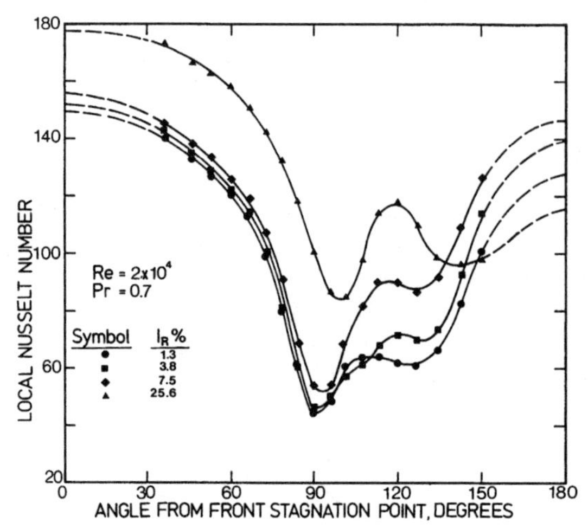

FIG. 10.12 Effect of intensity of turbulence on the local Nusselt number for a sphere in an air stream at Re $\doteq 2 \times 10^4$. Data of Galloway and Sage (G5).

Nu with intensity up to $I_R \approx 0.01$ (E2, R2). As I_R increases further, but still $I_R < I_{Rc}$, Nu increases roughly linearly, but more slowly. Similar effects have been observed for cylinders (M10). For spheres at higher Re, the average Nusselt number increases linearly with I_R for $I_R < I_{Rc}$ (R2). Few reliable data are available for $I_R > I_{Rc}$. Figure 10.13 presents a tentative correlation for the effect of turbulence on the average Nusselt number for spheres. The ordinate is Nu/Nu_0, the ratio of the Nusselt number at I_R to the value in the absence of turbulence, while the abscissa is the ratio of I_R, the intensity, to I_{Rc}, the critical intensity. The value of Nu_0 was calculated from the correlations in Table 5.4 and I_{Rc} from Eqs. (10-44) and (10-45). The correlation is divided into

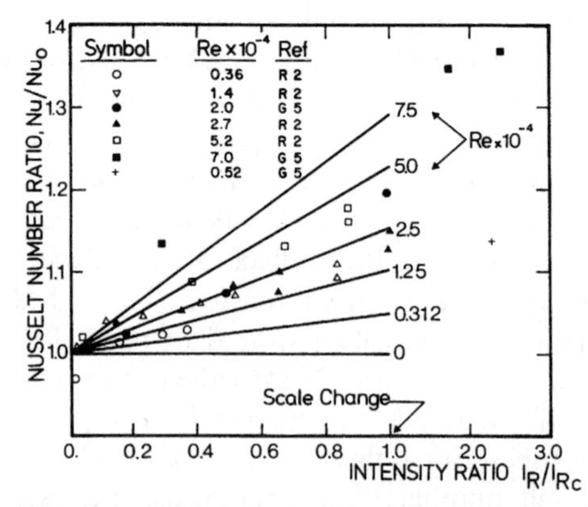

FIG. 10.13 Effect of intensity of turbulence on the average Nusselt number for a sphere in an air stream. (Note scale change on abscissa axis.)

a linear portion for $I_R/I_{Rc} \leq 1$ and a largely unexplored region for $I_R/I_{Rc} > 1$. For $I_R/I_{Rc} \leq 1$ a reasonable fit to the data is obtained by straight lines drawn from $Nu/Nu_0 = 1.0$ to the value of this ratio at $I_R = I_{Rc}$, denoted by Nu_c/Nu_0 and given by

$$Nu_c/Nu_0 = 1 + 4.8 \times 10^{-4} Re^{0.57}. \tag{10-48}$$

The linear relationship between Nu and I_R at a fixed Re proposed in Fig. 10.13 has also been found experimentally for stagnation point transfer from cylinders (K4) and spheres (G6). In addition, it has received some theoretical confirmation from predictions of turbulence models for stagnation point transfer (G1, S29, T13, W2).

Few reliable data are available on the effect of turbulence on transfer at high Pr or Sc. The data of Henry and Epstein (H8) for transfer from spheres in turbulent gas streams with $Sc \leq 5$ and the data of Mizushina *et al.* (M10) for cylinders with $Sc \doteq 10^3$ suggest that the representation in Fig. 10.13 should be independent of Pr or Sc.

VII. COMPRESSIBILITY AND NONCONTINUUM EFFECTS

Elsewhere in this book attention is focused on particles whose Mach and Knudsen numbers are small. The Mach number is defined as the ratio of the relative velocity between the particle and the fluid to the speed of sound in the fluid:

$$Ma = U_R/c. \tag{10-49}$$

For all practical purposes, isothermal flows with $Ma \dot{<} 0.2$ can be treated as incompressible, i.e., density variations in the fluid around the particle are negligible. Compressibility effects become important as Ma is increased, especially for Ma approaching and in excess of unity. The Knudsen number is defined as the ratio of the molecular mean free path in the fluid to some characteristic particle dimension. For a spherical particle

$$Kn = \lambda/d. \tag{10-50}$$

For Kn less than a value of order 10^{-3}, the particle is large by comparison with the scale of molecular processes in the continuous phase and the fluid can be treated as a continuum. Above this range, slip effects become significant. For liquids, c is sufficiently large and λ sufficiently small that these effects are virtually never significant. This is not the case for gases. The mean free path in centimeters for air at temperature T(K) and pressure p(kP) may be calculated (B10) from

$$\lambda = 2.15\mu T^{1/2}/p, \tag{10-51}$$

where μ is the viscosity in kg/ms. For air at 293 K and 100 kP, c is 343 m/s and λ is 6.53×10^{-6} cm. Compressibility effects are significant then for relative velocities greater than about 70 m/s, while noncontinuum effects are significant for particles smaller than about 10 μm.

Compressibility and noncontinuum effects are related. For an ideal gas, kinetic theory leads to the relationship (S11):

$$\mathrm{Kn} = \sqrt{\frac{\pi k}{2}} \frac{\mathrm{Ma}}{\mathrm{Re}}, \tag{10-52}$$

where k is the ratio of specific heats for the gas forming the continuous phase.[†] Hence the drag coefficient can be treated as a function of any two of Re, Ma, and Kn and in place of a unique relationship between C_D and Re (the "standard drag curve" discussed in Chapter 5), we require a family of curves. The discussion is simplified by distinguishing two distinct kinds of system showing noncontinuum and/or compressibility effects. Particles entrained in gases may be small enough for Kn to lie above the continuum range, although Re and Ma are very small. Flow around such particles shows noncontinuum effects but no appreciable compressibility, and this is discussed in Section A. The particles are subject to Brownian motion, so that for nonspherical particles the orientation is random and the hydrodynamic property of interest is the mean resistance to flow (see Chapter 4). The other class of flows corresponds to larger particles with high U_R, often in rarefied gases, and is of interest in connection with high-altitude flight and rocket propulsion. In this case, Ma is large and Re may be above the creeping flow range. Conventionally, this kind of system is described in terms of Re and Ma. We discuss this case from the viewpoint of compressibility effects in Section B. Effects of rarefaction on heat transfer are treated in Section C.

A. Noncontinuum Flow

It is conventional to distinguish three noncontinuum flow regimes. At small but not negligible Kn, where λ is up to 10% of d (or of the boundary layer thickness at higher Re), the gas adjacent to the particle surface has a significant tangential velocity or "slip." This range is termed the "slip flow regime," and flow is normally calculated by conventional methods with modified boundary conditions [see Pich (P10)]. When the mean free path and body dimensions are comparable, this approach breaks down, since both surface collisions and molecular collisions in the free stream are significant. This range, which is poorly understood, is termed the "transition regime." At high Kn, molecules moving away from the particle only undergo collision at large distances from the surface, so that the flow is dominated by molecule–particle interactions (D1, P9). This range is known as the "free molecule regime." The boundaries between these regimes are somewhat arbitrary. Table 10.2 summarizes two attempts at classification, but these are clearly not entirely consistent.

[†] For air at 293°K and 100 kP: Kn = 1.49 Ma/Re.

TABLE 10.2

Classification of Continuum and Noncontinuum Flow Regimes.[a]

	Schaaf and Chambré			Devienne (low Ma)
	General		Air	
Continuum	Re < 1:	$Ma < 0.01Re$	$Kn < 1.5 \times 10^{-2}$	$Kn < 10^{-3}$
	Re > 1:	$Ma < 0.01\sqrt{Re}$	$Kn < 1.5 \times 10^{-2} Re^{-1/2}$	
Slip flow	Re < 1:	$0.01 < \frac{Ma}{Re} < 0.1$	$1.5 \times 10^{-2} < Kn < 0.15$	$10^{-3} < Kn < 0.25$
	Re > 1:	$0.01 < \frac{Ma}{\sqrt{Re}} < 0.1$	$1.5 \times 10^{-2} < Kn\sqrt{Re} < 0.15$	
Transitional flow	Re < 1:	$0.1 < \frac{Ma}{Re} < 3$	$0.15 < Kn < 4.5$	$0.25 < Kn < 10$
	Re > 1:	$0.1 < \frac{Ma}{\sqrt{Re}} < 3\sqrt{Re}$	$\frac{0.15}{\sqrt{Re}} < Kn < 4.5$	
Free molecule flow	$Ma > 3Re$		$Kn > 4.5$	$Kn > 10$

[a] After Devienne (D8) and Schaaf and Chambré (S11).

Drag on a particle at nonnegligible Kn, but low Ma and Re, is conveniently expressed by the "slip correction factor":

$$C = \frac{\text{drag in continuum flow at same Re}}{\text{drag on particle}} \tag{10-53}$$

In the creeping flow range, C is equal to the ratio of the terminal velocity to the terminal velocity in continuum flow. The value of C is sensitive to the nature of molecular reflections from the surface of the particle (E5). The "accommodation coefficient," σ_R, may be interpreted as the fraction of molecules undergoing diffuse reflection, the balance being specularly reflected. Typical values for σ_R lie between 0.8 and unity. For near-continuum flow, Basset (B9) showed that

$$C = [1 - (2 - \sigma_R)\text{Kn}/\sigma_R + O(\text{Kn}^2)]^{-1}, \tag{10-54}$$

while Epstein's result (E5) for the opposite extreme of free-molecule flow may be written

$$C = 18\,\text{Kn}/(8 + \pi\sigma_R). \tag{10-55}$$

Phillips (P9) proposed an approximate theoretical solution which yields Eqs. (10-54) and (10-55) in the limits of low and high Kn, and appears to give a close fit to available data in the transitional range:

$$C = \frac{15 + 12c_1\,\text{Kn} + 9(c_1{}^2 + 1)\text{Kn}^2 + 18c_2(c_1{}^2 + 2)\text{Kn}^3}{15 - 3c_1\,\text{Kn} + c_2(8 + \pi\sigma_\text{R})(c_1{}^2 + 2)\text{Kn}^2}, \qquad (10\text{-}56)$$

where

$$c_1 = (2 - \sigma_\text{R})/\sigma_\text{R} \qquad \text{and} \qquad c_2 = (2 - \sigma_\text{R})^{-1}. \qquad (10\text{-}57)$$

For particles whose accommodation coefficient is known, Eq. (10-56) appears to give the most accurate estimate for drag. Since σ_R is rarely known to sufficient accuracy, C may instead be estimated for spheres over the whole range of Kn by a semiempirical expression whose form was first proposed by Knudsen and Weber (K6). With the numerical values due to Davies (D2):

$$C = 1 + \text{Kn}[2.514 + 0.8\exp(-0.55/\text{Kn})]. \qquad (10\text{-}58)$$

The constants in Eq. (10-58) depend upon σ_R, and other authors (D1, P10, S11) give slightly different values.

For nonspherical particles, values for the slip correction factor are available in slip flow (M11) and free-molecule flow (D1). To cover the whole range of Kn and arbitrary body shapes, it is common practice to apply Eq. (10-58) for nonspherical particles. The familiar problem then arises of selecting a dimension to characterize the particle. Some workers [e.g. (H2, P14)] have used the diameter of the volume-equivalent sphere; this procedure may give reasonable estimates for particles only slightly removed from spherical, or in near-continuum flow, but gives the wrong limit at high Kn. An alternative approach

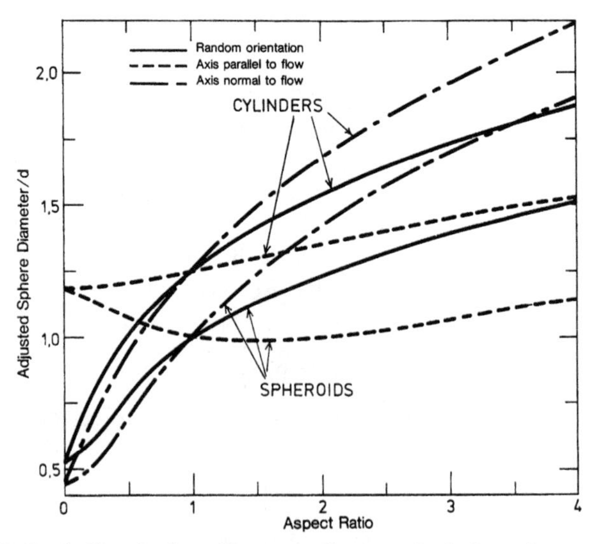

FIG. 10.14 Ratio of adjusted sphere diameter to diameter of cylinder and equatorial diameter of spheroids. After Dahneke (D1).

(D1) uses the "adjusted sphere," i.e., the sphere with the same value of C in free-molecule flow. The justification for this approach is that it gives the correct result in the limits of high and low Kn, and is therefore likely to be a good approximation in slip and transition flow as well. Figure 10.14 shows values for the diameter of the "adjusted sphere" for spheroids and cylinders, taken from Dahneke's tabulation. For cubes the "adjusted sphere diameter" to be used in defining Kn in Eq. (10-50) is 1.43 times the length of a side.

B. Compressibility Effect on Drag

The following discussion is concentrated on rigid spherical particles. There are many treatises [e.g. (K8, S13, S25, T10)] which consider aerodynamics and compressible external flows in a general manner, particularly with reference to aerofoils, wings, sharp cones, and other flight-related geometries. High-velocity flows are commonly accompanied also by significant aerodynamic heating and the temperature field can interact strongly with the velocity field. For simplicity, we restrict our attention to cases where the ratio of particle to gas absolute temperature, T_p/T_∞, is approximately unity. [For a review of data showing the effect of T_p/T_∞, see (B1).]

Early work in this area was reviewed by Hoerner (H14). [For more recent reviews of data, see (B1, B2, C10).] Many of the data in the literature for Ma $\gtrdot$ 0.2 are unreliable, just as for the low Ma results discussed in Chapter 5, because of high levels of freestream turbulence, interference by a support, wall effects, etc. The curves given in Fig. 10.15 have been prepared largely using data given in (B1) obtained in ballistic ranges where particle decelerations were measured over relatively short flight distances so that particle heating effects were small. The ratio of particle to gas density was sufficiently great that added mass and history effects (see Chapter 11) can be safely neglected in calculating C_D. The estimated accuracy of the data is $\pm 2\%$.

For subsonic velocities where $0.2 < \text{Ma} < 0.9$, the drag may be greater or less than that at low Ma depending on the value of Re. For $10^4 < \text{Re} < 3 \times 10^5$, a curve containing a dip at Ma $\doteq$ 0.85 given by Hoerner (H14) (see Fig. 10.15) was generally accepted for many years, but more recent work shows no evidence of this dip. Instead the drag increases monotonically with Ma at high Re. At very low Re, Kn becomes large [see Eq. (10-52)], and drag approaches the free-molecule flow limit. For intermediate Re (e.g., Re $\doteq$ 20) there is relatively little effect of Ma on drag.

For transonic velocities ($0.9 < \text{Ma} < 1.1$), C_D changes sharply with Ma, which accounts for the difficulty in obtaining reliable measurements in this range. Bailey and Hiatt (B1) give photographs showing the formation of wake and bow shock waves in this Ma range for Re $\doteq 10^3$. Sharp pressure increases occur across shock waves, and strong interactions with boundary layers therefore occur which tend to promote boundary layer separation (C4, S35). In the supersonic range ($1.1 < \text{Ma} < 6$), C_D becomes almost independent of Ma (see

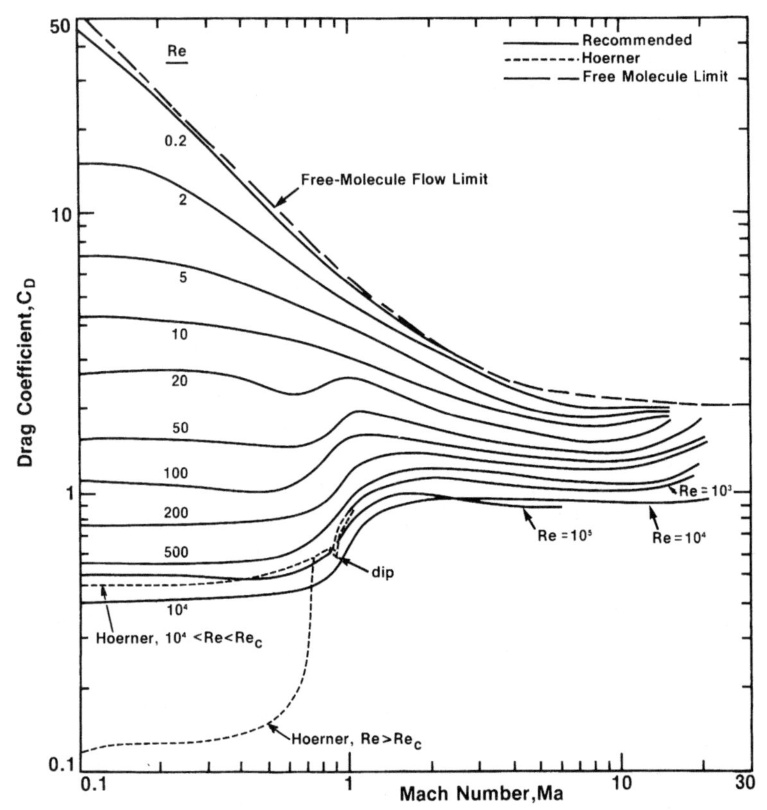

FIG. 10.15 Drag coefficient for rigid spherical particles in air as a function of Mach number with Reynolds number as parameter, for the case where the absolute temperatures of the particle and fluid are essentially the same.

Fig. 10.15) and depends primarily on Re, except for Re $\dot{<}$ 10, where C_D continues to fall as Ma increases. For hypersonic velocities (Ma > 6), temperature variations become more important with high local temperatures leading to ionization of gases. Available results (B2, K11) show that drag coefficients increase toward the free-molecule flow limit as Ma increases in the hypersonic range, while there is a monotonic decrease in drag coefficient with increasing Re in this range, at least up to Re $= 10^4$. Wake characteristics become particularly important and complex at super- and hypersonic speeds (C4).

In the free-molecule flow range, the drag on a sphere is given (S11), for diffuse molecular reflection† ($\sigma_R = 1$), by:

$$C_{\mathrm{Dfm}} = \frac{(1 + 2s^2)\exp(-s^2/2)}{\sqrt{\pi}s^3} + \frac{(4s^4 + 4s^2 - 1)\,\mathrm{erf}(s)}{2s^4} + \frac{2}{3s}\sqrt{\frac{\pi T_p}{T_\infty}}, \quad (10\text{-}59)$$

† The last term is absent for specular reflection. However, Eq. (10-59) is normally the applicable form.

where s is the "molecular speed ratio":

$$s = \mathrm{Ma}\sqrt{k/2}. \tag{10-60}$$

Crowe *et al.* (C10) developed a semiempirical procedure for estimating sphere drag coefficients as a function of Re and Ma, using C_{Dfm} and C_{D0}, the standard drag coefficient for continuum flow at the appropriate Re, calculated as in Chapter 5. A so-called "inviscid drag coefficient" is first estimated, corresponding to the hypothetical drag at large (but subcritical) Re:

$$C_{\mathrm{DI}} = 0.66 + 0.26\tanh(2\ln \mathrm{Ma}) + 0.17\exp[-2.5(\ln \mathrm{Ma}/1.4)^2]. \tag{10-61}$$

The drag coefficient for the case in question is then estimated as

$$C_{\mathrm{D}} = fC_{\mathrm{Dfm}} + (1 - f)C_{\mathrm{DI}}, \tag{10-62}$$

where the ratio f is:

$$f = \frac{B}{1+B}\{1 - \exp[-\mathrm{Re}\,\mathrm{Kn}^{0.6}e^{\mathrm{Kn}}(C_{\mathrm{D0}} - 0.4)/8]\} \tag{10-63}$$

and

$$B = \mathrm{Kn}^{0.4}\exp(1.2\sqrt{\mathrm{Kn}}). \tag{10-64}$$

This procedure has some advantage over the results presented graphically in Fig. 10.15 in that it allows for values of $T_{\mathrm{p}}/T_{\infty}$ other than unity via its effect on C_{Dfm}. Walsh (W1) and Henderson (H7) have also proposed empirical equations for C_{D} as a function of Ma and Re.

The results presented above are for cases where the fluid may be considered isothermal, except as outlined in the preceding paragraph where there may be gradients near the particle associated with aerodynamic heating of the particle itself. Another situation in which compressibility effects are important arises when there are substantial property gradients in the fluid through which a particle is moving. This case has been considered theoretically and experimentally by several workers [e.g. (C3, K1, S24)] for low Re, low Ma motion of spheres in gases with marked temperature gradients leading to simultaneous momentum and heat transfer. For example, Seymour (S24) carried out a numerical and experimental study of the motion of small spheres in an argon plasma at temperatures up to 13000 K for a sphere surface temperature of 2000 K and Re varying between 0.3 and 1.5. Not surprisingly, characterization of gas properties and experimental drag determination become very difficult under such extreme conditions.

Wortman (W8) has given an approximate method of extending results for spheres to other shapes which undergo random tumbling. The method requires calculation of an effective radius of curvature, using kinetic theory (H11) to define an equivalent cross section. The only restriction, aside from Ma > 1, is that the flight must last for a sufficiently long period that there is no statistically preferred orientation. This approach gave good agreement with supersonic data for cubes, especially in the hypersonic range, but it has not been confirmed for other shapes. Nevertheless, it appears to be the best method available for estimating the drag at supersonic velocities for shapes for which no direct measurements are available, especially if the particle is not far removed from spherical in shape.

C. Effect on Heat Transfer

Analogous to the slip velocity between gas and particle at Kn above the continuum flow range discussed in Section A above, a temperature discontinuity exists close to the surface at high Kn. Such a discontinuity represents an additional resistance to transfer. Hence, transfer rates are generally lowered by compressibility and noncontinuum effects. The temperature jump occurs over a distance $1.996k\lambda(2 - \sigma_t)/\mathrm{Pr}\,\sigma_t(k + 1)$ (K2, S11) where σ_t is the "thermal accommodation coefficient," interpreted as the extent to which the thermal energy of reflected molecules has adjusted to the surface temperature.

For small particles, subject to noncontinuum effects but not to compressibility, Re is very low; see Eq. (10-52). In this case, nonradiative heat transfer occurs purely by conduction. This situation has been examined theoretically in the near-free-molecule limit (S14) and in the near-continuum limit (T8). The following equation interpolates between these limits for a sphere in a motionless gas:

$$\mathrm{Nu} = 2b/(1 + b), \tag{10-65}$$

where

$$b = \frac{\sigma_t(1 + k)}{9k - 5}\,\mathrm{Kn}^{-1}. \tag{10-66}$$

Equations (10-65) and (10-66) give a good fit to data of Takao (T1) for heat transfer from a brass sphere to air with $\sigma_t = 0.8$. Natural convection at low pressures has been studied for cylinders and spheres (D9, K12).

At high Re and Ma in the free-molecule regime, transfer rates for spheres have been calculated by Sauer (S4). These results, together with others for cylinders and plates, have been summarized by Schaaf and Chambré (S11). The particles are subject to aerodynamic heating and the heat transfer coefficients are based upon the difference between the particle surface temperature and the recovery temperature (see standard aerodynamics texts). In the transitional region, the semiempirical result of Kavanau (K2),

$$\frac{\mathrm{Nu}^*}{\mathrm{Nu}} = 1 + 3.42\,\frac{\mathrm{Ma}\,\mathrm{Nu}^*}{\mathrm{Re}\,\mathrm{Pr}} = 1 + 2.73\,\frac{\mathrm{Kn}\,\mathrm{Nu}^*}{\mathrm{Pr}\sqrt{k}}, \tag{10-67}$$

may be used to relate the Nusselt number Nu to its value Nu* in continuum flow ($\mathrm{Kn} \to 0$) at the same Re and Pr for $0.1 < \mathrm{Ma} < 0.7$ and $2 < \mathrm{Re} < 120$.

Alternatively, one can use the interpolation formula suggested by Sherman (S26),

$$\mathrm{Nu_{fm}/Nu = 1 + (Nu_{fm}/Nu^*)}, \tag{10-68}$$

where $\mathrm{Nu_{fm}}$ is the Nusselt number in the free-molecule limit at the given Ma [see (S11)]. The latter equation has been found to give good predictions for cylinders as well as spheres. Equation (10-65) is a special case of Eq. (10-68) for Re = 0. A more complete review of heat transfer in rarefied gases is given by Springer (S32) who also covers the transition region at Ma greater than unity.

REFERENCES

A1. Achenbach, E., *J. Fluid Mech.* **46**, 321–335 (1971).
A2. Achenbach, E., *J. Fluid Mech.* **65**, 113–125 (1974).
A3. Achenbach, E., *Heat Transfer 1974* **2**, 229–233 (1974).
A4. Acrivos, A., *AIChE J.* **4**, 285–289 (1958).
A5. Acrivos, A., *AIChE J.* **6**, 584–590 (1960).
A6. Ahmadi, G., and Goldschmidt, V. W., *J. Appl. Mech.* **93**, 561–563 (1971).
A7. Amato, W. S., and Tien, C., *Int. J. Heat Mass Transfer* **15**, 327–339 (1972).
A8. Aufdermaur, A. N., and Joss, J., *Z. Angew. Math. Phys.* **18**, 852–866 (1967); **19**, 377 (1968).
A9. Austin, L. J., Ying, W. E., and Sawistowski, H., *Chem. Eng. Sci.* **21**, 1109–1110 (1966).
A10. Austin, L. J., Banczyk, L., and Sawistowski, H., *Chem. Eng. Sci.* **26**, 2120–2121 (1971).
B1. Bailey, A. B., and Hiatt, J., *Rep.* AEDC-TR-70-291. Arnold Eng. Dev. Cent., Tenn. (1971).
B2. Bailey, A. B., and Hiatt, J., *AIAA J.* **10**, 1436–1440 (1972).
B3. Baker, J. L. L., and Chao, B. T., *AIChE J.* **11**, 268–273 (1965).
B4. Bakker, C. A. P., Fentener, F. H., van Vlissingen, F. H. F., and Beek, W. J., *Chem. Eng. Sci.* **22**, 1349–1355 (1967).
B5. Bakker, C. A. P., van Buytenen, P. M., and Beek, W. J., *Chem. Eng. Sci.* **21**, 1039–1046 (1966).
B6. Barkla, H. M., *Ann. Sci.* **30**, 107–122 (1973).
B7. Barkla, H. M., and Auchterlonie, L. J., *J. Fluid Mech.* **47**, 437–447 (1971).
B8. Bartok, W., and Mason, S. G., *J. Colloid Sci.* **12**, 243–262 (1957).
B9. Basset, A. B., "A Treatise on Hydrodynamics," Vol. 2, Chap. 22. Deighton Bell, Cambridge, England, 1888. (Republished: Dover, New York, 1961.)
B10. Beard, K. V., *J. Atmos. Sci.* **33**, 851–864 (1976).
B11. Beitel, A., and Heideger, W. J., *Chem. Eng. Sci.* **26**, 711–717 (1971).
B12. Berg, J. C., and Morig, C. R., *Chem. Eng. Sci.* **24**, 937–946 (1969).
B13. Börner, H., *VDI-Forschungsh.* No. 512 (1965).
B14. Bogue, B. A., Myerson, A., and Kirwan, D. J., *Ind. Eng. Chem., Fundam.* **14**, 282 (1975).
B15. Bretherton, F. P., *J. Fluid Mech.* **14**, 284–304 (1962).
B16. Brian, P. L. T., *AIChE J.* **17**, 765–772 (1971).
B17. Brian, P. L. T., and Ross, J. R., *AIChE J.* **18**, 582–591 (1972).
B18. Brian, P. L. T., and Smith, K. A., *AIChE J.* **18**, 231–233 (1972).
B19. Brimacombe, J. K., Graves, A. D., and Inman, D., *Chem. Eng. Sci.* **25**, 1817–1818 (1970).
C1. Catton, I., and Schwartz, S. H., *J. Spacecr. Rockets* **9**, 468–471 (1972).
C2. Chandrasekhar, S., and Hoelscher, H. E., *AIChE J.* **21**, 103–109 (1975).
C3. Chang, I. D., *Z. Angew. Math. Phys.* **16**, 449–469 (1965).
C4. Chang, P. K., "Separation of Flow." Pergamon, New York, 1970.
C5. Churchill, S. W., and Churchill, R. U., *AIChE J.* **21**, 604–607 (1975).

C6. Clamen, A., and Gauvin, W. H., *AIChE J.* **15**, 184–189 (1969).
C7. Clift, R., and Gauvin, W. H., *Proc. Chemeca '70* Butterworths, Melbourne **1**, 14–28 (1970).
C8. Clift, R., and Gauvin, W. H., *Can. J. Chem. Eng.* **49**, 439–448 (1971).
C9. Cremers, C. J., and Finley, D., *Heat Transfer 1970* **4**, Paper NC 1.5, (1970).
C10. Crowe, C. T., Babcock, W. R., and Willoughby, P. G., *Prog. Heat Mass Transfer* **6**, 419–431 (1972).
D1. Dahneke, B. E., *Aerosol Sci.* **4**, 139–170 (1973).
D2. Davies, C. N., *Proc. Phys. Soc., London* **57**, 259–270 (1945).
D3. Davies, J. M., *J. Appl. Phys.* **20**, 821–828 (1949).
D4. Davies, J. T., *Adv. Chem. Eng.* **4**, 1–50 (1963).
D5. Davies, J. T., "Turbulence Phenomena." Academic Press, New York, 1972.
D6. Davies, J. T., and Rideal, E. K., "Interfacial Phenomena." Academic Press, New York, 1963.
D7. Davies, J. T., and Wiggill, J. B., *Proc. R. Soc., Ser. A* **255**, 277–291 (1960).
D8. Devienne, M., "Frottement et échanges thermiques dans las gaz raréfiés." Gauthier-Villars, Paris, 1958.
D9. Drake, R. M., Smith, C. G., and Stathakis, G. J., *in* "Heat Transfer, Thermodynamics and Education: Boelter Anniversary Volume" (H. A. Johnson, ed.). McGraw-Hill, New York, 1964.
D10. Drazin, M. P., *Proc. Cambridge Philos. Soc.* **47**, 142–145 (1951).
D11. Dryden, H. L., Schubauer, G. B., Mock, W. G., and Skramstad, H. K., *NACA Rep.* No. 581 (1937).
E1. Eastop, T. D., *Int. J. Heat Mass Transfer* **16**, 1954–1957 (1973).
E2. Eastop, T. D., Ph.D. Thesis, C.N.A.A., London, 1971.
E3. Elenbaas, W., *Physica* **9**, 285–296 (1942).
E4. Ellis, S. R. M., and Biddulph, M., *Chem. Eng. Sci.* **21**, 1107–1109 (1966).
E5. Epstein, P. S., *Phys. Rev.* **23**, 710–733 (1924).
F1. Fage, A., and Warsap, J. H., *Aero. Res. Counc. R. M. No.* 1283 (1930).
F2. Fand, R. M., and Keswani, K. K., *Int. J. Heat Mass Transfer* **16**, 1175–1191 (1973).
F3. Forgacs, O. L., and Mason, S. G., *J. Colloid Sci.* **14**, 457–472 (1959).
F4. Frankel, N. A., and Acrivos, A., *Phys. Fluids* **11**, 1913–1918 (1968).
F5. Friedlander, S. K., *AIChE J.* **3**, 381–385 (1957).
F6. Furuta, T., Jimbo, T., Okazaki, M., and Toei, R., *J. Chem. Eng. Jpn.* **8**, 456–462 (1975).
G1. Galloway, T. R., *AIChE J.* **19**, 608–617 (1973).
G2. Galloway, T. R., and Sage, B. H., *AIChE J.* **13**, 563–570 (1967).
G3. Galloway, T. R., and Sage, B. H., *Int. J. Heat Mass Transfer* **10**, 1195–1210 (1967).
G4. Galloway, T. R., and Sage, B. H., *Int. J. Heat Mass Transfer* **11**, 539–549 (1968).
G5. Galloway, T. R., and Sage, B. H., *AIChE J.* **18**, 287–293 (1972).
G6. Galloway, T. R., and Seid, D. M., *Lett. Heat Mass Transfer* **2**, 247–257 (1975).
G7. Garner, F. H., and Hoffman, J. M., *AIChE J.* **6**, 579–584 (1960).
G8. Garner, F. H., and Hoffman, J. M., *AIChE J.* **7**, 148–152 (1961).
G9. Garstang, T. E., *Proc. R. Soc., Ser. A* **142**, 491–508 (1933).
G10. Gebhart, B., *Adv. Heat Transfer* **9**, 273–348 (1973).
G11. Goldsmith, H. L., and Mason, S. G., *J. Fluid Mech.* **12**, 88–96 (1962).
G12. Goldsmith, H. L., and Mason, S. G., *in* "Rheology, Theory and Applications" (F. R. Eirich, ed.), Vol. 4, pp. 85–250. Academic Press, New York, 1967.
G13. Goldstein, R. G., Sparrow, E. M., and Jones, D. C., *Int. J. Heat Mass Transfer* **16**, 1025–1035 (1973).
G14. Goldstein, S., "Modern Developments in Fluid Dynamics." Oxford Univ. Press, London and New York, 1957.
G15. Goodridge, F., and Robb, I. D., *Ind. Eng. Chem., Fundam.* **4**, 49–55 (1965).
H1. Handlos, A. E., and Baron, T., *AIChE J.* **3**, 127–136 (1957).

H2. Hanel, G., *Atmos. Environ.* **4**, 280–300 (1970).
H3. Happel, J., and Brenner, H., "Low Reynolds Number Hydrodynamics," 2nd ed. Noordhoff, Leyden, Netherlands, 1973.
H4. Harper, E. Y., and Chang, J.-D., *J. Fluid Mech.* **33**, 209–225 (1968).
H5. Hatton, A. P., James, D. D., and Swire, H. W., *J. Fluid Mech.* **42**, 17–31 (1970).
H6. Heines, H., and Westwater, J. W., *Int. J. Heat Mass Transfer* **15**, 2109–2117 (1972).
H7. Henderson, C. B., *AIAA J.* **14**, 707–708 (1976).
H8. Henry, H. C., and Epstein, N., *Can. J. Chem. Eng.* **48**, 595–616 (1970).
H9. Hieber, C. A., and Gebhart, B., *J. Fluid Mech.* **38**, 137–159 (1969).
H10. Hinze, J. O., "Turbulence," 2nd ed. McGraw-Hill, New York, 1975.
H11. Hirschfelder, J. O., Curtiss, C. F., and Bird, R. B., "Molecular Theory of Gases and Liquids." Wiley, New York, 1954.
H12. Hjelmfelt, A. T., and Mockros, D. F., *Appl. Sci. Res. A* **16**, 149–161 (1966).
H13. Ho, B. P., and Leal, L. G., *J. Fluid Mech.* **65**, 365–400 (1974).
H14. Hoerner, S. F., "Fluid Dynamic Drag." Published by the author, Midland Park, New Jersey, 1958.
H15. Huang, H. S., and Winnick, J., *AIChE J.* **22**, 205–206 (1976).
J1. Jaluria, Y., and Gebhart, B., *Int. J. Heat Mass Transfer* **18**, 415–431 (1975).
J2. Jeffery, G. B., *Proc. R. Soc., Ser. A* **102**, 161–179 (1922).
J3. Joss, J., and Aufdermaur, A. N., *Int. J. Heat Mass Transfer* **13**, 213–215 (1970).
K1. Kassoy, D. R., Adamson, T. C., and Messiter, A. F., *Phys. Fluids* **9**, 671–681 (1966).
K2. Kavanau, L. L., *Trans. ASME* **77**, 617–624 (1955).
K3. Kestin, J., *Adv. Heat Transfer* **3**, 1–32 (1966).
K4. Kestin, J., and Wood, R. T., *J. Heat Transfer* **93**, 321–326 (1971).
K5. King, W. J., *Mech. Eng.* **54**, 347–353 (1932).
K6. Knudsen, M., and Weber, S., *Ann. Phys.* **36**, 981–994 (1911).
K7. Kojima, E., Akehata, T., and Shirai, T., *J. Chem. Eng. Jpn.* **8**, 108–113 (1975).
K8. Krasnov, N. F., "Aerodynamics of Bodies of Revolution." Elsevier, Amsterdam, 1970.
K9. Kuboi, R., Komasawa, I., and Otake, T., *Chem. Eng. Sci.* **29**, 651–657 (1974).
K10. Kudryashev, L. I., and Ipatenko, A. Y., *Sov. Phys.-Tech. Phys.* **4**, 275–284 (1959).
K11. Kussoy, M. I., Steward, D. A., and Horstmann, C. C., *AIAA J.* **9**, 1434–1435 (1971).
K12. Kyte, J. R., Madden, A. J., and Piret, E. L., *Chem. Eng. Prog.* **49**, 653–662 (1953).
L1. LaMer, V. K., ed., "Retardation of Evaporation by Monolayers." Academic Press, New York, 1962.
L2. Levins, D. M., and Glastonbury, J. R., *Trans. Inst. Chem. Eng.* **50**, 32–41 (1972).
L3. Lin, C. J., Peery, J. H., and Schowalter, W. R., *J. Fluid Mech.* **44**, 1–17 (1970).
L4. Lin, F. N., and Chao, B. T., *J. Heat Transfer* **96**, 435–442 (1974).
L5. Linde, H., *Proc. Int. Congr. Surf. Active Substances, 4th, Brussels, 1964* **2**, 1301–1309 (1967).
L6. Linde, H., and Schwarz, E., *Z. Phys. Chem.* **224**, 331–352 (1963).
L7. Linde, H., Schwarz, E., and Gröger, K., *Chem. Eng. Sci.* **22**, 823–836 (1967).
L8. Liu, V.-C., *J. Meteorol.* **13**, 399–405 (1956).
L9. Lloyd, J. R., and Moran, W. R., *J. Heat Transfer* **96**, 443–447 (1974).
L10. Lloyd, J. R., and Sparrow, E. M., *J. Fluid Mech.* **42**, 465–470 (1970).
L11. Lode, T., and Heideger, W. J., *Chem. Eng. Sci.* **25**, 1081–1090 (1970).
L12. Lumley, J., Ph.D. Thesis, John Hopkins Univ., Baltimore, Maryland, 1957.
L13. Luthander, S., and Rydberg, A., *Phys. Z.* **36**, 552–558 (1935).
M1. Maccoll, J. W., *J. R. Aeronaut. Soc.* **32**, 777–798 (1928).
M2. Magnus, G., *Poggendorffs Ann. Phys. Chem.* **88**, 1 (1853).
M3. Maroudas, N. G., and Sawistowski, H., *Chem. Eng. Sci.* **19**, 919–931 (1964).
M4. Mason, P. J., *J. Fluid Mech.* **71**, 577–599 (1975).
M5. Mason, S. G., and Manley, R. S., *Proc. R. Soc., Ser. A* **238**, 117–131 (1956).
M6. Mathers, W. G., Madden, A. J., and Piret, E. L., *Ind. Eng. Chem.* **49**, 961–968 (1957).

M7. Maxworthy, T., *J. Fluid Mech.* **31**, 643–655 (1968).
M8. Miles, J. W., *J. Fluid Mech.* **53**, 689–700 (1972).
M9. Miles, J. W., *J. Fluid Mech.* **72**, 363–371 (1975).
M10. Mizushina, T., Ueda, H., and Umemiya, N., *Int. J. Heat Mass Transfer* **15**, 769–780 (1972).
M11. Moe, M. M., and Tsang, L. C., *AIAA J.* **11**, 396–399 (1973).
M12. Moll, H. G., *Ing.-Arch.* **42**, 215–224 (1973).
M13. Moore, D. W., and Saffman, P. G., *J. Fluid Mech.* **31**, 635–642 (1968).
M14. Mudge, L. K., and Heideger, W. J., *AIChE J.* **16**, 602–608 (1970).
M15. Mujumdar, A. S., Ph.D. Thesis, McGill Univ., Montreal, 1971.
N1. Narasimhan, C., and Gauvin, W. H., *Can. J. Chem. Eng.* **45**, 181–188 (1967).
N2. Newman, L. B., Sparrow, E. M., and Eckert, E. R. G., *J. Heat Transfer* **94**, 7–15 (1972).
N3. Noordzij, M. P., and Rotte, J. W., *Chem. Eng. Sci.* **23**, 657–660 (1968).
O1. Oosthuizen, P. H., and Madan, S., *J. Heat Transfer* **93**, 240–242 (1971).
O2. Orell, A., and Westwater, J. W., *AIChE J.* **8**, 350–356 (1962).
O3. Ostrovskii, M. V., Frumin, G. T., Kremnev, L. Y., and Abramzon, A. A., *J. Appl. Chem. USSR* **40**, 1267–1274 (1967).
O4. Otto, W., Streicher, R., and Schügerl, K., *Chem. Eng. Sci.* **28**, 1777–1788 (1973).
P1. Pearson, R. S., and Dickson, P. F., *AIChE J.* **14**, 903–908 (1968).
P2. Pei, D. C. T., *Chem. Eng. Prog., Symp. Ser.* **61**(59), 57–63 (1965).
P3. Pera, L., and Gebhart, B., *Int. J. Heat Mass Transfer* **15**, 175–177 (1972).
P4. Pera, L., and Gebhart, B., *Int. J. Heat Mass Transfer* **16**, 1147–1163 (1974).
P5. Pérez de Ortiz, E. S., and Sawistowski, H., *Chem. Eng. Sci.* **28**, 2051–2061 (1973); **30**, 2063–2069 (1975).
P6. Pérez de Ortiz, E. S., and Sawistowski, H., *Chem. Eng, Sci.* **30**, 1527–1528 (1975).
P7. Petrak, D., *Chem. Tech.* (*Leipzig*) **28**, 591–595 (1976).
P8. Petrie, A. M., and Simpson, H. C., *Int. J. Heat Mass Transfer* **15**, 1497–1513 (1972).
P9. Phillips, W. F., *Phys. Fluids* **18**, 1089–1093 (1975).
P10. Pich, J., *in* "Aerosol Science" (C. N. Davies, ed.), pp. 223–285. Academic Press, New York, 1966.
P11. Plevan, R. E., and Quinn, J. A., *AIChE J.* **12**, 894–902 (1966).
P12. Poe, G. G., and Acrivos, A., *J. Fluid Mech.* **72**, 605–623 (1975).
P13. Poe, G. G., and Acrivos, A., *Int. J. Multiphase Flow* **2**, 365–377 (1976).
P14. Preining, O., *Atmos. Environ.* **1**, 271–286 (1967).
R1. Raithby, G. D., *Int. J. Heat Mass Transfer* **14**, 1875 (1971).
R2. Raithby, G. D., and Eckert, E. R. G., *Int. J. Heat Mass Transfer* **11**, 1233–1252 (1968).
R3. Rayleigh, Lord, *Sci. Pap.* **1**, No. 53 (1877).
R4. Redekopp, L. G., and Charwat, A. F., *J. Hydronaut.* **6**, 34–39 (1972).
R5. Rosenthal, D. K., *J. Fluid Mech.* **12**, 358–366 (1962).
R6. Rubinow, S. I., and Keller, J. B., *J. Fluid Mech.* **11**, 447–459 (1961).
S1. Sada, E., and Himmelblau, D. M., *AIChE J.* **13**, 860–865 (1967).
S2. Saffman, P. G., *J. Fluid Mech.* **22**, 385–400 (1965); **31**, 624 (1968).
S3. Sarma, L. V. K. V., and Rao, C. V. R., *Bull. Acad. Pol. Sci., Ser. Sci. Tech.* **21**, 451–459 (1973).
S4. Sauer, F. M., *J. Aerosol. Sci.* **18**, 353–354 (1951).
S5. Savery, C. W., and Borman, G. L., *Can. J. Chem. Eng.* **48**, 335–336 (1970).
S6. Saville, D. A., and Churchill, S. W., *AIChE J.* **16**, 268–273 (1970).
S7. Sawistowski, H., *in* "Recent Advances in Liquid-Liquid Extraction" (C. Hanson, ed.), pp. 293–366. Pergamon, Oxford, 1971.
S8. Sawistowski, H., and Goltz, G. E., *Trans. Inst. Chem. Eng.* **41**, 174–181 (1963).
S9. Sawistowski, H., and James, B. R., *Chem.-Ing.-Tech.* **35**, 175–179 (1963).
S10. Sawistowski, H., and James, B. R., *Proc. Int. Conf. Surf. Active Mater., 3rd* pp. 757–772 (1967).

S11. Schaaf, S. A., and Chambré, P. L., *in* "Fundamentals of Gas Dynamics" (H. W. Emmons, ed.), High Speed Aerodynamics and Jet Propulsion, Vol. 3, pp. 687–739. Princeton Univ. Press, Princeton, New Jersey, 1958.
S12. Schenkels, F. A. M., and Schenk, J., *Chem. Eng. Sci.* **24**, 585–593 (1969).
S13. Schlichting, H., "Boundary Layer Theory," 6th ed. McGraw-Hill, New York, 1968.
S14. Schmulian, R. J., *Rarefied Gas Dyn., Proc. Int. Symp., 6th* pp. 735–738 (1969).
S15. Schuepp, P. H., *J. Appl. Meteorol.* **10**, 1018–1025 (1971).
S16. Schuepp, P. H., and List, R., *J. Appl. Meteorol.* **8**, 254–263 (1969).
S17. Schuepp, P. H., and List, R., *J. Appl. Meteorol.* **8**, 743–746 (1969).
S18. Schütz, G., *Int. J. Heat Mass Transfer* **6**, 873–879 (1963).
S19. Scoggins, J. R., *J. Geophys. Res.* **69**, 591–598 (1964).
S20. Scoggins, J. R., *NASA Tech. Note* **D-3994** (1967).
S21. Seeley, L. E., Ph.D. Thesis, Univ. of Toronto, 1972.
S22. Sehrt, B., and Linde, H., Proc. *Int. Conf. Surf. Active Mater., 3rd* pp. 734–747 (1967).
S23. Selberg, B. P., and Nicholls, J. A., *AIAA J.* **6**, 401–408 (1968).
S24. Seymour, E. V., *J. Appl. Mech.* **93**, 739–748 (1971).
S25. Shapiro, A. H., "The Dynamics and Thermodynamics of Compressible Fluid Flow." Ronald Press, New York, 1953.
S26. Sherman, F. S., *Rarefied Gas Dyn., Proc. Int. Symp., 3rd* pp. 228–260 (1963).
S27. Sherwood, T. K., and Wei, J. C., *Ind. Eng. Chem.* **49**, 1030–1034 (1957).
S28. Sivier, K. R., and Nicholls, J. A., *NASA Contract. Rep.* **CR-1392** (1969).
S29. Smith, M. C., and Kuethe, A. M., *Phys. Fluids* **9**, 2337–2344 (1966).
S30. Soo, S. L., *Chem. Eng. Sci.* **5**, 57–67 (1956).
S31. Sparrow, E. M., and Gregg, J. L., *Trans ASME* **80**, 879–886 (1958).
S32. Springer, G. S., *Adv. Heat Transfer* **7**, 163–218 (1971).
S33. Sternling, C. V., and Scriven, L. E., *AIChE J.* **5**, 514–523 (1959).
S34. Stewart, W. E., *Int. J. Heat Mass Transfer* **14**, 1013–1031 (1971).
S35. Stewartson, K., "The Theory of Laminar Boundary Layers in Compressible Fluids." Oxford Univ. Press (Clarendon), London and New York, 1964.
T1. Takao, K., *Rarefied Gas Dyn., Proc. Int. Symp., 3rd* pp. 102–111 (1963).
T2. Tanaka, H., and Tago, O., *Kagaku Kogaku* **37**, 151–157 (1973).
T3. Taneda, S., *Rep. Res. Inst. Appl. Mech.* (*Kyushu*) **5**, 123–128 (1957).
T4. Taunton, J. W., Lightfoot, E. N., and Stewart, W. E., *Chem. Eng. Sci.* **25**, 1927–1938 (1970).
T5. Taylor, Sir G. I., *Proc. R. Soc., Ser. A* **100**, 114–121 (1922).
T6. Taylor, Sir G. I., *Proc. R. Soc., Ser. A* **104**, 213–218 (1923).
T7. Taylor, Sir G. I., *Proc. R. Soc., Ser. A* **156**, 307–317 (1936).
T8. Taylor, T. D., *Phys. Fluids* **6**, 987–992 (1963).
T9. Thomas, W. J., and Nicholl, E. McK., *Chem. Eng. Sci.* **22**, 1877–1878 (1967).
T10. Thompson, P. A., "Compressible Fluid Dynamics." McGraw-Hill, New York, 1971.
T11. Torobin, L. B., and Gauvin, W. H., *Can. J. Chem. Eng.* **38**, 142–153 (1960).
T12. Torobin, L. B., and Gauvin, W. H., *AIChE J.* **7**, 615–619 (1961).
T13. Traci, R. M., and Wilcox, D. C., *AIAA J.* **13**, 890–896 (1975).
T14. Traher, A., and Kirwan, D. J., *Ind. Eng. Chem., Fundam.* **12**, 244–245 (1973).
T15. Trevelyan, B. J., and Mason, S. G., *J. Colloid Sci.* **6**, 354–367 (1951).
U1. Uhlherr, P. H. T., and Sinclair, C. G., *Proc. Chemeca '70* Butterworths, Melbourne, **1**, 1–13 (1970).
V1. Van der Hegge Zijnen, B. G., *Appl. Sci. Res., Sect. A* **7**, 205–233 (1958).
W1. Walsh, M. J., *AIAA J.* **13**, 1526–1528 (1975).
W2. Wassel, A. T., and Denny, V. E., *Lett. Heat Mass Transfer* **3**, 375–386 (1976).
W3. West, F. B., Herrman, A. J., Chong, A. T., and Thomas, L. E. K., *Ind. Eng. Chem.* **44**, 625–631 (1952).
W4. Wilhelm, R., *VDI-Forschungsh.* No. 531 (1969).

W5. Wilhelm, R., *Chem.-Ing.-Tech.* **43**, 47–49 (1971).

W6. Willis, J. T., Browning, K. A., and Atlas, D., *J. Atmos. Sci.* **21**, 103–108 (1964).

W7. Woo, S. W., Ph.D. Thesis, McMaster Univ., Hamilton, Ontario, 1971.

W8. Wortman, A., *J. Spacecr. Rockets* **6**, 205–207 (1969).

Y1. Yuge, T., *J. Heat Transfer* **82**, 214–220 (1960).

Z1. Zarin, N. A., *NASA Contract. Rep.* No. **CR-1585** (1970).

Z2. Zarin, N. A., and Nicholls, J. A., *Combust. Sci. Technol.* **3**, 273–285 (1971).

Chapter 11

Accelerated Motion without Volume Change

I. INTRODUCTION

Prediction of fluid motion, drag, and transfer rates becomes much more complex when the motion is unsteady. Dimensional analysis gives an indication of the problems. A rigid sphere moving with steady velocity in a gravitational field is governed by an equation of the general form

$$f[g\,\Delta\rho, \rho, \mu, d, U_{\mathrm{R}}] = 0. \tag{11-1}$$

Since there are three dimensions, two dimensionless groups, e.g., N_{D} and N_{U} defined in Chapter 5, suffice to describe the motion. If the motion is unsteady, it is necessary to introduce the particle density explicitly, since it determines the particle inertia as well as the net gravity force. Also, since U_{R} varies with time and position, a further parameter must be introduced. This may be the distance x moved since the start of the motion. Equation (11-1) is then replaced by

$$f[g, \mu, \rho, d, U_{\mathrm{R}}, \rho_{\mathrm{p}}, x] = 0. \tag{11-2}$$

Two additional dimensionless groups are therefore required. These can be chosen as the density ratio[†] γ and displacement modulus, $M_{\mathrm{D}} = x/d$. Hence

$$f[N_{\mathrm{D}}, N_{\mathrm{U}}, \gamma, M_{\mathrm{D}}] = 0. \tag{11-3}$$

The additional complexity is not limited to introduction of two new groups. For example, Eq. (11-3) takes different forms for a particle accelerating from rest and a particle projected in a stagnant fluid. In principle, all time derivatives

[†] It has already been noted in Chapters 5 to 7, that $\gamma = \rho_{\mathrm{p}}/\rho$ must be included when secondary motion is superimposed on steady particle translation.

of U_R should be included in Eq. (11-3), leading to a series of new groups, of which the first is the acceleration modulus, $M_A = (dU_R/dt)d/U_R^2$.

In this chapter we first discuss the equations of motion for particles at low Re. Semiempirical extensions beyond the creeping flow regime are then considered. It is useful to distinguish two general kinds of unsteady motion:

Type 1 Characterized by rapid change of Re with M_D.
Type 2 Characterized by slow change of Re with M_D with the instantaneous flow pattern similar to that in steady motion at the instantaneous Re.

For Type 2 motions, drag is insensitive to acceleration, and calculation of the motion is greatly simplified. For Type 1 motions, the instantaneous drag may differ radically from the corresponding steady drag. In practice, Type 2 usually corresponds to particles moving through gases (high γ), whereas Type 1 generally describes motion in liquids.

Although a number of workers [e.g. (C5, D1, D2)] have considered flow around particles started impulsively from rest at constant nonzero velocities, this situation is of little practical interest. Attention is concentrated on free fall from rest and oscillatory motion.

II. INITIAL MOTION

A. General Considerations

As for steady motion, analytic solutions for unsteady motion of rigid and fluid particles are available only in creeping flow. The solution was developed by Basset (B3); the outline given here follows the treatment of Landau and Lifshitz (L4). The governing equation is the unsteady form of Eq. (1-36), i.e., for axisymmetric flow,

$$\partial(E^2\psi)/\partial t = E^4\psi. \tag{11-4}$$

The boundary conditions are the same as for steady motion considered in Chapters 1, 3, and 4, i.e., uniform flow remote from the particle, no slip and no normal flow at the particle boundary, and, for fluid particles, continuity of tangential stress at the interface. For a sphere the normal stress condition at the interface is again formally redundant, but indicates whether a fluid particle will remain spherical.

Consider a rigid sphere of radius a, executing rectilinear oscillatory motion relative to remote fluid with its velocity given by†:

$$U(t) = U_\omega e^{-i\omega t}. \tag{11-5}$$

By analogy with the solution for steady creeping flow, we assume that the stream function relative to the particle takes the form

$$\psi = f(r)e^{-i\omega t}\sin^2\theta. \tag{11-6}$$

† In Eqs. (11-5) to (11-10), only the real part is to be considered.

The form of $f(r)$ is obtained from Eq. (11-4) and the boundary conditions leading to

$$\psi = -U\frac{\sin^2\theta}{2}\left[r^2 - \frac{a^3}{r} - \frac{3a\delta}{2}\left\{\frac{(1+i)a + i\delta}{r} - \left(1 + i + \frac{i\delta}{r}\right)e^{(i-1)(r-a)/\delta}\right\}\right], \tag{11-7}$$

where

$$\delta = \sqrt{2\nu/\omega}. \tag{11-8}$$

For $\omega \to 0$, Eq. (11-7) reduces to the stream function for steady creeping flow past a rigid sphere, i.e., Eq. (3-7) with $\kappa = \infty$. The parameter δ may be regarded as a characteristic length scale for diffusion of vorticity generated at the particle surface into the surrounding fluid. When ω is very large, δ is small, and the flow can be considered irrotational except in the immediate vicinity of the particle. In the limit $\omega \to \infty$, Eq. (11-7) reduces to Eq. (1-29), the result for potential flow past a stationary sphere.

The total drag on the sphere may be obtained, as in steady flow, by integrating the normal and shear stresses over the surface. In terms of the instantaneous velocity U the result is (L4):

$$-F_{\mathrm{D}} = 6\pi\mu a U + \frac{V}{2}\rho\frac{dU}{dt} + 3\pi a^2\left[\frac{2\mu U}{\delta} + \delta\rho\frac{dU}{dt}\right], \tag{11-9}$$

where V is the volume of the sphere. Although the above results refer to pure oscillatory motion, they may be generalized to arbitrary rectilinear motion by representing the velocity as a Fourier integral:

$$U(t) = \int_{-\infty}^{\infty} U_\omega e^{-i\omega t}\, d\omega. \tag{11-10}$$

Equation (11-9) then becomes (L4)

$$-F_{\mathrm{D}} = 6\pi\mu a U + \frac{V\rho}{2}\frac{dU}{dt} + 6a^2\sqrt{\pi\rho\mu}\int_{-\infty}^{t}\left(\frac{dU}{dt}\right)_{t=s}\frac{ds}{\sqrt{t-s}}. \tag{11-11}$$

The first term of Eq. (11-11) is the Stokes drag for steady motion at the instantaneous velocity. The second term is the "added mass" or "virtual mass" contribution which arises because acceleration of the particle requires acceleration of the fluid. The volume of the "added mass" of fluid is $0.5V$, the same as obtained from potential flow theory. In general, the instantaneous drag depends not only on the instantaneous velocities and accelerations, but also on conditions which prevailed during development of the flow. The final term in Eq. (11-11) includes the "Basset history integral," in which past acceleration is included, weighted as $(t-s)^{-1/2}$, where $(t-s)$ is the time elapsed since the past acceleration. The form of the history integral results from diffusion of vorticity from the particle.

Equation (11-11) depends on neglect of inertial terms in the Navier–Stokes equation. Neglect of inertia terms is often less serious for unsteady motion than for steady flow since the convective acceleration term is small both for $\mathrm{Re} \to 0$ (Chapters 3 and 4), and for small amplitude motion or initial motion from rest. The second case explains why the error in Eq. (11-11) can remain small up to high Re, and why an empirical extension to Eq. (11-11) (see below) describes some kinds of high Re motion. Note also that the limited diffusion of vorticity from the particle at high ω or small t implies that the effects of a containing wall are less critical for accelerated motion than for steady flow at low Re.

Similar analyses have been developed for fluid spheres (S8, S9) and for rigid spheroids (L3) moving parallel to their axes. The main conclusions are discussed below.

B. Rigid Particles

1. *Spheres*

For a spherical particle released from rest at $t = 0$ in a stagnant fluid, the equation of motion follows from Eq. (11-11) as

$$(\rho_p + \rho/2)V\frac{dU}{dt} = g\,\Delta\rho\,V - 6\pi\mu aU - 6a^2\sqrt{\pi\rho\mu}\int_0^t \frac{\dot{U}(s)ds}{\sqrt{t-s}} \tag{11-12}$$

where the overdot denotes a time-derivative. Introducing dimensionless times

$$\tau = \nu t/a^2, \qquad \sigma = \nu s/a^2, \tag{11-13}$$

and the ratio, W_s, of the instantaneous velocity U to the steady velocity in Stokes flow, i.e.,

$$W_s = U/U_{\mathrm{TS}} = 9\mu U/2a^2 g\,\Delta\rho = \mathrm{Re}/\mathrm{Re}_{\mathrm{TS}}, \tag{11-14}$$

we can rewrite Eq. (11-12) in dimensionless form as

$$\left(\frac{2\gamma+1}{9}\right)\frac{dW_s}{d\tau} = 1 - W_s - \frac{1}{\sqrt{\pi}}\int_0^\tau \frac{\dot{W}_s(\sigma)d\sigma}{\sqrt{\tau-\sigma}}, \tag{11-15}$$

which may be transformed (B5, K1) to an ordinary differential equation:

$$\frac{d^2W_s}{d\tau^2} + B(2-B)\frac{dW_s}{d\tau} + B^2W_s = B^2\left(1 - \frac{1}{\sqrt{\pi\tau}}\right), \tag{11-16}$$

where B is the dimensionless acceleration, $dW_s/d\tau$, at $\tau = 0$:

$$B = 9/(2\gamma+1). \tag{11-17}$$

Equations (11-15) to (11-17) correspond to the general results for spheroids given in Table 11.1. The velocity and displacement can be calculated as functions of time, either by direct numerical integration (K1) of Eq. (11-16) with

TABLE 11.1

General Results for Rigid Spheroids Accelerating Freely from Rest in Creeping Flow[a,b]

Governing equations: $\frac{1}{B}\cdot\frac{dW_s}{d\tau} = 1 - W_s - \frac{\Delta_a}{\sqrt{\pi}}\int_0^{\tau}\frac{W_s(\sigma)\cdot d\sigma}{\sqrt{\tau-\sigma}}$ and $\frac{d^2W_s}{d\tau^2} + B(2 - B\Delta_a{}^2)\cdot\frac{dW_s}{d\tau} + B^2W_s = B^2(1 - \Delta_a/\sqrt{\pi\tau})$

Results:	$W_s(\tau)$	$M_D(\tau)/Re_{TS}$
1.[c] $\gamma < \gamma_c$:	$1 - \left(\frac{\alpha}{\alpha-\beta}\right)e^{\beta^2\tau}\,\mathrm{erfc}(\beta\sqrt{\tau}) - \left(\frac{\beta}{\beta-\alpha}\right)e^{\alpha^2\tau}\,\mathrm{erfc}(\alpha\sqrt{\tau})$	$\frac{\tau}{4} - \frac{\alpha}{4\beta^2(\alpha-\beta)}[e^{\beta^2\tau}\,\mathrm{erfc}(\beta\sqrt{\tau}) - 1 + 2\beta\sqrt{\tau/\pi}] - \frac{\beta}{4\alpha^2(\beta-\alpha)}[e^{\alpha^2\tau}\,\mathrm{erfc}(\alpha\sqrt{\tau}) - 1 + 2\alpha\sqrt{\tau/\pi}]$
2.[d] $\gamma > \gamma_c$:	$1 - R_W - \frac{X}{Y}I_W$	$\frac{\tau}{4} - \frac{(3X^2 - Y^2)}{4B^2}\left[R_W - 1 + 2X\sqrt{\frac{\tau}{\pi}}\right] - \frac{X(X^2 - 3Y^2)}{4YB^2}\left[I_W - 2Y\sqrt{\frac{\tau}{\pi}}\right]$
3.[e] $\gamma = \gamma_c$:	$1 + (2B - 1)e^{B\tau}\,\mathrm{erfc}(\sqrt{B\tau}) - 2\sqrt{B\tau/\pi}$	$\frac{\tau}{4} + \frac{3}{4B} - \frac{3}{2}\sqrt{\frac{\tau}{B\pi}} + \frac{1}{2}\left[\tau - \frac{3}{2B}\right]e^{B\tau}\,\mathrm{erfc}(\sqrt{B\tau})$

[a] After Boggio (B5), Hjelmfelt and Mockros (H5), and Lai (L1).

[b] Dimensionless time, $\tau = \nu t/a^2$. Dimensionless velocity, $W_s = U/U_{ST} = g\Delta_a U/2g\,\Delta\rho\,Ea^2$. Displacement modulus, $M_D = x/2a$; $B = (dW_s/d\tau)_{\tau=0} = 9\Delta_a/E(2\gamma + \Delta_A)$. For sphere, $E = \Delta_a = \Delta_A = 1$, and critical density ratio, $\gamma_c = \frac{5}{8}$.

[c] In this case $\alpha, \beta = \frac{1}{2}B\Delta_a[1 \pm \sqrt{1 - 4/B\Delta_a{}^2}]$ and $\mathrm{erfc}(z) = (2/\sqrt{\pi})\int_z^{\infty} e^{-x^2}\,dx = 1 - \mathrm{erf}(z)$.

[d] In this case $X = \frac{1}{2}B\Delta_a$, $Y = \frac{1}{2}B\Delta_a\sqrt{(4/B\Delta_a{}^2) - 1}$, and R_W, I_W are the real and imaginary parts of the complex error function:

$$W[(Y + iX)\sqrt{\tau}] = W(Z) = e^{-Z^2}\left[1 + \frac{2i}{\sqrt{\pi}}\int_0^Z e^{\xi^2}\,d\xi\right].$$

[e] In this case $\gamma_c = 9\Delta_a{}^3/8E - \frac{1}{2}\Delta_A$ and $B = 4/\Delta_a{}^2$.

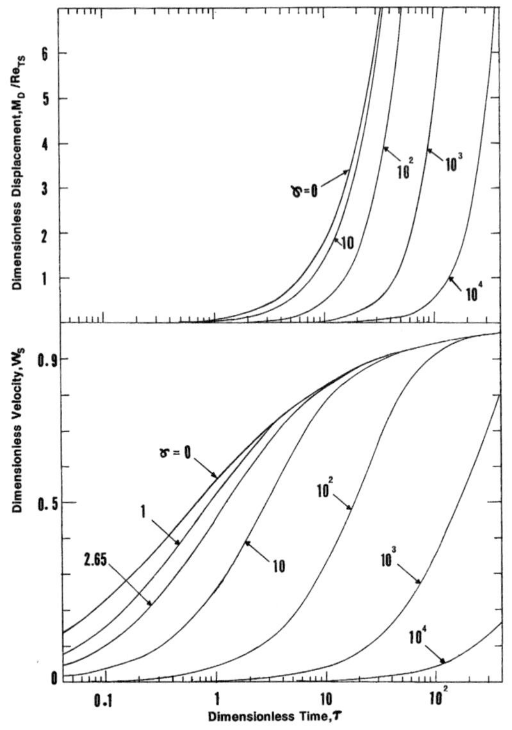

FIG. 11.1 Variation of dimensionless velocity and position with time for spheres of various density ratios accelerating freely from rest in creeping flow.

initial conditions

$$\text{at } \tau = 0: \qquad W_s = 0, \qquad \frac{dW_s}{d\tau} = B, \tag{11-18}$$

or from the analytic solutions in Table 11.1. The displacement is expressed as:

$$\frac{M_D}{Re_{TS}} = \frac{9\mu^2 x}{8a^4 g\rho\,\Delta\rho} = \frac{1}{4}\int_0^\tau W_s\,d\tau. \tag{11-19}$$

The general solution in Table 11.1 was first given for a sphere by Boggio (B5). For γ greater than a critical value denoted by γ_c, α and β are complex. This does not imply an oscillatory component in the motion, since the imaginary part of the expression is identically zero (H15). However, it may be more convenient to use the purely real forms (H7) given in Table 11.1.[†] Figure 11.1 shows

[†] Tabulations of erfc(z) and $W(Z)$ are available (A2). Brush *et al.* (B9) gave alternate results, including a series for $W_s(\tau)$ which converges rather slowly.

predictions for spheres of various density ratios. The validity of these results for creeping flow is well established (M6, M10), and the range of applicability is discussed below.

It is of interest to compare the predictions from Eq. (11-15) with simplified treatments which are often employed. If the troublesome history term is neglected,

$$W_s = 1 - \exp(-B\tau) \quad \text{and} \quad \frac{M_D}{\mathrm{Re_{TS}}} = \frac{1}{4}\left(\tau - \frac{W_s}{B}\right). \tag{11-20}$$

If added mass is also neglected, leaving only "steady drag,"

$$W_s = 1 - \exp\left(-\frac{9\tau}{2\gamma}\right) \quad \text{and} \quad \frac{M_D}{\mathrm{Re_{TS}}} = \frac{1}{4}\left(\tau - \frac{2\gamma W_s}{9}\right). \tag{11-21}$$

Figure 11.2 shows these approximate results for $\gamma = 2.65$ together with the full solution. Neglect of the unsteady drag terms is clearly unjustified, with the history term being more important than added mass. Comparisons for other values of γ are given elsewhere (H7, H15). Figure 11.3 shows the times and distances required for a particle to reach various fractions of its steady terminal velocity. Neglect of the unsteady terms becomes less serious at high γ, but discrepancies from the exact solution are still apparent as $W_s \to 1$.

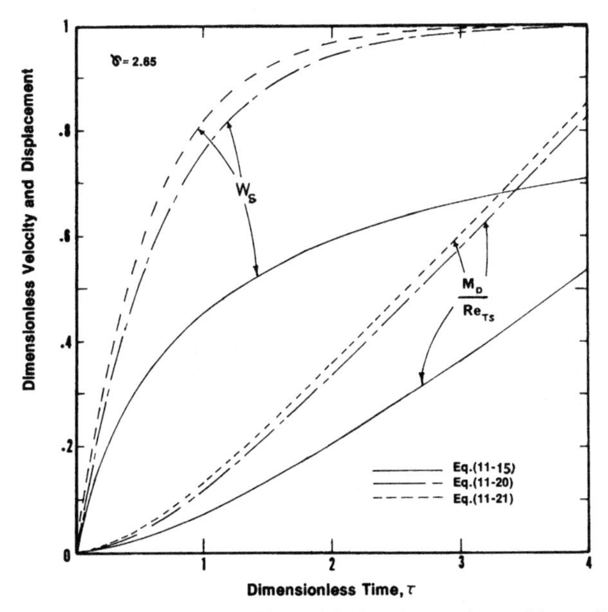

FIG. 11.2 Variation of velocity and position with time for a sphere with $\gamma = 2.65$ accelerating freely from rest in creeping flow.

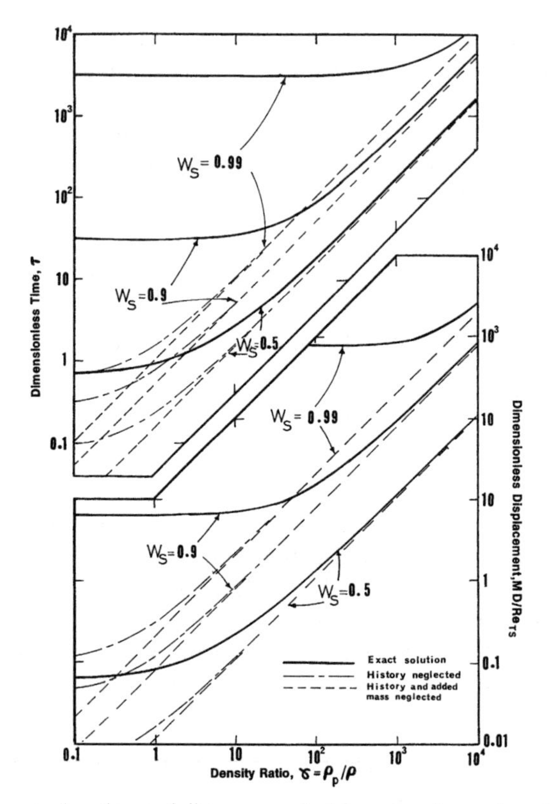

FIG. 11.3 Dimensionless time and distance required for sphere to reach various fractions of its steady terminal velocity in creeping flow.

2. *Spheroids*

Lai and Mockros (L3) generalized Eq. (11-11) for a spheroid moving parallel to its axis of symmetry to give the approximation:

$$-F_{\mathrm{D}} = \Delta_a 6\pi\mu a U + \Delta_{\mathrm{A}} \frac{\rho V}{2}\frac{dU}{dt} + \Delta_{\mathrm{H}} 6a^2\sqrt{\pi\rho\mu}\int_{-\infty}^{t}\frac{\dot{U}(s)ds}{\sqrt{t-s}}, \tag{11-22}$$

where a is the equatorial radius defined in Fig. 4.2 and Δ_a is the ratio of steady drag to that on a sphere of radius a, as discussed in Chapter 4; its value is given by Eq. (4-20) with the axial resistance, c_1, from Table 4.1. The ratio of the history term to that on a sphere of radius a is

$$\Delta_{\mathrm{H}} = \Delta_a^{\;2}. \tag{11-23}$$

The added mass coefficient, Δ_{A}, is given by:

Oblate ($E < 1$)
$$\Delta_{\mathrm{A}} = \frac{2[E\cos^{-1}E - \sqrt{1-E^2}]}{E^2\sqrt{1-E^2} - E\cos^{-1}E}, \tag{11-24}$$

$$\text{Prolate } (E > 1) \qquad \Delta_A = \frac{2[E\ln(E+\sqrt{E^2-1}) - \sqrt{E^2-1}]}{E^2\sqrt{E^2-1} - E\ln(E+\sqrt{E^2-1})}. \tag{11-25}$$

The coefficient of the added mass term, $\frac{1}{2}\Delta_A$, is identical with the ratio of form drag to friction drag in steady creeping flow given in Table 4.1. The ratio of the added mass on a spheroid to that on a sphere of radius a is $E\Delta_A$. For a spheroid falling from rest with its axis vertical, Eq. (11-22) leads to a result equivalent to Eq. (11-15) for a sphere, as summarized in Table 11.1. If the history term is neglected, Eq. (11-20) applies. If both history and added mass are neglected, the trajectory is given by,

$$W_s = 1 - \exp\left\{-\frac{9\Delta_a\tau}{2\gamma E}\right\} \quad \text{and} \quad \frac{M_D}{\mathrm{Re}_{TS}} = \frac{1}{4}\left(\tau - \frac{2\gamma E W_s}{9\Delta_a}\right) \tag{11-26}$$

Figure 11.4 shows the velocity-time curves from the full solution for weightless rigid spheroids ($\gamma = 0$) and for density ratios typical of particles in liquids ($\gamma = 2.65$) and gases ($\gamma = 10^3$). Figure 11.5 shows the ratio of the value of τ for which $W_s = 0.5$ to the corresponding value for a sphere. The effect of spheroid

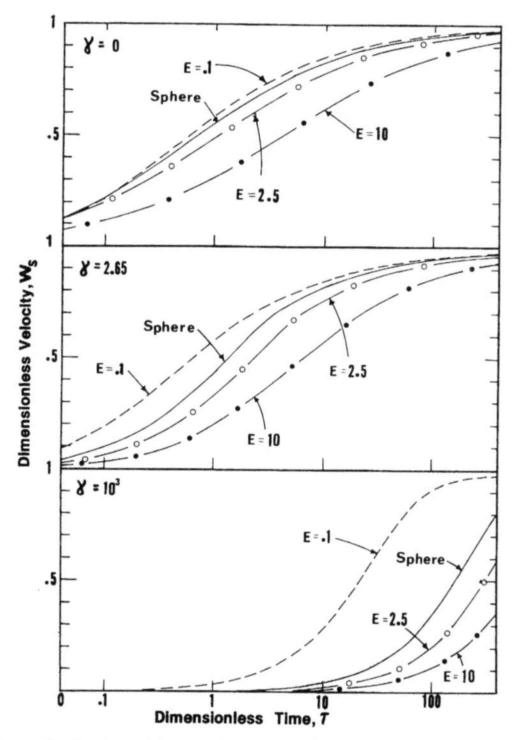

FIG. 11.4 Variation of velocity with time for spheroids accelerating freely from rest in creeping flow.

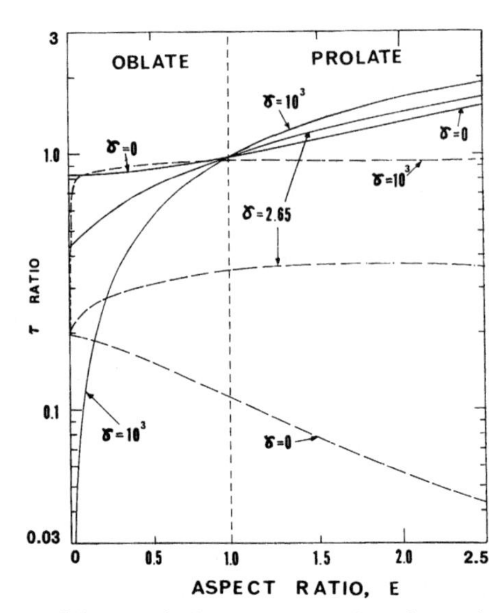

FIG. 11.5 Dependence of time required to reach 50% of steady terminal velocity ($W_s = 0.5$) on aspect ratio for rigid spheroids. Solid lines give ratio of τ for spheroids to that for a sphere of the same equatorial diameter. Broken lines give the ratio of τ from Eq. (11-20) neglecting history terms to the value from the complete solution. Analysis is for creeping flow.

shape is generally weak,† except for disk-like particles. As the spheroid becomes more prolate, the time and distance required to achieve a given fraction of the terminal velocity increase. Figure 11.5 also shows the effect of neglecting the history term. As for spheres, the errors become smaller as γ increases. For very small γ, neglect of the history term predicts the wrong effect of aspect ratio.

3. *Disks*

The drag on a thin disk moving normal to its faces is obtained by setting $E = 0$ in Eq. (11-22) (L3), i.e.,

$$-F_D = 16a\mu U + \frac{8a^3\rho}{3}\frac{dU}{dt} + \frac{128a^2}{3\pi}\sqrt{\frac{\rho\mu}{\pi}}\int_{-\infty}^{t}\frac{\dot{U}(s)ds}{\sqrt{t-s}}. \tag{11-27}$$

The motion of a disk of small aspect ratio accelerating freely from rest is then given (L2) by Table 11.1, with $\Delta_a = 8/3\pi$ and $B = 24/(3\pi E\gamma + 4)$. For an "ideal" disk ($E = 0$), $B = 6$ and the variation of W_s and M_D/Re_{TS} with τ is independent of the density ratio. The corresponding velocity-time curve is indistinguishable from that for $\gamma = 0$, $E = 0.1$ in Fig. 11.4.

† If τ is based on volume-equivalent radius, rather than equatorial radius as used here, E has almost no effect on the trajectory for prolate spheroids (L1). However, this definition of τ obscures the effect of shape for oblate particles.

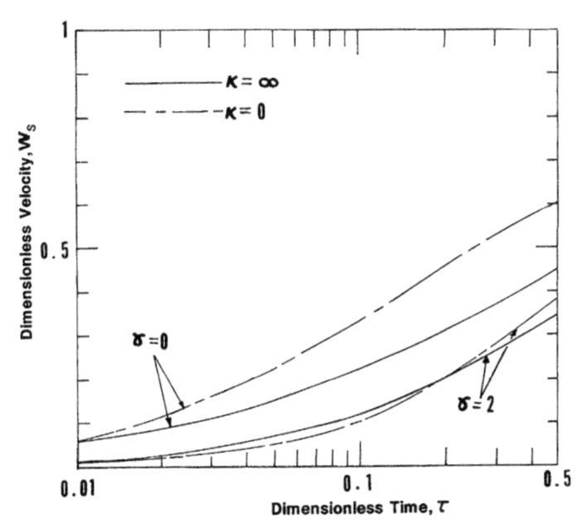

FIG. 11.6 Variation of velocity with time for rigid ($\kappa = \infty$) and circulating inviscid ($\kappa = 0$) spheres accelerating freely from rest.

C. Fluid Particles

Sy *et al.* (S8, S9) and Morrison and Stewart (M12) analyzed the initial motion of fluid spheres with creeping flow in both phases. For bubbles ($\gamma = 0$, $\kappa = 0$), the condition that internal and external Reynolds numbers remain small is sufficient to ensure a spherical shape. However, for other κ and γ, the Weber number must also be small to prevent significant distortion (S9). For $\kappa = 0$, the equation governing the particle velocity may be transformed to an ordinary differential equation (K1), to give a result corresponding to Eq. (11-16), i.e.,

$$\frac{d^3 W_s}{d\tau^3} + (4B - 9)\frac{d^2 W_s}{d\tau^2} + 4B(B - 3)\frac{dW_s}{d\tau} - 4B^2 W_s = 4B^2\left[\frac{2}{3\sqrt{\pi\tau}} - 1\right] \tag{11-28}$$

The initial conditions are (K1):

$$\text{at } \tau = 0: \qquad W_s = 0, \qquad \frac{dW_s}{d\tau} = \frac{2B}{3}, \qquad \frac{d^2 W_s}{d\tau^2} = -\frac{4B^2}{3}. \tag{11-29}$$

As for a rigid sphere, the initial acceleration is $g(1 - \gamma)/(\gamma + 1/2)$, i.e., the added mass is half the displaced mass regardless of the nature of the sphere.

Explicit forms for $W_s(\tau)$ and $M_D(\tau)$ are not available, while numerical solution of Eq. (11-28) is complicated by stability problems (C7). Sy *et al.* gave analytic solutions for $\gamma = \kappa = 0$ (S8), and numerical results for other cases (S9). Figure 11.6 shows the velocity-time curves for two density ratios and for $\kappa = 0$ and $\kappa = \infty$. For short times, rigid particles show higher W_s than circulating particles (because the dimensional initial acceleration is the same for the two cases while the rigid particle has a lower terminal velocity). A reduction in the dispersed phase viscosity reduces the time and distance required to attain a given W_s at longer times. Curves for intermediate κ generally fall between these curves, but near the intersection of the $\kappa = 0$ and $\kappa = \infty$ curves, a drop with finite κ shows a higher W_s than either a rigid or an inviscid sphere.

III. RECTILINEAR ACCELERATION AT HIGHER Re

A. Rigid Particles

1. *General Considerations*

The only rigid particle for which accelerated motion beyond the creeping flow range has been considered in detail is the sphere. Odar and Hamilton (O6) suggested that Eq. (11-11) be extended to higher Re as:

$$-F_{\mathrm{D}} = C_{\mathrm{D}} \frac{\pi d^2 \rho}{8} U^2 + \Delta_{\mathrm{A}} \frac{V\rho}{2} \frac{dU}{dt} + \Delta_{\mathrm{H}} \frac{3d^2}{2} \sqrt{\pi\rho\mu} \int_{-\infty}^{t} \dot{U}(s) \frac{ds}{\sqrt{t-s}}. \tag{11-30}$$

The first term again represents drag in steady motion at the instantaneous velocity, with C_{D} an empirical function of Re as in Chapter 5. The other terms represent contributions from added mass and history, with empirical coefficients, Δ_{A} and Δ_{H}, to account for differences from creeping flow. From measurements of the drag on a sphere executing simple harmonic motion in a liquid, Δ_{A} and Δ_{H} appeared to depend only on the acceleration modulus according to:

$$\Delta_{\mathrm{A}} = 2.1 - 0.132 M_{\mathrm{A}}^2/(1 + 0.12 M_{\mathrm{A}}^2), \tag{11-31}$$

$$\Delta_{\mathrm{H}} = 0.48 + 0.52 M_{\mathrm{A}}^3/(1 + M_{\mathrm{A}})^3. \tag{11-32}$$

For initial motion from rest, i.e., as $U \to 0$ and $M_{\mathrm{A}} \to \infty$, then $\Delta_{\mathrm{A}} = \Delta_{\mathrm{H}} = 1$, and Eq. (11-30) reduces to Eq. (11-11).

Equations (11-30) to (11-32) give a good description of the motion of particles released from rest (C8, O1, O2). In dimensionless form, the equation of motion is (C8):

$$\frac{(2\gamma + \Delta_{\mathrm{A}})}{9} \mathrm{Re}' = \mathrm{Re}_{\mathrm{TS}} - \frac{A_{\mathrm{D}}}{24} - \frac{\Delta_{\mathrm{H}}}{\sqrt{\pi}} \int_0^{\tau} \frac{\mathrm{Re}'(\sigma)\, d\sigma}{\sqrt{\tau - \sigma}}, \tag{11-33}$$

where

$$\mathrm{Re}_{\mathrm{TS}} = g d^3 \rho\, \Delta\rho / 18\mu^2, \tag{11-34}$$

$$A_{\mathrm{D}} = C_{\mathrm{D}} \mathrm{Re}^2, \tag{11-35}$$

and

$$\mathrm{Re}' = d(\mathrm{Re})/d\tau = \mathrm{Re}^2 M_{\mathrm{A}}/4. \tag{11-36}$$

Here A_{D} is a function of Re, analogous to N_{D} used for steady motion in Chapter 5, and may be evaluated using the correlations in Table 5.2. Since $N_{\mathrm{D}} = 24\mathrm{Re}_{\mathrm{TS}}$ for a spherical particle at its terminal velocity, $\mathrm{Re}_{\mathrm{TS}}$ fixes the terminal Reynolds number Re_{T} via the correlations in Table 5.3. The relationship between Re_{T} and $\mathrm{Re}_{\mathrm{TS}}$ is shown by the uppermost curve in Fig. 11.11. In view of the complex dependence of Δ_{A} and Δ_{H} on Re and Re′, Eq. (11-33) must be

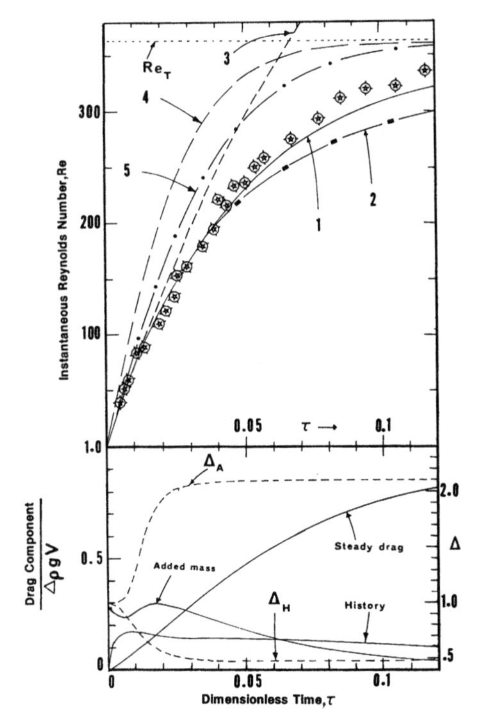

FIG. 11.7 Variation of Re with dimensionless time for sphere falling from rest; $\gamma = 1.22$, $Re_{TS} = 3371$, $Re_T = 364$. Data are from Moorman (M10), run 33. (1) Eq. (11-33); Δ_A and Δ_H from Eqs. (11-31) and (11-32); (2) Eq. (11-33); $\Delta_A = \Delta_H = 1$; (3) Creeping flow solution, Table 11.1; (4) Steady drag only ($\Delta_A = \Delta_H = 0$); (5) Steady drag with $\Delta_A = 1$. Bottom part of figure, giving drag components and coefficients, corresponds to curve 1.

solved numerically. Suitable integration procedures† are described by Odar (O2) and Clift (C7). For fall from rest, the initial conditions are

$$\text{at } \tau = 0: \qquad \text{Re} = 0, \qquad \text{Re}' = 9\text{Re}_{TS}/(2\gamma + 1), \tag{11-37}$$

and the particle displacement follows by numerical solution of

$$M_D = \frac{1}{4}\int_0^\tau \text{Re}\, d\tau. \tag{11-38}$$

There is no *a priori* justification for Eq. (11-30) since the form of Eq. (11-11) (and not simply the coefficients) depends on the assumption of creeping flow. Moreover, the form of the equation is open to criticism; for example, momentum arguments suggest that the added mass term be written $(\rho V/2)d(\Delta_A U)/dt$. However, Eq. (11-30) is the form for which Eqs. (11-31) and (11-32) were determined (O5), and appears to give an accurate description of the motion of spheres from rest as demonstrated in Fig. 11.7. Curve 1 shows the predictions

† Kuo (K3) proposed a procedure based on piecewise application of the results in Table 11.1, taking C_D, Δ_A, and Δ_H as constant over each interval. For extensive calculations, this procedure is more laborious than numerical integration.

for the instantaneous Reynolds number, $\mathrm{Re}(\tau)$, corresponding to one run from the extensive data of Moorman (M10). Similar comparisons have been made for all of Moorman's runs, covering the range $1.22 < \gamma < 9.55$, $5.8 < \mathrm{Re_{TS}} < 7.6 \times 10^4$ (O1, O2), for the data of Mockros and Lai (M6) where $1.15 < \gamma < 500$, $6.4 < \mathrm{Re_{TS}} < 2 \times 10^4$ (C8), and for data of Richards (R3) with $\gamma < 0.1$ and $9 < \mathrm{Re_{TS}} < 1.5 \times 10^7$ (C7, C8). Richards' results provide a particularly critical test since the added mass and history terms become more significant as $\gamma \rightarrow 0$. In each case, Eqs. (11-31) to (11-38) give a good description of the motion except for limitations due to secondary motion noted below.

Richards' results agree with the predictions for Re as high as 10^4 (C7). This is not only far removed from creeping flow, but is also well above the range for which Odar and Hamilton derived Eqs. (11-31) and (11-32). The fact that Δ_A and Δ_H values determined for oscillatory motion work well for unidirectional motion at much higher Re probably arises because the trajectory is relatively insensitive to these coefficients. This is demonstrated by curve 2 in Fig. 11.7, obtained with $\Delta_A = \Delta_H = 1$. In the early part of the motion, changes in these terms tend to compensate, so that curves 1 and 2 only differ when approaching $\mathrm{Re_T}$. Similar conclusions apply for other particle characteristics (H1), although the errors introduced by assuming $\Delta_A = \Delta_H = 1$ are generally larger for lower $\mathrm{Re_{TS}}$ (C8).

Although Eqs. (11-30) to (11-32) give a good description of free acceleration from rest, this does not necessarily mean that they apply to all types of unsteady motion. Fall or rise from rest is a particular case in which the creeping flow assumptions apply initially. The approach is less realistic if Re is initially large (e.g., for a particle released in a flowing fluid).

2. *Solutions for Particles in Liquids*

Unless $\gamma \gg 1$, all terms in Eq. (11-33) must be retained. Since Eq. (11-30) has no formal justification, the individual terms cannot definitely be ascribed to added mass or history effects. Even so, the relative magnitudes of the terms are of interest. Figure 11.7 shows the three terms for specific values of γ and $\mathrm{Re_{TS}}$, expressed as fractions of the immersed particle weight. "Added mass" dominates initially; "history" passes through a maximum and decays slowly; "steady drag" increases monotonically to become the sole component at the terminal velocity. Both Δ_A and Δ_H depart from unity early in the motion. For smaller $\mathrm{Re_{TS}}$, "history" may be the dominant drag component for a brief period (O2).

Figure 11.8 shows typical curves for $\mathrm{Re/Re_T}$ as functions of τ and M_D, calculated from Eqs. (11-31) to (11-36) for $\gamma = 2.65$. Even for low $\mathrm{Re_T}$ (curve 2), the velocity approaches the terminal value more rapidly than predicted by the creeping flow solution. At higher $\mathrm{Re_T}$ the steady terminal velocity is approached more rapidly, but the M_D value required to achieve a given fraction of $\mathrm{Re_T}$ increases with $\mathrm{Re_T}$. The trajectory is generally more sensitive to $\mathrm{Re_T}$ than to γ as shown by Fig. 11.9, where we have plotted the τ and $M_D/\mathrm{Re_T}$ required to

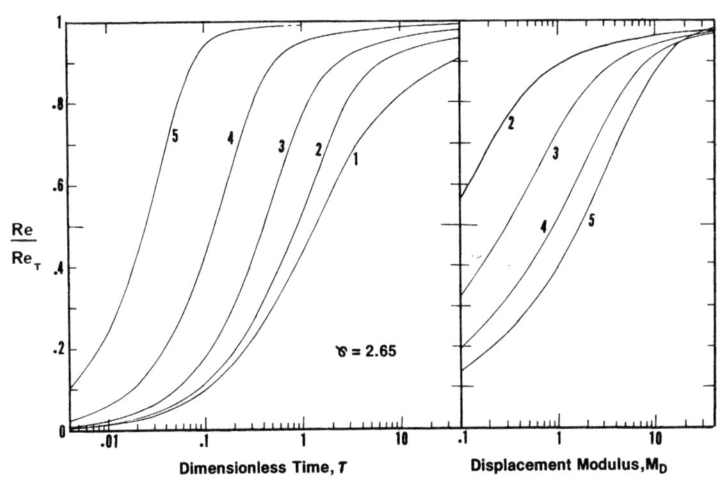

FIG. 11.8 Variation of velocity with time and distance for spheres with $\gamma = 2.65$. (1) $Re_T = Re_{TS}$ (creeping flow); (2) $Re_T = 1.0$, $Re_{TS} = 1.132$; (3) $Re_T = 10.0$, $Re_{TS} = 17.7$; (4) $Re_T = 100.0$, $Re_{TS} = 453$; (5) $Re_T = 10^3$, $Re_{TS} = 19600$.

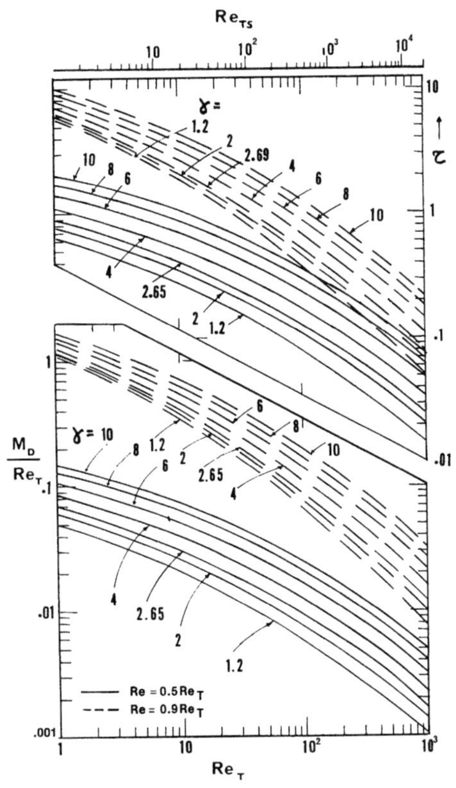

FIG. 11.9 Values of τ and M_D/Re_T required for spheres of different γ and Re_T to reach 50% and 90% of their steady terminal velocities.

achieve 50 and 90% of U_T. Figure 11.9 enables rapid estimates to be made for particles in liquids. All drag components are significant and the motion is considered "Type 1."

3. *Approximate Solutions*

The creeping flow solutions for ($\mathrm{Re}/\mathrm{Re}_{TS}$) and ($M_D/\mathrm{Re}_{TS}$) give a close approximation for early motion even for quite high Re (M6). Curve 3 in Fig. 11.7 shows a typical case: due to compensating errors in the drag terms, the creeping flow predictions are close to the observed velocities up to Re of order 100. Figure 11.10 shows the M_D at which the difference between the numerical and creeping flow values for M_D reaches 5%. Beyond this point the error increases rapidly, since the limiting Re for the creeping flow solution is Re_{TS} rather than Re_T. The Re at which the error reaches a given value is almost independent of γ in this range. Figure 11.11 shows the Re at which the error in the creeping flow solution for $\mathrm{Re}(\tau)$ reaches 1 and 5%. Mockros and Lai (M6) determined a range of validity empirically, without specifying what error they considered to constitute disagreement. Their curve is shown in Fig. 11.11 and typically corresponds to 20% error in Re.

Neglect of added mass and history simplifies calculation of unsteady motion considerably. However, for γ characteristic of particles in liquids, this introduces substantial errors as illustrated by curve 4 in Fig. 11.7. The accuracy of the simplification improves as γ and Re increase, but even for γ as high as 10 trajectories calculated neglecting history and added mass substantially underpredict the duration of accelerated motion. Neglect of added mass causes the predicted trajectory to be in error from the start of the motion. Since it is the

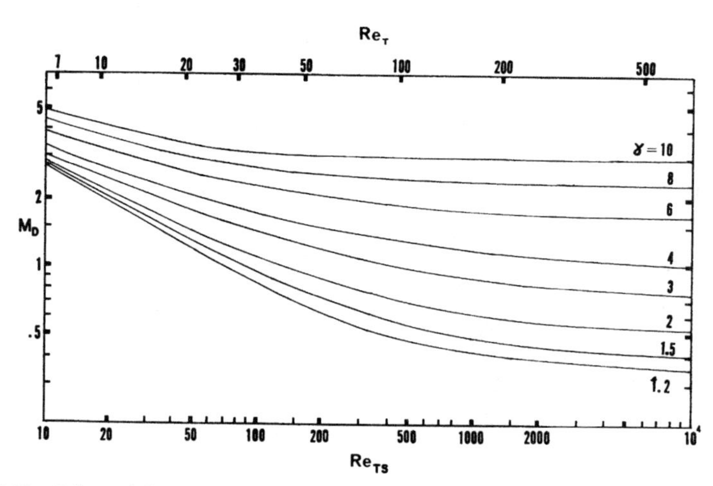

FIG. 11.10 Values of displacement modulus at which error in creeping flow solution for $M_D(\tau)$ reaches 5%.

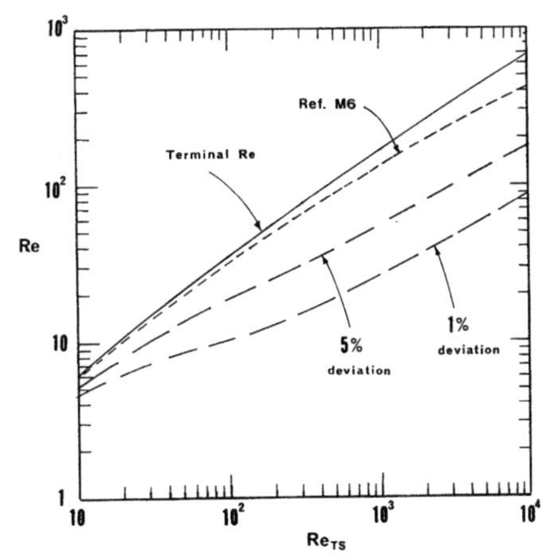

FIG. 11.11 Values of Re at which error in creeping flow solution for Re(τ) becomes significant.

history integral which complicates Eq. (11-33), it is tempting to neglect this term alone. Curve 5 in Fig. 11.7 shows the resulting trajectory with $\Delta_A = 1$. The prediction is somewhat improved, since the initial acceleration is correct, but the range of validity is much shorter than for the creeping flow approximation. Thus there is no justification for neglecting the history term and retaining added mass, since history is only negligible if γ is so large that both terms are negligible.

4. *Development of Flow Field*

Numerical solutions have been reported for fluid motion around spheres falling freely from rest (H4, L5, L7). The value of Re at which wake separation first occurs may be much higher than in steady motion (Re = 20; see Chapter 5) and increases with Re′ (H4). Figure 11.12 shows typical results for the case where $\gamma = 1.72$, $Re_T = 145$ ($Re_{TS} = 770$). Lateral wake development occurs quickly so that the separation circle rapidly approaches its steady position. Downstream growth is considerably slower. Similar trends are predicted for a sphere started impulsively (R1, R4).

Free-fall experiments with $Re_T > 10^3$ show that a sphere released from rest initially accelerates vertically, and then moves horizontally while its vertical velocity falls sharply (R3, S2, S3, V2). As for "steady" motion discussed in Chapter 5, secondary motion results from asymmetric shedding of fluid from the wake (S3, V2). Wake-shedding limits applicability of the equations given above. Data on the point at which wake-shedding occurs are scant, but lateral motion has been detected for M_D in the range 4–5 (C7). Deceleration occurs for $Re > 0.9\ Re_T$. The first asymmetric shedding occurs at much higher Re than in "steady" motion ($Re \doteq 200$; see Chapter 5), due to the relatively slow downstream development, as shown in Fig. 11.12.

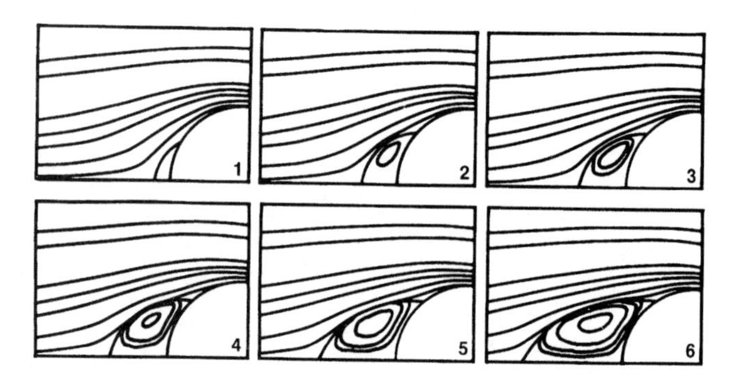

	τ	Re	Separation angle (θ_s)	Wake length/ sphere diameter (L_W/d)
1	0.123	102.6	150°	0.095
2	0.15	114.5	138°	0.24
3	0.164	119.3	134.5°	0.31
4	0.185	125.4	131°	0.41
5	0.20	129.1	128°	0.50
6	0.24	136.4	125°	0.66
At steady state:		145	121°	1.1

FIG. 11.12 Fluid streamlines relative to sphere falling from rest showing development of wake, after (L5), $\gamma = 1.72$, $\mathrm{Re_T} = 145$. Conditions as above.

5. *Solutions for Particles in Gases*

As noted above, added mass and history contributions can be neglected for large γ, especially at high $\mathrm{Re_T}$ or $\mathrm{Re_{TS}}$. The motion is then of "Type 2," with the fluid responding rapidly to changes in particle velocity. If the history term is neglected and $\gamma \gg 1$, Eq. (11-33) becomes

$$\frac{2\gamma}{9}\frac{d\,\mathrm{Re}}{d\tau} = \mathrm{Re_{TS}} - \frac{A_\mathrm{D}}{24}. \tag{11-39}$$

It is convenient to rewrite Eq. (11-39) in terms of $W = U/U_\mathrm{T}$ and a new dimensionless time:

$$T = gt/U_\mathrm{T}. \tag{11-40}$$

Equation (11-39) then becomes

$$\frac{dW}{dT} = 1 - \frac{A_\mathrm{D}}{24\mathrm{Re_{TS}}} = 1 - \frac{C_\mathrm{D}\,\mathrm{Re}^2}{N_\mathrm{D}}. \tag{11-41}$$

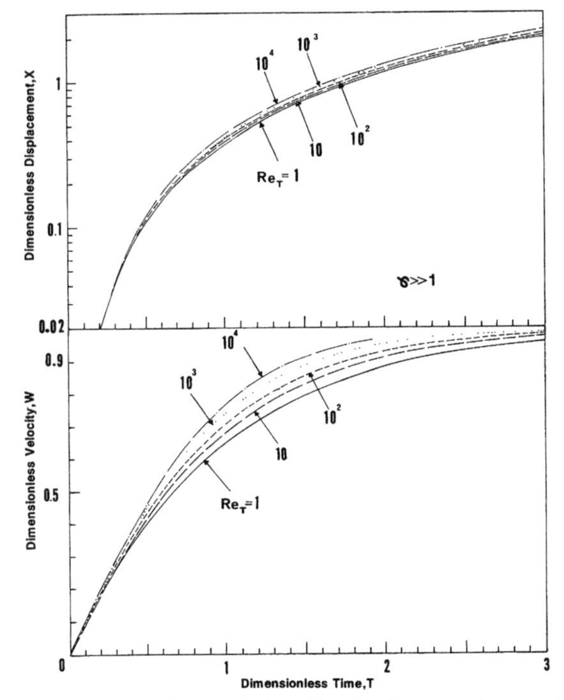

FIG. 11.13 Values of dimensionless velocity, $W = U/U_T$, and dimensionless displacement, $X = gx/U_T^2$, as functions of dimensionless time, $T = gt/U_T$ for spheres released from rest in stagnant gases ($\gamma \gg 1$).

The dimensionless displacement is given by

$$X = gx/U_T^2 = \int_0^T W\,dT. \tag{11-42}$$

With these simplifications, W and X can be generated as functions of T, with the particle characterized by a single dimensionless parameter, either Re_{TS}, N_D or Re_T. Figure 11.13 shows predictions for a particle released from rest ($W = X = 0$ at $T = 0$), while Fig. 11.14 gives trajectories for particles projected vertically upwards such that the particle comes to rest at $T = 0$. Figures 11.13 and 11.14 enable rapid estimations for many problems involving unsteady motion of particles in gases.

6. *Heat and Mass Transfer*

Hatim (H2) obtained numerical solutions for heat transfer from a sphere of constant temperature accelerating from rest. The trajectory was calculated from Eq. (11-33), and the time-dependent Navier–Stokes and energy equations

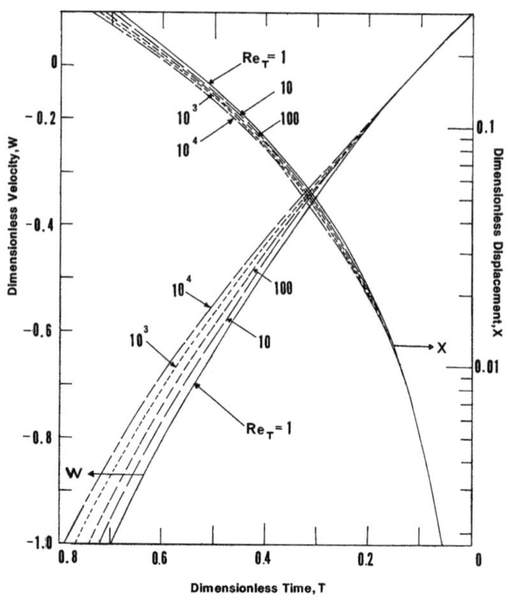

FIG. 11.14 Dimensionless velocity and displacement as functions of dimensionless time for spheres projected vertically upwards into stagnant gas ($\gamma \gg 1$).

were solved numerically.† Two distinct cases were considered. If steady-state conduction was assumed to be established before the sphere was released (i.e., Nu = 2 at $t = 0$), the Nusselt number was calculated to rise monotonically towards the final value corresponding to the terminal velocity. If the sphere was assumed to be released into a fluid of uniform initial temperature, Nu was initially very large, but fell rapidly to a minimum somewhat in excess of Nu = 2, and subsequently rose to the final steady value. In each case, the minimum local transfer rate at the sphere surface at any time following separation occurred downstream of the separation point.

B. FLUID PARTICLES

As for steady motion, shape changes and oscillations may complicate the accelerated motion of bubbles and drops. Here we consider only acceleration of drops and bubbles which have already been formed; formation processes are considered in Chapter 12. As for solid spheres, initial motion of fluid spheres is controlled by added mass, and the initial acceleration under gravity is $g(\gamma - 1)/(\gamma + \frac{1}{2})$ (E1, H15, W2). Quantitative measurements beyond the initial stages are scant, and limited to falling drops with intermediate Re_T, and rising

† Instantaneous overall drag coefficients determined from the Navier–Stokes equation were within about 10% of values obtained from Eqs. (11-30)–(11-33). This provides an additional justification of the approach discussed above.

spherical- and circular-cap bubbles. It is not uncommon for fluid particles to achieve a maximum rise or fall velocity soon after release and then decelerate somewhat. The deceleration thereafter is caused by accumulation of surface-active impurities at the interface [e.g. (A4, W4)] and is considered briefly in Chapter 7.

1. *Drops*

Edge *et al.* (E1) studied the motion of oscillating drops immediately after formation in carefully purified aqueous systems. Their results generally confirm the observations of other workers (S5, W5). Drops fall vertically at first. Internal circulation may be strong at first, depending on the mode of formation (G1). Shape oscillations may be initiated by deformation immediately prior to detachment (E1, S5) and occur at Lamb's theoretical frequency; see Eq. (7-30). Thereafter the frequency decreases, and oscillations are associated with growth and shedding of the wake. Except for drops which show large-scale asymmetric wake shedding in "steady" motion, a steady velocity close to the terminal value is attained before macroscopic shedding occurs, as for rigid spheres discussed above. Drops typically travel 5–10 equivalent diameters before shedding occurs, somewhat further than rigid spheres. Once shedding starts, wake shedding coupled with oscillations of shape and velocity is observed. A drop whose "steady fall" is accompanied by asymmetric shedding and a zig-zag trajectory passes through the intermediate regime of oscillating rectilinear motion with shedding of a vortex chain.

Drops accelerated by an air stream may split, as described in Chapter 12. For drops which do not split, measured drag coefficients are larger than for rigid spheres under steady-state conditions (R2). The difference is probably associated more with shape deformations than with the history and added mass effects discussed above. For micron-size drops where there is no significant deformation, trajectories may be calculated using steady-state drag coefficients (S1).

2. *Large Bubbles*

Walters and Davidson (W1, W2) investigated the motion of large spherical and circular bubbles, initially at rest in a stagnant fluid. As predicted by irrotational flow theory, the initial acceleration following release is $2g$ for a spherical bubble and g for a cylindrical bubble (the difference being caused by the different added mass in the two cases). In each case, a tongue of liquid moves upward into the rear of the bubble, so that it immediately begins to deform towards a spherical-cap or circular-cap shape.† Rapid generation of vorticity in the wake may cause two small satellite bubbles to detach from a two-dimensional bubble

† Similar initial motion occurs for bubbles in fluidized beds, where the final shape is attained after rising through a distance of the order of the initial radius (C10, M14).

(W1), or a ring of such satellites to form behind a three-dimensional bubble (W2).

3. *Toroidal Bubbles*

If a relatively large bubble, typically 5 ml or more, is formed rapidly (e.g., by injection of gas at high velocity through a tube, or bursting of a submerged balloon in water), a toroidal vortex ring may be formed with the gas bubble as its "core" (W2). The toroidal bubble is inherently unstable. As time progresses, circulation inside the bubble decreases, the bubble slows down and the toroid diameter increases while the core diameter decreases (B8, W2). Eventually, the bubble becomes unstable and breaks up into a number of approximately equal segments which may retain their relative positions in the toroid (B8). Theoretical predictions for toroidal bubbles by Pedley (P1) are in qualitative agreement with experimental results (B8).

IV. OSCILLATORY MOTION

Periodic fluctuations of fluid velocity usually increase the mean drag and transfer rates for entrained particles, and this has led to applications of pulsations in industrial contacting equipment. Certain natural phenomena may also be affected. For example, modification of drag by flow oscillations may be important for various flying and swimming organisms (H11, H13). Similarly, pulsations promote the onset of movement of particles originally at rest on the bottom of a duct containing a flowing fluid (C2). In addition, fluid oscillations are related to the motion of particles in turbulent fluids, as discussed in Chapter 10.

A. Rigid Particles

1. *General Considerations*

The instantaneous drag on a rigid spherical particle moving with velocity U_p in a fluid whose instantaneous velocity in the vicinity of the particle is U_f follows from an extension to Eq. (11-30):

$$-F_\mathrm{D} = \frac{\pi d^2 \rho}{8} C_\mathrm{D} U_\mathrm{R}|U_\mathrm{R}| + \frac{\rho V \Delta_\mathrm{A}}{2}\frac{dU_\mathrm{R}}{dt} + \frac{3d^2}{2}\Delta_\mathrm{H}\sqrt{\pi\rho\mu}\int_0^t \frac{\dot{U}_\mathrm{R}(s)\,ds}{\sqrt{t-s}} - \rho V \frac{dU_\mathrm{f}}{dt}, \tag{11-43}$$

where $U_\mathrm{R} = U_\mathrm{p} - U_\mathrm{f}$ is the velocity of the particle relative to the fluid. Only rectilinear motion is considered. The first three components are discussed above. The last component, the "pressure gradient term," represents the force required to accelerate the fluid which would occupy V if the particle were absent. Like Eq. (11-30), Eq. (11-43) is subject to the objection that it is an empirical modification of a result which is justified only for creeping flow.

Although its practical applicability is not so well established as that of Eq. (11-30) for motion from rest, it represents a convenient starting point for a discussion of oscillatory motion. If the fluid oscillates in the vertical direction, and velocities are positive downwards, the equation of motion for a freely moving particle follows from Eq. (11-43) as:

$$\left(\gamma + \frac{\Delta_A}{2}\right)\frac{dU_R}{dt} = (\gamma - 1)g - \frac{3C_D U_R|U_R|}{4d} - \frac{9\Delta_H}{d}\sqrt{\frac{\nu}{\pi}}\int_0^t \frac{\dot{U}_R(s)ds}{\sqrt{t-s}} - (\gamma - 1)\frac{dU_f}{dt}. \tag{11-44}$$

2. *Creeping Flow*

If the particle Re is always small, the creeping flow assumptions apply. Equation (11-44) then becomes:

$$(\gamma + 1/2)\frac{dU_p}{dt} = (\gamma - 1)g - \frac{18\nu U_R}{d^2} - \frac{9}{d}\sqrt{\frac{\nu}{\pi}}\int_0^t \frac{\dot{U}_R(s)ds}{\sqrt{t-s}} + \frac{3}{2}\frac{dU_f}{dt}. \tag{11-45}$$

The linearity of Eq. (11-45) implies that the mean terminal velocity is unaffected by oscillation (M7). The velocities may then be written as sums of mean values and variations from the mean, i.e.,

$$U_p = \bar{U}_p + u_p; \qquad U_f = \bar{U}_f + u_f; \qquad U_R = \bar{U}_R + u_R, \tag{11-46}$$

where

$$\bar{U}_R = \Delta\rho g d^2/18\mu \tag{11-47}$$

is the Stokes terminal velocity, Eq. (3-18). Equation (11-45) becomes:

$$(\gamma + 1/2)\frac{du_p}{dt} = \frac{3}{2}\frac{du_f}{dt} - \frac{9}{d}\sqrt{\frac{\nu}{\pi}}\int_0^t \frac{\dot{u}_R(s)ds}{\sqrt{t-s}} - \frac{18\nu u_R}{d^2}. \tag{11-48}$$

Molerus (M7) and Hjelmfelt and Mockros (H6) have developed complete solutions to Eq. (11-48). Velocities can be expressed as Fourier integrals. It therefore suffices to consider pure sinusoidal oscillations:

$$u_f = A\omega \cos \omega t. \tag{11-49}$$

The particle velocity is related to u_f by an amplitude ratio η and phase shift β:

$$u_p = \eta A\omega \cos(\omega t + \beta). \tag{11-50}$$

Table 11.2 gives expressions for η and β in terms of γ and a dimensionless period†:

$$\tau_0 = \nu/\omega a^2. \tag{11-51}$$

Figure 11.15 shows predictions for density ratios typical of bubbles and of particles in liquids and gases. For low frequency (high τ_0), the particle follows

† This group is sometimes called a "Stokes number." It is also $\delta^2/2a^2$, with δ defined by Eq. (11-8).

TABLE 11.2

Amplitude Ratio and Phase Shift for Spheres Entrained in Oscillating Fluids at Low Reynolds Number

$$\text{Amplitude ratio} = \eta = \sqrt{(1 + h_1)^2 + h_2^2}$$
$$\text{Phase shift} = \beta = \tan^{-1}[h_2/(1 + h_1)]$$

	h_1	h_2
Rigid spheres		
Full solution	$H_1(1 + H_2)$	$H_1 H_2(1 + 2\tau_0)$
History neglected	$\dfrac{2(1-\gamma)(2\gamma+1)}{81\tau_0^2 + (2\gamma+1)^2}$	$\dfrac{9\tau_0 h_1}{2\gamma+1}$
Steady drag only	$\dfrac{-4\gamma^2}{81\tau_0^2 + 4\gamma^2}$	$\dfrac{9\tau_0 h_1}{2\gamma}$
History and added mass neglected	$\dfrac{4\gamma(1-\gamma)}{81\tau_0^2 + 4\gamma^2}$	$\dfrac{9\tau_0 h_1}{2\gamma}$
Fluid spheres (mobile interface, $\gamma = 0$)	$\dfrac{G_1}{G_1^2 + G_2^2}$	$\dfrac{G_2}{G_1^2 + G_2^2}$

where

$$H_1 = \frac{2(1-\gamma)/(2\gamma+1)}{H_2^2(1+\sqrt{2\tau_0})^2 + (1+H_2)^2}; \qquad H_2 = \frac{9}{(2\gamma+1)}\sqrt{\frac{\tau_0}{2}};$$

$$G_1 = \frac{1}{2} + \frac{18\sqrt{2}\tau_0^{3/2}}{1+(1+3\sqrt{2\tau_0})^2}; \qquad G_2 = 18\tau_0\left[\frac{1}{2} - \frac{6\tau_0 + \sqrt{2\tau_0}}{1+(1+3\sqrt{2\tau_0})^2}\right]$$

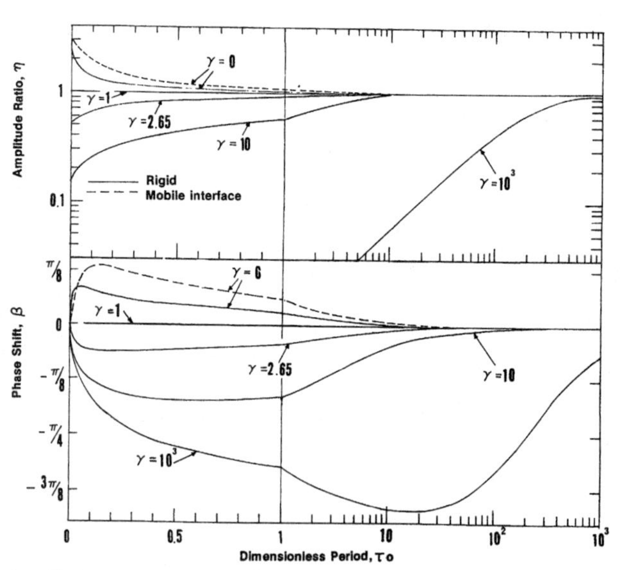

FIG. 11.15 Amplitude ratio and phase shift for particles entrained in oscillating fluids at low Re. (Note change to logarithmic scale at $\tau_0 = 1$.)

the fluid motion. This also occurs for all τ_0 at $\gamma = 1$. Deviation of the particle from the fluid motion increases at higher frequency, and is most marked for high γ. It has been shown theoretically (M5) and experimentally (M9) that these results can be applied to particles in bounded fluids provided that true local values are used for u_f. Application of these results to particles in turbulent fluids is discussed in Chapter 10.

Various levels of simplification are also available with corresponding results given in Table 11.2. Hjelmfelt and Mockros (H6) have given detailed comparisons. The approximate results agree closely with the exact solutions at high τ_0 (low ω), but are inaccurate when the particle does not follow the fluid closely. As in motion from rest, neglect of the history term generally introduces larger errors than neglect of added mass. Both terms are less critical for high γ. Neglect of the pressure gradient term introduces errors at low τ_0.

3. *Higher Reynolds Numbers*

Outside the creeping flow range, Eq. (11-43) becomes nonlinear, making solutions more difficult to obtain. Although the nonlinearity implies that it is no longer sufficient to express an arbitrary oscillatory motion as a Fourier integral, most treatments have considered purely sinusoidal variations in U_f. Particle motion can be obtained numerically from Eqs. (11-44), (11-31), and (11-32). Since these equations were derived for oscillatory motion, this approach should have some validity. Oscillations reduce the mean terminal velocity outside the creeping flow range, due to the convex form of the "steady drag" term (B6, H12, M7).

A common simplification is to assume constant C_D (equivalent to assuming that Re is always in the Newton's law range). It is then convenient to define a dimensionless frequency and amplitude:

$$N_\omega = \omega(2\gamma + \Delta_A)\sqrt{\rho d/3gC_D\,\Delta\rho}, \tag{11-52}$$

$$N_A = 3C_D A\,\Delta\rho/(2\gamma + \Delta_A)^2\rho d. \tag{11-53}$$

Equation (11-44) now becomes†:

$$N_\omega \frac{dW_R}{dt} = 1 - W_R|W_R| + N_A N_\omega{}^2 \sin t^+ - N_H \int_0^{t^+} \frac{\dot{W}_R(s^+)\,ds^+}{\sqrt{t^+ - s^+}} \tag{11-54}$$

where

$$t^+ = t\omega \qquad \text{and} \qquad s^+ = s\omega, \tag{11-55}$$

$$W_R = U_R/U_T; \qquad W_f = U_f/U_T; \qquad U_T = \sqrt{4gd\,\Delta\rho/3\rho C_D}, \tag{11-56}$$

and

$$N_H = 6\Delta_H\sqrt{3\omega\mu/\pi g d C_D\,\Delta\rho}. \tag{11-57}$$

† The first term on the right of Eq. (11-54) is -1 for $\gamma < 1$.

If the history term is neglected, Eq. (11-54) can be solved if W_R does not change sign (H10). The mean relative velocity is then approximately (H12)

$$\bar{W}_R^{\,2} = \frac{1}{2} + \frac{N_\omega^{\,2}}{8}\left[\sqrt{\left(1 + \frac{4}{N_\omega^{\,2}}\right)^2 - 8N_A^{\,2}} - 1\right]. \tag{11-58}$$

Al-Taweel and Carley (A3) gave corresponding expressions for the amplitude ratio and phase shift. For $N_A^{\,2}N_\omega^{\,2} \ll 1$, Eq. (11-58) reduces approximately (H14) to

$$\bar{W}_R = 1 - N_A^{\,2}N_\omega^{\,4}/4(4 + N_\omega^{\,2}). \tag{11-59}$$

Figure 11.16 shows $\bar{W}_R$ as a function of N_ω, calculated from Eq. (11-58) for various values of N_A. Chan *et al.* (C3) extended this approach to a sphere in a horizontally oscillating fluid.

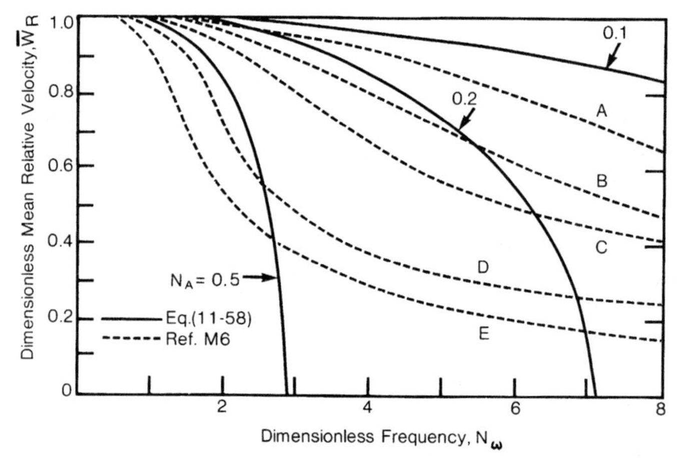

FIG. 11.16 Ratio of mean terminal velocity to terminal velocity in absence of oscillations for particles in sinusoidally oscillating fluids. Unbroken lines are predictions from Eq. (11-58); broken lines are numerical predictions (M8) for 2 mm spheres in water with $\gamma = 2.5$ and N_A values as follows: curve A—0.28; B—0.42; C—0.56; D—1.11; E—1.67.

A number of authors (B1, B2, H3, H10, M8) have solved Eq. (11-44) numerically, often neglecting the history term and using empirical approximations for C_D. Typical predictions are shown in Fig. 11.16.[†] Qualitatively, the trends are the same as predicted by Eq. (11-58), but the numerical approach predicts less retardation since C_D decreases as Re increases (T4). Here $\bar{W}_R$ is predicted to become zero only for $N_\omega \to \infty$ at finite N_A. Thus assumption of constant C_D leads to significant error. A rather different approach was initiated by Bailey

[†] For the specific particle considered (M8), the terminal Reynolds number is approximately 520, corresponding to $C_D \doteq 0.55$, but N_ω or N_A were evaluated using the "Newton's law" value, $C_D = 0.445$.

(B1), based on a time-averaged form of Eq. (11-44) and applicable to any form of oscillation if the period is long compared with the response time of the fluid (see below). At the other extreme of simplification, Rschevkin (R5) showed that if all effects except added mass and pressure gradient are omitted from Eq. (11-44), a particle moves in phase with the fluid oscillations (i.e., $\beta = 0$) with amplitude ratio:

$$\eta = (2 + \Delta_A)/(2\gamma + \Delta_A). \tag{11-60}$$

The validity of the various simplifications has been the subject of considerable discussion [e.g. (A3, B2, H3, T4)]. Schöneborn (S4) showed that in the range where periodic wake shedding normally occurs ($Re \dot{>} 200$; see Chapter 5), the effect of fluid oscillations depends on the relationship between the forced fluid frequency and the natural wake frequency:

(a) If the fluid frequency is low, typically $0.02U_T/d$ or less, it is reasonable to use a "quasi-steady" model in which history is neglected and $\Delta_A = 1$. Equation (11-58) can then be used for particles in the Newton's law range. Particles with lower Re_T generally settle more rapidly (T4), as shown by the numerical results in Fig. 11.16.

(b) At higher frequencies, $\bar{W}_R$ is lower than the value predicted by the "quasi-steady" approach, providing that A/d is sufficiently great for vortex shedding to occur (H3). The difference is most marked when the fluid frequency corresponds to the frequency of wake shedding (see Fig. 5.9). Then $\bar{W}_R$ is typically 20–30% lower than predicted by Eq. (11-54) (B2, S4), due to resonance between the fluid and wake frequencies. A falling particle then shows substantial secondary motion, usually following a zig-zag trajectory (S4), and shedding large vortices at the driving frequency (B2, C1).

(c) For forcing frequencies greater than about twice the natural wake frequency, wake shedding is suppressed and the particle trajectory is close to rectilinear. The mean settling velocity is then predicted well by Eq. (11-44) with the history term included (S4). Schöneborn found little difference between predictions using $\Delta_A = \Delta_H = 1$, and using the Odar and Hamilton values. The difference was most apparent at lower Re, where Eqs. (11-31) and (11-32) were more reliable. As for free fall from rest, these results applied for Re much higher than the range covered in Odar and Hamilton's experiments. For particles not in the wake-shedding range (i.e., $Re \dot{<} 130$), Eqs. (11-31), (11-32), and (11-44) gave the best prediction of $\bar{W}_R$. This approach also gives the best values for the amplitude ratio, for all cases.

Other simplifications apply under specific conditions. Bailey (B1) showed that history can be neglected in calculating $\bar{W}_R$ at high frequencies, i.e., $N_\omega \gg 1$. Al-Taweel and Carley (A3) found that the amplitude ratio for particles in liquids is given by the creeping flow results in Table 11.2 at low frequency ($\tau_0 \dot{>} 0.2$), while the ideal fluid result, Eq. (11-60) with $\Delta_A = 1$, applies at high frequency ($\tau_0 \dot{<} 0.002$ to 0.01, dependent on γ).

4. *Levitation*

Levitation is defined as a stable condition in which a particle responds to vertically oscillating fluid so that net gravity forces are completely neutralized and the particle merely oscillates about a fixed position (H12). In terms of the preceding analysis, this means $\overline{W}_R = 0$ (cf. Fig. 11.16). Contrary to the predictions of the numerical solutions, Feinman (F1) found that levitation can be caused by sinusoidal oscillations. Equation (11-58) predicts that $\overline{W}_R$ becomes zero if:

$$N_A N_\omega = \sqrt{2}. \tag{11-61}$$

Levitation is inconsistent with the assumptions leading to Eq. (11-58), since W_R must change sign during a cycle. Even so, Eq. (11-61) appears to represent a lower bound on the $N_A N_\omega$ needed to initiate levitation. Krantz *et al.* (K2) found that levitation conditions were correlated closely by

$$N_A N_\omega = 1.86. \tag{11-62}$$

Inclusion of the dimensionless history coefficient N_H gave no improvement in the correlation.

As noted above, the nonlinearity of Eq. (11-44) implies that the form of the oscillations in U_f will also affect the particle motion. If the oscillations have larger upward than downward velocities, the mean upward drag is increased. Van Oeveren and Houghton (V1) confirmed that particles could be made to levitate or move against gravity more readily by "sawtooth" oscillations in which $|U_f|$ is greater on the upward stroke. The reverse wave form, in which the downstroke is more rapid, should increase $\overline{W}_R$ above unity provided that $N_A N_\omega$ is sufficiently large (B1, B6).

5. *Heat and Mass Transfer*

There is conflicting evidence regarding the extent to which imposed vibrations increase particle to fluid heat and mass transfer rates (G2), with some authors even claiming that transfer rates are decreased. For sinusoidal velocity variations superimposed on steady relative motion, enhancement of transfer depends on a scale ratio A/d and a velocity ratio Af/U_T (G3). These quantities are rather like the scale and intensity of turbulence (see Chapter 10). For $Af/U_T < 1/2\pi$, the vibrations do not cause reversal in the relative motion and the enhancement of mass transfer has been correlated (G3) by

$$(\mathrm{Sh}_v - \mathrm{Sh}_t)/\mathrm{Sh}_t = 1.05[N - 0.06]^{1.26} \qquad (0.06 \le N \dot{<} 0.65), \tag{11-63}$$

where

$$N = \frac{4Af}{U_T}\left[\frac{2A}{d}\right]^{-0.45}. \tag{11-64}$$

Vibration then has less than 10% effect on Sh for $N \dot{<} 0.2$, which explains why earlier workers (B4) failed to detect an effect of vibration normal to the axis of translation.

B. Fluid Particles

1. *Bubbles*

Sinusoidal oscillations of the continuous phase cause levitation or counter-gravity motion much more readily for gas bubbles, due to changes in bubble volume which cause a steady component in the pressure gradient drag term (J1, J2). If the fluid motion is given by Eq. (11-49), the pressure in the vicinity of the bubble also varies sinusoidally. For normal experimental conditions, the resulting volume oscillations are isothermal (P2), and given by (J1):

$$V = \bar{V}/(1 - \varepsilon \sin \omega t), \tag{11-65}$$

where the bubble is at depth H below the liquid surface, the mean pressure around the bubble is $\bar{p}$, and

$$\varepsilon = \rho A \omega^2 H/\bar{p}. \tag{11-66}$$

For small ε, bubbles have zero mean velocity if

$$\omega^2 A = \sqrt{2g\bar{p}/\rho H}. \tag{11-67}$$

This result is valid for Re < 2. At higher Re, problems again arise from the nonlinearity of Eq. (11-43), and larger values of $\omega^2 A$ are required to cause levitation. For Re $\doteq$ 100, conditions for the bubble to have no mean motion relative to the continuous phase were correlated (B10) by

$$\omega^2 A = g + \sqrt{3g^2 + 2g\bar{p}/\rho H}. \tag{11-68}$$

Jameson (J1) also derived expressions for the amplitude ratio and phase shift in creeping flow, based on the assumption that fluid stresses due to the velocity and pressure fields do not affect one another, i.e., the drag terms are unaffected by volume oscillations. Two limiting cases were considered, corresponding to zero tangential velocity and zero shear stress at the bubble boundary. Resulting values for the amplitude ratio and phase shift are given in Table 11.2 and shown in Fig. 11.15, with τ_0 based on the mean bubble radius. In the range where the bubble departs most from the fluid motion, mobility of the interface should increase the amplitude ratio and phase shift. As a result, a fluid particle follows the motion of ambient fluid less faithfully than a solid particle (M12). In practice, η is larger than either theoretical value, possibly due to shape changes not considered in the analysis (J1) and other simplifications (M2).

The above results apply to spherical bubbles, and analysis for nonspherical bubbles is considerably more complex (P3). Marmur and Rubin (M3) have given an approximate analysis of the motion of spherical bubbles in a radially oscillating liquid. Large oscillating bubbles produced by underwater explosions have been reviewed by Holt (H9).

2. *Drops*

Chonowski and Angelino (C6) studied chlorobenzene drops in an oscillating water column ($\gamma \doteq 1.1$, $\kappa \doteq 0.8$). Over the range of the experiments ($50 < \text{Re} < 240$; $0.5 < A < 2$ cm; $1 < f < 3\ \text{s}^{-1}$), W_R was a decreasing function of $A^{1/3}f$. An equation of motion was developed in which history is ignored[†] [cf. Eq. (11-44)]:

$$\frac{\gamma\, dU_R}{dt} + \frac{|U_R|}{U_R}\left|\frac{dU_R}{dt}\right| = (\gamma - 1)g - \frac{3C_D U_R|U_R|}{4d} - (\gamma - 1)\frac{dU_f}{dt}, \qquad (11\text{-}69)$$

with C_D estimated from the Hu and Kintner correlation (see Chapter 7). The amplitude ratio and mean velocity were well predicted by analytic solutions to Eq. (11-69), but the general applicability of this result is untested. For high-amplitude, high-frequency oscillations, the mean drag on drops may actually decrease (A1).

3. *Mass and Heat Transfer*

Oscillations are often used to improve the efficiency of transfer processes in industrial phase-contacting equipment [e.g., see (B7, T3)]. Substantial improvements in mass transfer rates occur (A3), but these are due to a combination of effects including increased hold-up of fluid particles in the column due to the increase in mean drag described above, break up of fluid particles to give smaller bubbles or drops with increased interfacial area, and improved transfer coefficients. Experimental measurements of transfer coefficients on single drops in pulsed fluids show an increase in mass transfer with increasing pulsation amplitude and frequency (A1), with enhancement increased if $\dot{U}_R$ changes direction during each cycle. Qualitatively, the greatest enhancement should occur when the imposed frequency is close to the natural frequency of the fluid particle [see Eq. (7-30) (M11)], or for bubbles and drops too small or too large to undergo shape dilations and secondary motion in steady translation through stagnant fluid (see Chapter 7).

V. ARBITRARY ACCELERATED MOTION

A. General Considerations

Apart from the specific classes of motion discussed above, understanding of unsteady fluid–particle interaction is not well advanced. Torobin and Gauvin

[†] The reason for writing the added mass term in this form is not clearly explained.

(T2) and Clift and Gauvin (C9) have reviewed the relevant literature. There have been two distinct approaches to unsteady drag. The first ignores the history term completely, and either neglects added mass or assumes $\Delta_A = 1$. The drag force is then correlated using an acceleration-dependent C_D. This approach has some justification for Type 2 motion (see below). However, it has been widely used for Type 1 motion also, and resulting correlations tend to be specific to the conditions used. Ensuing problems in interpretation are exemplified by the well-known work of Lunnon (L9) on spheres accelerating from rest. Similar problems were encountered by Schwartzberg and Beckerman (S6), who determined the drag on particles following spiral trajectories in horizontally gyrating liquids. They assumed $\Delta_A = 1$ and $\Delta_H = 0$, and found that the apparent C_D was increased by a factor dependent on $(1/\nu)|dU_R/dt|$, whose form suggests that the unsteady terms would better account for the additional drag. Acceleration-dependent C_D values are generally higher than the "standard" values discussed in Chapter 5 [e.g., see (M1, R6)]. Some workers report anomalously low drag for water drops in air (I1, O7) but this appears to result from freestream turbulence and difficulty in accurately determining particle accelerations (M4, T2). Small liquid drops in gases can be treated as rigid spheres for trajectory calculations (S1), provided that evaporation rates are low and that acceleration is insufficient to cause break up or significant deformation.

The second general approach is based on extension of the creeping flow result, as in earlier sections. Corrsin and Lumley's modification (C11) of the equation proposed by Tchen (T1) allows Eq. (11-43) to be generalized as

$$-F_{Di} = \frac{\pi d^2 \rho}{8} C_D U_{Ri}|U_R| + \frac{\rho V \Delta_A}{2}\frac{dU_{Ri}}{dt} + \frac{3d^2\Delta_H}{2}\sqrt{\pi\rho\mu}\int_0^t \left(\frac{dU_{Ri}}{dt}\right)_{t=s}\frac{dt}{\sqrt{t-s}} - \rho V\left[\frac{DU_{fi}}{Dt} - \nu\nabla^2 U_{fi}\right], \quad (11\text{-}70)$$

where F_{Di}, U_{Ri}, and U_{fi} are orthogonal components of the total drag force and of velocities $\mathbf{U}_R$ and $\mathbf{U}_f$, respectively, and

$$\frac{dU_{Ri}}{dt} \equiv \frac{dU_{pi}}{dt} - \frac{\partial U_{fi}}{\partial t} - \sum_j U_{pj}\frac{\partial U_{fi}}{\partial x_j}. \quad (11\text{-}71)$$

The pressure gradient term has been extended to its full form from the Navier–Stokes equation. Equation (11-70) has been discussed by Corrsin and Lumley (C11), Hinze (H5), and Soo (S7). It is applicable only if a particle is small compared to the scale of velocity variations in the fluid (L8), i.e., if

$$\frac{d^2}{\nu}\cdot\frac{dU_f}{dx} \ll 1 \quad \text{and} \quad \frac{d^2}{U_f}\cdot\frac{d^2U_f}{dx^2} \ll 1. \quad (11\text{-}72)$$

Effects such as lift due to particle rotation or fluid velocity gradients can readily be included in Eq. (11-70) if appropriate. The resulting equation of motion is the usual starting point for analysis of particle motion in turbulent fluids, as discussed in Chapter 10, and is also the basis of particle trajectory calculations. For rectilinear motion in which Re is initially small, Δ_A and Δ_H can be estimated from Eqs. (11-31) and (11-32), provided that wake shedding does not occur. Lewis and Gauvin (L6) found that this approach gives a good description of the motion of particles in a decelerating plasma jet.† If Re is high throughout the unsteady motion, the history term can be neglected, since the flow pattern never resembles creeping flow.

Useful trajectory estimates for high Re can frequently be obtained by simply assuming that C_D takes its standard value and that $\Delta_A = 1$. For example, Guthrie *et al.* (G4) found that this approach gave useful predictions of the motion of a sphere projected into a stagnant liquid, even though a cavity formed in the particle wake. Prediction of unsteady drag is generally most difficult in ranges where the flow pattern changes markedly with Re. Acceleration delays the laminar/turbulent transition in the boundary layer (C4) and thus increases the critical Reynolds number (W3); deceleration has the reverse effect.

Additional difficulties arise when the motion is not rectilinear. Odar (O2, O4) measured the drag on a sphere following a circular path in a stagnant liquid. If the path diameter was at least 7 times the particle diameter, the drag was virtually unaffected by acceleration or curvature of the trajectory, in agreement with the approach based on Eq. (11-70). However, the added mass and history coefficients were both greater than unity (cf. $\Delta_H < 1$ for motion from rest) and dependent on the curvature of the trajectory. In addition, a particle moving steadily on a circular path experiences "lift" in the outward radial direction (O3), correlated by a "normal drag coefficient":

$$C_N = \text{Radial force}/(\pi d^2 \rho U_R{}^2/8). \tag{11-73}$$

At Re $\doteq$ 20, C_N increased sharply to pass through a maximum of approximately 0.22 at Re $\doteq$ 40, declining to be very small for Re $\gtrsim$ 150. Large normal drag is probably related to wake development, and similar effects may be expected whenever the flow pattern changes markedly with Re. In the critical range, lateral acceleration would tend to produce asymmetric boundary layer transition, so that significant lift can be anticipated.

B. Calculation of Particle Trajectories

From the examples of experimental studies discussed, it is clear that it is impossible to predict with any confidence either the magnitude or the direction of the drag on a particle when the relative velocity and acceleration are not

† This is an unusual case in which ν is so high that the history term is significant even though γ is large. It demonstrates the advisability of evaluating the unsteady drag components before assuming that motion is of Type 2.

parallel. Even when motion is rectilinear, accurate predictions are expected only for the particular cases discussed. The following general guidelines are proposed for calculation of particle trajectories:

a. *Classify Motion as "Type 1" or "Type 2"* This implies estimating whether terms in Eq. (11-70) dependent on relative acceleration are significant. This can be done beforehand if rough estimates are available for acceleration and velocity during the motion. Alternatively, the trajectory can be calculated according to the relatively simple "Type 2" procedure, with the magnitudes of the unsteady terms evaluated in the course of the calculations. As a general rule, motion of a particle in a gas is usually of Type 2, while that in a liquid is almost always of Type 1.

b. *Type 1 Motion* If the acceleration-dependent drag terms are significant, Eq. (11-70) is recommended for estimating the instantaneous drag. Numerical solution is then necessary (C7). The difficulty lies in estimation of Δ_A and Δ_H. If the motion is similar to a case for which experimental results are available, reported data should be used [e.g., Eqs. (11-31) and (11-32) for motion which is close to rectilinear, or the results of Odar or Schwartzberg and Beckerman for circular or spiral motion]. Otherwise there may be no ready alternative to the assumption $\Delta_A = \Delta_H = 1$. Errors in the predicted trajectories are especially serious when such complications as shedding or asymmetry of the wake are present. If Re is high throughout the motion, it is reasonable to ignore the history term. The calculation procedure then becomes the same as for Type 2 motion.

c. *Type 2 Motion* If the history term in Eq. (11-70) can be neglected, and the added mass is either negligible or constant, estimation of particle trajectories is simpler. The equations for particle velocity form a set of at most three first-order ordinary differential equations, given by the scalar components of

$$V\left(\frac{\rho\Delta_A}{2} + \rho_p\right)\frac{d\mathbf{U}_p}{dt} = \mathbf{F}_g - \frac{\pi d^2 \rho}{8} C_D \mathbf{U}_R |U_R|, \tag{11-74}$$

where $\mathbf{F}_g$ contains body forces (such as gravity), the pressure gradient term,† and any significant lift terms. The particle displacement is given by the components of

$$d\mathbf{x}/dt = \mathbf{U}_p. \tag{11-75}$$

Much has been written on the solution of Eqs. (11-74) and (11-75) [see, e.g., (H8, K4, M13)], frequently without serious consideration of the validity of Type 2 simplifications. Approximate methods, avoiding numerical integration, are also available for specific types of motion, such as a particle projected with arbitrary velocity in a gravitational field (H8, K4).

† The pressure gradient contribution can be significant, even for high γ [see (D3)].

The drag components contain $C_D U_{Ri}|U_R|$, with C_D evaluated for $\text{Re} = dU_R/\nu$. Some authors have used $C_D U_{Ri}^2$, or even $C_{Di} U_{Ri}^2$, where C_{Di} corresponds to $\text{Re}_i = dU_{Ri}/\nu$. These simplifications are only valid in Stokes flow, and can lead to substantial errors at higher Re [see, e.g., (R7)]. The effect of freestream turbulence can be included, via the correlations in Chapter 10, provided that the turbulence intensity can be estimated. Alternatively, one of the available correlations for drag in accelerated motion through a turbulent fluid can be used [see (C9)], although these are only applicable for limited ranges of experimental conditions.

REFERENCES

A1. Abdelrazek, I. D., and Gomaa, H. G., *Indian J. Technol.* **14**, 271–276 (1976).

A2. Abramowitz, M., and Stegun, I. A., eds., "Handbook of Mathematical Functions," Appl. Math. Ser. No. 55. Natl. Bur. Stand., Washington, D.C., 1964.

A3. Al-Taweel, A. M., and Carley, J. F., *AIChE Symp. Ser.* **67**, No. 116, 114–131 (1971).

A4. Aybers, N. M., and Tapucu, A., *Wärme- Stoffübertrag.* **2**, 118–128 (1969).

B1. Bailey, J. E., *Chem. Eng. Sci.* **29**, 767–773 (1974).

B2. Baird, M. H. I., Senior, M. G., and Thompson, R. J., *Chem. Eng. Sci.* **22**, 551–557 (1967).

B3. Basset, A. B., "A Treatise on Hydrodynamics," Vol. 2, Ch. 22. Deighton Bell, Cambridge, England, 1888. (Republished: Dover, New York, 1961.)

B4. Baxi, C. B., and Ramachandran, A., *J. Heat Transfer* **91**, 337–344 (1969).

B5. Boggio, T., *Rendiconti R. Accad. Naz. Lincei* **16**, Ser. 5, 613–620, 730–737 (November 1907).

B6. Boyadzhiev, L., *J. Fluid Mech.* **57**, 545–548 (1973).

B7. Bretsznajder, S., Jaszczak, M., and Pasiuk, M., *Int. Chem. Eng.* **3**, 496–502 (1963).

B8. Brophy, J., Course 302-495B Final Rep., McGill Univ., Montreal, 1973.

B9. Brush, L. M., Ho, H. W., and Yen, B. C., *J. Hydraul. Div., Am. Soc. Civ. Eng.* **90**, 149–160 (1964).

B10. Buchanan, R. H., Jameson, G. J., and Oedjoe, D., *Ind. Eng. Chem., Fundam.* **1**, 82–86 (1962).

C1. Carstens, M. R., *Trans. Am. Geophys. Union* **33**, 713–721 (1952).

C2. Chan, K. W., Baird, M. H. I., and Round, G. F., *Proc. R. Soc., Ser. A* **330**, 537–559 (1972).

C3. Chan, K. W., Baird, M. H. I., and Round, G. F., *Chem. Eng. Sci.* **29**, 1585–1592 (1974).

C4. Chang, P. K., "Separation of Flow." Pergamon, Oxford, 1970.

C5. Chen, J. L. S., *J. Appl. Mech.* **41**, 873–878 (1974).

C6. Chonowski, A., and Angelino, H., *Can. J. Chem. Eng.* **50**, 23–30 (1972).

C7. Clift, R., unpublished analysis, 1974.

C8. Clift, R., Adamji, F. A., and Richards, W. R., *Int. Symp. Part. Technol., IIT Res. Inst., Chicago*, 1973.

C9. Clift, R., and Gauvin, W. H., *Can. J. Chem. Eng.* **49**, 439–448 (1971).

C10. Collins, R., *Chem. Eng. Sci.* **26**, 995–997 (1971).

C11. Corrsin, S., and Lumley, J. L., *Appl. Sci. Res., Sect. A* **6**, 114–116 (1956).

D1. Dennis, S. C. R., and Walker, J. D. A., *J. Eng. Math.* **5**, 263–278 (1971).

D2. Dennis, S. C. R., and Walker, J. D. A., *Phys. Fluids* **15**, 517–527 (1972).

D3. Drew, D. A., *Phys. Fluids* **17**, 1688–1691 (1974).

E1. Edge, R. M., Flatman, A. T., Grant, C. D., and Kalafatoglu, I. E., *Symp. Multi-Phase Flow Syst., Inst. Chem. Eng., London* Paper C3 (1974).

F1. Feinman, J., Ph.D. Thesis, Univ. of Pittsburgh, 1964.

G1. Garner, F. H., and Lane, J. J., *Trans. Inst. Chem. Eng.* **37**, 162–172 (1959).

G2. Gibert, H., and Angelino, H., *Can. J. Chem. Eng.* **51**, 319–325 (1973).
G3. Gibert, H., and Angelino, H., *Int. J. Heat Mass Transfer* **17**, 625–632 (1974).
G4. Guthrie, R. I. L., Clift, R., and Henein, H., *Metall. Trans.*, *B* **6**, 321–329 (1975).
H1. Hamilton, W. S., and Lindell, J. E. *J. Hydraul. Div., Am. Soc. Civ. Eng.* **97**, 805–817 (1971).
H2. Hatim, B. M. H., Ph.D. Thesis, Imperial College, London, 1975.
H3. Herringe, R. A., *Chem. Eng. J.* **11**, 89–99 (1976).
H4. Hilprecht, L., *VDI Forschungsheft* 577 (1976).
H5. Hinze, J. O., "Turbulence," 2nd Ed. McGraw-Hill, New York, 1975.
H6. Hjelmfelt, A. T., and Mockros, L. F., *Appl. Sci. Res.* **16**, 149–161 (1966).
H7. Hjelmfelt, A. T., and Mockros, L. F., *J. Eng. Mech. Div., Am. Soc. Civ. Eng.* **93**, 87–102 (1967).
H8. Holland-Batt, A. B., *Trans. Inst. Chem. Eng.* **50**, 156–167 (1972).
H9. Holt, M. *Annu. Rev. Fluid Mech.* **9**, 187–214 (1977).
H10. Houghton, G., *Proc. R. Soc., Ser. A* **272**, 33–43 (1963).
H11. Houghton, G., *Nature (London)* **204**, 666–668 (1964).
H12. Houghton, G., *Can. J. Chem. Eng.* **44**, 90–95 (1966).
H13. Houghton, G., *Proc. Am. Philos. Soc.* **110**, 165–173 (1966).
H14. Houghton, G., *Chem. Eng. Sci.* **23**, 287–288 (1968).
H15. Hughes, R. R., and Gilliland, E. R., *Chem. Eng. Prog.* **48**, 497–504 (1952).
I1. Ingebo, R. D., *NACA* **TN 3762** (1956).
J1. Jameson, G. J., *Chem. Eng. Sci.* **21**, 35–48 (1966).
J2. Jameson, G. J., and Davidson, J. F., *Chem. Eng. Sci.* **21**, 29–34 (1966).
K1. Konopliv, N., *AIChE J.* **17**, 1502–1503 (1971).
K2. Krantz, W. B., Carley, J. F., and Al-Taweel, A. M., *Ind. Eng. Chem., Fundam.* **12**, 391–396 (1973).
K3. Kuo, C. Y., *J. Eng. Mech. Div., Am. Soc. Civ. Eng.* **96**, 177–180 (1970).
K4. Kurten, H., Raasch, J., and Rumpf, H., *Chem.-Ing.-Tech.* **38**, 941–948 (1966).
L1. Lai, R. Y. S., *J. Hydraul. Div., Am. Soc. Civ. Eng.* **99**, 939–957 (1973).
L2. Lai, R. Y. S., *Appl. Sci. Res., Ser. A* **27**, 440–450 (1973).
L3. Lai, R. Y. S., and Mockros, L. F., *J. Fluid Mech.* **52**, 1–15 (1972).
L4. Landau, L. D., and Lifshitz, E. M., "Fluid Mechanics." Pergamon, Oxford, 1959.
L5. LeClair, B. P., and Hamielec, A. E., *Fluid Dyn. Symp., McMaster Univ., Hamilton, Ont.* 1970.
L6. Lewis, J. A., and Gauvin, W. H., *AIChE J.* **19**, 982–990 (1973).
L7. Lin, C. L., and Lee, S. C., *Comput. Fluids* **1**, 235–250 (1973).
L8. Lumley, J., Ph.D. Thesis, John Hopkins University, Baltimore, 1957.
L9. Lunnon, R. G., *Proc. R. Soc., Ser. A* **110**, 302–326 (1926).
M1. Marchildon, E. K., Ph.D. Thesis, McGill Univ., Montreal, 1965.
M2. Marmur, A., and Rubin, E., *Can. J. Chem. Eng.* **53**, 560–562 (1975).
M3. Marmur, A., and Rubin, E., *J. Fluids Eng.* **98**, 483–487 (1976).
M4. Mathews, L. A., and Salt, D. L., *Q. J. R. Meterol. Soc.* **96**, 860 (1972).
M5. Mazur, P., and Bedeaux, D., *Physica* **76**, 235–246 (1974).
M6. Mockros, L. F., and Lai, R. Y. S., *J. Eng. Mech. Div., Am. Soc. Civ. Eng.* **95**, 629–640 (1969).
M7. Molerus, O., *Chem.-Ing.-Tech.* **36**, 866–870 (1964).
M8. Molerus, O., and Werther, J., *Chem.-Ing.-Tech.* **40**, 522–524 (1968).
M9. Molinier, J., Kuychoukov, G., and Angelino, H. *Chem. Eng. Sci.* **26**, 1401–1412 (1971).
M10. Moorman, R. W., Ph.D. Thesis, Univ. of Iowa, Iowa City, 1955.
M11. Mori, Y., Imabayashi, M., Hijikata, K., and Yoshida, Y., *Int. J. Heat Mass Transfer* **12**, 571–585 (1969).
M12. Morrison, F. A., and Stewart, M. B., *J. Appl. Mech.* **43**, 399–403 (1976).
M13. Morsi, S. A., and Alexander, A. J., *J. Fluid Mech.* **55**, 193–208 (1972).

M14. Murray, J. D., *J. Fluid Mech.* **28**, 417–428 (1967).
O1. Odar, F., *J. Fluid Mech.* **25**, 591–592 (1966).
O2. Odar, F., *U.S. Army, Corps Eng., Cold Reg. Res. Eng. Lab., Res. Rep.* **190** (1966).
O3. Odar, F., *J. Appl. Mech.* **90**, 238–241 (1968).
O4. Odar, F., *J. Appl. Mech.* **90**, 652–654 (1968).
O5. Odar, F., personal communication (1973).
O6. Odar, F., and Hamilton, W. S., *J. Fluid Mech.* **18**, 302–314 (1964).
O7. Ogden, T. L., and Jayaweera, K. O. L. F., *Q. J. R. Meterol. Soc.* **97**, 571–574 (1971); **98**, 861 (1972).
P1. Pedley, T. J., *J. Fluid Mech.* **32**, 97–112 (1968).
P2. Plesset, M. S., and Hsieh, D. Y., *Phys. Fluids* **3**, 882–892 (1960).
P3. Plesset, M. S., and Prosperetti, A., *Annu. Rev. Fluid Mech.* **9**, 145–185 (1977).
R1. Rafique, K., Ph.D. Thesis, Imperial College, London, 1971.
R2. Reichman, J. M., and Temkin, S., *Bull. Am. Phys. Soc.* **17**, 1117 (1972).
R3. Richards, W. D., Course 302-495B Final Rep., McGill Univ., Montreal, 1973.
R4. Rimon, Y., and Cheng, S. I., *Phys. Fluids* **12**, 949–959 (1969).
R5. Rschevkin, S. N., "The Theory of Sound." Academic Press, New York, 1963.
R6. Rudinger, G., *J. Basic Eng.* **92**, 165–172 (1970).
R7. Rudinger, G., *AIAA J.* **12**, 1138–1140 (1974).
S1. Sartor, J. D., and Abbott, C. E., *J. Appl. Meteorol.* **14**, 232–239 (1975).
S2. Schmiedel, J., *Phys. Z.* **29**, 593–610 (1928).
S3. Schmidt, F. S., *Ann. Phys. (Leipzig)* **61**, 633–664 (1920).
S4. Schöneborn, P.-R., Doktor-Ing. Dissertation, Univ. Erlangen, 1973; *Chem.-Ing.-Tech.* **47**, 305 (1975).
S5. Schroeder, R. R., and Kintner, R. C., *AIChE J.* **11**, 5–8 (1965).
S6. Schwartzberg, H. G., and Beckerman, G., *Chem. Eng. Prog., Symp. Ser.* **66**, No. 105, 127–132 (1970).
S7. Soo, S. L., "Fluid Dynamics of Multiphase Systems." Ginn (Blaisdell), Boston, Massachusetts, 1967.
S8. Sy, F., Taunton, J. W., and Lightfoot, E. N., *AIChE J.* **16**, 386–391 (1970).
S9. Sy, F., and Lightfoot, E. N., *AIChE J.* **17**, 177–181 (1971).
T1. Tchen, C. M., Ph.D. Thesis, Univ. Delft, 1947.
T2. Torobin, L., and Gauvin, W. H., *Can. J. Chem. Eng.* **37**, 224–236 (1959).
T3. Tudose, R. Z., *Int. Chem. Eng.* **4**, 664–666 (1964).
T4. Tunstall, E. B., and Houghton, G., *Chem. Eng. Sci.* **23**, 1067–1081 (1968).
V1. Van Oeveren, R. M., and Houghton, G., *Chem. Eng. Sci.* **26**, 1958–1961 (1971).
V2. Viets, H., *AIAA J.* **9**, 2087–2089 (1971).
W1. Walters, J. K., and Davidson, J. F., *J. Fluid Mech.* **12**, 408–416 (1962).
W2. Walters, J. K., and Davidson, J. F., *J. Fluid Mech.* **17**, 321–336 (1963).
W3. Wang, C. C., M. Eng. Thesis, McGill Univ., Montreal, 1969.
W4. Weiner, A., Ph.D. Thesis, Univ. of Pennsylvania, Philadelphia, 1974.
W5. Winnikow, S., and Chao, B. T., *Phys. Fluids* **9**, 50–61 (1966).

Chapter 12

Formation and Breakup of Fluid Particles

I. INTRODUCTION

In earlier chapters, we have considered steady and unsteady motion and transfer processes for fluid and rigid particles without treating the initiation or termination of these processes. This final chapter is concerned with formation and breakup of fluid particles. The problems are distinct from those encountered in Chapter 11 for the unsteady motion of rigid particles. A single rigid particle can be launched, dropped, or suspended in a fluid flow,† but a fluid particle must be formed at the same time as it is launched. In keeping with earlier chapters, we consider here only cases where single particles or very dilute dispersions are generated. This division is necessarily arbitrary, since many techniques can be used to produce dilute or concentrated clouds of particles. Space limitations have severely restricted treatment, but the reader is referred to relevant reviews where these exist.

II. FORMATION OF BUBBLES AND DROPS

Generation of small bubbles and drops is essential in a wide range of phase-contacting equipment. In bubble columns, fermentation vessels, extraction equipment, etc., bubbles and drops are usually formed by forcing the dispersed phase through orifices or a porous sparging device into the continuous phase, frequently with mechanical agitation to aid dispersion. A variety of atomizers, spray nozzles, and sprinklers have been devised for dispersing liquids into gases. In most of these applications, the objective is to produce a cloud of

† See (H16, L14) for reviews on dispersal of solid powders.

small fluid particles. Since we are concerned here only with single particles, discussion is limited to simple techniques, capable of producing single or widely spaced drops or bubbles. Mechanically agitated or rotated delivery devices are not considered, nor are devices using impinging jets, impaction, ultrasonic or mechanical vibrations, electrical forces, etc. For general reviews of spraying, atomization, and injection devices, see (G2, G6, H16, L7, O1, S19).

A. Formation at an Orifice

A number of workers (H11, J1, K2, K6, K15, L7, S25, V1) have reviewed the formation of bubbles or drops by flow through orifices or nozzles. Here, we consider only injection at modest flow rates through single orifices of diameter d_{or} less than about 0.65 cm. At high velocities and for large orifices, significant jetting and multiple particle formation occur (J2, S25). Sections 1 and 2 are concerned only with single orifices of circular cross section facing in the flow direction, i.e., upward for $\rho_p < \rho$ and downward for $\rho_p > \rho$. The continuous phase is stagnant except for motion caused by flow of the dispersed phase. More complex situations are treated briefly in Sections 3 and 4, and mass transfer during formation is discussed in Section 5.

1. *Bubble Formation*

As a bubble is formed by flow of gas through an upward-facing orifice, the pressure within the bubble decreases due to upward displacement of its centroid and to decrease in the capillary pressure, $2\sigma/r$. Thus, the gas flow rate may vary with time. If there is a high pressure drop restriction, such as a long capillary, between the gas reservoir and the orifice, the pressure fluctuations due to forming bubbles are much smaller than the pressure drop between the gas reservoir and the orifice. In this case, the gas flow rate can be taken as constant. Otherwise, account must be taken of both the "line" pressure drop and the reservoir volume. If the volume of the reservoir or "plenum chamber" upstream of the orifice is very large by comparison with the volume of bubbles being formed, the varying gas efflux will not significantly change the pressure in the chamber. This corresponds to the other limiting case of bubble formation under constant pressure conditions. For conditions intermediate between the limits of constant flow and constant pressure, the chamber volume V_{ch} must be taken into account. Unfortunately, some workers have failed to report the characteristics of their orifices and chambers so that their results are hard to interpret.

Thus bubble formation at an orifice is a surprisingly complex phenomenon. For intermediate conditions and a perfectly wetted orifice, the volume of the bubble formed may be written[†]:

$$V = f(\bar{Q}, d_{or}, \rho, \mu, \sigma, \rho_p, K, V_{ch}, g, H), \tag{12-1}$$

[†] Even this list is not always complete. For example, if the orifice is a tube projecting into and poorly wetted by the liquid, the outer tube diameter is also important (S2).

where $\bar{Q}$ is the time-mean flow rate, K the "orifice constant," and H the submergence. For most practical purposes, $\mu_p \ll \mu$ and μ_p can be omitted. Usually ρ_p can also be removed (D5), since $\rho_p \ll \rho$ for bubble formation except for cases, such as high pressure formation, where the momentum of the incoming gas must be considered (L2).

a. *Theoretical Models* The many models proposed to describe bubble formation in liquids are summarized in Table 12.1. All are mechanistic in the sense that they are based on a sequence of events suggested by photographic observation. All depend on some form of force balance for predicting one or more stages in bubble growth. Almost all approximate the bubble as spherical throughout the growth period. The simplest group may be termed "one-stage models." In these, bubbles originating at the orifice are assumed to grow smoothly until detachment, which occurs when the rear of the bubble passes

TABLE 12.1

Theoretical Models for Bubble Formation at a Submerged Orifice

Ref.	Conditions	Number of stages	Forces included[a]	Comments
(H8)	Constant flow	1	B, Dd, I, P, S	
(S22)	Constant flow	1	B, Dd, I, P, S	Same as (H8) but includes cross flow
(D5)	Constant flow	1	B, Da	Low flow rates
(D5)	Constant flow	1	B, Da, Ia	Higher flow rates
(D6)	Constant flow	1	B, Ia	High flow rate, low-viscosity liquid
(D3)	Constant flow	1	B, Ib	High flow rate, low-viscosity liquid
(W4)	Constant flow	1	B, Iab, S, W	Low-viscosity fluid
(K13)	Constant flow	2	B, Da, S	
(K16)	Constant flow	2	B, Da, Ia	
(R2)	Constant flow	2	B, Da, Ia, S	
(C1)	Constant flow	2	B, Dc, Ib, S, W	Also extended to coflowing stream
(W8)	Constant flow	2	B, Ia, M	Hemispherical in first stage
(R11)	Constant flow	2	B, Dc, Ia, S	
(T7)	Constant flow	1	B, Dbc, S	
(D5)	Constant pressure	1	B, Da	Low flow
(D5)	Constant pressure	1	B, Da, Ia	High flow
(D6)	Constant pressure	1	B, Ia	High flow, low-viscosity liquids
(S3)	Constant pressure	2	B, Da, Ia, S	
(L3)	Constant pressure	1	B, Ia	Extension of (D6)
(K4)	Intermediate	2	B, D, Ia, S	
(K17)	Intermediate	3	B, I	
(M5)	Intermediate	—	B, Ib, W	Shape originally hemispherical
(S24)	Intermediate	1	B, Db, M, P, S	
(L2)	Intermediate	1	B, Ia, M	
(M2)	Intermediate	1	B, Ic, S	Shape varies; numerical solution

[a] B: buoyancy; D: drag (a: Stokes; b: Hadamard; c: empirical expression; d: kept as constant to fit to data); I: inertia (a: $C_A = 11/16$; b: $C_A = 1/2$; c: C_A kept as constant to fit to data); M: gas momentum; P: excess pressure term; S: surface tension force; W: wake effect from previous bubble.

the orifice or when buoyancy exceeds the retarding forces. In "multiple-stage models," it is assumed that there is a basic change in the growth mechanism at one or more points in the growth process. Typically, it is assumed that the bubble resides on the orifice during the first stage, and that the second stage begins at "lift-off," with the bubble subsequently fed by a tongue of gas from the orifice. There is some photographic evidence for this sequence of events [e.g. (D5, K17)].

In view of the complexity of the bubble formation process, it is not surprising that the models are successful only under restricted conditions. The simplest models, and the only ones to give simple analytic expressions for the volume of the bubble produced, apply for constant flow formation. All the models have inherent limitations:

(i) The assumption that bubbles remain spherical is reasonable for most low M systems, but can be significantly in error for large M systems (M2, W8).
(ii) Assumptions regarding the sequence of events, in particular the criteria for such events as lift-off and detachment, are often arbitrary. In some models, force balances are applied throughout, while in others they are applied only as a means of predicting the volume at the end of one stage in growth.
(iii) When surface tension forces and contact angles are included, the σ and θ_c used are invariably determined under static conditions, even though bubble formation is a dynamic process.
(iv) Expressions used for drag and added mass are at best approximate. No allowance is made in any of the models for history effects, which may well be important since $\rho \gg \rho_p$ (see Chapter 11).
(v) Terms such as the updraught due to the wake of the preceding bubble are generally ignored, but may be important. Individual models also ignore other terms (see Table 12.1), often without adequate justification.

b. *Constant Flow Conditions* Despite the shortcomings noted in many of the models, useful results can be obtained for constant flow conditions by judicious combination of dimensional analysis, force balances, and empirical results. Neglecting μ_p and ρ_p for the reasons given above and noting that K, V_{ch} and H are not required when Q is constant, we write the reduced form of Eq. (12-1) in terms of a dimensionless bubble volume as

$$V' = Vg\,\Delta\rho/d_{or}\sigma = f(Q', \mu', M), \tag{12-2}$$

where Q' is a dimensionless flow rate:

$$Q' = (\rho/d_{or}\sigma)^{5/6}g^{1/3}Q \tag{12-3}$$

and μ' is a dimensionless viscosity:

$$\mu' = \mu/\sqrt{\rho d_{or}\sigma}. \tag{12-4}$$

For bubble formation with $\rho_p \ll \rho$, $\Delta\rho$ is taken as ρ. However, V' is defined in terms of $\Delta\rho$ to facilitate its interpretation as the magnitude of gravitational forces relative to surface tension forces, and to aid comparison with drop formation. If it is assumed that bubbles remain essentially spherical throughout formation, then σ and the orifice diameter d_{or} enter only as $d_{or}\sigma$ which can be treated as a single quantity. Hence M can be omitted, so that all the simple models take the form

$$V' = f(Q', \mu'). \tag{12-5}$$

For very low gas rates, i.e., $Q' \to 0$,

$$V' = \pi\psi_H, \tag{12-6}$$

where ψ_H is the Harkins correction factor (H5) given in Fig. 12.4 and discussed below. This correction factor makes allowance for dispersed phase fluid retained at the orifice when detachment occurs. Equation (12-6)† is frequently employed in measuring surface and interfacial tensions by the bubble forming or "drop-weight" method [e.g., see (D8)].

At high flow rates with liquids of low viscosity (i.e., relatively large Q', small μ'), a simple equation developed by Davidson and Schüler (D6) is commonly used, i.e.,

$$V' = 1.378(Q')^{1.2}, \tag{12-7}$$

or, in dimensional form,

$$V = 1.378Q^{1.2}g^{-0.6}. \tag{12-8}$$

The numerical coefficient in Eq. (12-7) is obtained using an added mass coefficient, C_A, of 11/16, for a spherical bubble forming at a perforation in a flat plate. For a nozzle protruding into a fluid $C_A = 1/2$ and the coefficient becomes 1.138 (D3, W2). An early empirical correlation (V3) gave a value of 1.72. Since the mean frequency of bubble formation is Q/V, Eq. (12-8) predicts that the frequency becomes only weakly dependent on flow rate at relatively high flow. In practice, the dependence becomes even weaker than the -0.2 power predicted, with the mean frequency of bubble formation becoming essentially independent of Q (e.g., see Fig. 12.2). Hence, for bubble formation in liquids of low viscosity like water, it is common [e.g. (V1)] to describe formation at low flow rates where Eq. (12-6) applies as "constant volume formation" and that at quite high Q as "constant frequency formation," with an "intermediate region" for the range where neither result applies. For viscous liquids (high μ') and intermediate Q', another equation developed by Davidson and Schüler (D5) gives

$$V' = 6.48(Q'\mu')^{0.75}. \tag{12-9}$$

Ruff (R11) has developed a semiempirical model which approximates to Eqs. (12-6), (12-7), and (12-9) in the appropriate limits. It may be regarded as

† For a projecting nonwetted nozzle, the outer diameter of the tube should be used in place of the inner (S2) in evaluating V'.

an improvement on the model of Kumar and Kuloor (K15) with a better expression for the drag coefficient and an empirical correlation rather than an arbitrary model to describe the second stage of growth. Two successive stages in bubble formation are considered, with

$$V' = V_1' + V_2'. \tag{12-10}$$

The bubble volume achieved in the first stage is predicted from a force balance:

$$V_1' - 0.0578(V_1')^{-2/3}(Q')^2 - 2.417(V')^{-1/3}Q'\mu' - 0.204Q'\sqrt{\mu' Q'/V_1'} = \pi, \tag{12-11}$$

where the first term arises from gravitational forces, the second results from inertia and drag, the third and fourth from drag, and that on the right-hand side from surface tension. The volume added in the second stage was correlated empirically as

$$V_2' = (Q')^{1/2} + 4.0(Q'\mu')^{3/4}. \tag{12-12}$$

Dimensionless bubble volumes predicted by solving Eq. (12-11) numerically and adding the second stage increment are plotted in Fig. 12.1 as functions of the dimensionless flow rate Q', with μ' as parameter. It is important to note that μ' is constant for a given orifice in a given gas-liquid system. Hence, Fig. 12.1 or Eqs. (12-10) to (12-12) give a convenient means of predicting bubble formation in any liquid–gas system whose properties are known. Over most of

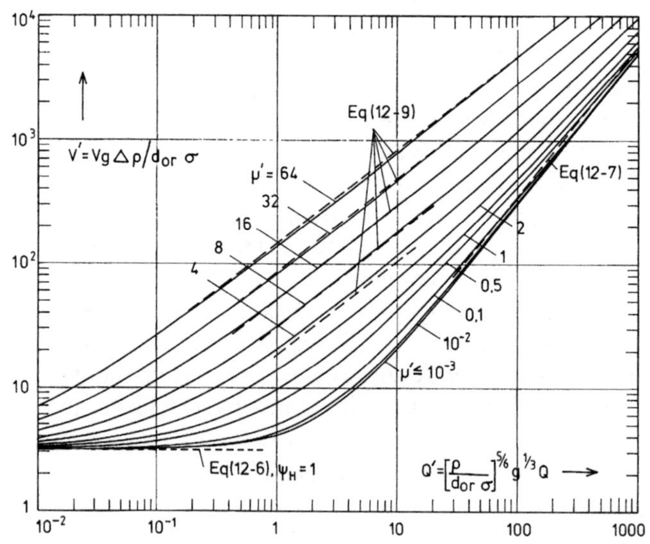

FIG. 12.1 Variation of dimensionless bubble volume, $V' = Vg\Delta\rho/d_{or}\sigma$, with dimensionless gas flow rate $Q' = (\rho/d_{or}\sigma)^{5/6}g^{1/3}Q$ for bubble formation under constant flow conditions: predictions of Ruff model with limiting cases also shown.

the range of Q', the correlation agrees within $\pm 15\%$ with most data appearing in the literature [e.g. (B5, D5, D6, K13, K16, R2, S11, S22, W8)]. Discrepancies for bubbles formed in liquid metals (A2, S2) are more serious, possibly because of experimental difficulties, surface effects, or bubble deformation at the low values of M characteristic of liquid metal–gas systems. Also shown in Fig. 12.1 are lines corresponding to Eqs. (12-6), (12-7), and (12-9). These simplified equations may be viewed as limiting cases, with the ranges of Q' and μ' for their application indicated by Fig. 12.1.

Predictions of Ruff's model for air bubbles forming in water are shown in dimensional form by the two solid lines in Fig. 12.2 for two commonly used orifice diameters, 0.63 and 0.32 cm (1/4 and 1/8 inch). Some data are also shown for the larger orifice and agreement is generally very favorable. Note that the orifice diameter plays an important role only at low Q, where surface tension provides the major restraining force.

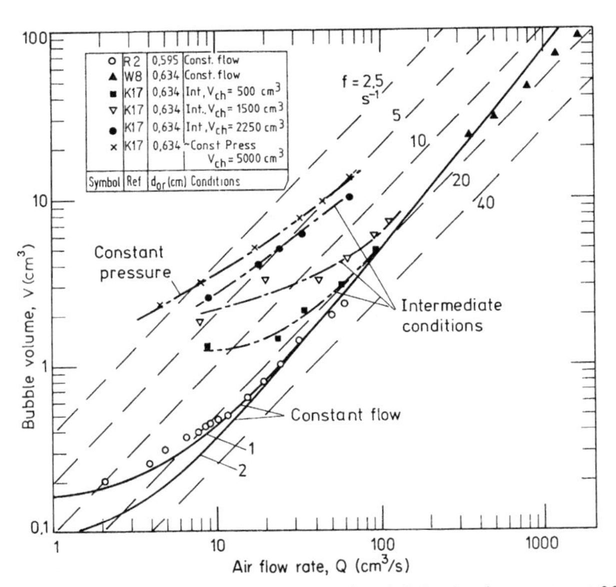

FIG. 12.2 Bubble volume as a function of flow rate for air injection into water at 20°C. Curves for constant flow obtained from Ruff model, Eqs. (12-10) to (12-12): (1) $d_{or} = 0.63$ cm, $\mu' = 1.5 \times 10^{-3}$; (2) $d_{or} = 0.32$ cm, $\mu' = 2.1 \times 10^{-3}$. Experimental results shown for constant flow, intermediate, and constant pressure conditions.

Deviations from the theories tend to occur at large Q where the frequency of bubble formation becomes essentially independent of Q, whereas theory predicts $f \propto Q^{-0.2}$. For example, the frequency in air–water systems levels out at about 17 s^{-1} as shown in Fig. 12.2. This almost certainly results from the updraught caused by preceding bubbles, ignored in almost all the models. At still higher flow rates, bubble pairing occurs at the orifice, when a bubble coalesces with that just formed before it can escape. Incipient pairing occurs for

$$Q > Cg^{1/2}d_{or}^{5/2}, \tag{12-13}$$

where the constant coefficient C has been given values from 1.3 to 6.2 (W2, W8).

c. *Constant Pressure Conditions* Bubble formation under constant pressure conditions is of practical interest for sieve trays and other multiorifice distributors. As noted above, the flow rate through the orifice varies with time. The "orifice equation" may be written

$$Q = K(p_{ch} - \rho gH + \rho ga - 2\sigma/a)^{1/2}, \tag{12-14}$$

where a is the instantaneous radius of the forming bubble and p_{ch} the pressure in the chamber behind the orifice. In general, Eq. (12-14), or a more complete orifice equation like that proposed by Potter (P10), must be solved simultaneously with a force balance equation to predict initial bubble volumes. Models are outlined in Table 12.1. Due to the added complexity of the formation process, analytic results cannot be summarized neatly. Dimensional quantities K and $p(= p_{ch} - \rho gH)$ are required in addition to those needed for the constant flow case, so that dimensionless presentation of the results also becomes cumbersome. Furthermore, fewer experimental studies have been reported for constant pressure than for constant flow conditions.

If accurate predictions of bubble volume are required, the original models should be consulted. The Marmur model (M2) appears to work well for low-viscosity liquids over a broad range of V_{ch} and Q, but is too complex to be useful for normal predictive purposes. Of the relatively simple models, the most reliable are those of Lanauze and Harris for low μ' (L3, L4), and those of Kumar and Kuloor (K15, S3) and Davidson and Schüler (D5) for viscous liquids.

For many purposes, approximate predictions suffice, and may be obtained from the results for constant flow formation using some simple guidelines. Bubbles obtained under constant pressure tend to be larger than under constant flow conditions at the same time-mean flow rate, $\bar{Q}$, because most of the flow with variable Q occurs during the latter stages of formation. It is convenient to define a ratio of bubble volumes formed under constant pressure and constant flow conditions as

$$Y = V_{CP}/V_{CF} = f_{CF}/f_{CP}. \tag{12-15}$$

For large Q', $\mu' \to 0$, and large values of the dimensionless orifice constant,

$$K' = Kg^{0.4}\rho^{0.5}/Q^{0.8}, \tag{12-16}$$

Y approaches approximately 1.5 (D6). As $K' \to 0$, constant flow conditions are approached and $Y \to 1$. For intermediate and low flow rates, Y may be as high as 10, as shown by the curve for approximately constant pressure conditions in Fig. 12.2.

d. *Intermediate Conditions* The importance of the chamber volume, V_{ch}, was first recognized by Hughes *et al.* (H23) and Davidson and Amick (D7). If V_{ch} is relatively small and the orifice constant relatively large, then both the flow rate through the orifice and the pressure in the chamber vary with time. Practical interest, for orifices in such devices as sieve trays in phase-contacting equipment, lies as much in reverse passage of the continuous phase ("weeping") as in bubble formation.

Models used to describe bubble formation under intermediate conditions are listed in Table 12.1. Generally, they must be solved numerically. Reasonable agreement has been obtained (K18, M2, M5) for low-viscosity liquids so long as "pairing" at the orifice did not occur. In addition to pairing, six other flow regimes have been identified, and charted for formation in water at three sizes of orifice (M6). Following Hughes *et al.* (H23), many workers have used a capacitance number

$$N_{ch} = 4g\,\Delta\rho\, V_{ch}/\pi d_{or}{}^2 \rho_p c^2, \tag{12-17}$$

where c is the velocity of sound in the gas, but in other studies [e.g. (K17)] this group was found to have no significance. In general, bubbles produced under intermediate conditions are intermediate in size between those formed at constant pressure and constant flow at the same $\bar{Q}$. The results in Fig. 12.2 for a 0.63 cm diameter orifice illustrate the effect of increasing chamber volume and thus going from constant flow to constant pressure.

e. *Bubble Formation in Fluidized Beds* As noted in Chapter 8, the surface tension between the bubble and dense phase of a fluidized bed is generally taken to be zero. Equations (12-10) to (12-12) can be rewritten as

$$V'' = V_1'' + V_2'', \tag{12-18}$$

$$V_1'' - 0.0578(Q'')^2(V'')^{-2/3} - 2.417Q''(V'')^{-1/3} - 0.204(Q'')^{1.5}(V'')^{-1/2} = 0, \tag{12-19}$$

and

$$V_2'' = (Q'')^{1.2} + 4.0(Q'')^{0.75}, \tag{12-20}$$

where

$$V'' = Vg/\nu^2 \tag{12-21}$$

and

$$Q'' = Qg^{1/3}/\nu^{5/3}. \tag{12-22}$$

Hence bubble formation in fluidized beds may be predicted in a manner similar to that employed for liquids using an effective dense phase kinematic viscosity ν. The equations reduce to the Davidson and Schüler forms, Eqs. (12-7) and (12-9), for large and small Q'', respectively, and show reasonable agreement with experimental results (H7). For bubble formation, it is best to take the effective kinematic viscosity of the bed as about 0.5 cm^2/s, a value an order of magnitude less than the value derived from bubble shape measurements (see Chapter 8).

At flow rates greater than about 100 cm^3/s, the frequency becomes essentially constant at a value of about 20 s^{-1}, close to the value for bubble formation in real liquids (B4, H7). The volume of bubbles formed, however, is generally less than Q/f due to leakage of gas into the dense phase (N3). The length of jets, when these occur, feeding forming bubbles is correlated (M12) by the equation

$$\frac{L_{\text{jet}}}{d_{\text{or}}} = 5.2\left(\frac{\rho_g d_{\text{or}}}{\rho_s d_s}\right)^{0.3}\left[1.3\left(\frac{u_{\text{or}}^2}{g d_{\text{or}}}\right)^{0.2} - 1\right]. \tag{12-23}$$

Some effect of chamber volume has been demonstrated (H21) for low pressure drop orifices, as for bubbles forming in liquids.

2. *Drop Formation*

a. *Regimes of Jet Formation* When a liquid of density ρ_p issues steadily from a horizontal orifice into an immiscible fluid of density ρ, drops may form at the orifice or at the end of a disintegrating cylindrical jet as shown schematically in Fig. 12.3. At low flow rates, formation occurs close to the orifice, as for gas bubbles. As Q is increased, a critical flow, Q_{jet}, is reached at which a jet forms. At higher flow rates, drops form by jet breakup. At flow rates between Q_{jet} and a value labeled $Q_{\max}$, break up occurs by Rayleigh instability (i.e., axisymmetric amplification of surface perturbations), while the jet length in-

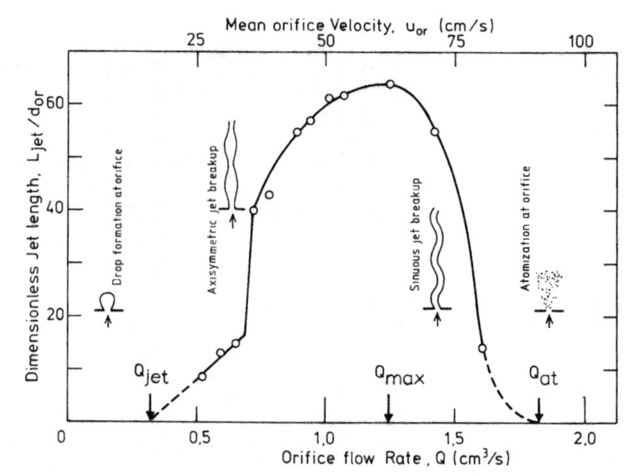

FIG. 12.3 Typical jet development for injection of a liquid through an orifice into another liquid: experimental results of Meister and Scheele (M10) for heptane injection into water with $d_{\text{or}} = 0.16$ cm, $\mu_p = 0.00393$ poise, $\rho_p = 0.683$ gm/cm^3, $\sigma = 36.2$ dynes/cm.

creases with increasing Q. Above Q_{max}, the jet length decreases again, and breakup results primarily from growth of asymmetric disturbances. The jet length decreases until, for $Q > Q_{at}$, the liquid shatters at or very near the orifice to give many droplets of nonuniform size. Atomization of liquids by forcing through an orifice ("pressure atomization") is treated in a number of reviews [e.g. (B6, F6, G2, G6, H16)]. Only drop formation at much lower flow rates is discussed here.

The condition for incipient jetting has been derived by Scheele and Meister (S4) as

$$Q_{jet} = 1.36\sqrt{\frac{\sigma d_{or}^{3}}{\rho_p}[1 - d_{or}/(1.24V^{1/3})]}, \tag{12-24}$$

where V is the volume of drops which would form if jetting did not occur, obtained from Eq. (12-28) below. The numerical coefficient, 1.36, applies when the velocity profile in the jet at the orifice is parabolic; a coefficient of 1.57 should be used for a flat velocity profile.

b. *Formation at Low Flow Rates* Drop formation with $Q < Q_{jet}$ occurs at the orifice, and is qualitatively similar to bubble formation. Quantitative differences arise because the momentum and viscosity of the entering fluid are often appreciable for drops, but rarely for bubbles. The momentum effect is particularly important, and causes drops to be smaller than those formed under near-static conditions ($Q' \to 0$) where

$$V = \psi_H \pi d_{or}\sigma/\Delta\rho\, g, \tag{12-25}$$

which is the dimensional form of Eq. (12-6) given above for bubbles. The Harkins factor, ψ_H, accounts for the fact that a residual drop remains at the orifice when detachment occurs, causing the volume of the detached drop, V, to be less than the volume at which the net gravity force exactly balances the interfacial tension forces. Smoothed values of ψ_H (H5) are shown in Fig. 12.4 as a function of $d_{or}/V^{1/3}$, together with the fitted equations (H13, L5):

$$\psi_H = \left\{0.92878 + 0.87638\frac{d_{or}}{V^{1/3}} - 0.261\left(\frac{d_{or}}{V^{1/3}}\right)^2\right\}^{-1} \quad (0.6 < d_{or}/V^{1/3} < 2.4) \tag{12-26}$$

$$\psi_H = 1.000 - 0.66023\frac{d_{or}}{V^{1/3}} + 0.33936\left(\frac{d_{or}}{V^{1/3}}\right)^2 \quad (0 \le d_{or}/V^{1/3} \le 0.6) \tag{12-27}$$

If the continuous fluid does not wet the orifice material, corrections must be made as for bubbles.

As Q' is increased slowly, modifications in shape occur (H3). As for bubble formation, many equations and models have been proposed for predicting

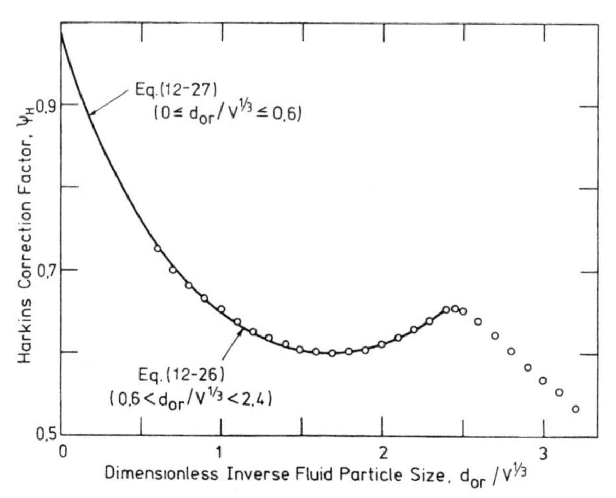

FIG. 12.4 Harkins' correction factor for formation of drops or bubbles, accounting for residual volume retained at the orifice when detachment occurs.

initial drop volumes. Some of these [e.g. (G8, H9, N1, N4)] are almost entirely empirical. Even though these empirical approaches are simple to use and have enjoyed some popularity, they should be employed with caution outside the range of variables investigated by the workers who derived them. For example, some correlations include no viscosity term, and are therefore very unlikely to apply to viscous liquids. Several more mechanistic, semiempirical models have also been proposed for drop formation in liquid–liquid systems [e.g. (H13, K9, R14, S4)]. Because liquids are essentially incompressible, drop formation corresponds closely to bubble formation under constant flow conditions. Many of the underlying ideas are the same as those in models for bubble formation, and the same criticisms apply. For example, the drop is generally assumed to remain spherical throughout formation. Formation is again usually treated as a two-stage process, as revealed by photographic observation (S4): the first stage terminates at "lift-off," predicted by a force balance, while the second stage corresponds to "necking" and eventual severing of the liquid filament. Kumar and co-workers (K14, K15, R5) have attempted to establish models general enough to cover both drop and bubble formation with constant flow. Grigar *et al.* (G7) give a method for calculating pressure drops across orifices or nozzles, accounting for both hydraulic and drop formation effects. Humphrey *et al.* (H25) give a detailed experimental analysis of flow patterns in forming drops [see also (S1)].

Unfortunately, none of the models for drop formation can be recommended with complete confidence. Most of the experimental results for liquid–liquid systems have been obtained with water as one phase and a low-viscosity organic liquid as the other. Under these conditions, especially at modest flow rates ($u_{or} \lessdot 15$ cm/s, a number of the models predict drop volumes within 15–20%.

Scheele and Meister (S4), for example, give

$$Vg\,\Delta\rho = \psi_{\mathrm{H}}\left[\pi\sigma d_{\mathrm{or}} + \frac{13\mu Q d_{\mathrm{or}}}{V^{2/3}} - \frac{16\rho_{\mathrm{p}} Q^2}{3\pi d_{\mathrm{or}}^2} + \frac{9}{2}\{g\rho_{\mathrm{p}}\,\Delta\rho\,\sigma d_{\mathrm{or}}^2 Q^2\}^{1/3}\right], \quad (12\text{-}28)$$

where the terms account, respectively, for buoyancy, interfacial tension, Stokes' drag, momentum, and volume added during necking. The factor 16/3 in the third term assumes a parabolic velocity profile in the orifice, and should be replaced by 4 for a flat velocity profile. Although appreciable discrepancies may exist (D9, K9) and the prominence given to ψ_{H}, derived under static conditions, is questionable, Eq. (12-28) appears to be the best compromise between accuracy and ease of use for drop formation with $Q < Q_{\mathrm{jet}}$, especially for low-viscosity systems.

c. *Formation by Jet Disintegration* With $Q_{\mathrm{jet}} < Q < Q_{\mathrm{at}}$, drop size is governed by jet stability. Rayleigh (R6) was the first to apply linearized stability analysis to the growth of small axisymmetric disturbances on a cylindrical jet, and his treatment has been extended to account for viscous and nonlinear effects [see, e.g. (L11, M9, P3, T8, Y1)]. For most purposes, the linearized theory is surprisingly accurate for $Q_{\mathrm{jet}} < Q < Q_{\mathrm{at}}$. It is assumed that the amplitude of disturbance grows as

$$a' = a'_0 \exp(\alpha t + 2\pi i x/\lambda), \quad (12\text{-}29)$$

where a'_0 is the initial amplitude, λ the wavelength, and x the distance along the jet. Values for the "growth rate," α, can be obtained by solving the fourth-order determinant equation derived by Tomotika (T8), or by an approximate procedure proposed by Meister and Scheele (M9). The drop size follows as

$$d_{\mathrm{e}} = (1.5\lambda_{\mathrm{m}}/d_{\mathrm{or}})^{1/3} d_{\mathrm{or}}, \quad (12\text{-}30)$$

where λ_{m} is the "most dangerous" wavelength, corresponding to disturbances with the fastest growth (i.e., maximum α). Rayleigh considered a water jet issuing into air, and by neglecting the viscosity of both phases obtained $\lambda_{\mathrm{m}} = 4.5 d_{\mathrm{or}}$ and $d_{\mathrm{e}} \approx 1.9 d_{\mathrm{or}}$, results which are in reasonable agreement with experimental findings. Other limiting cases can also be obtained (M9) from the Tomotika equation. The linearized analysis has been extended to breakup of stationary liquid threads (R13) and threads undergoing extensional flow (M14).

Meister and Scheele (M10) examined phenomena determining the jet length, L_{jet}. For Q somewhat greater than Q_{jet}, L_{jet} can be predicted from the linearized stability theory as the distance required for a symmetric disturbance to grow to an amplitude equal to the jet radius.† For the apparatus and conditions

† The analysis also explains why jets are not normally observed for bubble formation at an orifice. When a gas is injected into a liquid, unstable disturbances amplify to the radius of the orifice within a very short distance (M10). Some data on jet lengths for gas flow into liquids have been published (P4).

investigated the initial amplitude was well approximated by

$$a_0' = (d_{or}/2)\exp\{-6.0\} = 0.00124 d_{or}. \tag{12-31}$$

At somewhat higher Q, drops in many liquid–liquid systems merge since their terminal velocity is less than the jet velocity; as a result, the jet lengthens abruptly. This occurs, for example, at a flow rate of about 0.7 cm^3/s for the experimental results shown in Fig. 12.3. Sinuous disturbances are shown to be unimportant at low Q, but at higher Q they account for the attainment of a maximum jet length since they are capable of ejecting drops sideways, out of the path of the oncoming jet. Mass transfer affects jet breakup if concentration variations are sufficient to cause significant gradients of interfacial tension (B9).

For low-viscosity liquid–liquid systems, tables presented by Hozawa and Tadaki (H19) offer an alternative means of predicting initial drop sizes for $Q_{jet} < Q < Q_{at}$. This method also has the advantage of giving an estimate of the spread in drop sizes caused by pairing and other complex interaction effects. Empirical expressions are also available for predicting the size of drops produced by jet break up [e.g. (P2, T1)].

3. *Influence of Orifice Shape and Orientation*

Kumar and Kuloor (K15) have surveyed work on the influence of orifice shape and orientation. For very low flow rates, where surface tension effects are dominant, bubbles appear to form from an equisided orifice, such as an equilateral triangle or regular hexagon, as from the inscribed circular orifice. At higher flow rates, an orifice with a shape not too far removed from circular gives roughly the same bubble volume as the circular orifice of the same area at the same flow rate. Irregular geometries, such as elongated rectangular slots, show more complex behavior.

Inclining an orifice may increase or decrease the volume of bubbles formed, although the effect of orientation is often rather small (S22). Kumar and Kuloor (K15) extended their two-stage model to an orifice inclined at an arbitrary angle to the horizontal. Equations are given both for constant flow and constant chamber pressure, and agreement with experiment is favorable. Analogous equations are suggested for drops forming at inclined orifices.

4. *Influence of Flow of Continuous Fluid*

If the continuous fluid has a net vertical velocity component, the additional drag causes earlier or later detachment and hence reduces or increases the volume of particle formed according to whether the drag force assists or impedes detachment. Significantly smaller bubbles or drops can be produced by causing the continuous fluid to flow cocurrently with the dispersed phase (C1).

Horizontal components of velocity also tend to affect the volume of bubbles and drops produced at an orifice. At low Q there is little effect, but larger bubbles tend to be produced as the horizontal mean velocity is increased at intermediate Q (S22), presumably because of a reduction in the updraught effect noted above.

5. *Mass Transfer during Formation*

Mass transfer during formation of drops or bubbles at an orifice can be a very significant fraction of the total mass transfer in industrial extraction or absorption operations. Transfer tends to be particularly favorable because of the exposure of fresh surface and because of vigorous internal circulation during the formation period. In discussing mass transfer in extraction, it has become conventional (H12) to distinguish four steps: (1) formation, (2) release, (3) free rise or fall, (4) coalescence. Free rise or fall has been treated in previous chapters. Steps 1 and 2 are considered here.

By making mass transfer during coalescence negligible and by varying column heights, one can determine mass transfer during formation and release by extrapolation to zero column height. However, it is difficult to apportion this transfer between formation and release. In the period immediately after detachment, mass transfer rates may be high due to internal circulation, shape oscillations, and acceleration. Theoretical models have been proposed for formation, but not for release. Empirical correlations also exist for the formation step, but usually include transfer during release. We consider here models for "slow formation," i.e., formation without internal circulation, and "fast formation" in which the momentum of the entering fluid causes circulation within the forming bubble or drop. Both situations correspond to flow rates below Q_{jet}.

a. *Slow Formation* The most realistic models for slow formation are based upon one of two assumptions: "surface stretch," in which the fluid at the interface is assumed to remain there throughout formation, and "fresh surface," in which fresh fluid elements are assumed to arrive at the interface to provide the increase in area. The interfacial area during formation is assumed to be

$$A = A_{\text{R}} + \beta t^{n}, \tag{12-32}$$

where A_{R} is the surface area of the residual drop or bubble left at the orifice after detachment. Values of n of 2/3 (R4) and 1 (H10) have been reported for drop formation at constant flow rate. Combining Eq. (12-32) with Eq. (7-51) or (7-52) yields

$$\overline{kA} = 2BA_{\text{f}}\sqrt{\mathscr{D}/\pi t_{\text{f}}}, \tag{12-33}$$

where A_{f} is the surface area of the drop or bubble at detachment and t_{f} is the duration of the formation process. Values for the constant B are given in Table 12.2. Equation (12-33) can be used when resistance in the dispersed phase controls if $\overline{kA}$ is replaced by $(\overline{kA})_{\text{p}}$ and $\mathscr{D}$ by $\mathscr{D}_{\text{p}}$. The predicted values of $\overline{kA}$ are not strongly sensitive to n, and differ little between the two models. Experimental data are not sufficiently accurate to differentiate between them, and good agreement has been found for drops with the controlling resistance in either phase as long as there is no circulation during formation (G9, H12, P8, P9, R4). Corrections have been given to account for changing concentration within the forming drop (W3) and for curvature of the interface for very small drops (N2).

TABLE 12.2

Models for Transfer during Slow Formation of Bubbles and Drops

Surface stretch model:

$$B = \frac{1}{(2n+1)^{1/2}}\left[1 + \frac{2n}{n+1}\left(\frac{A_R}{A_f}\right)\left\{1 + n\left(\frac{A_R}{A_f}\right)\right\}\right]^{1/2}$$

For $n = 1$, $A_R = 0$: $B = 0.577$. For $n = \frac{2}{3}$, $A_R = 0$: $B = 0.655$.

Fresh surface model:

$$B = b + (1-b)\frac{A_R}{A_f}, \qquad b = \frac{\sqrt{\pi}}{2}\frac{\Gamma(n+1)}{(n+\frac{1}{2})\Gamma(n+\frac{1}{2})}$$

For $n = 1$, $A_R = 0$: $B = 0.667$. For $n = \frac{2}{3}$, $A_R = 0$: $B = 0.737$.

b. *Fast Formation* On the basis of flow visualization, Humphrey *et al.* (H26) proposed that circulation occurs in a forming drop if

$$u_{or}^3 \rho_p^2 d_{or}^4 / \sigma \mu_e d_e^2 \gtrsim 0.7, \tag{12-34}$$

where d_e is the equivalent diameter of the drop at detachment and the effective viscosity is

$$\mu_e = \mu_p + \mu[1 - \exp\{-1/(1+\kappa)\}]. \tag{12-35}$$

Circulation increases transfer, and two models have been proposed. Zheleznyak (Z2) assumed that the fluid from the orifice moves along the axis of the drop to the stagnation point, and then travels with the orifice velocity, u_{or}, back to the base of the drop and into the interior. If the drop is spherical, the instantaneous mass transfer product with the external resistance controlling is

$$kA = 2\pi(u_{or}\mathscr{D})^{1/2}a^{3/2}. \tag{12-36}$$

If it is assumed further that the surface area of the residual drop is negligible and that formation occurs at constant flow, the time-average product is

$$\overline{kA} = \frac{4A_f}{d_{or}}\sqrt{\frac{\mathscr{D}}{3}}\left(\frac{Q}{36\pi\sqrt{t_f}}\right)^{1/3}. \tag{12-37}$$

Equation (12-37) correctly predicts the effect of d_{or} and Q, and Zheleznyak found that it agreed closely with his own data. Although proposed for the external resistance, Eq. (12-37) should also apply to the internal resistance if written in terms of dispersed phase properties.

Siskovic and Narsimhan (S14) modified the model of Handlos and Baron discussed in Chapter 7, using an estimate for internal circulation, to obtain

$$(\overline{kA})_p = 0.01 u_{or} A_f \tag{12-38}$$

for the controlling resistance in the dispersed phase. Agreement with their data was satisfactory when the constant was reduced from 0.01 to 0.0062.

c. *Empirical Correlations* Skelland and co-workers proposed empirical equations for transfer during formation and release, with the mass transfer coefficient based on the arithmetic mean of the driving forces at the beginning and end of the whole process. For the continuous phase (S16):

$$\overline{kA} = 0.386 A_e \left(\frac{\mathscr{D}}{t_f}\right)^{0.5} \left(\frac{\rho\sigma}{\Delta\rho\, g\mu t_f}\right)^{0.407} \left(\frac{g t_f^2}{d_e}\right)^{0.148}, \tag{12-39}$$

while for the dispersed phase (S17)

$$(\overline{kA})_p = 0.0432 \frac{A_e d_e}{t_f} \left(\frac{u_{or}^2}{g d_e}\right)^{0.089} \left(\frac{t_f \mathscr{D}_p}{d_e^2}\right)^{0.334} \left(\frac{\sqrt{\rho_p d_e \sigma}}{\mu_p}\right)^{0.601}, \tag{12-40}$$

where d_e and A_e refer to a sphere with the same volume as the drop at detachment. Equations (12-39) and (12-40) should be used with caution outside the limited range of properties covered by the original experiments.

d. *Effect of Surface Active Agents* Skelland and Caenepeel (S15) added surface-active materials to examine their effect on transfer during formation and coalescence. By comparison with Eq. (12-40), addition of surfactant reduced $(\overline{kA})_p/A_e$ by a factor of five during formation with dispersed phase resistance controlling, but increasing surfactant concentration returned the transfer rate almost to its value in the pure system. No quantitative explanation for this behavior is available. As in the work of Rajan and Heideger (R1), surfactants had relatively little effect when the resistance in the continuous phase controlled, reducing $\overline{kA}/A_e$ by at most 50%.

B. Formation with Phase Change

Bubbles, drops, and solid particles are of importance in many processes, such as boiling, condensation, sublimation, crystallization, cavitation, electrolysis, and effervescence, in which a change of phase occurs. A detailed review of these subjects is beyond the scope of this book, but a few basic points and useful references will be given.

All the above processes involve an initiation stage, called nucleation, followed by particle growth. Both homogeneous and heterogeneous nucleation are possible, although the latter is generally more important, except in certain processes in the atmosphere or in ultrapure systems with large driving forces. The surface tensions of pure liquids are so high that preexisting nuclei must generally be present for vapor bubbles to form in cavitation, boiling, or electrolysis. Frequently microscopic scratches, pits, or crevices on solid surfaces trap gas pockets at which vapor bubble growth may begin. The equilibrium diameter

d_{eq} of a spherical bubble at a point in a liquid where the local pressure is p_l is

$$d_{eq} = 4\sigma/(p_v + p_g - p_l), \tag{12-41}$$

where p_g is the partial pressure of any noncondensable gas and p_v the vapor pressure of the liquid at the given temperature. Bubbles smaller than d_{eq} decrease in size while larger pockets grow. Growth may be promoted by lowering the local p_l, as in cavitation or effervescence, or by raising p_v as in nucleate boiling. In atmospheric nucleation, nuclei are commonly provided by sodium chloride or potassium iodide crystals or by airborne dust or droplets. Microscopic solid particles may also be important for heterogeneous nucleation in liquids or at liquid–liquid interfaces. [For discussions and reviews of nucleation, see (B1, B10, C4, H15, H16, H18, K5, R9, Z1).]

Bubble growth in vaporization is usually controlled by diffusion of mass or heat, although chemical steps can be rate-controlling for electrolytic processes (D2). Thorough reviews of diffusion-controlled bubble growth are available (B1, H20). Theoretical treatments [e.g. (F4, P7, S8)] generally consider spherical symmetry with spherical or hemispherical bubbles in a liquid of large extent and fluid properties assumed uniform within each phase, although extensions have been made, e.g., to show the effect of other contact angles (B8) and bubble translation (R10). Sideman (S10) reviewed studies of heat transfer to drops and bubbles undergoing simultaneous change of phase.

The growth and collapse of cavitation bubbles is commonly described by considering irrotational expansion of a spherical cavity in an incompressible liquid of infinite extent, subject to the unsteady form of Bernoulli's equation (B3, P5). Effects of compressibility and bubble migration must also be considered for oscillating bubbles produced by underwater explosions (B3, C5).

C. Other Means of Forming Bubbles and Drops

A horizontal interface between two fluids such that the lower fluid is the less dense tends to deform by the process known as Rayleigh–Taylor instability (see Section III.A). Spikes of the denser fluid penetrate downwards, until the interface is broken up and one fluid is dispersed into the other. This is observed, for example, in formation of drops from a wet ceiling, and of bubbles in film boiling. For low-viscosity fluids, the equivalent diameter of the particle formed is of order $\sqrt{\sigma/g\,\Delta\rho}$.

Various experimental techniques have been devised for introducing relatively large single drops or bubbles into liquids. The most common method is inversion of a cup, immersed in the fluid for $\rho_p < \rho$ or at the surface for $\rho_p > \rho$. Other variants are opening a shutter (C6), withdrawal of a solid cylindrical tube (W1), and bursting of a stretched balloon containing the dispersed phase (W2). The mode of injection sometimes plays an important role, for example, in affecting wake shedding from large bubbles (C9). Smaller bubbles may be generated by a focused laser beam (L10), enabling the exact bubble position to be predetermined. Flow visualization using hydrogen bubbles generated by electrolysis is described by Schraub *et al.* (S6) and by Tory and Haywood (T9). Bubbles are entrained when a jet of liquid enters a pool from above (L13, V2). Formation of bubbles at the surface of liquids in turbulent flow, associated with "white water," has also received some attention (F3).

Kintner *et al.* (K7) and Damon *et al.* (D1) have discussed photographic techniques applicable to the study of bubbles and drops. Sometimes it is desirable to hold a bubble or drop stationary, to study internal or external flow patterns and transfer processes. To prevent the particle from migrating to the wall, it is desirable to establish a minimum in the velocity profile at the position where the particle is to reside, and various techniques have been devised (D4, F1, G1, P11, M15, R15, S20) to do this. Vertical wandering of such particles may occur (W7), and may be reduced by using a duct tapered so that the area decreases towards the top (D4). Acoustic levitation of liquid drops may also be used (A3).

III. BREAKUP OF DROPS AND BUBBLES

In multiphase flow equipment, the size distribution of drops and bubbles is commonly determined by the dynamics of break up and coalescence. Coalescence involves multiple fluid–particle systems and hence is beyond the scope of this book. A number of processes may cause breakup and these are discussed here.

A. Breakup in Stagnant Media

When one fluid overlays a less dense fluid, perturbations at the interface tend to grow by Rayleigh–Taylor instability (L1, T4). Surface tension tends to stabilize the interface while viscous forces slow the rate of growth of unstable surface waves (B2). The leading surface of a drop or bubble may therefore become unstable if the wavelength of a disturbance at the surface exceeds a critical value

$$\lambda_{cr} = 2\pi\sqrt{\sigma/g\,\Delta\rho}. \tag{12-42}$$

For rising bubbles and drops, instability manifests itself as an indentation at the upper surface which grows deeper as time advances. Splitting tends to occur if the disturbance grows sufficiently quickly relative to the velocity at which it is swept around to the equator by tangential movement along the interface. A typical sequence of events is shown in Fig. 12.5. This mechanism of splitting applies to bubbles in liquids and in fluidized beds (C2, C3, H14) and to drops in gases and liquids (G5, K8, K10, P12, R15). For unstable drops falling in air, an indentation develops at the front leading to a hollow rim with an "inverted bag" of liquid attached (H1, H4, L6). This mode of breakup is shown in Fig. 12.6. The bag is inflated as time progresses with the penetration velocity given

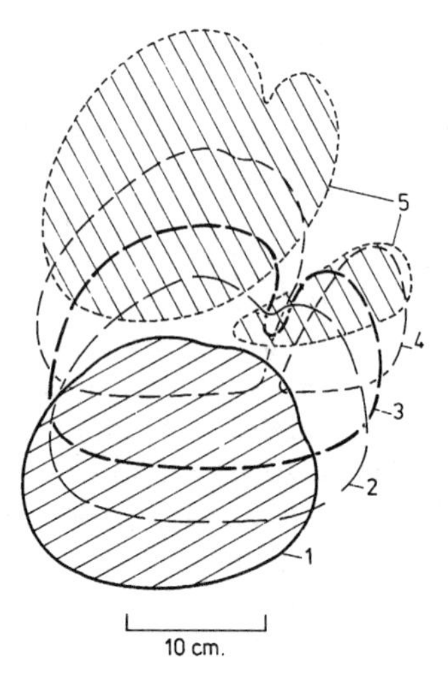

FIG. 12.5 Breakup of a large two-dimensional bubble in viscous sugar solution, traced from photographs by Clift and Grace (C2). (1) 0 s; (2) 0.16 s; (3) 0.32 s; (4) 0.56 s; (5) 0.84 s.

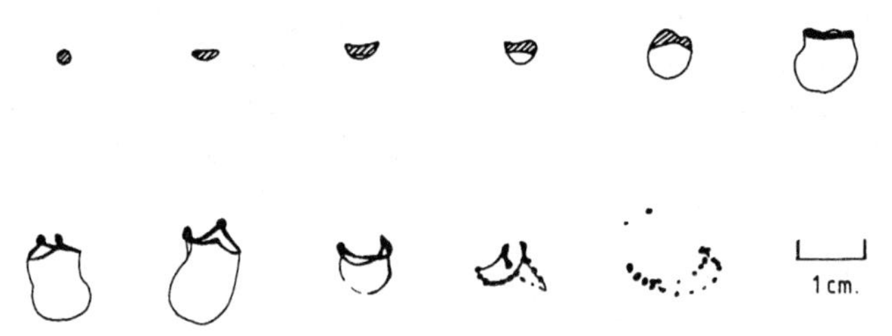

FIG. 12.6 Breakup of a water drop in an air stream moving downwards relative to a particle, traced from photographs by Lane (L6).

approximately (F2) by $0.3\sqrt{\dot{U}\lambda}$ similar to Eq. (9-37) for bubbles in tubes, where $\dot{U}$ is the drop acceleration. The wall of the bag thins until the film shatters into as many as several hundred fragments (M1, M3). The toroidal rim also breaks up into larger droplets containing about 75% of the original drop mass (B6).

The time available for a disturbance to grow is approximately (G5)

$$t_a = \frac{d_e}{2U}\left(\frac{2 + 3\kappa}{2}\right)\ln\{\cot(\lambda/4d_e)\}. \tag{12-43}$$

The time required for growth, $t_g = 1/\alpha$, may be estimated, to a first approximation, from the linearized theory which leads to the equation

$$[\sigma k'^3 - gk'\,\Delta\rho + \alpha^2(\rho + \rho_p)][k' + m + \kappa(k' + m_p)] + 4\alpha k'\mu(k' + \kappa m_p)(\kappa k' + m) = 0 \tag{12-44}$$

where $k' = 2\pi/\lambda$ is the wavenumber and

$$m = (k'^2 + \alpha/\nu)^{1/2},$$
$$m_p = (k'^2 + \alpha/\nu_p)^{1/2}. \tag{12-45}$$

On intuitive grounds λ cannot be larger than about half the circumference. Maximum values of t_g may be obtained by solving Eq. (12-44) for $\lambda_{cr} < \lambda \leq \pi d_e/2$. If the density and/or viscosity of one of the phases is much larger than that of the other phase, simpler approximate forms of Eq. (12-44) may be solved (P6, W5).

Comparison of the computed values of t_a and t_g gives an indication of the likelihood of splitting. Values of U may be obtained using the correlations given in Chapters 7 and 8. Experimental evidence shows that splitting occurs when $t_a > 1.4t_g$ for liquid drops, and when $t_a > 3.8t_g$ for gas bubbles (G5). Maximum stable drop and bubble sizes predicted with this procedure are given in Table 12.3. For $\kappa \dot{>} 0.5$, i.e., for liquid drops falling in gases and for many liquid–liquid systems, the maximum stable diameter is given approximately (G5, L12) by

$$(d_e)_{max} \approx 4\sqrt{\sigma/g\,\Delta\rho}. \tag{12-46}$$

Equation (12-46) implies that the Eotvos number cannot exceed a value of about 16. Since the spherical-cap regime requires $Eo \geq 40$ (see Fig. 2.5), stability considerations explain why drops falling in gases and drops in many liquid–liquid systems never attain the spherical-cap regime. Moreover, since $We = 4Eo/3C_D$ and C_D is nearly constant for large drops in air, it is also possible to

TABLE 12.3

Maximum Stable Drop and Bubble Sizes for Systems at Room Temperature

System	Experimental $(d_e)_{max}$ (cm)	Ref.	Predicted $(d_e)_{max}$ (cm)
Water drops in air	1.0	(M11)	1.21
Isobutanol drops in air	0.45	(F1)	0.69
Carbon tetrachloride drops in air	0.36	(M11)	0.65
Carbon tetrachloride drops in water	1.04	(H22)	1.34
Nitrobenzene drops in water	1.54	(H22)	1.56
Chlorobenzene drops in water	3.16	(K12)	3.00
Bromoform drops in water	0.56	(K12)	0.69
Diphenyl ether drops in water	3.25	(K12)	3.17
Carbon tetrachloride drops in 31p aq. sucrose solution	13.5	(G5)	13.5
Air bubbles in 0.5p aq. sucrose solution	4.6	(G5)	5.0
Air bubbles in 2p paraffin oil	6.3	(G5)	6.7
Air bubbles in water			4.9
Air bubbles in mercury			3.4
Air bubbles in fluidized bed			8.4

use a critical We criterion for breakup [e.g., see (T11)]. Experimental maximum stable bubble and drop sizes for stagnant media are given in Table 12.3 for some systems. [For other systems, see (F1, G5, H22, K12, M11).]

B. Breakup Due to Resonance

There is some evidence that isolated drops may shake themselves apart if shape oscillations become sufficiently violent (L7). It has been suggested (E1, G11, H22) that breakup occurs when the exciting frequency of eddy shedding matches the natural frequency of the drop. However, other workers (S7) have found that oscillations give way to random wobbling before breakup occurs. While it is possible that resonance may produce breakup in isolated cases, this mechanism appears to be less important than the Taylor instability mechanism described above.

Sevik and Park (S9) suggested that resonance can cause bubble and drop breakup in turbulent flow fields when the characteristic turbulence frequency matches the lowest or natural frequency mode of an entrained fluid particle. Breakup in turbulent flow fields is discussed below.

C. Breakup Due to Velocity Gradients

A drop or bubble in a shear field tends to rotate and deform. If the velocity gradients are large enough, interfacial tension forces are no longer able to maintain the fluid particle intact, and it ruptures into two or more smaller particles (A1, K1, R12, T3, T10). Observations of drop and bubble breakup have also been obtained in hyperbolic flows (R12, T3). Figure 12.7 shows tracings of photographs showing the effect of increasing shear rate; further sequences appear in (R12, T3, T10).

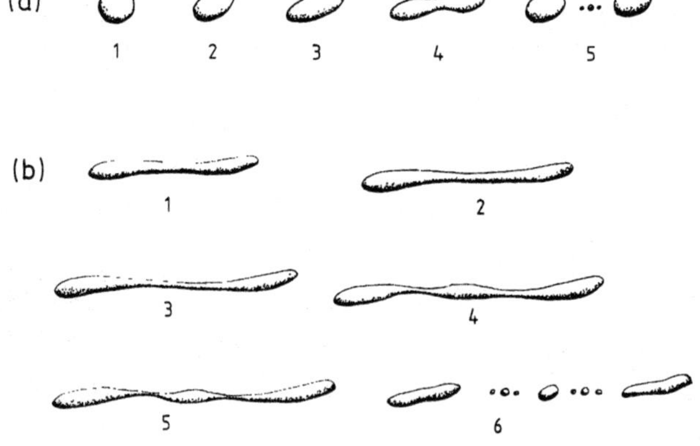

Fig. 12.7 Breakup of liquid drops in simple shear. Velocity gradient G increases in each sequence. (a) Rumscheidt and Mason (R12): $\kappa = 1$, $\sigma = 4.8$ dynes/cm; (b) Torza *et al.* (T10): $\kappa = 1.1$, $\sigma = 1.3$ dynes/cm.

Theoretical predictions relating to the orientation and deformation of fluid particles in shear and hyperbolic flow fields are restricted to low Reynolds numbers and small deformations (B7, C8, T3, T10). The fluid particle may be considered initially spherical with radius a_0. If the surrounding fluid is initially at rest, but at time $t = 0$, the fluid is impulsively given a constant velocity gradient G, the particle undergoes damped shape oscillations, finally deforming into an ellipsoid (C8, T10) with axes in the ratio $E^{-1/2}:1:E^{1/2}$, where

$$\frac{1-E}{1+E} = \frac{5(19\kappa + 16)}{4(\kappa + 1)\sqrt{(19\kappa)^2 + (20/N)^2}}, \tag{12-47}$$

with

$$N = a_0 G\mu/\sigma. \tag{12-48}$$

The corresponding orientation at large time is given by

$$\theta = \frac{\pi}{4} + \frac{1}{2}\tan^{-1}\left(\frac{19\kappa N}{20}\right). \tag{12-49}$$

The relaxation time for the oscillations is approximately

$$\tau_r = a_0\mu_p/\sigma. \tag{12-50}$$

For the limiting cases, $N \to 0$ and $\kappa \to \infty$, Eqs. (12-47) and (12-49) reduce to equations derived by Taylor (T2, T3) for fluid particles in steady-state shear or hyperbolic flows, i.e.,

$$\text{as } N \to 0, \qquad \theta \simeq \frac{\pi}{4} \quad \text{and} \quad \frac{1-E}{1+E} = N\left(\frac{1 + 19\kappa/16}{1+\kappa}\right), \tag{12-51}$$

$$\text{as } \kappa \to \infty, \qquad \theta \simeq \frac{\pi}{2} \quad \text{and} \quad \frac{1-E}{1+E} = \frac{5}{4\kappa}. \tag{12-52}$$

Rumscheidt and Mason (R12) proposed that breakup occurs if N exceeds a critical value

$$N_{\text{crit}} = \frac{1}{2}\left(\frac{1+\kappa}{1 + 19\kappa/16}\right), \tag{12-53}$$

which varies only between 0.5 and 0.42 as κ varies from zero to infinity. The corresponding value of E would be 1/3 if Eq. (12-47) continued to apply up to such aspect ratios.

Experimental results show reasonable agreement with the above equations even for deformations considerably larger than those for which the theory might be expected to apply. In practice, breakup occurs for $E \leq 0.26 \pm 0.05$ (R12), whereas observed relaxation times are longer than predicted from Eq. (12-50) (T10). Experimentally no breakup occurs when $\kappa > 3$; instead a drop becomes aligned and elongated in the flow direction with its aspect ratio given approximately by Eq. (12-52). There is also a lower limit of κ below which no breakup occurs, $\kappa \doteq 0.005$ for simple shear fields (K1).

The mode of breakup depends upon the rate at which the shear rate is applied (T10). If dG/dt is too large, a fluid particle develops pointed ends for κ less than about 0.2 and fragments break off both ends. On the other hand, if G is increased gradually, necking occurs in the center until rupture produces two large droplets of nearly equal size separated by tiny satellite droplets (Fig. 12.7a). With large dG/dt and $0.2 < \kappa < 3$, a drop is pulled out into a long thread which eventually breaks up due to Rayleigh instability (Fig. 12.7b).

In the experiments referred to above, the systems were relatively free of surface-active agents and Reynolds numbers were small. Care must be exercised, therefore, when applying these results to drops or bubbles under other conditions.

D. Breakup in Turbulent Flow Fields

Two-phase systems are often exposed to turbulent flow conditions in order to maximize the interfacial area of the fluids being contacted. In addition, turbulence is often present in wind tunnels and other laboratory equipment, as well as in nature where it can influence breakup processes (F5). Prediction of drop or bubble sizes in turbulent contacting equipment for any geometry and operating conditions is a formidable problem, primarily because of the inherent theoretical and experimental difficulties in treating turbulent flows. To these difficulties, which exist in single phase systems, must be added the complexity of interaction of dispersed particles with turbulent flow fields.

Work in this field tends to follow directly from two simple concepts proposed by Hinze (H17):

(a) The total local shear stress, τ, imposed by the continuous phase acts to deform a drop or bubble, and to break it if the counterbalancing surface tension forces and viscous stresses inside the fluid particle are overcome. The condition for breakup is then:

$$\tau > (\sigma + \mu_p\sqrt{\tau/\rho_p})/d_e. \tag{12-54}$$

(b) Only the energy associated with eddies with length scales smaller than d_e is available to cause splitting; larger eddies merely transport the drop or bubble. Hence, the turbulent energy available to cause breakup of a fluid particle of diameter d_e is given by

$$\frac{\pi}{6}\rho {d_e}^3 \int_{1/d_e}^{\infty} E(k')dk'$$

where k' is the wavenumber and $E(k')$ the energy spectrum.

Equation (12-54) leads to a prediction of a critical (or "maximum stable") size if τ can be evaluated. For example, Hughmark (H24) applied Eq. (12-54) with some success to experimental data obtained in fully developed turbulent pipe flow by Paul and Sleicher (P1, S18). These workers arbitrarily specified the maximum stable size as that for which 20% breakage occurs. It was found that drops usually split into two daughters of approximately equal size, though much smaller drops were stripped off larger ones in some cases. There is some evidence (C7, S18) that most of the breakup occurs in the wall region of the pipe, where it is possible for the time-mean velocity gradients to cause distortion and breakup as discussed in the previous subsection. For developing pipe flow, on the other hand, breakup tends to occur near the center of the pipe, and may be due to pressure fluctuations which cause one or two small fluid particles to detach from the original particle (S23).

Concept (b) is less useful, except in rare cases where the energy spectrum has been measured. It is common to assume that the turbulence is homogeneous and isotropic and that the eddies in question are in the inertial ($-5/3$ power) subrange. This assumption is unlikely to be valid in an overall sense though it may be reasonable locally (G10) or for the high wavenumber (small) eddies which are of primary interest. For an example of the application of the theory, see Middleman (M13).

There is little evidence showing the mode of breakup in turbulent flow fields. Hinze (H17) speaks of a "bulgy" mode of breakup. Published photographs (C7, T12) show highly deformed bubbles and necking drops, protuberances and cell-like surface structures (see Fig. 12.8). Experimental evidence regarding single bubbles and drops in well-characterized turbulent fields would be most welcome.

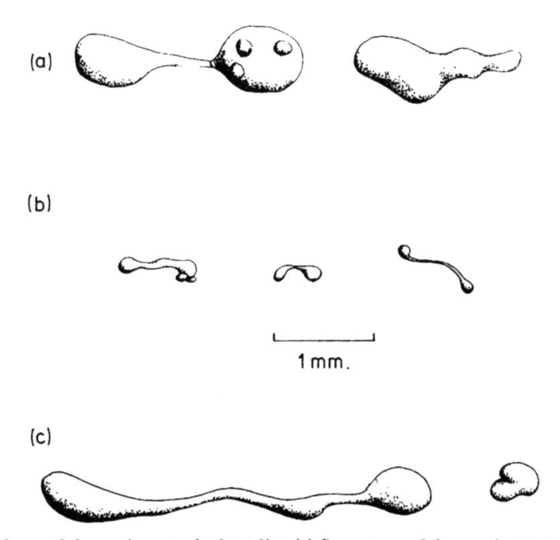

FIG. 12.8 Breakup of drops in a turbulent liquid flow, traced from photographs by Collins and Knudsen (C7). (a) $\gamma = 0.79$, $\kappa = 1.2$, $\sigma = 40.3$ dynes/cm; (b) $\gamma = 0.85$, $\kappa = 9$, $\sigma = 13.0$ dynes/cm; (c) $\gamma = 0.87$, $\kappa = 16$, $\sigma = 17.6$ dynes/cm.

E. Breakup of Drops in Air Blasts

Break up of drops accelerated by air blasts (including shock waves) can occur by an "inverted bag" mechanism similar to that described in Section A above, for $\mathrm{Eo}_{ac} = \dot{U}\,\Delta\rho\, d_e^{\,2}/\sigma$ between about 16 and 10^5 (H1, H2, H4, L6). Reichman and Temkin (R7) give a detailed description of four stages of bag-type breakup. Under some circumstances, deformation preceding breakup appears more like a parasol than an inflating bag (S12). The distance x moved by the drop is given approximately by

$$x/a_0 = 1.6T^2, \tag{12-55}$$

where a_0 is the radius of the drop before exposure to the air blast,

$$T = tu/2a_0\gamma^{1/2} \tag{12-56}$$

is a dimensionless time, and u is the air velocity (R8, S13). Equation (12-55), corresponds to constant acceleration with a C_D value of 2.1. A criterion for breakup has been derived based on a critical thickness beyond which deformation is irreversible (R7). For water drops accelerated by a shock wave of velocity u, breakup occurs for

$$\mathrm{We} = u^2 d_e \rho/\sigma \geq 6.5. \tag{12-57}$$

For higher accelerations with $10^2 < \mathrm{Eo}_{ac} < 10^5$, liquid tends to be stripped from the surface of the drop as a spray (F2, R3, T5), this phenomenon usually being called "boundary layer stripping." The time for the drop to disintegrate completely by this stripping mechanism is of order $T = 3.5$ (R8). Breakup becomes increasingly chaotic with increasing Eo_{ac}. For Eo_{ac} of order 10^5 and higher, the drop tends to shatter due to development of a series of indentations on the windward surface (R8), in a time interval given approximately by

$$T_{\mathrm{crit}} = 45\mathrm{We}^{-1/4}. \tag{12-58}$$

For theoretical analyses of instability of accelerating drops, see (H6, K11, T5).

F. Other Causes of Breakup

Breakup of water drops due to strong electrical forces has been studied in connection with rain phenomena [e.g. (A4, L8, L9, M4, M7)]. As a strong electrical field is imposed on a freely falling drop, marked elongation occurs in the direction of the field and can lead to stripping of charge-bearing liquid. A simple criterion derived by Taylor (T6) can be used to predict the critical condition for instability. It has also been shown (W6) that soap bubbles can be rendered unstable by electric fields.

Raindrop breakup also occurs when drops collide, and this has been studied by a number of workers [see (M8)]. It is probable that the collision mode of breakup is much more significant in determining the size distribution of raindrops in the atmosphere than the Rayleigh–Taylor instability mode discussed above.

Impaction of water drops on solid surfaces has been studied (G3), and under some circumstances smaller drops are detached and leave the surface. Impingement of drops on thin liquid films may also cause breakup (K3, S5). Breakup of bubbles in fluidized beds due to impingement on fixed horizontal cylinders has also been observed (G4). Sound waves may lead to instability of bubbles in liquids (S21).

REFERENCES

A1. Acrivos, A., *Proc. Int. Colloq. Drops Bubbles, Calif. Inst. Technol. Jet Propul. Lab.* **2**, 390–404 (1974).

A2. Andreini, R. J., Foster, J. S., and Callan, R. W., *Metall. Trans.* **8B**, 625–631 (1977).

A3. Apfel, R. E., *Proc. Int. Colloq. Drops Bubbles, Calif. Inst. Technol. Jet Propul. Lab.* **1**, 246–265 (1974).

A4. Ausman, E. L., and Brook, M., *J. Geophys. Res.* **72**, 6131–6135 (1967).

B1. Bankoff, S. G., *Adv. Chem. Eng.* **6**, 1–60 (1966).

B2. Bellman, R., and Pennington, R. H., *Q. Appl. Math.* **12**, 151–162 (1954).

B3. Birkhoff, G., and Zarantonello, E. H., "Jets, Wakes, and Cavities." Academic Press, New York, 1957.

B4. Bloore, P. D., and Botterill, J. S. M., *Nature (London)* **190**, 250–251 (1961).

B5. Bondarev, G. S., and Romanov, V. F., *Khim. Tekhnol. Topl. Masel* No. 2, 40–43 (1973).

B6. Brodkey, R. S., "The Phenomena of Fluid Motions." Addison-Wesley, Reading, Massachusetts, 1967.

B7. Buckmaster, J. D., *J. Appl. Mech.* **95**, 18–24 (1973).

B8. Buehl, W. M., and Westwater, J. W., *AIChE J.* **12**, 571–576 (1966).

B9. Burkholder, H. C., and Berg, J. C., *AIChE J.* **20**, 863–880 (1974).

B10. Byers, H. R., *Ind. Eng. Chem.* **57**(11), 32–40 (1965).

C1. Chuang, S. C., and Goldschmidt, V. W., *J. Basic Eng.* **92**, 705–711 (1970).

C2. Clift, R., and Grace, J. R., *Chem. Eng. Sci.* **27**, 2309–2310 (1972).

C3. Clift, R., Grace, J. R., and Weber, M. E., *Ind. Eng. Chem., Fundam.* **13**, 45–51 (1974).

C4. Cole, R., *Adv. Heat Transfer* **10**, 86–166 (1974).

C5. Cole, R. H., "Underwater Explosions." Dover, New York, 1965.

C6. Collins, R., *Chem. Eng. Sci.* **20**, 851–855 (1965).

C7. Collins, S. B., and Knudsen, J. G., *AIChE J.* **16**, 1072–1080 (1970).

C8. Cox, R. G., *J. Fluid Mech.* **37**, 601–623 (1969).

C9. Crabtree, J. R., and Bridgwater, J., *Chem. Eng. Sci.* **22**, 1517–1520 (1967).

D1. Damon, K. G., Angelo, J. B., and Park, R. W., *Chem. Eng. Sci.* **21**, 813 (1966).

D2. Darby, R., and Haque, M. S., *Chem. Eng. Sci.* **28**, 1129–1138 (1973).

D3. Davidson, J. F., and Harrison, D., "Fluidised Particles." Cambridge Univ. Press, London and New York, 1963.

D4. Davidson, J. F., and Kirk, F. A., *Chem. Eng. Sci.* **24**, 1529–1530 (1969).

D5. Davidson, J. F., and Schüler, B. O. G., *Trans. Inst. Chem. Eng.* **38**, 144–154 (1960).

D6. Davidson, J. F., and Schüler, B. O. G., *Trans. Inst. Chem. Eng.* **38**, 335–342 (1960).

D7. Davidson, L., and Amick, E. H., *AIChE J.* **2**, 337–342 (1956).

D8. Davies, J. T., and Rideal, E. K., "Interfacial Phenomena." Academic Press, New York, 1963.

D9. De Chazal, L. E. M., and Ryan, J. T., *AIChE J.* **17**, 1226–1229 (1971).
E1. Elzinga, E. R., and Banchero, J. T., *AIChE J.* **7**, 394–399 (1961).
F1. Finlay, B. A., Ph.D. Thesis, Univ. of Birmingham, 1957.
F2. Fishburn, B. D., *Acta Astron.* **1**, 1267–1284 (1974).
F3. Flack, J. E., Kveisengen, J. I., and Nath, J. H., *Science* **134**, 392–393 (1961).
F4. Forster, H. K., and Zuber, N., *J. Appl. Phys.* **25**, 474–478 (1954).
F5. Fournier D'Albe, E. M., and Hidayetulla, M. S., *Q. J. R. Meteorol. Soc.* **81**, 610–613 (1955).
F6. Fraser, R. P., and Eisenklam, P., *Trans. Inst. Chem. Eng.* **34**, 294–319 (1956).
G1. Garner, F. H., and Kendrick, P., *Trans. Inst. Chem. Eng.* **37**, 155–163 (1959).
G2. Giffen, E., and Muraszen, A., "Atomization of Liquid Fuels." Chapman & Hall, London, 1953.
G3. Gillespie, T., and Rideal, E., *J. Colloid Sci.* **10**, 281–298 (1955).
G4. Glass, D. H., Ph.D. Thesis, Cambridge Univ., 1967.
G5. Grace, J. R., Wairegi, T., and Brophy, J., *Can. J. Chem. Eng.* **56**, 3–8 (1978).
G6. Green, H., and Lane, W., "Particulate Clouds: Dusts, Smokes and Mists," 2nd ed. Van Nostrand, Princeton, New Jersey, 1964.
G7. Grigar, K., Prochazka, J., and Landau, J., *Chem. Eng. Sci.* **25**, 1773–1783 (1970).
G8. Grigar, K., Prochazka, J., and Landau, J., *Collect. Czech. Chem. Commun.* **37**, 1943–1951 (1972).
G9. Groothuis, H., and Kramers, H., *Chem. Eng. Sci.* **4**, 17–25 (1955).
G10. Guenkel, A., Ph.D. Thesis, McGill University, Montreal, 1973.
G11. Gunn, R., *J. Geophys. Res.* **54**, 383–385 (1949).
H1. Haas, F. C., *AIChE J.* **10**, 920–924 (1964).
H2. Habler, G., *Forsch. Ingenieures.* **38**, 183–192 (1972).
H3. Halligan, J. E., and Burkhart, L. E., *AIChE J.* **14**, 411–414 (1968).
H4. Hanson, A. R., Domich, E. G., and Adams, H. S., *Phys. Fluids* **6**, 1070–1080 (1963).
H5. Harkins, W. D., and Brown, F. E., *J. Am. Chem. Soc.* **41**, 499–524 (1919).
H6. Harper, E. Y., Grube, G. W., and Chang, I., *J. Fluid Mech.* **52**, 565–591 (1972).
H7. Harrison, D., and Leung, L. S., *Trans. Inst. Chem. Eng.* **39**, 409–414 (1961).
H8. Hayes, W. B., Hardy, B. W., and Holland, C. D., *AIChE J.* **5**, 319–324 (1959).
H9. Hayworth, C. B., and Treybal, R. E., *Ind. Eng. Chem.* **42**, 1174–1181 (1950).
H10. Heertjes, P. M., and De Nie, L. H., *Chem. Eng. Sci.* **21**, 755–768 (1966).
H11. Heertjes, P. M., and De Nie, L. H., *in* "Recent Advances in Liquid-Liquid Extraction" (C. Hanson, ed.), pp. 367–406. Pergamon, Oxford, 1971.
H12. Heertjes, P. M., and De Nie, L. H., *Chem. Eng. Sci.* **26**, 697–703 (1971).
H13. Heertjes, P. M., De Nie, L. H., and DeVries, H. J., *Chem. Eng. Sci.* **26**, 441–459 (1971).
H14. Henriksen, H. K., and Ostergaard, K., *Chem. Eng. Sci.* **29**, 626–629 (1974).
H15. Hidy, G. M., ed., "Aerosols and Atmospheric Chemistry." Academic Press, New York, 1972.
H16. Hidy, G. M., and Brock, J. R., "The Dynamics of Aerocolloidal Systems." Pergamon, Oxford, 1970.
H17. Hinze, J. O., *AIChE J.* **1**, 289–295 (1955).
H18. Holl, J. W., *J. Basic Eng.* **92**, 681–688 (1970).
H19. Hozawa, M., and Tadaki, T., *Kagaku Kogaku* **37**, 827–833 (1973).
H20. Hsieh, D. Y., *J. Basic Eng.* **87**, 991–1005 (1965).
H21. Hsiung, T. P., M. Eng. Thesis, McGill University, Montreal, 1974.
H22. Hu, S., and Kintner, R. C., *AIChE J.* **1**, 42–50 (1955).
H23. Hughes, R. R., Handlos, A. E., Evans, H. D., and Maycock, R. L., *Chem. Eng. Prog.* **51**, 557–563 (1955).
H24. Hughmark, G. A., *AIChE J.* **17**, 1000 (1971).
H25. Humphrey, J. A. C., Hummel, R. L., and Smith, J. W., *Can. J. Chem. Eng.* **52**, 449–456 (1974).

H26. Humphrey, J. A. C., Hummel, R. L., and Smith, J. W., *Chem. Eng. Sci.* **29**, 1496–1499 (1974).

J1. Jackson, R., *Chem. Engr.* No. 178, 107–118 (1964).

J2. Jeffreys, G. V., *Chem. and Process Eng.* (*London*) **49**(11), 111–122 (1968).

K1. Karam, H. J., and Bellinger, J. C., *Ind. Eng. Chem., Fundam.* **7**, 576–581 (1968).

K2. Kehat, E., and Sideman, S., *in* "Recent Advances in Liquid-Liquid Extraction" (C. Hanson, ed.), pp. 455–494. Pergamon, Oxford, 1971.

K3. Kennedy, T. D. A., and Collier, J. G., *Symp. Multi-Phase Flow Syst., Paper J4, I. Chem. E. Symp. Ser.* No. 38 (1974).

K4. Khurana, A. K., and Kumar, R., *Chem. Eng. Sci.* **24**, 1711–1723 (1969).

K5. Kiang, C. S., Stauffer, D., Mohnen, V. A., Hamill, P., and Walker, G. H., *Proc. Int. Colloq. Drops Bubbles, Calif. Inst. Technol. Jet Propul. Lab.* **2**, 506–528 (1974).

K6. Kintner, R. C., *Adv. Chem. Eng.* **4**, 51–94 (1963).

K7. Kintner, R. C., Horton, T. J., Graumann, R. E., and Amberkar, S., *Can. J. Chem. Eng.* **39**, 235–241 (1961).

K8. Klett, J. D., *J. Atmos. Sci.* **28**, 646–647 (1971).

K9. Kohli, A., Ph.D. Thesis, Univ. of Maryland, College Park, 1974.

K10. Komabayasi, M., Gonda, T., and Isono, K., *J. Meteorol. Soc. Jpn.* **42**, 330–340 (1964).

K11. Korner, W., *Acta Mech.* **13**, 87–115 (1972).

K12. Krishna, P. M., Venkateswarlu, D., and Narasimhamurty, G. S. R., *J. Chem. Eng. Data* **4**, 336–343 (1959).

K13. Krishnamurthi, S., Kumar, R., and Kuloor, N. R., *Ind. Eng. Chem., Fundam.* **7**, 549–554 (1968).

K14. Kumar, R., *Chem. Eng. Sci.* **26**, 177–184 (1971).

K15. Kumar, R., and Kuloor, N. R., *Adv. Chem. Eng.* **8**, 256–368 (1970).

K16. Kumar, R., and Kuloor, N. R., *Can. J. Chem. Eng.* **48**, 383–388 (1970).

K17. Kupferberg, A., and Jameson, G. J., *Trans. Inst. Chem. Eng.* **47**, 241–250 (1969).

K18. Kupferberg, A., and Jameson, G. J., *Trans. Inst. Chem. Eng.* **48**, 140–150 (1970).

L1. Lamb, H., "Hydrodynamics," 6th ed. Cambridge Univ. Press, London, 1932.

L2. Lanauze, R. D., and Harris, I. J., *Trans. Inst. Chem. Eng.* **52**, 337–348 (1974).

L3. Lanauze, R. D., and Harris, I. J., *Chem. Eng. Sci.* **27**, 2102–2105 (1972).

L4. Lanauze, R. D., and Harris, I. J., *Chem. Eng. Sci.* **29**, 1663–1668 (1974).

L5. Lando, J. H., and Oakley, H. T., *J. Colloid Interface Sci.* **25**, 526–530 (1967).

L6. Lane, W. R., *Ind. Eng. Chem.* **43**, 1312–1317 (1951).

L7. Lane, W. R., and Green, H. L., *in* "Surveys in Mechanics" (G. K. Batchelor and R. M. Davies, eds.), pp. 162–215. Cambridge Univ. Press, London and New York, 1956.

L8. Latham, J., *Q. J. R. Meteorol. Soc.* **91**, 87–90 (1965).

L9. Latham, J., and Myers, V., *J. Geophys. Res.* **75**, 515–520 (1970).

L10. Lauterborn, W., and Bolle, H., *Proc. Int. Colloq. Drops Bubbles, Calif. Inst. Technol. Jet Propul. Lab.* **2**, 322–337 (1974).

L11. Lee, H. C., *IBM J. Res. Dev.* **18**, 364–369 (1974).

L12. Lehrer, I. H., *Isr. J. Technol.* **13**, 246–252 (1975).

L13. Lin, T. J., and Donnelly, H. G., *AIChE J.* **12**, 563–571 (1966).

L14. Liu, B. Y. H., ed., "Fine Particles: Aerosol Generation, Measurement, Sampling and Analysis." Academic Press, New York, 1976.

M1. Magarvey, R. H., and Taylor, B. W., *J. Appl. Phys.* **27**, 1129–1135 (1956).

M2. Marmur, A., and Rubin, E., *Chem. Eng. Sci.* **31**, 453–463 (1976).

M3. Mason, B. J., *Endeavour* **23**, 136–141 (1964).

M4. Matthews, B. J., *J. Geophys. Res.* **72**, 3007–3013 (1967).

M5. McCann, D. J., and Prince, R. G. H., *Chem. Eng. Sci.* **24**, 801–814 (1969).

M6. McCann, D. J., and Prince, R. G. H., *Chem. Eng. Sci.* **26**, 1505–1512 (1971).

M7. McDonald, J. E., *J. Meteorol.* **11**, 478–494 (1954).

M8. McTaggart-Cowan, J. D., and List, R., *J. Atmos. Sci.* **32**, 1401–1411 (1975).
M9. Meister, B. J., and Scheele, G. F., *AIChE J.* **13**, 682–688 (1967).
M10. Meister, B. J., and Scheele, G. F., *AIChE J* **15**, 689–699 (1969).
M11. Merrington, A. C., and Richardson, E. G., *Proc. Phys. Soc., London* **59**, 1–13 (1947).
M12. Merry, J.M.D., *AIChE J.* **21**, 507–510 (1975).
M13. Middleman, S., *Ind. Eng. Chem., Process Des. Dev.* **13**, 78–84 (1974).
M14. Mikami, T., Cox, R. G., and Mason, S. G., *Int. J. Multiphase Flow* **2**, 113–138 (1975).
M15. Moo-Young, M., Fulford, G., and Cheyne, I., *Ind. Eng. Chem., Fundam.* **10**, 157–160 (1971).
N1. Narayana, S., Basu, A., and Roy, N. K., *Chem. Eng. Sci.* **25**, 1950–1951 (1970).
N2. Newman, J., "Electrochemical Systems." Prentice-Hall, Englewood Cliffs, New Jersey, 1973.
N3. Nguyen, X. T., and Leung, L. S., *Chem. Eng. Sci.* **27**, 1748–1750 (1972).
N4. Null, H. R., and Johnson, H. F., *AIChE J.* **4**, 273–281 (1958).
O1. Orr, C., "Particulate Technology." Macmillan, New York, 1966.
P1. Paul, H. I., and Sleicher, C. A., *Chem. Eng. Sci.* **20**, 57–59 (1965).
P2. Perrut, M., and Loutaty, R., *Chem. Eng. J.* **3**, 286–293 (1972).
P3. Pimbley, W. T., *IBM J. Res. Dev.* **20**, 148–156 (1976).
P4. Pinczewski, W. V., Yeo, H. K., and Fell, C. J. D., *Chem. Eng. Sci.* **28**, 2261–2263 (1973).
P5. Plesset, M. S., and Prosperetti, A., *Annu. Rev. Fluid Mech.* **9**, 145–185 (1977).
P6. Plesset, M. S., and Whipple, C. G., *Phys. Fluids* **17**, 1–7 (1974).
P7. Plesset, M. S., and Zwick, S. A., *J. Appl. Phys.* **25**, 493–500 (1954).
P8. Popovich, A. T., Jervis, R. E., and Trass, O., *Chem. Eng. Sci.* **19**, 357–365 (1964).
P9. Popovich, A. T., and Lenges, J., *Wärme-Stoffübertrag.* **4**, 87–92 (1971).
P10. Potter, O. E., *Chem. Eng. Sci.* **24**, 1733–1734 (1969).
P11. Pruppacher, H. R., and Beard, K. V., *Q. J. R. Meteorol. Soc.* **96**, 247–256 (1970).
P12. Pruppacher, H. R., and Pitter, R. L., *J. Atmos. Sci.* **28**, 86–94 (1971).
R1. Rajan, S. M., and Heideger, W. J., *AIChE J.* **17**, 202–206 (1971).
R2. Ramakrishnan, S., Kumar, R., and Kuloor, N. R., *Chem. Eng. Sci.* **24**, 731–747 (1969).
R3. Ranger, A. A., and Nicholls, J. A., *AIAA J.* **7**, 285–290 (1969).
R4. Ranz, W. E., and Marshall, W. R., *Chem. Eng. Prog.* **48**, 141–146, 173–180 (1952).
R5. Rao, E. V. L. N., Kumar, R., and Kuloor, N. R., *Chem. Eng. Sci.* **21**, 867–878 (1966).
R6. Rayleigh, Lord, *Proc. London Math. Soc.* **10**, 4–13 (1878).
R7. Reichman, J. M., and Temkin, S., *Proc. Int. Colloq. Drops Bubbles, Calif. Inst. Technol. Jet Propul. Lab.* **2**, 446–464 (1974).
R8. Reinecke, W. G., and Waldman, G. D., *S.A.M.S.O. Tech. Rep.* No. 70–142, Norton Air Force Base, Calif. (1970).
R9. Rohsenow, W. M., *Ind. Eng. Chem.* **58**(1), 40–47 (1966).
R10. Ruckenstein, E., and Davis, E. J., *Int. J. Heat Mass Transfer* **14**, 939–952 (1971).
R11. Ruff, K., *Chem.-Ing.-Tech.* **44**, 1360–1366 (1972).
R12. Rumscheidt, F. D., and Mason, S. G., *J. Colloid Sci.* **16**, 238–261 (1961).
R13. Rumscheidt, F. D., and Mason, S. G., *J. Colloid Sci.* **17**, 260–269 (1962).
R14. Ryan, J. T., Ph.D. Thesis, Univ. of Missouri, Columbia, 1966.
R15. Ryan, R. T., *J. Appl. Meteorol.* **15**, 157–165 (1976).
S1. Sandry, T. D., Ph.D. Thesis, Iowa State Univ., Ames, 1973.
S2. Sano, M., and Mori, K., *J. Inst. Metals* (*Jpn.*) **17**, 344–352 (1976).
S3. Satyanarayan, A., Kumar, R., and Kuloor, N. R., *Chem. Eng. Sci.* **24**, 749–761 (1969).
S4. Scheele, G. F., and Meister, B. J., *AIChE J.* **14**, 9–19 (1968).
S5. Schotland, R. M., *Discuss. Faraday Soc.* **30**, 72–77 (1960).
S6. Schraub, F. A., Kline, S. J., Henry, J., Runstadler, P. W., and Littel, A., *J. Basic Eng.* **87**, 429–444 (1965).
S7. Schroeder, R. R., and Kintner, R. C., *AIChE J.* **11**, 5–8 (1965).
S8. Scriven, L. E., *Chem. Eng. Sci.* **10**, 1–13 (1959).

S9. Sevik, M., and Park, S. H., *J. Fluids Eng.* **95**, 53–60 (1973).
S10. Sideman, S., *Adv. Chem. Eng.* **6**, 207–286 (1966).
S11. Siemes, W., and Kauffmann, J. F., *Chem. Eng. Sci.* **5**, 127–139 (1956).
S12. Simpkins, P. G., *Proc. Int. Colloq. Drops Bubbles, Calif. Inst. Technol. Jet Propul. Lab.* **2**, 372–389 (1974).
S13. Simpkins, P. G., and Bales, E. L., *J. Fluid Mech.* **55**, 629–639 (1972).
S14. Siskovic, N., and Narsimhan, G., *Proc. Chemeca '70*, Butterworths, Melbourne, **1**, 63–75 (1970).
S15. Skelland, A. H. P., and Caenepeel, C. L., *AIChE J.* **18**, 1154–1163 (1972).
S16. Skelland, A. H. P., and Hemler, C. L., Unpublished work, Univ. of Notre Dame, South Bend, Indiana; cited in Skelland, A. H. P., "Diffusional Mass Transfer," Ch. 8. Wiley, New York, 1974.
S17. Skelland, A. H. P., and Minhas, S. S., *AIChE J.* **17**, 1316–1324 (1971).
S18. Sleicher, C. A., *AIChE J.* **8**, 471–477 (1962).
S19. Soo, S. L., "Fluid Dynamics of Multiphase Systems." Ginn (Blaisdell), Boston, Massachusetts, 1967.
S20. Srikrishna, M., and Narasimhamurty, G. S. R., *Indian Chem. Eng.* **13**, 4–11 (1971).
S21. Strube, H. W., *Acustica* **25**, 289–303 (1971).
S22. Sullivan, S. L., Hardy, B. W., and Holland, C. D., *AIChE J.* **10**, 848–854 (1964).
S23. Swartz, J. E., and Kessler, D. P., *AIChE J* **16**, 254–260 (1970).
S24. Swope, R. D., *Can. J. Chem. Eng.* **49**, 169–174 (1971).
S25. Szekely, J., and Themelis, N. J., "Rate Phenomena in Process Metallurgy." Wiley (Interscience), New York, 1971.
T1. Takahashi, T., and Yoshiro, K., *Kagaku Kogaku* **36**, 527–533 (1972).
T2. Taylor, Sir G. I., *Proc. R. Soc., Ser. A* **138**, 41–48 (1932).
T3. Taylor, Sir G. I., *Proc. R. Soc., Ser. A* **146**, 501–523 (1934).
T4. Taylor, Sir G. I., *Proc. R. Soc., Ser. A* **201**, 192–196 (1950).
T5. Taylor, Sir G. I., *in* "The Scientific Papers of G. I. Taylor" (G. K. Batchelor ed.), pp. 457–464. Cambridge Univ. Press, London and New York, 1963.
T6. Taylor, Sir G. I., *Proc. R. Soc., Ser. A* **280**, 383–397 (1964).
T7. Tien, R. H., and Turkdogan, E. T., *AIME Annu. Meet., 104th, New York, 1975.*
T8. Tomotika, S., *Proc. R. Soc., Ser. A* **150**, 322–337 (1935).
T9. Tory, A. C., and Haywood, K. H., *Mech. Eng.* **93**, 53 (1971).
T10. Torza, S., Cox, R. G., and Mason, S. G., *J. Colloid Interface Sci.* **38**, 395–411 (1972).
T11. Tovbin, M. V., Panasyuk, O. A., and Oleinik, L. N., *Colloid J. USSR* **27**, 516–519 (1965).
T12. Towell, G. D., Strand, C. P., and Ackerman, G. H., *AIChE—I. Chem. E. Symp. Ser.* No. 10, 97–105 (1965).
V1. Valentin, F. H. H., "Absorption in Gas-Liquid Dispersions: Some Aspects of Bubble Technology." Spon, London, 1967.
V2. Van de Sande, E., and Smith, J. M., *Chem. Eng. Sci.* **28**, 1161–1168 (1973).
V3. Van Krevelen, D. W., and Hoftijzer, P. J., *Chem. Eng. Prog.* **46**, 29–35 (1950).
V4. Vijayan, S., Ponter, A. B., and Jeffreys, G. V., *Chem. Eng. J.* **10**, 145–154 (1975).
W1. Walia, D. S., and Vir, D., *Chem. Eng. Sci.* **31**, 525–534 (1976).
W2. Walters, J. K., and Davidson, J. F., *J. Fluid Mech.* **12**, 408–416 (1962).
W3. Walters, J. K., and Davidson, J. F., *J. Fluid Mech.* **17**, 321–336 (1963).
W4. Weber, M. E., unpublished calculations, McGill Univ., Montreal (1975).
W5. Willson, A. J., *Proc. Cambridge Philos. Soc.* **61**, 595–607 (1965).
W6. Wilson, C. T. R., and Taylor, Sir G. I., *Proc. Cambridge Philos. Soc.* **22**, 728–730 (1925).
W7. Woodward, M. C., and Stow, C. D., *J. Appl. Meteorol.* **14**, 571–577 (1975).
W8. Wraith, A. E., *Chem. Eng. Sci.* **26**, 1659–1671 (1971).
Y1. Yuen, M. C., *J. Fluid Mech.* **33**, 151–163 (1968).
Z1. Zettlemoyer, A. C., ed., "Nucleation." Dekker, New York, 1969.
Z2. Zheleznyak, A. S., *J. Appl. Chem. USSR* **40**, 834–837 (1967).

Appendices

APPENDIX

Log_{10} Re as Function of

$Log\ N_D^{1/3}$	0.00	0.01	0.02	0.03	0.04
-0.2	-1.981	-1.951	-1.921	-1.891	-1.861
-0.1	-1.681	-1.651	-1.621	-1.591	-1.561
0.0	-1.382	-1.352	-1.322	-1.292	-1.263
0.1	-1.084	-1.054	-1.024	-0.995	-0.965
0.2	-0.787	-0.758	-0.729	-0.699	-0.670
0.3	-0.495	-0.466	-0.437	-0.408	-0.379
0.4	-0.208	-0.180	-0.152	-0.125	-0.097
0.5	0.066	0.092	0.119	0.145	0.171
0.6	0.322	0.347	0.372	0.398	0.425
0.7	0.582	0.607	0.632	0.657	0.682
0.8	0.825	0.849	0.872	0.894	0.917
0.9	1.048	1.069	1.090	1.114	1.138
1.0	1.279	1.302	1.325	1.348	1.371
1.1	1.505	1.527	1.549	1.570	1.592
1.2	1.720	1.741	1.761	1.782	1.803
1.3	1.924	1.944	1.964	1.984	2.004
1.4	2.120	2.139	2.158	2.177	2.196
1.5	2.308	2.326	2.345	2.363	2.381
1.6	2.489	2.506	2.524	2.542	2.559
1.7	2.663	2.681	2.698	2.715	2.732
1.8	2.833	2.850	2.867	2.884	2.900
1.9	3.000	3.016	3.032	3.049	3.065
2.0	3.163	3.179	3.195	3.211	3.228
2.1	3.324	3.340	3.356	3.372	3.389
2.2	3.485	3.501	3.517	3.533	3.549
2.3	3.646	3.662	3.678	3.694	3.710
2.4	3.807	3.820	3.833	3.846	3.859
2.5	3.939	3.953	3.966	3.980	3.994
2.6	4.077	4.092	4.106	4.120	4.134
2.7	4.220	4.235	4.249	4.264	4.279
2.8	4.367	4.382	4.397	4.411	4.426
2.9	4.516	4.531	4.546	4.561	4.576
3.0	4.666	4.681	4.696	4.711	4.726
3.1	4.817	4.832	4.847	4.862	4.876
3.2	4.966	4.981	4.996	5.010	5.025
3.3	5.113	5.128	5.142	5.157	5.171
3.4	5.257	5.272	5.286	5.300	5.314

A

Log$_{10}$ $N_D^{1/3}$ for Rigid Spheres

0.05	0.06	0.07	0.08	0.09
-1.831	-1.801	-1.771	-1.741	-1.711
-1.532	-1.502	-1.472	-1.442	-1.412
-1.233	-1.203	-1.173	-1.143	-1.114
-0.935	-0.906	-0.876	-0.847	-0.817
-0.640	-0.611	-0.582	-0.553	-0.524
-0.350	-0.322	-0.293	-0.265	-0.237
-0.069	-0.042	-0.015	0.012	0.039
0.196	0.222	0.247	0.272	0.297
0.452	0.478	0.504	0.530	0.556
0.706	0.730	0.754	0.778	0.802
0.939	0.962	0.984	1.005	1.027
1.162	1.185	1.209	1.232	1.256
1.393	1.416	1.438	1.461	1.483
1.614	1.635	1.656	1.678	1.699
1.823	1.844	1.864	1.884	1.904
2.023	2.043	2.062	2.082	2.101
2.215	2.234	2.252	2.271	2.289
2.399	2.417	2.435	2.453	2.471
2.577	2.594	2.612	2.629	2.646
2.749	2.766	2.783	2.800	2.817
2.917	2.934	2.950	2.967	2.983
3.081	3.098	3.114	3.130	3.147
3.244	3.260	3.276	3.292	3.308
3.405	3.421	3.437	3.453	3.469
3.565	3.581	3.597	3.614	3.630
3.727	3.743	3.759	3.775	3.792
3.872	3.886	3.899	3.912	3.926
4.008	4.022	4.035	4.049	4.063
4.148	4.163	4.177	4.192	4.206
4.293	4.308	4.323	4.337	4.352
4.441	4.456	4.471	4.486	4.501
4.591	4.606	4.621	4.636	4.651
4.741	4.756	4.771	4.786	4.802
4.891	4.906	4.921	4.936	4.951
5.040	5.055	5.069	5.084	5.099
5.186	5.200	5.215	5.229	5.243
5.328	5.342	5.356	5.370	5.383

APPENDIX

$\text{Log}_{10}\, N_U{}^{1/3}$ as a Function of

Log $N_D{}^{1/3}$	0.00	0.01	0.02	0.03	0.04
-0.2	-1.781	-1.761	-1.741	-1.721	-1.701
-0.1	-1.581	-1.561	-1.541	-1.521	-1.501
0.0	-1.382	-1.362	-1.342	-1.322	-1.303
0.1	-1.184	-1.164	-1.144	-1.125	-1.105
0.2	-0.987	-0.968	-0.949	-0.929	-0.910
0.3	-0.795	-0.776	-0.757	-0.738	-0.719
0.4	-0.608	-0.590	-0.572	-0.555	-0.537
0.5	-0.434	-0.418	-0.401	-0.385	-0.369
0.6	-0.278	-0.263	-0.248	-0.232	-0.215
0.7	-0.118	-0.103	-0.088	-0.073	-0.058
0.8	0.025	0.039	0.052	0.064	0.077
0.9	0.148	0.159	0.170	0.184	0.198
1.0	0.279	0.292	0.305	0.318	0.331
1.1	0.405	0.417	0.429	0.440	0.452
1.2	0.520	0.531	0.541	0.552	0.563
1.3	0.624	0.634	0.644	0.654	0.664
1.4	0.720	0.729	0.738	0.747	0.756
1.5	0.808	0.816	0.825	0.833	0.841
1.6	0.889	0.896	0.904	0.912	0.919
1.7	0.964	0.971	0.978	0.985	0.992
1.8	1.034	1.040	1.047	1.054	1.060
1.9	1.100	1.106	1.112	1.119	1.125
2.0	1.163	1.169	1.175	1.181	1.188
2.1	1.224	1.230	1.236	1.243	1.249
2.2	1.285	1.291	1.297	1.303	1.309
2.3	1.346	1.352	1.358	1.364	1.371
2.4	1.407	1.410	1.413	1.416	1.419
2.5	1.439	1.443	1.447	1.450	1.454
2.6	1.478	1.482	1.486	1.490	1.494
2.7	1.521	1.525	1.530	1.534	1.539
2.8	1.567	1.572	1.577	1.582	1.586
2.9	1.616	1.621	1.626	1.631	1.636
3.0	1.666	1.671	1.676	1.681	1.686
3.1	1.717	1.722	1.727	1.732	1.737
3.2	1.766	1.771	1.776	1.781	1.785
3.3	1.813	1.818	1.823	1.827	1.832
3.4	1.858	1.862	1.866	1.870	1.874

B

$\text{Log}_{10}\, N_D{}^{1/3}$ for Rigid Spheres

0.05	0.06	0.07	0.08	0.09
-1.681	-1.661	-1.641	-1.621	-1.601
-1.482	-1.462	-1.442	-1.422	-1.402
-1.283	-1.263	-1.243	-1.223	-1.204
-1.085	-1.066	-1.046	-1.027	-1.007
-0.890	-0.871	-0.852	-0.833	-0.814
-0.700	-0.682	-0.663	-0.645	-0.627
-0.519	-0.502	-0.485	-0.468	-0.451
-0.354	-0.338	-0.323	-0.308	-0.293
-0.198	-0.182	-0.166	-0.150	-0.134
-0.044	-0.030	-0.016	-0.002	0.012
0.089	0.102	0.114	0.125	0.137
0.212	0.225	0.239	0.252	0.266
0.343	0.356	0.368	0.381	0.393
0.464	0.475	0.486	0.498	0.509
0.573	0.584	0.594	0.604	0.614
0.673	0.683	0.692	0.702	0.711
0.765	0.774	0.782	0.791	0.799
0.849	0.857	0.865	0.873	0.881
0.927	0.934	0.942	0.949	0.956
0.999	1.006	1.013	1.020	1.027
1.067	1.074	1.080	1.087	1.093
1.132	1.138	1.144	1.150	1.157
1.194	1.200	1.206	1.212	1.218
1.255	1.261	1.267	1.273	1.279
1.315	1.321	1.328	1.334	1.340
1.377	1.383	1.389	1.395	1.402
1.422	1.426	1.429	1.432	1.436
1.458	1.462	1.466	1.470	1.474
1.499	1.503	1.507	1.512	1.516
1.543	1.548	1.553	1.558	1.562
1.591	1.596	1.601	1.606	1.611
1.641	1.646	1.651	1.656	1.661
1.692	1.697	1.702	1.707	1.712
1.742	1.747	1.751	1.756	1.761
1.790	1.795	1.800	1.804	1.809
1.836	1.840	1.845	1.849	1.853
1.878	1.882	1.886	1.890	1.894

Nomenclature

A	surface area of particle; cross-sectional area of duct; amplitude of oscillation
A_D	$C_D\mathrm{Re}^2$ as defined by Eq. (11-35)
A_e	surface area of volume-equivalent sphere
A_f	true surface area of bubble or drop at detachment
A_n	coefficient in Kronig–Brink series, Eq. (3-82)
$A_{p'}$	surface area of perimeter-equivalent sphere
A_p	projected area of particle
A_R	surface area of residual bubble or drop left at orifice after detachment of parent fluid particle
A_ω	amplitude of turbulent oscillations with angular frequency ω
A_0	minimum surface area of particle undergoing shape oscillations
Ar	Archimedes number $= g\rho\,\Delta\rho\, d_e{}^3/\mu^2$
a	radius of sphere, disk, or spherical-cap; equatorial radius of spheroid; semimajor axis of fluid particle
a'	amplitude of oscillation or disturbance
$\bar{a}$	average radius of curvature over leading portion of large fluid particle
a_R	radius of curvature of rear indented surface of large fluid particle
a_0	initial radius of bubble or drop
$a_0{}'$	initial amplitude of disturbance
a_1, a_2, a_3, a_4	constants in fitted polynomials
B	function of Kn defined by Eq. (10-64); dimensionless initial acceleration [Eq. (11-17) and Table 11.1]; numerical parameter in models for transfer during slow formation
B_1, B_2	functions of $\kappa\gamma$ in Table 5.7
B_1, B_2, B_3	constants in Eq. (4-34)
Bi	Biot number $= ka/H\mathscr{D}_p$, $kd_e/2H\mathscr{D}_p$, or ha/K_{tp}
b	semiaxis of spheroidal particle along axis of symmetry; breadth of particle; vertical semiaxis of spheroidal cap; distance from centre of particle to axis of tube; group defined by Eq. (10-66); numerical parameter in models for transfer during slow formation
b_1, b_2	semiminor axes of particle (see Fig. 7-10)
C	coefficient in Eq. (3-20); slip correction factor, defined by Eq. (10-53); constants in Eqs. (9-10) and (12-13)

C_D	drag coefficient $= 2F_D/\rho U_R{}^2 A_p$; $= 2F_D/\pi\rho U_R{}^2 a^2$ for sphere; $= 4\Delta\rho\, g d_e/3\rho U_T{}^2$ for fluid particle at its terminal velocity
C_{DF}	drag coefficient corresponding to skin friction alone
C_{Dfm}	drag coefficient for free-molecule flow
C_{DI}	inviscid drag coefficient given by Eq. (10-61)
C_{Di}	drag coefficient under steady conditions corresponding to the instantaneous Reynolds number, Re_i
C_{D0}	drag coefficient for incompressible flow ($\mathrm{Ma} \to 0$) at the same Re
C_{DP}	drag coefficient corresponding to pressure distribution (form drag) alone
C_{DSt}	drag coefficient given by Stokes law, $=24/\mathrm{Re}$
C_{D1}, C_{D2}, C_{D3}	coefficients of pressure drag, drag due to deviatoric normal stress, and drag due to shear stress
$C_{D\infty}$	drag coefficient in unbounded fluid
$C_D{}'$	drag coefficient for sphere subject to secondary motion; drag coefficient for cylinder defined by Eq. (6-23)
C_e	capacitance
C_L	lift coefficient, $=2F_L/\pi a^2 \rho U_R{}^2$ for sphere
C_N	normal drag coefficient for circular motion, Eq. (11-73)
C_o	orbit constant
C_s	constant in Eq. (9-43) and Table 9.4
C_t, C_{tp}	heat capacity at constant pressure of continuous, dispersed phase
c	concentration in continuous phase; numerical constant in Table 3.1; maximum particle dimension; speed of sound in same medium
$\bar{c}$	mean resistance, Eq. (4-6)
c_i	principal translational resistances, Eq. (4-4) (c_1 is axial resistance)
c_{is}	translational resistance of sphere in creeping flow $= 6\pi a$
c_p	concentration in dispersed phase; bulk concentration of solute in dispersed phase
$\bar{c}_p$	average concentration in dispersed phase
c_{p0}	initial concentration in dispersed phase
c_s	concentration in continuous phase at surface of particle
c_1, c_2	constants defined by Eq. (10-57)
c_∞	concentration in continuous phase remote from particle
¢	degree of circularity, defined by Eq. (2-6)
$¢_{op}$	operational circularity, defined by Eq. (2-9)
D	diameter of containing vessel or tube; hydraulic diameter of duct
D_i, D_o	inner and outer diameters of annular section
D_1, D_2	length and width of cross section of rectangular duct
$\mathscr{D}$, $\mathscr{D}_p$	molecular diffusivity in continuous, dispersed phase
d	characteristic dimension of particle; diameter of sphere, cylinder, or disk; equatorial diameter of spheroid
d_A	diameter of sphere with same projected area as particle in its orientation of maximum stability on a horizontal surface, $=\sqrt{4A_p/\pi}$
d_B	maximum bubble width
d_e	diameter of volume-equivalent sphere, $= (6V/\pi)^{1/3}$
$(d_e)_{max}$	maximum stable volume-equivalent diameter
d_{eq}	equilibrium diameter of bubble
d_{or}	orifice diameter
d_s	diameter of solid particles in fluidized bed
E	aspect ratio, $= b/a$ for spheroid or L/d for cylinder
$\bar{E}$	aspect ratio averaged over shape oscillations
E^2	operator defined by Eq. (1-37)
Eo	Eotvos number, $= g\,\Delta\rho\, d^2/\sigma$ or $g\,\Delta\rho\, d_e{}^2/\sigma$

Eo_D	Eotvos number based on diameter of duct, $= g\,\Delta\rho\,D^2/\sigma$
Eo'	modified Eotvos number, based on difference between surface tension of pure fluids and at equilibrium with surface active contaminant
Eo^*	modified Eotvos number, based on difference between surface tension of pure fluids and minimum value at which surface film collapses
Eo_{ac}	acceleration Eotvos number, $= \dot{U}\,\Delta\rho\,d_e^2/\sigma$
e	eccentricity, $= \sqrt{1 - E^2}$
e_i	eccentricity, $= \sqrt{1 - b_i^2/a^2}$
e_1	flatness ratio, Eq. (2-3)
e_2	elongation ratio, Eq. (2-4)
F	fractional approach to equilibrium ("extraction efficiency"), $= (c_{p0} - \bar{c}_p)/(c_{p0} - Hc_\infty)$
F_D, $\mathbf{F_D}$	net drag force (scalar, vector)
F_{Di}	component of drag force in $\mathbf{i}$-direction
$F_{D\parallel}$	drag component parallel to velocity component U_2 for orthotropic particle (see Fig. 4.11)
$F_{D\perp}$	drag component normal to velocity component U_2 for orthotropic particle (see Fig. 4.11)
$F_{D\infty}$	drag force in unbounded fluid
$\mathbf{F_g}$	net force on particle in type 2 accelerating motion (see Ch. 11), excluding steady drag, history, and added mass
F_L	net lift force
F_M	value of extraction efficiency, F, for mobile portion of particle
F_S	value of extraction efficiency, F, for stagnant portion of particle
F_{St}	drag force in Stokes flow, $= 6\pi a\mu U_R$ for sphere
Fr_D	Froude number, $= U_T\sqrt{\rho/\Delta\rho\,gD}$
f	frequency of oscillation of particle or fluid; fraction defined by Eq. (10-63); frequency of formation of bubbles or drops at orifice
$\bar{f}$	$(f_W + f_N)/2$
f_A	fraction of particle surface with buoyancy directed outwards at an angle less than 45° to the vertical
f_{AI}	area of contact between mobile and stagnant parts of spherical fluid particle, divided by surface area of particle, $4\pi a^2$
f_{AS}	fraction of particle surface occupied by stagnant cap
f_{CF}, f_{CP}	frequency of formation at constant flow, constant pressure
f_N	natural frequency of shape oscillation, Eq. (7-30)
f_{VS}	stagnant fraction of particle volume
f_W	frequency of wake shedding
G	shear rate, Eq. (10-30); velocity gradient in continuous phase
G_1, G_2	functions defined in Table 11.2
Gr	Grashof number (either Gr_t or Gr_w)
Gr_t, Gr_w	thermal and composition Grashof numbers, defined by Eqs. (10-13) and (10-14)
g, $\mathbf{g}$	gravitational acceleration (scalar, vector)
$g(\theta_W)$	function defined by Eq. (8-21)
H	distribution coefficient, Eq. (1-39); dimensionless group defined by Eq. (7-7); distance below free surface
H_1, H_2	functions defined in Table 11.2
h	heat transfer coefficient
h_v	coordinate directed vertically upwards
h_0	heat transfer coefficient for particle in stagnant medium
h_1, h_2	functions defined in Table 11.2
I^*	dimensionless moment of inertia defined by Eq. (6-11)
I_R	relative intensity of turbulence defined by Eq. (10-43)

I_{Rc}	value of I_R required to induce critical transition at given Reynolds number, Re
I_W	imaginary part of complex error function, Table 11.1
i, j, k	orthogonal unit vectors; **i** generally in direction of particle motion
J	dimensionless group defined by Eq. (7-8)
K	factor for spheroids using equatorial diameter as characteristic length, Eq. (4-61); velocity correction factor defined by Eq. (6-29); K_F, K_U, or K_μ in creeping flow; orifice constant, Eq. (12-14)
K'	factor for spheroids using L' as characteristic length, Eq. (4-68); constant defined by Eq. (9-13); dimensionless orifice constant, Eq. (12-16)
K_A	terminal velocity of particle divided by terminal velocity of sphere of diameter d_A
K_e	terminal velocity of particle divided by terminal velocity of volume-equivalent sphere
K_F	drag factor defined by Eq. (9-6)
K_{MD}	Sherwood number factor defined by Eq. (9-22)
K_{Me}	velocity ratio defined by Eq. (9-25)
K_{MU}	mass transfer factor defined by Eq. (9-21)
K_t, K_{tp}	thermal conductivity of continuous, dispersed phase
K_U	velocity ratio defined by Eq. (9-7)
K_μ	viscosity ratio defined by Eq. (9-8)
Kn	Knudsen number, $= \lambda/d$
$(\overline{KA})_p$	time-average product of surface area and overall transfer coefficient based on dispersed phase concentrations
k	instantaneous or steady external mass transfer coefficient; volumetric shape factor $= V/d_A{}^3$; ratio of specific heats, $= C_t$/heat capacity at constant volume
k'	wave number, $= 2\pi/\lambda$
$\bar{k}$	time-average external mass transfer coefficient
k_e	volumetric shape factor of isometric particle of similar form
$\bar{k}_p$	time-average internal mass transfer coefficient
k_0	mass transfer coefficient for particle in a stagnant medium
$\overline{kA}$, $(\overline{kA})_p$	time-average product of interfacial area and external, internal mass transfer coefficient
$(kA)_F$	mass transfer product for front surface
$(kA)_R$	mass transfer product for rear surface
L	reference length; characteristic dimension of particle; length of cylinder, slender body, or slug
L'	characteristic length defined by Eq. (4-67)
L'_{aft}	characteristic length defined by Eq. (6-32)
L_c	height of closed cylindrical column
L_{jet}	length of jet from orifice to point of break up
L_s	scale of turbulence
L_W	wake length measured from rear of particle
l	length of particle; side of parallelepiped or cube; shortest distance from center of particle to wall
l'	characteristic length defined by Eq. (10-24)
M	Morton number, $= g\mu^4 \Delta\rho/\rho^2\sigma^3$
M_A	acceleration modulus, $= (d/U_R{}^2)(dU_R/dt)$
M_D	displacement modulus, $= x/d$
Ma	Mach number, = characteristic velocity/c
m	constant in Table 3.1; mass transfer flux; group defined by Eq. (12-45)
m^*	mass transfer flux in absence of interfacial motion
m_p	group defined by Eq. (12-45)
N	group defined by Eq. (11-64); dimensionless velocity gradient, $= r_0 G\mu/\sigma$

N_A	dimensionless amplitude defined by Eq. (11-53)
N_{ch}	capacitance number defined by Eq. (12-17)
N_{crit}	critical value of N for bubble or drop breakup
N_D	dimensionless diameter group, $C_D Re_T^2 = 4g\,\Delta\rho\, d^3/3\rho\nu^2$ for sphere, $4g\,\Delta\rho\, d_e^3/3\rho\nu^2$ for fluid particle at terminal velocity
N_H	dimensionless history group defined by Eq. (11-57)
N_t	dimensionless thermal group defined by Eq. (10-29)
N_U	dimensionless terminal velocity group, $Re_T/C_D = 3\rho U_T^3/4g\,\Delta\rho\,\nu$
N_U'	value of N_U for sphere subject to secondary motion
N_ω	dimensionless frequency defined by Eq. (11-52)
Nu	Nusselt number, $=hL/K_t$
Nu*	Nu for continuum flow (Kn $\rightarrow$ 0) at same Re and Pr
Nu_c	Nu corresponding to critical transition, Eq. (10-48)
Nu_{fm}	Nu in free molecule limit (Kn $\rightarrow \infty$) at given Ma
Nu_{loc}	local Nusselt number
Nu_0	Nu in absence of turbulence at same Re
n	integer; coordinate normal to particle surface; constant defined by Eq. (5-40); index in Eq. (12-32)
P_A	perimeter of projected- area-equivalent sphere
P'	perimeter of an axisymmetric body projected normal to the axis
P_p	projected perimeter of particle
Pe	Peclet number in continuous phase, $= dU/\mathscr{D}$, $LU/\mathscr{D}$ or $d_e U/\mathscr{D}$
Pe′	Pe based on L', $=L'U/\mathscr{D}$
Pe_G	shear Peclet number, $=Re_G Sc$
Pe_p	Peclet number in dispersed phase, $=dU/\mathscr{D}_p$
Pr	Prandtl number, $=\mu C_t/K_t$
p	pressure or modified pressure in continuous phase
$\bar{p}$	mean ambient pressure
p_{ch}	pressure in chamber for constant pressure bubble formation
p_g	partial pressure of noncondensable gas
p_{HD}, $(p_{HD})_p$	hydrodynamic surface pressures in continuous, dispersed fluid
p_l	local pressure
p_m, p_m'	modified pressure, dimensionless modified pressure
p_p	modified pressure in dispersed phase
p_s	pressure or modified pressure in continuous phase at particle surface
p_v	vapour pressure of liquid
p_0	reference pressure; modified pressure at front stagnation point; constant in Eq. (3-9)
p_{0p}	constant in Eq. (3-10)
p_∞	pressure or modified pressure remote from particle
Q	function defined in Table 5.4; volumetric flow rate
$\bar{Q}$	time-mean volumetric flow rate
Q'	dimensionless flow rate defined by Eq. (12-3)
Q_{at}	volumetric flow rate at onset of atomization
Q_{jet}	volumetric flow rate at onset of jet formation
Q_{max}	volumetric flow rate for maximum jet length
R	distance to surface of axisymmetric particle, Fig. 1.1; ratio of form drag to skin friction; dimensionless radial coordinate, $=r/a$; radius of cylindrical tube
R_W	real part of complex error function, Table 11.1
R_0	radius of curvature at nose or lowest point of particle
R_1, R_2	principal radii of curvature
$\mathscr{R}$	dimensionless distance to particle surface, $=2R/d_e$ or R/L
Ra	Rayleigh number, $=Gr_t Pr$ or $Gr_w Sc$

Ra' — Rayleigh number based on l'
Ra_c' — value of Ra' above which $\frac{1}{3}$-power relationship applies
Re — particle Reynolds number, normally $= Ud/\nu$ or Ud_e/ν
Re' — Reynolds number based on L'; dimensionless acceleration $= d(\mathrm{Re})/d\tau$
Re_c — critical Reynolds number
Re_G — shear Reynolds number, $= Gd^2/\nu$
Re_i — instantaneous Reynolds number, $= U_{Ri}d/\nu$
Re_L — limiting Reynolds number for creeping flow
Re_M — metacritical Reynolds number at which $C_D = 0.3$ $(\mathrm{Re}_M > \mathrm{Re}_c)$
Re_m — Reynolds number corresponding to minimum C_D, Table 10.1
Re_p — Reynolds number based on dispersed phase properties, $= Ud/\nu_p$ or Ud_e/ν_p
Re_s — Reynolds number corresponding to onset of separation
Re_T — Reynolds number at terminal velocity $= U_T d/\nu$
Re_{TS} — Reynolds number corresponding to Stokes terminal velocity, Eq. (11-34)
Re_v — vibration Reynolds number, $= 4a'fd/\nu$
Re'_{aft} — Reynolds number based on L'_{aft}
r — radial coordinate in spherical or cylindrical coordinates
S — settling factor, Eq. (4-2)
S_e — settling factor based on volume-equivalent sphere
Sc, Sc_p — Schmidt number in continuous, dispersed phase, $= \nu/\mathscr{D}$, $\nu_p/\mathscr{D}_p$
Sh — external Sherwood number, $= kd/\mathscr{D}$; area-free Sherwood number for external resistance, $= kd_e/\mathscr{D}$ resistance, $= kd_e/\mathscr{D}$
Sh' — Sherwood number based on L' or l'
Sh_e — Sherwood number for external resistance based on volume-equivalent sphere, $= (kA/A_e)d_e/\mathscr{D}$, $(\overline{kA}/A_e)d_e/\mathscr{D}$
Sh_I — value of Sh at surface of contact between mobile and stagnant parts of particle
Sh_{loc} — local Sherwood number
Sh_M — Sherwood number over mobile portion of interface
Sh_p — area-free Sherwood number for internal resistance, $= \bar{k}_p d_e/\mathscr{D}_p$
Sh_{pe} — Sherwood number for internal resistance based on volume-equivalent sphere, $= [(\overline{kA})_p/A_e]d_e/\mathscr{D}$
Sh_r — Sherwood number in the presence of rotation
Sh_S — Sherwood number over stagnant portion of interface
Sh_{sphere} — Sherwood number for sphere at the same Re
Sh_v — Sherwood number with vibrations superimposed on translation
Sh_0 — Sherwood number for diffusion into stagnant medium, $= 2$ for sphere
Sh_∞ — Sherwood number in unbounded fluid
Sh'_{aft} — Sherwood number based on L'_{aft}
Sh_0' — Sherwood number for diffusion into stagnant medium, based on L' or l'
$\mathrm{Sh}_0'{}_{aft}$ — Sherwood number for diffusion into stagnant medium, based on L'_{aft}
Sr — Strouhal number, $= fd/U_T$, fb/U_T, or $f_W w/U_T$
s — molecular speed ratio defined by Eq. (10-60); dummy time-coordinate
s^+ — dimensionless time defined by Eq. (11-55)
T — temperature in continuous phase; dimensionless times defined by Eqs. (11-40) and (12-56)
T' — dimensionless temperature, $= (T - T_\infty)/(T_s - T_\infty)$
T_{crit} — value of dimensionless time at which droplet shatters
T_p — temperature in dispersed phase
T_s — surface temperature
T_∞ — temperature in continuous phase remote from particle
Ta — dimensionless group used by Tadaki and Maeda, $= \mathrm{Re}M^{0.23}$

t	time; particle thickness, i.e. minimum distance between two parallel planes tangential to opposite surfaces
t^+	dimensionless time defined by Eq. (11-55)
t^*	dimensionless time defined by Eq. (5-7)
t_a	time available for disturbance to grow
t_c	coalescence time
t_f	duration of formation process
t_g	time required for disturbance to grow, $=\alpha^{-1}$
t_x	time required for Sh to come within $100x\%$ of steady value
U, $\mathbf{U}$	velocity of particle, normally relative to remote continuous phase, or of continuous phase relative to particle (scalar, vector)
U_e	relative velocity giving same Sherwood number in an unbounded fluid
U_f, $\bar{U}_f$	instantaneous and time-average continuous phase velocities
U_{fi}	i-component of U_f
U_p, $\bar{U}_p$	instantaneous and time-average particle velocity
U_{pi}	i-component of U_p
U_R, $\bar{U}_R$	instantaneous and time-average velocity of particle relative to continuous phase
U_{Ri}	i-component of U_R
U_s	spreading velocity of surface tension-lowering material at interface
U_{sphere}	terminal velocity of equivalent diameter spherical particle
U_T	terminal velocity of particle
U_{TS}	terminal velocity of spherical particle given by Stokes law, $=gd^2\,\Delta\rho/18\mu$
$U_{T\infty}$	terminal velocity in unbounded fluid
$(U_T)_{pure}$	terminal velocity in surfactant-free system
U_ω	maximum velocity of oscillating particle corresponding to frequency ω
U_0	reference velocity; fluid velocity on axis of cylinder far from particle
$\dot{U}$, $\dot{U}_R$	time derivative of U, U_R, $= dU/dt$, dU_R/dt
$\mathbf{u}$	local fluid velocity vector
$\mathbf{u}'$	dimensionless velocity vector, $=\mathbf{u}/U_0$
u	undisturbed local fluid velocity; velocity of air stream or shock wave
u'	fluctuating component of velocity
u_f	fluctuating velocity in continuous phase, $= U_f - \bar{U}_f$
u_n, $(u_n)_p$	velocity in continuous, dispersed phase normal to surface
u_{or}	velocity of fluid through orifice
u_p	fluctuating velocity of particle, $= U_p - \bar{U}_p$
u_R	fluctuating relative velocity, $= U_R - \bar{U}_R$
u_r	r-component of $\mathbf{u}$
u_t, $(u_t)_p$	velocity in continuous, dispersed phase tangential to surface
u_x, u_y	velocity in continuous phase parallel, normal to surface
u_x, u_y, u_z	Cartesian components of $\mathbf{u}$
u_θ	θ-component of $\mathbf{u}$
V	volume of particle
V', V''	dimensionless volumes defined by Eqs. (12-2) and (12-21)
V_{CF}, V_{CP}	volume of bubble formed at orifice under constant flow, constant pressure conditions
V_{ch}	chamber volume
V_d	drift volume
V_R	volume of indentation at rear of particle
V_s	sphere volume
V_W	volume of closed wake
V_1', V_2'	dimensionless volumes contributed by successive stages of bubble formation

V_1'', V_2''	dimensionless volumes contributed by successive stages of bubble formation in a fluidized bed
$\mathbf{v}$	velocity vector representing deviation of continuous phase from uniform stream
v_s	equatorial surface velocity due to rotation
W	$\log_{10} N_D$, Table 5.3; $\log_{10} (C_D \mathrm{Re_T}^2)^{1/3}$; ratio of instantaneous particle velocity to terminal velocity, $=U/U_T$
$W(Z)$	complex error function, Table 11.1
W_f, W_R	dimensionless fluid and relative velocities defined by Eq. (11-56)
W_S	dimensionless particle velocity, Eq. (11-14) and Table 11.1
We	Weber number, $=U_T^2 d_e \rho/\sigma$ or $u^2 d_e \rho/\sigma$
$\mathrm{We_a}$	Weber number based on radius of spherical cap, $=U_T^2 a\rho/\sigma$
w	$\log_{10}$ Re; maximum width of particle; component mass fraction
w'	dimensionless mass fraction, $=(w - w_\infty)/(w_s - w_\infty)$
w_s, w_∞	surface, remote component mass fraction
X	dimensionless boundary-layer coordinate parallel to surface, $=2x/d_e$ or x/L; dimensionless displacement, Eq. (11-42); function of B and Δ_a, Table 11.1
X_M	maximum value of dimensionless boundary-layer coordinate, X
x	Cartesian boundary layer coordinate parallel to surface; index defined by Eq. (6-2); distance travelled by particle; distance along jet, Eq. (12-29)
x_1, x_1'	dimensional, dimensionless Cartesian coordinate
Y	ratio of terminal velocity to terminal velocity in Stokes flow, $=U_T/U_{TS}$; function of B and Δ_a in Table 11.1; ratio of bubble volumes formed under constant pressure and constant flow conditions
Y_M, Y_S	weighting factors for mass transfer at mobile and stagnant portions of an interface, Eq. (3-91)
y	Cartesian boundary layer coordinate normal to surface; distance measured vertically upwards from point 0 or lowest point on particle surface
y_1, y_1'	dimensional, dimensionless Cartesian coordinate
Z	total degree of circulation, Eq. (3-25); distance of particle from end of closed column; complex function in Table 11.1, $=(X + iY)\sqrt{\tau}$
z_1, z_1'	dimensional, dimensionless Cartesian coordinate
α	thermal diffusivity, $=K_t/\rho C_t$; fraction of particle surface area aft of maximum perimeter in a plane normal to flow, $=L'_{\mathrm{aft}}/L'$; angle between surface normal and direction of gravity; function in Table 11.1; exponential growth rate of disturbance
β	dimensionless displacement from axis, $=b/R$; phase shift, Eq. (11-50); function in Table 11.1; constant in Eq. (12-32)
β_t, β_w	thermal and composition compressibility coefficients, Eqs. (10-10) and (10-11)
Γ	vortex strength for circulating sphere divided by strength of corresponding Hill's vortex; retardation coefficient defined by Eq. (7-10)
γ	density ratio, $=\rho_p/\rho$; Euler's constant, $=0.5772157\ldots$
γ_c	value of density ratio above which functions α and β are complex
Δ	drag ratio, Eq. (4-1); skirt thickness
Δ'	conductance factor, Eq. (4-56)
Δ_A	added mass coefficient, Eqs. (11-22) and (11-30)
$\bar{\Delta}_a$	mean drag on spheroid in steady motion ÷ drag on sphere of radius a at same Re (based on d)
Δ_{ai}	drag on spheroid in i direction in steady motion ÷ drag on sphere of radius a at same Re (based on d)
$\bar{\Delta}_e$	mean drag ratio based on volume-equivalent sphere
Δ_{ei}	drag ratio (i-direction drag) based on volume-equivalent sphere
Δ_H	history coefficient, Eqs. (11-22) and (11-30)
Δ_P'	drag ratio based on projected perimeter-equivalent sphere

Δc	factor introduced by O'Brien, Eq. (4-38)
Δf	frequency difference, $=(f_W - f_N)/2$
ΔP^+	excess pressure drop due to particle
ΔV	factor introduced by O'Brien, Eq. (4-39)
$\Delta\rho$	absolute value of density difference between particle and continuous phase, $=\lvert\rho_p - \rho\rvert$
$\Delta\rho'$	absolute value of surface to bulk density difference, $=\lvert\rho_s - \rho_\infty\rvert$
$\Delta\sigma$	surface or interfacial tension difference
δ	thickness of fictitious mass transfer film, Eq. (4-64); thickness of disk; cell depth; length scale defined by Eq. (11-8)
δ_{pen}	penetration depth, $=\sqrt{\pi\mathscr{D}t}$
ε	parameter in Table 3.1; permittivity; ratio of amplitude of area oscillation to minimum area; effective height of surface roughness elements; dimensionless group defined by Eq. (11-66)
Z	dimensionless vorticity, $=\zeta a/U_0$
$\zeta, \boldsymbol{\zeta}$	vorticity (scalar, vector)
ζ_s	vorticity at particle surface
η	spheroidal angular coordinate; amplitude ratio, Eq. (11-50)
θ	angular coordinate, normally measured away from front stagnation point; angle between axis of symmetry and direction of motion; angle defined in Fig. 4.11; angle of inclination of tube from vertical; coordinate of axis of rotating particle; angle between direction of flow and major axis of particle deformed by shear field
θ_c	contact angle
θ_R	included angle for rear indented surface, Fig. 8.9
θ_s	separation angle measured from front stagnation point
θ_W	included wake angle for front surface, Figs. 8.1 and 8.9
θ_0	angle from front stagnation point to leading edge of stagnant cap
κ	viscosity ratio, $=\mu_p/\mu$
λ	angle between vertical and direction of motion; dimensionless group defined by Eq. (7-18); diameter ratio, $=d/D$ or d_e/D; molecular mean free path; wavelength of disturbance
λ_{cr}	critical wavelength
λ_m	"most dangerous wavelength", corresponding to maximum α
λ_n	eigenvalue in Kronig–Brink series, Eq. (3-82)
μ, μ_p	viscosity of continuous, dispersed phase
μ'	dimensionless viscosity defined by Eq. (12-4)
μ_e	effective viscosity defined by Eq. (12-35)
μ_s	continuous phase viscosity yielding the actual terminal velocity when calculated from Stokes' law, Eq. (9-9)
μ_w	viscosity of water in Braida's experiments, 9×10^{-4} Ns/m^2
ν, ν_p	kinematic viscosity of continuous, dispersed phase
ξ	dummy time variable, Table 11.1
ρ, ρ_p	density of continuous, dispersed phase
ρ_g	density of gas in fluidized bed
ρ_s	continuous phase density at particle surface; density of solid particles in fluidized bed
ρ_∞	density of continuous phase remote from particle
$\sum$	perimeter-equivalent factor defined by Eq. (2-12)
σ	interfacial or surface tension; dimensionless time defined by Eq. (11-13)
σ_R	accommodation coefficient for molecular collisions at particle surface
σ_{SF}	surface tension in absence of surface-active contaminants
σ_t	thermal accommodation coefficient

τ dimensionless time ("Fourier number"), $=\mathscr{D}t/a^2$, $\alpha t/a^2$, or $\nu t/a^2$; local shear stress
τ_{nn} deviatoric normal stress at interface
τ_{nt} shear stress at interface
τ_p Fourier number based on dispersed phase, $=4\mathscr{D}_p t/d_e^2$
τ_r period of rotation; relaxation time
τ_x value of Fourier number at time t_x
τ_0 dimensionless period of oscillation, $=\nu/\omega a^2$
Φ velocity potential
Φ_v viscous dissipation function
ϕ dimensionless concentration in continuous phase, $=(c - c_\infty)/(c_s - c_\infty)$ or $(c - c_\infty)/(c_{p0}/H - c_\infty)$; angle of inclination of particle axis from vertical; spherical polar coordinate of axis of rotating particle
ϕ_p dimensionless concentration in dispersed phase, $=(c_p - Hc_\infty)/(c_{p0} - Hc_\infty)$
ϕ_ω phase lag
ϕ_0 initial phase angle for particle undergoing rotation
χ modified circularity defined by Eq. (4-28)
Ψ dimensionless stream function, $=\psi/U_0L^2$
ψ Stokes stream function for continuous phase relative to particle; sphericity, $=A_e/A$
ψ' stream function for particle motion through stagnant fluid
ψ_H Harkins' correction factor, Fig. 12.4
ψ_{op} operational sphericity, Eq. (2-8)
ψ_p stream function in dispersed phase
ψ_W working sphericity, Eq. (2-10)
$\mathbf{\Omega}$ angular velocity of rotating particle
ω angular frequency of oscillation

Index

C

Z

Errata

p. 43, caption of Fig. 3.8: ζ_5 should be ζ_s.
p.75, Table 4.1: Add brackets around $\cos^{-1} E$ in the first two Exact expressions under Oblate.
p.103, line 3 of d: Change "pockets" to "packets".
p.111, Table 5.1 and p.140. ref. S1: Naumann (2 n's at the end).
p.113, eq. (5-14): Coefficient 1.73 in the Re_T relation (i.e. second equation) should be 1.50.
p.116: Caption of Fig. 5.14 should read: Terminal velocities of rigid sphere in water at 20°C and in air at 20°C and 101.3 kPa pressure.
p.120, 4 lines up: Change "maximum" to "maxima" and "reflects" to "reflect".
p.126, line 11 of 2nd paragraph: (A1, R7) should be (A1, R9).
p.153, caption of Fig. 6.11: Replace "heat transfer" by "mass transfer" and Pr by Sc.
p.155, Table 6.1: In line 3, change to ; in line 4 change to ; in line 5 change to twice (at beginning of the line and before the last equal sign.)
p.187, line 5: Change "amd" to "and".
p.191, Equation (7-39): Last term should be , i.e. lower case k.
p.230, 2 line above equation (9.25): $K_{\chi e}$ should be K_{Me}.
pp.233 and 234: The footnote at the bottom of page 234 should be at the bottom of page 233.
p.236: equation (9.37): The symbol d should be D.
p.243, Z1: Initials for Zukoski should be S.S., not E.E.
p.267, Table 10.1, last line: First term in C_D expression should be 3990(log Re)$^{-6.1}$. In line 2 of the footnote, Re_m should be Re_M. In final line of footnote, Re_χ should be Re_M.
p.292, Fig. 11.3: On right-hand ordinate scale, the upper label 10 should be 10^2. Also MD in the label on the right-hand ordinate axis should be M_D, i.e. the D should be a subscript.
p.304, Fig. 11.14: On left-hand ordinate axis. The top label "0" should move to the very top.
p.332, last line: Add ")" after 15 cm/s.